SIXTH EDITION

Historical Geology

Evolution of Earth and Life Through Time

Reed Wicander
Central Michigan University

James S. Monroe
Emeritus, Central Michigan University

BROOKS/COLE
CENGAGE Learning™

Australia • Brazil • Japan • Korea • Mexico • Singapore • Spain • United Kingdom • United States

Historical Geology: Evolution of Earth and Life Through Time, Sixth Edition
Reed Wicander, James S. Monroe

Earth Sciences Editor: Yolanda Cossio

Development Editor: Liana Monari

Assistant Editor: Samantha Arvin

Editorial Assistant: Jenny Hoang

Technology Project Manager: Alexandria Brady

Marketing Manager: Nicole Mollica

Marketing Communications Manager:
Belinda Krohmer

Project Manager, Editorial Production:
Hal Humphrey

Art Director: John Walker

Print Buyer: Karen Hunt

Permissions Editor: Bob Kauser

Production Service: Amanda Maynard,
Pre-Press PMG

Text Designer: Lisa Buckley

Photo Researcher: Terri Wright

Copy Editor: Christine Hobberlin

Illustrators: Accurate Art, Precision Graphics,
Rolin Graphics, Pre-Press PMG

Cover Designer: Denise Davidson

Cover Image: © Naturfoto Honal/Corbis

Compositor: Pre-Press PMG

For product information and technology assistance, contact us at
Cengage Learning Customer & Sales Support, 1-800-354-9706

For permission to use material from this text or product, submit all requests online at **www.cengage.com/permissions**
Further permissions questions can be e-mailed to
permissionrequest@cengage.com

Library of Congress Control Number: 2008943952

Student Edition:
ISBN-13: 978-0-495-56007-4

ISBN-10: 0-495-56007-3

Brooks/Cole
10 Davis Drive
Belmont, CA 94002-3098
USA

Cengage Learning is a leading provider of customized learning solutions with office locations around the globe, including Singapore, the United Kingdom, Australia, Mexico, Brazil, and Japan. Locate your local office at **www.cengage.com/international**

Cengage Learning products are represented in Canada by Nelson Education, Ltd.

To learn more about Brooks/Cole, visit **www.cengage.com/brookscole**

Purchase any of our products at your local college store or at our preferred online store **www.ichapters.com**

Printed in China by China Translation & Printing Services Limited
2 3 4 5 6 7 12 11 10

About the Authors

REED WICANDER

REED WICANDER is a geology professor at Central Michigan University where he teaches physical geology, historical geology, prehistoric life, and invertebrate paleontology. He has co-authored several geology textbooks with James S. Monroe. His main research interests involve various aspects of Paleozoic palynology, specifically the study of acritarchs, on which he has published many papers. He is a past president of the American Association of Stratigraphic Palynologists and a former councillor of the International Federation of Palynological Societies. He is the current chairman of the Acritarch Subcommission of the Commission Internationale de Microflore du Paléozoique.

Photo courtesy of Melanie Wicander

Photo courtesy of Sue Monroe

JAMES S. MONROE

JAMES S. MONROE is professor emeritus of geology at Central Michigan University where he taught physical geology, historical geology, prehistoric life, and stratigraphy and sedimentology since 1975. He has co-authored several textbooks with Reed Wicander and has interests in Cenozoic geology and geologic education. Now retired, he continues to teach geology classes for Osher Lifelong Learning Institute, an affiliate of California State University, Chico, and leads field trips to areas of geologic interest.

Brief Contents

Chapter 1   The Dynamic and Evolving Earth   1

Chapter 2   Minerals and Rocks   18

Chapter 3   Plate Tectonics: A Unifying Theory   37

Chapter 4   Geologic Time: Concepts and Principles   65

Chapter 5   Rocks, Fossils, and Time—Making Sense of the Geologic Record   85

Chapter 6   Sedimentary Rocks—The Archives of Earth History   109

Chapter 7   Evolution—The Theory and Its Supporting Evidence   131

Chapter 8   Precambrian Earth and Life History—The Archean Eon   153

Chapter 9   Precambrian Earth and Life History—The Proterozoic Eon   173

Chapter 10   Early Paleozoic Earth History   196

Chapter 11   Late Paleozoic Earth History   217

Chapter 12   Paleozoic Life History: Invertebrates   240

Chapter 13   Paleozoic Life History: Vertebrates and Plants   260

Chapter 14   Mesozoic Earth History   281

Chapter 15   Life of the Mesozoic Era   303

Chapter 16   Cenozoic Earth History: The Paleogene and Neogene   328

Chapter 17   Cenozoic Geologic History: The Pleistocene and Holocene Epochs   353

Chapter 18   Life of the Cenozoic Era   376

Chapter 19   Primate and Human Evolution   400

             Epilogue   415

Appendix A   Metric Conversion Chart   421

Appendix B   Classification of Organisms   422

Appendix C   Mineral Identification   427

             Glossary   430

             Answers to Multiple-Choice Review Questions   437

             Index   438

Contents

Chapter 1
The Dynamic and Evolving Earth 1

Introduction 2

What is Geology? 4

Historical Geology and the Formulation
of Theories 4

Origin of the Universe and Solar System, and Earth's
Place in Them 5

Origin of the Universe—Did It Begin with a Big Bang? 5

Our Solar System—Its Origin and Evolution 6

Perspective *The Terrestrial and Jovian Planets* 8

Earth—Its Place in Our Solar System 10

Why is Earth a Dynamic and Evolving Planet? 10

Organic Evolution and the History of Life 13

Geologic Time and Uniformitarianism 13

How Does the Study of Historical
Geology Benefit Us? 14

Summary 15

Courtesy of NASA Goddard Space Flight Center/Reto Stockli

Chapter 2
Minerals and Rocks 18

Introduction 19

Matter and Its Composition 19

Elements and Atoms 19

Bonding and Compounds 20

Minerals—The Building Blocks of Rocks 21

How Many Minerals Are There? 22

Rock-Forming Minerals and the Rock Cycle 23

Igneous Rocks 24

Texture and Composition 24

Classifying Igneous Rocks 26

Sedimentary Rocks 26

Sediment Transport, Deposition, and Lithification 28

Classification of Sedimentary Rocks 28

Metamorphic Rocks 30

What Causes Metamorphism? 32

Metamorphic Rock Classification 32

Plate Tectonics and the Rock Cycle 34

Summary 35

Sue Monroe

Chapter 3
Plate Tectonics: A Unifying Theory 37

Introduction 38

Early Ideas About Continental Drift 39
  Alfred Wegener and the Continental Drift Hypothesis 39
  Additional Support for Continental Drift 42

Paleomagnetism and Polar Wandering 42

Magnetic Reversals and Seafloor Spreading 43

Plate Tectonics and Plate Boundaries 46

Perspective *Tectonics of the Terrestrial Planets* 48
  Divergent Boundaries 50
  An Example of Ancient Rifting 50
  Convergent Boundaries 50
  Recognizing Ancient Convergent Boundaries 54
  Transform Boundaries 55

Hot Spots and Mantle Plumes 55

How Are Plate Movement and Motion Determined? 56

The Driving Mechanism of Plate Tectonics 58

How Are Plate Tectonics and Mountain Building Related? 59
  Terrane Tectonics 59

Plate Tectonics and the Distribution of Life 60

Plate Tectonics and the Distribution of Natural Resources 61

Summary 62

Chapter 4
Geologic Time: Concepts and Principles 65

Introduction 66

How is Geologic Time Measured? 66

Early Concepts of Geologic Time and Earth's Age 67

Relative Dating Methods 68

Establishment of Geology as a Science—The Triumph of Uniformitarianism over Neptunism and Catastrophism 69
  Neptunism and Catastrophism 69
  Uniformitarianism 70
  Modern View of Uniformitarianism 71

Lord Kelvin and a Crisis in Geology 71

Absolute Dating Methods 72
  Atoms, Elements, and Isotopes 72
  Radioactive Decay and Half-Lives 72
  Long-Lived Radioactive Isotope Pairs 76
  Fission-Track Dating 76
  Radiocarbon and Tree-Ring Dating Methods 77

Geologic Time and Climate Change 78

Perspective *Denver's Weather—280 Million Years Ago!* 81

Summary 83

NASA Johnson Space Center—Earth Sciences and Image Analysis

Copyright and photograph by Dr. Parvinder S. Sethi

Chapter 5
Rocks, Fossils, and Time—Making Sense of the Geologic Record 85

Introduction 86

Stratigraphy 87
  Vertical Stratigraphic Relationships 87
  Lateral Relationships—Facies 91

Marine Transgressions and Regressions   91
Extent, Rate, and Causes of Marine Transgressions
and Regressions   93

Fossils and Fossilization   93
How Do Fossils Form?   94
Fossils and Telling Time   97

The Relative Geologic Time Scale   98

Stratigraphic Terminology   98
Lithostratigraphic Units and Biostratigraphic Units   98
Time Stratigraphic Units and Time Units   100

Correlation   100

Perspective *Monument Valley Navajo Tribal Park*   104

Absolute Dates and the Relative Geologic
Time Scale   105

Summary   107

Sue Monroe

Chapter 6
Sedimentary Rocks—The Archives of Earth History   109

Introduction   110

Sedimentary Rock Properties   110
Composition and Texture   111
Sedimentary Structures   112
Geometry of Sedimentary Rocks   116
Fossils—The Biologic Content of Sedimentary Rocks   116

Depositional Environments   116
Continental Environments   116
Transitional Environments   119
Marine Environments   122

Perspective *Evaporites—What We Know
and Don't Know*   125

Interpreting Depositional Environments   126

Paleogeography   127

Summary   128

Chapter 7
Evolution—The Theory and Its Supporting Evidence   131

Introduction   132

Evolution: What Does It Mean?   132
Jean-Baptiste de Lamarck and His Ideas on Evolution   133
The Contributions of Charles Darwin
and Alfred Wallace   134
Natural Selection—What Is Its Significance?   135

Perspective *The Tragic Lysenko Affair*   136

Mendel and the Birth of Genetics   136
Mendel's Experiments   137
Genes and Chromosomes   138

The Modern View of Evolution   138
What Brings About Variation?   139
Speciation and the Rate of Evolution   140
Divergent, Convergent, and Parallel Evolution   141
Microevolution and Macroevolution   142
Cladistics and Cladograms   142
Mosaic Evolution and Evolutionary Trends   144
Extinctions   144

What Kinds of Evidence Support
Evolutionary Theory?   145
Classification—A Nested Pattern of Similarities   146
How Does Biological Evidence Support Evolution?   146
Fossils: What Do We Learn from Them?   149
Missing Links—Are They Really Missing?   150
The Evidence—A Summary   150

Summary   151

Chapter 8
Precambrian Earth and Life History— The Archean Eon   153

Introduction   154

What Happened During the Eoarchean?   155

Continental Foundations—Shields, Platforms,
and Cratons   156
Archean Rocks   157
Greenstone Belts   157
Evolution of Greenstone Belts   160

Perspective *Earth's Oldest Rocks*   162

Archean Plate Tectonics and the Origin
of Cratons   163

The Atmosphere and Hydrosphere   163
How Did the Atmosphere Form and Evolve?   164
Earth's Surface Waters—The Hydrosphere   165

The Origin Of Life   165
Experimental Evidence and the Origin of Life   166
Submarine Hydrothermal Vents and the Origin of Life   167
The Oldest Known Organisms   168

Archean Mineral Resources   169

Summary   170

Chapter 9

Precambrian Earth and Life History—The Proterozoic Eon   173

Introduction   174

Evolution of Proterozoic Continents   175

Paleoproterozoic History of Laurentia   175

Perspective *The Sudbury Meteorite Impact and Its Aftermath*   178

Mesoproterozoic Accretion and Igneous Activity   178

Mesoproterozoic Orogeny and Rifting   179

Meso- and Neoproterozoic Sedimentation   179

Proterozoic Supercontinents   180

Ancient Glaciers and Their Deposits   182

Paleoproterozoic Glaciers   183

Glaciers of the Neoproterozoic   183

The Evolving Atmosphere   184

Banded Iron Formations (BIFs)   184

Continental Red Beds   185

Life of the Proterozoic   185

Eukaryotic Cells Evolve   186

Endosymbiosis and the Origin of Eukaryotic Cells   187

The Dawn of Multicelled Organisms   188

Neoproterozoic Animals   189

Proterozoic Mineral Resources   191

Summary   192

Chapter 10

Early Paleozoic Earth History   196

Introduction   197

Continental Architecture: Cratons and Mobile Belts   197

Paleozoic Paleogeography   199

Early Paleozoic Global History   199

Early Paleozoic Evolution of North America   201

The Sauk Sequence   202

Perspective *The Grand Canyon—A Geologist's Paradise*   204

The Cambrian of the Grand Canyon Region: A Transgressive Facies Model   205

The Tippecanoe Sequence   206

Tippecanoe Reefs and Evaporites   206

The End of the Tippecanoe Sequence   211

The Appalachian Mobile Belt and the Taconic Orogeny   212

Early Paleozoic Mineral Resources   213

Summary   214

Chapter 11

Late Paleozoic Earth History   217

Introduction   218

Late Paleozoic Paleogeography   218

The Devonian Period   218

The Carboniferous Period   220

The Permian Period   222

Late Paleozoic Evolution of North America   222

The Kaskaskia Sequence   222

Reef Development in Western Canada   222

Perspective *The Canning Basin, Australia—A Devonian Great Barrier Reef*   223

Black Shales   224

The Late Kaskaskia—A Return to Extensive Carbonate Deposition   225

The Absaroka Sequence   226

What Are Cyclothems, and Why Are They Important?   227

Cratonic Uplift—The Ancestral Rockies   229

The Middle Absaroka—More Evaporite Deposits and Reefs   229

History of the Late Paleozoic Mobile Belts   231

Cordilleran Mobile Belt   231

Ouachita Mobile Belt   233

Appalachian Mobile Belt   233

What Role Did Microplates and Terranes Play in the Formation of Pangaea?   235

Late Paleozoic Mineral Resources   235

Summary   236

Chapter 12

Paleozoic Life History: Invertebrates   240

Introduction   241

What Was the Cambrian Explosion?   241

The Emergence of a Shelly Fauna   242

Paleozoic Invertebrate Marine Life   244

The Present Marine Ecosystem   244

Cambrian Marine Community   246

Perspective *Trilobites—Paleozoic Arthropods*   248

The Burgess Shale Biota   250

Ordovician Marine Community   251

Silurian and Devonian Marine Communities   252

Carboniferous and Permian Marine Communities   254

Mass Extinctions   255

The Permian Mass Extinction   256

Summary   257

Chapter 13
Paleozoic Life History: Vertebrates and Plants 260

Introduction 261

Vertebrate Evolution 261

Fish 262

Amphibians—Vertebrates Invade the Land 267

Evolution of the Reptiles—The Land Is Conquered 269

Plant Evolution 271

Perspective *Palynology: A Link between Geology and Biology* 273

Silurian and Devonian Floras 274

Late Carboniferous and Permian Floras 276

Summary 278

Chapter 14
Mesozoic Earth History 281

Introduction 282

The Breakup of Pangaea 282

The Effects of the Breakup of Pangaea on Global Climates and Ocean Circulation Patterns 285

Mesozoic History of North America 286

Continental Interior 286

Eastern Coastal Region 287

Gulf Coastal Region 288

Western Region 290

Mesozoic Tectonics 290

Mesozoic Sedimentation 293

Perspective *Petrified Forest National Park* 297

What Role Did Accretion of Terranes Play in the Growth of Western North America? 298

Mesozoic Mineral Resources 298

Summary 300

Chapter 15
Life of the Mesozoic Era 303

Introduction 304

Marine Invertebrates and Phytoplankton 304

Aquatic and Semiaquatic Vertebrates 307

The Fishes 307

Amphibians 307

Plants—Primary Producers on Land 308

The Diversification of Reptiles 309

Archosaurs and the Origin of Dinosaurs 310

Dinosaurs 310

Warm-Blooded Dinosaurs? 314

Flying Reptiles 315

Mesozoic Marine Reptiles 315

Crocodiles, Turtles, Lizards, and Snakes 317

From Reptiles to Birds 317

Perspective *Mary Anning and Her Contributions to Paleontology* 318

Origin and Evolution of Mammals 319

Cynodonts and the Origin of Mammals 319

Mesozoic Mammals 321

Mesozoic Climates and Paleobiogeography 322

Mass Extinctions—A Crisis in Life History 323

Summary 324

© Eivind Bovor

Chapter 16
Cenozoic Earth History: The Paleogene and Neogene 328

Introduction 329

Cenozoic Plate Tectonics—An Overview 330

Cenozoic Orogenic Belts 332

Alpine–Himalayan Orogenic Belt 333

Circum-Pacific Orogenic Belt 335

North American Cordillera 335

Laramide Orogeny 336

Cordilleran Igneous Activity 339

Basin and Range Province 341

Colorado Plateau 342

Rio Grande Rift 342

Pacific Coast 344

The Continental Interior   345
Cenozoic History of the Appalachian
    Mountains   346
Perspective *The Great Plains*   347
North America's Southern and Eastern Continental
    Margins   348
    Gulf Coastal Plain   348
    Atlantic Continental Margin   349
Paleogene and Neogene Mineral Resources   350
Summary   351

Chapter 17
Cenozoic Geologic History: The Pleistocene and Holocene Epochs   353

Introduction   354
Pleistocene and Holocene Tectonism and
    Volcanism   355
    Tectonism   355
    Volcanism   356
Pleistocene Stratigraphy   356
    Terrestrial Stratigraphny   357
    Deep-Sea Stratigraphy   359
Onset of the Ice Age   360
    Climates of the Pleistocene and Holocene   360
    Glaciers—How Do They Form?   361
Glaciation and Its Effects   361
    Glacial Landforms   362
    Changes in Sea Level   363

Sue Monroe

Perspective *Waterton Lakes National Park, Alberta,
and Glacier National Park, Montana*   366
    Glaciers and Isostasy   367
    Pluvial and Proglacial Lakes   367
What Caused Pleistocene Glaciation?   370
    The Milankovitch Theory   371
    Short-Term Climatic Events   372
Glaciers Today   372
Pleistocene Mineral Resources   373
Summary   373

Chapter 18
Life of the Cenozoic Era   376

Introduction   377
Marine Invertebrates and Phytoplankton   378
Perspective *The Messel Pit Fossil Site in Germany*   380
Cenozoic Vegetation and Climate   381
Cenozoic Birds   382
The Age of Mammals Begins   383
    Monotremes and Marsupial Mammals   383
    Diversification of Placental Mammals   383
Paleogene and Neogene Mammals   386
    Small Mammals—Insectivores, Rodents, Rabbits, and Bats   387
    A Brief History of the Primates   387
    The Meat Eaters—Carnivorous Mammals   388
    The Ungulates or Hoofed Mammals   389
    Giant Land-Dwelling Mammals—Elephants   392
    Giant Aquatic Mammals—Whales   393
Pleistocene Faunas   394
    Ice Age Mammals   394
    Pleistocene Extinctions   395
Intercontinental Migrations   396
Summary   398

Chapter 19
Primate and Human Evolution   400

Introduction   401
What Are Primates?   401
Prosimians   402
Anthropoids   403
Hominids   404
    Australopithecines   407

Perspective *Footprints at Laetoli*   409
   The Human Lineage   410
   Neanderthals   411
   Cro-Magnons   412
Summary   413

Epilogue   415

Appendix A
Metric Conversion Chart   421

Appendix B
Classification of Organisms   422

Appendix C
Mineral Identification   427

Glossary   430

**Answers to Multiple-Choice Review
Questions   437**

Index   438

Preface

Earth is a dynamic planet that has changed continuously during its 4.6 billion years of existence. The size, shape, and geographic distribution of the continents and ocean basins have changed through time, as have the atmosphere and biota. As scientists and concerned citizens, we have become increasingly aware of how fragile our planet is and, more importantly, how interdependent all of its various systems and subsystems are.

We have also learned that we cannot continually pollute our environment and that our natural resources are limited and, in most cases, nonrenewable. Furthermore, we are coming to realize how central geology is to our everyday lives. For these and other reasons, geology is one of the most important college or university courses a student can take.

Historical geologists are concerned with all aspects of Earth and its life history. They seek to determine what events occurred during the past, place those events into an orderly chronological sequence, and provide conceptual frameworks for explaining such events. Equally important is using the lessons learned from the geologic past to understand and place in context some of the global issues facing the world today, such as depletion of natural resources, global climate warming, and decreasing biodiversity. Thus, what makes historical geology both fascinating and relevant is that, like the dynamic Earth it seeks to understand, it is an exciting and ever-changing science in which new discoveries and insights are continually being made.

Historical Geology: Evolution of Earth and Life Through Time, sixth edition, is designed for a one-semester geology course and is written with students in mind. One of the problems with any science course is that students are overwhelmed by the amount of material that must be learned. Furthermore, most of the material does not seem to be linked by any unifying theme and does not always appear to be relevant to their lives. This book, however, is written to address that problem in that it shows, in its easy-to-read style, that historical geology is an exciting science, and one that is increasingly relevant in today's world.

The goal of this book is to provide students with an understanding of the principles of historical geology and how these principles are applied in unraveling Earth's history. It is our intent to present the geologic and biologic history of Earth, not as a set of encyclopedic facts to memorize, but rather as a continuum of interrelated events reflecting the underlying geologic and biologic principles and processes that have shaped our planet and life upon it. Instead of emphasizing individual, and seemingly unrelated events, we seek to understand the underlying causes of why things happened the way they did and how all of Earth's systems and subsystems are interrelated. Using this approach, students will gain a better understanding of how everything fits together, and why events occurred in a particular sequence.

Because of the nature of the science, all historical geology textbooks share some broad similarities. Most begin with several chapters on concepts and principles, followed by a chronological discussion of Earth and life history. In this respect we have not departed from convention. We have, however, attempted to place greater emphasis on basic concepts and principles, their historical development, and their importance in deciphering Earth history; in other words, how do we know what we know. By approaching Earth history in this manner, students come to understand Earth's history as part of a dynamic and complex integrated system, and not as a series of isolated and unrelated events.

New and Retained Features in the Sixth Edition
Just as Earth is dynamic and evolving, so too is *Historical Geology: Evolution of Earth and Life Through Time*. The sixth edition has undergone considerable rewriting and updating, resulting in a book that is still easy to read and has a high level of current information, many new photographs and figures, as well as several new Perspectives, all of which are designed to help students maximize their learning and understanding of their planet's history. Drawing on the comments and suggestions of reviewers and users of the fifth edition, we have

incorporated a number of new features into this edition, as well as keeping the features that were successful in the previous edition.

- The *Chapter Objectives* outline at the beginning of each chapter has been retained to alert students to the key points that the chapter will address.
- An expanded coverage of the origin of the universe and solar system, as well as a new *Perspective* on the planets have been added to Chapter 1.
- Chapter 2 (Earth Materials), which provides the necessary background for those students who are unfamiliar with minerals, rocks, and the rock cycle has been retained.
- Chapter content has been rewritten to help clarify concepts and make the material more exploratory.
- An emphasis is placed on global climate warming throughout the text.
- Many of the *Perspectives* contain either new topics or have been updated.
- The art program has undergone revision to better illustrate the material covered in the text. In addition, the figure captions have been expanded and improved to help explain what students are seeing.
- Many of the figures throughout the book are now designated as "active figures" in which students can access the Geology Resource Center to view animations of these figures.
- The format of about 10 multiple-choice questions and 10 short-answer questions has been retained in the *Review Questions* section at the end of each chapter. Answers to all of the multiple-choice questions are provided at the back of the book. In addition, the *Apply Your Knowledge* questions that include both thought-provoking and quantitative questions that require students to apply what they have learned to solving geologic problems, has also been retained.
- The *Epilogue* has been expanded to include not only global climate warming issues, but sections on acid rain and ozone depletion. The *Epilogue* is designed to tie together current issues with the historical perspective of geology presented in the previous 19 chapters.

It is our strong belief that the rewriting and updating done in the text, as well as the revision of the art program, have greatly improved the sixth edition of *Historical Geology: Evolution of Earth and Life Through Time*. We think that these changes and enhancements make the textbook easier to read and comprehend, as well as a more effective teaching tool that engages students in the learning process, thereby fostering a better understanding of the material.

Text Organization

As in the previous editions, we develop three major themes in this textbook that are essential to the interpretation and appreciation of historical geology, introduce these themes early, and reinforce them throughout the book. These themes are *plate tectonics* (Chapter 3), a unifying theory for interpreting much of Earth's physical history and, to a large extent, its biological history; *time* (Chapter 4), the dimension that sets historical geology apart from most of the other sciences; and *evolutionary theory* (Chapter 7), the explanation for inferred relationships among living and fossil organisms. Additionally, we have emphasized the intimate interrelationship existing between physical and biological events, and the fact that Earth is a complex, dynamic, and evolving planet whose history is best studied by using a systems approach.

This book was written for a one-semester course in historical geology to serve both majors and non-majors in geology and in the Earth sciences. We have organized *Historical Geology: Evolution of Earth and Life Through Time*, sixth edition, into the following informal categories:

- Chapter 1 reviews the principles and concepts of geology and the three themes this book emphasizes. The text is written at an appropriate level for those students taking historical geology with no prerequisites, but the instructor may have to spend more time expanding some of the concepts and terminology discussed in Chapter 1.
- Chapter 2 "Earth Materials—Minerals and Rocks," can be used to introduce those students who have not had an introductory geology course to minerals and rocks, or as a review for those students that have had such a course.
- Chapter 3 explores plate tectonics, which is the first major theme of this book. Particular emphasis is placed on the evidence substantiating plate tectonic theory, why this theory is one of the cornerstones of geology, and why plate tectonic theory serves as a unifying paradigm in explaining many apparently unrelated geologic phenomena.
- The second major theme of this book, the concepts and principles of geologic time, is examined in Chapter 4.
- Chapter 5 expands on that theme by integrating geologic time with rocks and fossils.
- Depositional environments are sometimes covered rather superficially (perhaps little more than a summary table) in some historical geology textbooks. Chapter 6, "Sedimentary Rocks—The Archives of Earth History," is completely devoted to this topic; it contains sufficient detail to be meaningful, but avoids an overly detailed discussion more appropriate for advanced courses.
- The third major theme of this book, organic evolution, is examined in Chapter 7. In this chapter, the theory of evolution is covered as well as its supporting evidence.
- Precambrian time—fully 88% of all geologic time— is sometimes considered in a single chapter in other historical geology textbooks. However, in this book, Chapters 8 and 9 are devoted to the geologic and

biologic histories of the Archean and Proterozoic eons, respectively, with much updating on this early period of Earth history, as well as a discussion on the Precambrian time subdivisions.

- Chapters 10 through 19 constitute our chronological treatment of the Phanerozoic geologic and biologic history of Earth. These chapters are arranged so that the geologic history of an era is followed by a discussion of the biologic history of that era. We think that this format allows easier integration of life history with geologic history.

- An Epilogue summarizes the major topics and themes of this book, with an emphasis on global climate warming.

In these chapters, there is an integration of the three themes of this book as well as an emphasis on the underlying principles of geology and how they helped decipher Earth's history. We have found that presenting the material in the order discussed above works well for most students. We know, however, that many instructors prefer an entirely different order of topics, depending on the emphasis in their course. We have, therefore, written this book so that instructors can present the chapters in whatever order that suits the needs of a particular course.

Chapter Organization
All chapters have the same organizational format as follows:

- Each chapter opens with a photograph relating to the chapter material, an *Outline* of the topics covered, and a *Chapter Objectives* outline that alerts the student to the learning outcome objectives of the chapter.

- An *Introduction* follows that is intended to stimulate interest in the chapter and show how the chapter material fits into the larger geologic perspective.

- The text is written in a clear informal style, making it easy for students to comprehend.

- Numerous color diagrams and photographs complement the text and provide a visual representation of the concepts and information presented.

- All of the chapters, except Chapter 2, contain a *Perspective* that presents a brief discussion of an interesting aspect of historical geology or geological research pertinent to that chapter, and many of them have been revised or replaced from the fifth edition.

- Each of the chapters on geologic history in the second half of this book contains a final section on mineral resources characteristic of that time period. These sections provide applied economic material of interest to students.

- The end-of-chapter materials begin with a concise review of important concepts and ideas in the *Summary*.

- The *Important Terms*, which are printed in boldface type in the chapter text, are listed at the end of each chapter for easy review, along with the page number where that term is first defined. A full *Glossary* of important terms appears at the end of the text.

- The *Review Questions* are another important feature of this book. They include multiple-choice questions with answers as well as short essay questions. Many new multiple-choice questions as well as short answer and essay questions have been added to each chapter of the sixth edition.

- The *Apply Your Knowledge* section includes questions that are thought-provoking and quantitative and include some of the questions posed in the *What Would You Do?* boxes in previous editions.

- The global paleogeographic maps that illustrate in stunning relief the geography of the world during various time periods have been retained in this edition. These maps enable students to visualize what the world looked like during the time period being studied and add a visualization dimension to the text material.

- As in the previous editions, end-of-chapter summary tables are provided for the chapters on geologic and biologic history. These tables are designed to give an overall perspective of the geologic and biologic events that occurred during a particular time interval and to show how these events are interrelated. The emphasis in these tables is on the geologic evolution of North America. Global tectonic events and sea-level changes are also incorporated into these tables to provide global insights and perspective.

Ancillary Materials

We are pleased to offer a full suite of text and multimedia products to accompany the sixth edition of *Historical Geology: Evolution of Earth and Life Through Time*.

For Instructors

Online Instructor's Manual with Test Bank This comprehensive manual is designed to help instructors prepare for lectures. It contains lecture outlines, teaching suggestions and tips, and an expanded list of references, plus a bank of test questions. (ISBN: 0-495-56013-8)

ExamView® Computerized Testing Create, deliver, and customize tests and study guides (both print and online) in minutes with this easy-to-use assessment and tutorial system. ExamView offers both a *Quick Test Wizard* and an *Online Test Wizard* that guide you step-by-step through the process of creating tests, while its unique capabilities allow you to see the test you are creating on the screen exactly as it will print or display online. You can build tests of up to 250 questions using up to 12 question types. Using *Exam-View's* complete word processing capabilities, you can enter an unlimited number of new questions or edit existing questions. (ISBN: 0-495-56015-4)

PowerLecture™: A 1-stop Microsoft® PowerPoint® Tool
PowerLecture instructor resources are a collection of

book-specific lecture and class tools on DVD. The fastest and easiest way to build powerful, customized media-rich lectures, PowerLecture assets include chapter-specific PowerPoint presentations, images, animations and video, instructor manual, test bank, and videos. PowerLecture media-teaching tools are an effective way to enhance the educational experience. (ISBN: 0-495-56012-X)

For Students

Geology Resource Center   This password-protected website features an array of resources to complement your experience with geology—animations, videos, current news feeds, and Google Earth activities. If access to the Geology Resource Center is not packaged with your textbook, you can purchase access electronically at www.iChapters.com using ISBN-10: 0-495-60570-0 ISBN-13: 978-0-495-60570-6.

Student Companion Site   This book-specific website contains resources to help you study—flashcards, glossary, quizzes, chapter objectives, and chapter summaries. Visit www.cengage.com/earthscience/wicander/historical6e.

Acknowledgments

As the authors, we are, of course, responsible for the organization, style, and accuracy of the text, and any mistakes, omissions, or errors are our responsibility. The finished product is the culmination of many years of work during which we received numerous comments and advice from many geologists who reviewed parts of the text. We wish to express our sincere appreciation to the reviewers who reviewed the fifth edition and made many helpful and useful comments that led to the improvements seen in this sixth edition.

Dr. Allen I. Benimoff, *College of Staten Island*
Stuart Birnbaum, *University of Texas, San Antonio*
Claudia Bolze, *Tulsa Community College*
James L. Carew, *College of Charleston*
Mitchell W. Colgan, *College of Charleston*
Bruce Corliss, *Duke University*
John F. Cottrell, *Monroe Community College*
Michael Dalman, *Blinn College*
Mike Farabee, *Estrella Mountain Community College*
Chad Ferguson, *Bucknell University*
Robert Fillmore, *Western State College of Colorado*
Michael Fix, *University of Missouri, St. Louis*
Sarah Fowell, *University of Alaska, Fairbanks*
Dr. Heather Gallacher, *Cleveland State University*
Richard Gottfrief, *Frederick Community College*
Richard D. Harnell, *Monroe Community College*
Rob Houston, *Northwestern Michigan College*
Amanda Julson, *Blinn College*
Lanna Kopachena, *Eastfield College*
Niranjala Kottachchi, *Fresno City College*
Steve LoDuca, *Eastern Michigan University*
Ntungwa Maasha, *Coastal Georgia Community College*

L. Lynn Marquez, *Millersville University*
Pamela Nelson, *Glendale Community College*
Christine O'Leary, *Wallace State Community College*
Dr. Mark W. Presley, *Eastfield College (Dallas County Community College District)*
Gary D. Rosenberg, *Indiana University-Purdue University*
Michael Rygel, *SUNY Potsdam*
Steven Schimmrich, *SUNY, Ulster County Community College*
Michelle Stoklosa, *Boise State University*
Edmund Stump, *Arizona State University*
David Sunderlin, *Lafayette College*
Donald Swift, *Old Dominion University*
Shawn Willsey, *College of Southern Idaho*
Mark A. Wilson, *College of Wooster*

We would like to thank the reviewers of the first four editions. Their comments and suggestion resulted in many improvements and pedagogical innovation, making this book a success for students and instructors alike.

Paul Belasky, *Ohlone College*
Thomas W. Broadhead, *University of Tennessee at Knoxville*
Mark J. Camp, *University of Toledo*
James F. Coble, *Tidewater Community College*
William C. Cornell, *University of Texas at El Paso*
Rex E. Crick, *University of Texas at Arlington*
Chris Dewey, *Mississippi State University*
Dean A. Dunn, *The University of Southern Mississippi*
Richard Fluegeman, Jr., *Ball State University*
Annabelle Foos, *University of Akron*
Susan Goldstein, *University of Georgia*
Joseph C. Gould, *University of South Florida*
Bryan Gregor, *Wright State University*
Thor A. Hansen, *Western Washington University*
Paul D. Howell, *University of Kentucky*
Jonathan D. Karr, *North Carolina State University*
William W. Korth, *Buffalo State College*
R. L. Langenheim, Jr., *University of Illinois at Urbana-Champaign*
Michael McKinney, *University of Tennessee at Knoxville*
George F. Maxey, *University of North Texas*
Glen K. Merrill, *University of Houston-Downtown*
Arthur Mirsky, *Indiana University, Purdue University at Indianapolis*
N. S. Parate, *HCC-NE College, Pinemont Center*
William C. Parker, *Florida State University*
Anne Raymond, *Texas A & M University*
G. J. Retallack, *University of Oregon*
Mark Rich, *University of Georgia*
Scott Ritter, *Brigham Young University*
Gary D. Rosenberg, *Indiana University-Purdue University, Indianapolis*
Barbara L. Ruff, *University of Georgia*
W. Bruce Saunder, *Bryn Mawr College*
William A. Smith, *Charleston Southern University*
Ronald D. Stieglitz, *University of Wisconsin, Green Bay*
Carol M. Tang, *Arizona State University*

Jane L. Teranes, *Scripps Institution of Oceanography*
Michael J. Tevesz, *Cleveland State University*
Matthew S. Tomaso, *Montclair State University*
Art Troell, *San Antonio College*
Robert A. Vargo, *California University of Pennsylvania*

We also wish to thank Kathy Benison, Richard V. Dietrich (Professor Emeritus), David J. Matty, Jane M. Matty, Wayne E. Moore (Professor Emeritus), and Sven Morgan of the Geology Department, and Bruce M. C. Pape (Emeritus) of the Geography Department of Central Michigan University, as well as Eric Johnson (Hartwick College, New York) and Stephen D. Stahl (St. Bonaventure, New York) for providing us with photographs and answering our questions concerning various topics. We are also grateful for the generosity of the various agencies and individuals from many countries who provided photographs.

Special thanks must go to Marcus Boggs, West Coast Editorial Director at Cengage Learning, who initiated this sixth edition and encouraged us throughout its revision, as well as Yolanda Cossio, Brooks/Cole sciences publisher for her help and encouragement. We are equally indebted to our content project manager Hal Humphrey for all his help, and to Amanda Maynard of Pre-Press PMG for her help and guidance as our project manager. Liana Monari, our development editor, assisted us in many aspects of this edition and we appreciate all of her help. We would also like to thank Christine Hobberlin for her copyediting skills. We appreciate her help in improving our manuscript. Lisa Buckley is the interior designer responsible for another visually-pleasing edition. We thank Bob Kauser, text permissions manager for securing the necessary permissions, and Terri Wright for her invaluable help in locating appropriate photos.

Because historical geology is largely a visual science, we extend thanks to the artists at Precision Graphics, Graphic World, and Pre-Press PMG who were responsible for much of the art program. We also thank the artists at Magellan Geographix, who rendered many of the maps, and Dr. Ron Blakey, who allowed us to use his global paleogeographic maps.

As always, our families were very patient and encouraging when most of our spare time and energy were devoted to this book. We thank them for their continued support and understanding.

Reed Wicander
James S. Monroe

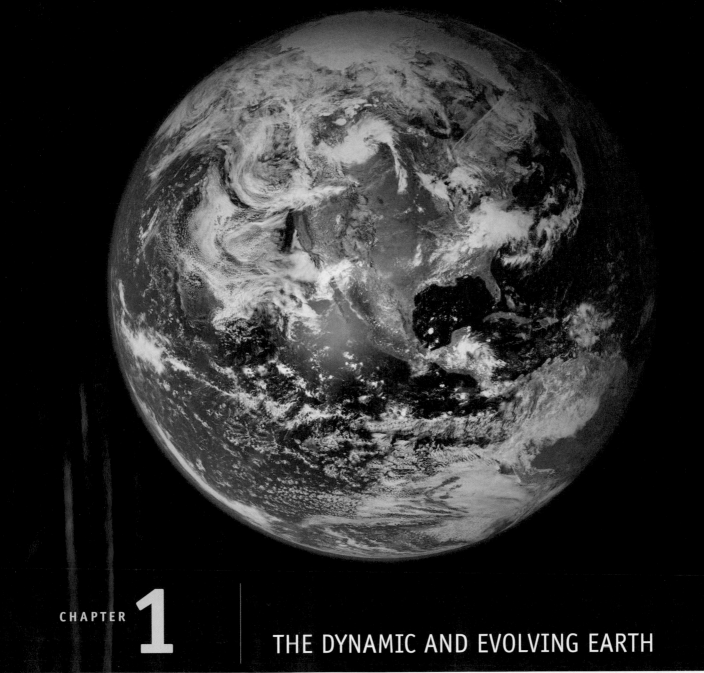

Courtesy of NASA Goddard Space Flight Center/Reto Stockli

CHAPTER **1**

THE DYNAMIC AND EVOLVING EARTH

Satellite-based image of Earth. North America is clearly visible in the center of this view, as well as Central America and the northern part of South America. The present locations of continents and ocean basins are the result of plate movements. The interaction of plates through time has affected the physical and biological history of Earth.

[OUTLINE]

Introduction

What Is Geology?

Historical Geology and the Formulation of Theories

Origin of the Universe and Solar System, and Earth's Place in Them

   Origin of the Universe—Did It Begin with a Big Bang?

   Our Solar System—Its Origin and Evolution

   Perspective *The Terrestrial and Jovian Planets*

   Earth—Its Place in Our Solar System

Why Is Earth a Dynamic and Evolving Planet?

Organic Evolution and the History of Life

Geologic Time and Uniformitarianism

How Does the Study of Historical Geology Benefit Us?

Summary

At the end of this chapter, you will have learned that

- Earth is a complex, dynamic planet that has continually evolved since its origin some 4.6 billion years ago.

- To help understand Earth's complexity and history, it can be viewed as an integrated system of interconnected components that interact and affect one another in various ways.

- Theories are based on the scientific method and can be tested by observation and/or experiment.

- The universe is thought to have originated approximately 14 billion years ago with a Big Bang, and the solar system and planets evolved from a turbulent, rotating cloud of material surrounding the embryonic Sun.

- Earth consists of three concentric layers—core, mantle, and crust—and this orderly division resulted during Earth's early history.

- Plate tectonics is the unifying theory of geology and this theory revolutionized the science.

- The theory of organic evolution provides the conceptual framework for understanding the evolution of Earth's fauna and flora.

- An appreciation of geologic time and the principle of uniformitarianism is central to understanding the evolution of Earth and its biota.

- Geology is an integral part of our lives.

Introduction

What kind of movie would we have if it were possible to travel back in time and film Earth's history from its beginning 4.6 billion years ago? It would certainly be a story of epic proportions, with incredible special effects, a cast of trillions, a plot with twists and turns—and an ending that is still a mystery!

Unfortunately, we can't travel back in time, but we *can* tell the story of Earth and its inhabitants, because that history is preserved in its geologic record. In this book, you will learn how to decipher that history from the clues Earth provides. Before we learn about the underlying principles of historical geology and how to use those principles and the clues preserved in Earth's geologic record, let's take a sneak preview of the full-length feature film *The History of Earth*.

In this movie, we would see a planet undergoing remarkable change as continents moved about its surface. As a result of these movements, ocean basins would open and close, and mountain ranges would form along continental margins or where continents collided with each other. Oceanic and atmospheric circulation patterns would shift in response to the moving continents, sometimes causing massive ice sheets to form, grow, and then melt away. At other times, extensive swamps or vast interior deserts would appear.

We would also witness the first living cells evolving from a primordial organic soup, sometime between 4.6 and 3.6 billion years ago. Somewhere around 1.5 billion years later, cells with a nucleus would evolve, and not long thereafter multi-celled soft-bodied animals would make their appearance in the world's oceans, followed in relatively short order by animals with skeletons, and then animals with backbones.

Up until about 450 million years ago, Earth's landscape was essentially barren. At that time, however, the landscape comes to life as plants and animals move from their home in the water to take up residency on land. Viewed from above, Earth's landmasses would take on new hues and colors as different life-forms began inhabiting the terrestrial environment. From that moment on, Earth would never be the same, as plants, insects, amphibians, reptiles, birds, and mammals made the land their home. Near the end of our film, humans evolve, and we see how their activities greatly impact the global ecosystem. It seems only fitting that the movie's final image is of Earth, a shimmering blue-green oasis in the black void of space and a voice-over saying, "To be continued."

Every good movie has a theme, and the major theme of *The History of Earth* is that Earth is a complex, dynamic planet that has changed continuously since its origin some 4.6 billion years ago. Because of its epic nature, three interrelated sub-themes run throughout *The History of Earth*. The first is that Earth's outermost part is composed of a series of moving plates (*plate tectonics*) whose interactions have affected the planet's physical and biological history. The second is that Earth's biota has evolved or changed throughout its history (*organic evolution*). The third is that the physical and biological changes that occurred did so over long periods of time (*geologic* or *deep time*). These three interrelated sub-themes are central to our understanding and appreciation of our planet's history.

As you study and read the various topics covered in this book, keep in mind that the themes and topics discussed in this chapter and throughout the book are like the interconnected components of a system, and not just isolated and unrelated pieces of information. By relating each chapter's topic to its place in the entire Earth system, you will gain a greater appreciation of Earth's evolution and the role of its various interacting internal and external systems, subsystems, and cycles.

By viewing Earth as a whole—that is, thinking of it as a system—we not only see how its various components are interconnected, but we can also better appreciate its complex and dynamic nature. The system concept makes it easier for us to study a complex subject, such as Earth, because it divides the whole into smaller components that we can easily understand, without losing sight of how the separate components fit together as a whole.

A **system** is a combination of related parts that interact in an organized manner. We can thus consider Earth as a system of interconnected components that interact and affect each other in many different ways. The principal subsystems of Earth are the *atmosphere, biosphere, hydrosphere, lithosphere, mantle,* and *core* (• Figure 1.1). The complex interactions among these subsystems result in a dynamically changing planet in which matter and energy

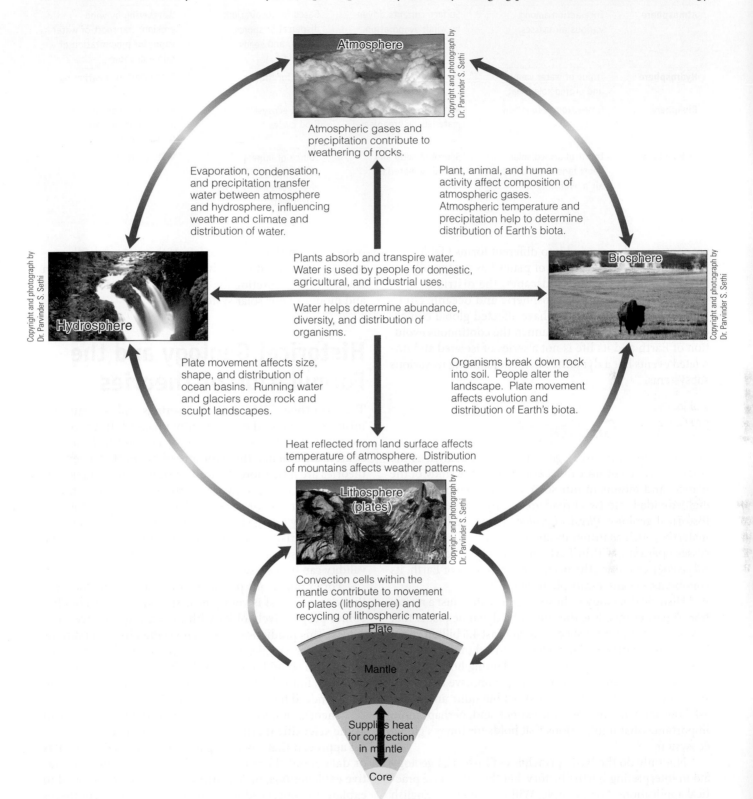

Atmospheric gases and precipitation contribute to weathering of rocks.

Evaporation, condensation, and precipitation transfer water between atmosphere and hydrosphere, influencing weather and climate and distribution of water.

Plant, animal, and human activity affect composition of atmospheric gases. Atmospheric temperature and precipitation help to determine distribution of Earth's biota.

Plants absorb and transpire water. Water is used by people for domestic, agricultural, and industrial uses.

Water helps determine abundance, diversity, and distribution of organisms.

Plate movement affects size, shape, and distribution of ocean basins. Running water and glaciers erode rock and sculpt landscapes.

Organisms break down rock into soil. People alter the landscape. Plate movement affects evolution and distribution of Earth's biota.

Heat reflected from land surface affects temperature of atmosphere. Distribution of mountains affects weather patterns.

Convection cells within the mantle contribute to movement of plates (lithosphere) and recycling of lithospheric material.

Supplies heat for convection in mantle

Atmosphere

Hydrosphere

Biosphere

Lithosphere (plates)

Plate

Mantle

Core

Copyright and photograph by Dr. Parvinder S. Sethi

• **Figure 1.1 Subsystems of Earth** The atmosphere, hydrosphere, biosphere, lithosphere, mantle, and core are all subsystems of Earth. This simplified diagram shows how these subsystems interact, with some examples of how materials and energy are cycled throughout the Earth system. The interactions between these subsystems make Earth a dynamic planet that has evolved and changed since its origin 4.6 billion years ago.

TABLE 1.1 Interactions among Earth's Principal Subsystems

	Atmosphere	Hydrosphere	Biosphere	Lithosphere
Atmosphere	Interaction among various air masses	Surface currents driven by wind; evaporation	Gases for respiration; dispersal of spores, pollen, and seeds by wind	Weathering by wind erosion; transport of water vapor for precipitation of rain and snow
Hydrosphere	Input of water vapor and stored solar heat	Hydrologic cycle	Water for life	Precipitation; weathering and erosion
Biosphere	Gases from respiration	Removal of dissolved materials by organisms	Global ecosystems; food cycles	Modification of weathering and erosion processes; formation of soil
Lithosphere	Input of stored solar heat; landscapes affect air movements	Source of solid and dissolved materials	Source of mineral nutrients; modification of ecosystems by plate movements	Plate tectonics

are continuously recycled into different forms (Table 1.1). For example, the movement of plates has profoundly affected the formation of landscapes, the distribution of mineral resources, and atmospheric and oceanic circulation patterns, that in turn have affected global climate changes. Examined in this manner, the continuous evolution of Earth and its life is not a series of isolated and unrelated events but a dynamic interaction among its various subsystems.

What Is Geology?

Geology, from the Greek *geo* and *logos,* is defined as the study of Earth, but now must also include the study of the planets and moons in our solar system. Geology is generally divided into two broad areas: physical geology and historical geology. *Physical geology* is the study of Earth materials, such as minerals and rocks, as well as the processes operating within Earth and on its surface. *Historical geology* examines the origin and evolution of Earth, its continents, oceans, atmosphere, and life.

Historical geology is, however, more than just a recitation of past events. It is the study of a dynamic planet that has changed continuously during the past 4.6 billion years. In addition to determining what occurred in the past, geologists are also concerned with explaining how and why past events happened. It is one thing to observe in the fossil record that dinosaurs went extinct but quite another to ask how and why they became extinct, and, perhaps more important, what implications that holds for today's global ecosystem.

Not only do the basic principles of historical geology aid in interpreting Earth's history, but they also have practical applications. For example, William Smith, an English surveyor and engineer, recognized that by studying the sequences of rocks and the fossils they contained, he could predict the kinds and thicknesses of rocks that would have to be excavated in the construction of canals. The same principles Smith used in the late 18th and early 19th centuries are still used today in mineral and oil exploration and also in interpreting the geologic history of the planets and moons of our solar system.

Historical Geology and the Formulation of Theories

The term **theory** has various meanings and is frequently misunderstood and consequently misused. In colloquial usage, it means a speculative or conjectural view of something—hence the widespread belief that scientific theories are little more than unsubstantiated wild guesses. In scientific usage, however, a theory is a coherent explanation for one or several related natural phenomena supported by a large body of objective evidence. From a theory, scientists derive predictive statements that can be tested by observations and/or experiments so that their validity can be assessed.

For example, one prediction of plate tectonic theory is that oceanic crust is young near spreading ridges but becomes progressively older with increasing distance from ridges. This prediction has been verified by observations (see Chapter 3). Likewise, according to the theory of evolution, fish should appear in the fossil record before amphibians, followed by reptiles, mammals, and birds—and that is indeed the case (see Chapter 7).

Theories are formulated through the process known as the **scientific method.** This method is an orderly, logical approach that involves gathering and analyzing facts or data about the problem under consideration. Tentative explanations, or **hypotheses,** are then formulated to explain the observed phenomena. Next, the hypotheses are tested to see if what they predicted actually occurs in a given situation. Finally, if one of the hypotheses is found, after repeated tests, to explain the phenomena, then that hypothesis is proposed as a theory. Remember, however, that in science even a theory is still subject to further testing and refinement as new data become available.

The fact that a scientific theory can be tested and is subject to such testing separates science from other forms of human inquiry. Because scientific theories can be tested, they have the potential for being supported or even proved wrong. Accordingly, science must proceed without any appeal to beliefs or supernatural explanations, not because such beliefs or explanations are necessarily untrue, but because we have no way to investigate them. For this reason, science makes no claim about the existence or nonexistence of a supernatural or spiritual realm.

Each scientific discipline has certain theories that are of particular importance. For example, the theory of organic evolution revolutionized biology when it was proposed in the 19th century. In geology, plate tectonic theory has changed the way geologists view Earth. Geologists now view Earth from a global perspective in which all of its subsystems and cycles are interconnected, and Earth history is seen as a continuum of interrelated events that are part of a global pattern of change.

Origin of the Universe and Solar System, and Earth's Place in Them

How did the universe begin? What has been its history? What is its eventual fate, or is it infinite? These are just some of the basic questions people have asked and wondered about since they first looked into the nighttime sky and saw the vastness of the universe beyond Earth.

Origin of the Universe—Did It Begin with a Big Bang?

Most scientists think that the universe originated about 14 billion years ago in what is popularly called the **Big Bang.** The Big Bang is a model for the evolution of the universe in which a dense, hot state was followed by expansion, cooling, and a less-dense state. According to modern *cosmology* (the study of the origin, evolution, and nature of the universe), the universe has no edge and therefore no center. Thus, when the universe began, all matter and energy were compressed into an infinitely small high-temperature and high-density state in which both time and space were set at zero. Therefore, there is no "before the Big Bang" but only what occurred after it. As demonstrated by Einstein's theory of relativity, space and time are unalterably linked to form a space–time continuum, that is, without space, there can be no time.

How do we know that the Big Bang took place approximately 14 billion years ago? Why couldn't the universe have always existed as we know it today? Two fundamental phenomena indicate that the Big Bang occurred. First, the universe is expanding, and second, it is permeated by background radiation.

When astronomers look beyond our own solar system, they observe that everywhere in the universe galaxies are moving away from each other at tremendous speeds. Edwin Hubble first recognized this phenomenon in 1929. By measuring the optical spectra of distant galaxies, Hubble noted that the velocity at which a galaxy moves away from Earth increases proportionally to its distance from Earth. He observed that the spectral lines (wavelengths of light) of the galaxies are shifted toward the red end of the spectrum; that is, the lines are shifted toward longer wavelengths. Galaxies receding from each other at tremendous speeds would produce such a redshift. This is an example of the *Doppler effect,* which is a change in the frequency of a sound, light, or other wave caused by movement of the wave's source relative to the observer (• Figure 1.2).

One way to envision how velocity increases with increasing distance is by reference to the popular analogy of a rising loaf of raisin bread in which the raisins are uniformly distributed throughout the loaf (• Figure 1.3). As the dough rises, the raisins are uniformly pushed away from each other at velocities directly proportional to the distance between any two raisins. The farther away a given raisin is to begin with, the farther it must move to maintain the regular spacing during the expansion, and hence the greater its velocity must be.

In the same way that raisins move apart in a rising loaf of bread, galaxies are receding from each other at a rate proportional to the distance between them, which is exactly what astronomers see when they observe the universe. By measuring this expansion rate, astronomers can calculate how long ago the galaxies were all together at a single point, which turns out to be about 14 billion years, the currently accepted age of the universe.

Arno Penzias and Robert Wilson of Bell Telephone Laboratories made the second important observation that provided evidence of the Big Bang in 1965. They

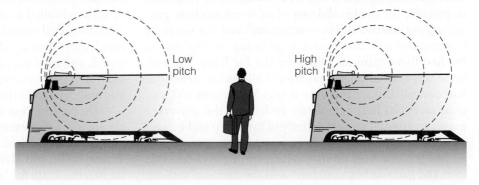

• Figure 1.2 **The Doppler Effect** The sound waves of an approaching whistle are slightly compressed so that the individual hears a shorter-wavelength, higher-pitched sound. As the whistle passes and recedes from the individual, the sound waves are slightly spread out, and a longer-wavelength, lower-pitched sound is heard.

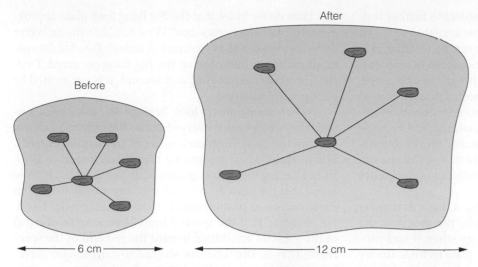

Before

After

6 cm

12 cm

• **Figure 1.3 The Expanding Universe** The motion of raisins in a rising loaf of raisin bread illustrates the relationship that exists between distance and speed and is analogous to an expanding universe. In this diagram, adjacent raisins are located 2 cm apart before the loaf rises. After one hour, any raisin is now 4 cm away from its nearest neighbor and 8 cm away from the next raisin over, and so on. Therefore, from the perspective of any raisin, its nearest neighbor has moved away from it at a speed of 2 cm per hour, and the next raisin over has moved away from it at a speed of 4 cm per hour. In the same way that raisins move apart in a rising loaf of bread, galaxies are receding from each other at a rate proportional to the distance between them.

discovered that there is a pervasive background radiation of 2.7 Kelvin (K) above absolute zero (absolute zero equals −273°C; 2.7 K = −270.3°C) everywhere in the universe. This background radiation is thought to be the fading afterglow of the Big Bang.

Currently, cosmologists cannot say what it was like at time zero of the Big Bang because they do not understand the physics of matter and energy under such extreme conditions. However, it is thought that during the first second following the Big Bang, the four basic forces—*gravity* (the attraction of one body toward another), *electromagnetic force* (the combination of electricity and magnetism into one force and binds atoms into molecules), *strong nuclear force* (the binding of protons and neutrons together), and *weak nuclear force* (the force responsible for the breakdown of an atom's nucleus, producing radioactive decay)—separated, and the universe experienced enormous cosmic inflation. By the end of the first three minutes following the Big Bang, the universe was cool enough that almost all nuclear reactions had ceased, and by the time it was 30 minutes old, nuclear reactions had completely ended and the universe's mass consisted of almost entirely of hydrogen and helium nuclei.

As the universe continued expanding and cooling, stars and galaxies began to form and the chemical makeup of the universe changed. Initially, the universe was 76% hydrogen and 24% helium, whereas today it is 70% hydrogen, 28% helium, and 2% all other elements by weight. How did such a change in the universe's composition

occur? Throughout their life cycle, stars undergo many nuclear reactions in which lighter elements are converted into heavier elements by nuclear fusion. When a star dies, often explosively, the heavier elements that were formed in its core are returned to interstellar space and are available for inclusion in new stars. In this way, the composition of the universe is gradually enhanced in heavier elements. In fact, it is estimated that when the universe is one trillion years old, it will consist of 20% hydrogen, 60% helium, and 20% all other elements.

Our Solar System—Its Origin and Evolution

Our solar system, which is part of the Milky Way Galaxy, consists of the Sun, eight planets, one dwarf planet (Pluto), 153 known moons or satellites (although this number keeps changing with the discovery of new moons and satellites surrounding the outer planets), a tremendous number of asteroids—most of which orbit the Sun in a zone between Mars and Jupiter—and millions of comets and meteorites, as well as interplanetary dust and gases (• Figure 1.4). Any theory formulated to explain the origin and evolution of our solar system must therefore take into account its various features and characteristics.

Many scientific theories for the origin of the solar system have been proposed, modified, and discarded since the French scientist and philosopher René Descartes first proposed, in 1644, that the solar system formed from a gigantic whirlpool within a universal fluid. Today, the **solar nebula theory** for the origin of our solar system involves the condensation and collapse of interstellar material in a spiral arm of the Milky Way Galaxy (• Figure 1.5).

The collapse of this cloud of gases and small grains into a counterclockwise-rotating disk concentrated about 90% of the material in the central part of the disk and formed an embryonic Sun, around which swirled a rotating cloud of material called a *solar nebula*. Within this solar nebula were localized eddies in which gases and solid particles condensed. During the condensation process, gaseous, liquid, and solid particles began accreting into ever-larger masses called *planetesimals,* which collided and grew in size and mass until they eventually became planets.

The composition and evolutionary history of the planets are a consequence, in part, of their distance from the Sun (see Perspective). The **terrestrial planets**—Mercury, Venus, Earth, and Mars—so named because they are similar to *terra,* Latin for "earth," are all small and composed of rock and metallic elements that condensed at the high temperatures of the inner nebula. The **Jovian planets**—Jupiter, Saturn, Uranus, and Neptune—so named because they resemble Jupiter (the Roman god was also named Jove) all

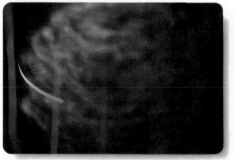

● Figure 1.4 **Diagrammatic Representation of the Solar System** This representation of the solar system shows the planets and their orbits around the Sun. On August 24, 2006, the International Astronomical Union downgraded Pluto from a planet to a dwarf planet. A dwarf planet has the same characteristics as a planet, except that it does not clear the neighborhood around its orbit. Pluto orbits among the icy debris of the Kuiper Belt, and therefore, does not meet the criteria for a true planet.

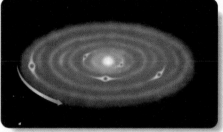

All NASA

a A huge rotating cloud of gas contracts and flattens

b to form a disk of gas and dust with the sun forming in the center,

c and eddies gathering up material to form planets.

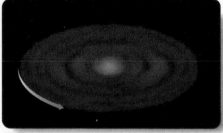

● Figure 1.5 **Solar Nebula Theory** According to the currently accepted theory for the origin of our solar system, the planets and the Sun formed from a rotating cloud of gas.

have small rocky cores compared to their overall size, and are composed mostly of hydrogen, helium, ammonia, and methane, which condense at low temperatures.

While the planets were accreting, material that had been pulled into the center of the nebula also condensed, collapsed, and was heated to several million degrees by gravitational compression. The result was the birth of a star, our Sun.

During the early accretionary phase of the solar system's history, collisions between various bodies were common, as indicated by the craters on many planets and moons. Asteroids probably formed as planetesimals in a localized eddy between what eventually became Mars and Jupiter in much the same way that other planetesimals

formed the terrestrial planets. The tremendous gravitational field of Jupiter, however, prevented this material from ever accreting into a planet. Comets, which are interplanetary bodies composed of loosely bound rocky and icy material, are thought to have condensed near the orbits of Uranus and Neptune.

The solar nebula theory of the formation of the solar system thus accounts for most of the characteristics of the planets and their moons, the differences in composition between the terrestrial and Jovian planets, and the presence of the asteroid belt. Based on the available data, the solar nebula theory best explains the features of the solar system and provides a logical explanation for its evolutionary history.

Perspective

The Terrestrial and Jovian Planets

The planets of our solar system are divided into two major groups that are quite different, indicating that the two underwent very different evolutionary histories (Figure 1). The four inner planets—Mercury, Venus, Earth, and Mars—are the terrestrial planets; they are small and dense (composed of a metallic core and silicate mantle-crust), ranging from no atmosphere (Mercury) to an oppressively thick one (Venus). The outer four planets (Pluto, is now considered a dwarf planet)—Jupiter, Saturn, Uranus, and Neptune—are the Jovian planets; they are large, ringed, low-density planets with liquid interior cores surrounded by thick atmospheres.

Mercury has a heavily cratered surface that has changed very little since its early history (Figure 2). Because Mercury is so small, its gravitational attraction is insufficient to retain atmospheric gases; any atmosphere that it may have held when it formed probably escaped into space quickly.

Venus is surrounded by an oppressively thick atmosphere that completely obscures its surface. However, radar images from orbiting spacecraft reveal a wide variety of terrains, including volcanic features, folded mountain ranges, and a complex network of faults (Figure 3).

Earth is unique among our solar system's planets in that it has a hospitable atmosphere, oceans of water, and a variety of climates, and it supports life (Figure 4).

The Moon is one-fourth the diameter of Earth, has a low density relative to the terrestrial planets, and is extremely dry (Figure 5). Its surface is divided into low-lying dark-colored plains and light-colored highlands that are heavily cratered, attesting to a period of massive meteorite bombardment in our solar system more than 4 billion years ago. The hypothesis, known as the *large-impact hypothesis*, that best accounts for the origin of the Moon has a giant planetesimal, the size of Mars or

larger, crashing into Earth 4.6 to 4.4 billion years ago, causing ejection of a large quantity of hot material that cooled and formed the Moon.

Mars has a thin atmosphere, little water, and distinct seasons (Figure 6). Its southern hemisphere is heavily cratered like the surfaces of Mercury and the Moon. The northern hemisphere has large smooth plains, fewer craters, and evidence of extensive volcanism. The largest volcano in the solar system is found in the northern hemisphere as are huge canyons, the largest of which, if present on Earth, would stretch from San Francisco to New York!

Jupiter is the largest of the Jovian planets (Figure 7). With its moons, rings, strong magnetic field, and intense radiation belts, Jupiter is the most complex and varied planet in our solar system. Jupiter's cloudy and violent atmosphere is divided into a series of different colored bands and a variety of spots (the Great

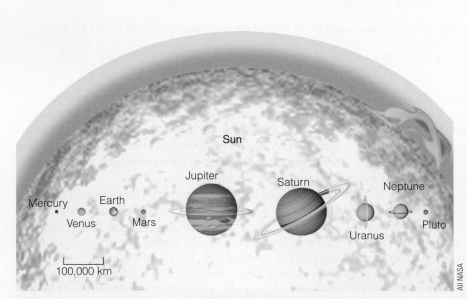

Sun
Jupiter
Saturn
Neptune
Mercury Earth
Venus Mars
Uranus
Pluto

100,000 km

All NASA

Figure 1 The relative sizes of the planets and the Sun. (Distances between planets are not to scale.)

Figure 2 Mercury

Red Spot) that interact in incredibly complex motions.

Saturn's most conspicuous feature is its ring system, consisting of thousands of rippling, spiraling bands of countless particles (Figure 8). The width of Saturn's rings would just reach from Earth to the Moon.

Uranus (Figure 9) is the only planet that lies on its side; that is, its axis of rotation nearly parallels the plane in which the planets revolve around the Sun. Some scientists thank that a collision with an Earth-sized body early in its history may have knocked Uranus on its side. Like the other Jovian planets, Uranus has a ring system, albeit a faint one.

Neptune is a dynamic stormy planet with an atmosphere similar to those of the other Jovian planets (Figure 10). Winds up to 2000 km/hr blow over the planet, creating tremendous storms, the largest of which, the Great Dark Spot, seen in the center of Figure 10, is nearly as big as Earth and is similar to the Great Red Spot on Jupiter.

Figure 3 Venus

Figure 4 Earth

Figure 5 The Moon

Figure 6 Mars

Figure 7 Jupiter

Figure 8 Saturn

Figure 9 Uranus

Figure 10 Neptune

Earth—Its Place in Our Solar System

Some 4.6 billion years ago, various planetesimals in our solar system gathered enough material together to form Earth and the other planets. Scientists think that this early Earth was probably cool, of generally uniform composition and density throughout, and composed mostly of silicates, compounds consisting of silicon and oxygen, iron and magnesium oxides, and smaller amounts of all the other chemical elements (• Figure 1.6a). Subsequently, when the combination of meteorite impacts, gravitational compression, and heat from radioactive decay increased the temperature of Earth enough to melt iron and nickel, this homogeneous composition disappeared (Figure 1.6b) and was replaced by a series of concentric layers of differing composition and density, resulting in a differentiated planet (Figure 1.6c).

This differentiation into a layered planet is probably the most significant event in Earth history. Not only did it lead to the formation of a crust and eventually continents, but it also was probably responsible for the emission of gases from the interior that eventually led to the formation of the oceans and atmosphere (see Chapter 8).

Why Is Earth a Dynamic and Evolving Planet?

Earth is a dynamic planet that has continuously changed during its 4.6-billion-year existence. The size, shape, and geographic distribution of continents and ocean basins have changed through time, the composition of the atmosphere has evolved, and life-forms existing today differ from those that lived during the past. Mountains and hills have been worn away by erosion, and the forces of wind, water, and ice have sculpted a diversity of landscapes. Volcanic eruptions and earthquakes reveal an active interior, and folded and fractured rocks indicate the tremendous power of Earth's internal forces.

Earth consists of three concentric layers: the core, the mantle, and the crust (• Figure 1.7). This orderly division results from density differences between the layers as a function of variations in composition, temperature, and pressure.

The **core** has a calculated density of 10–13 grams per cubic centimeter (g/cm^3) and occupies about 16% of Earth's total volume. Seismic (earthquake) data indicate that the core consists of a small, solid, inner region and a larger, apparently liquid, outer portion. Both are thought to consist largely of iron and a small amount of nickel.

The **mantle** surrounds the core and comprises about 83% of Earth's volume. It is less dense than the core (3.3–5.7 g/cm^3) and is thought to be composed largely of *peridotite*, a dark, dense igneous rock containing abundant iron and magnesium (see Figure 2.10). The mantle is divided into three distinct zones based on physical characteristics. The lower mantle is solid and forms most of the volume of Earth's interior. The **asthenosphere** surrounds the lower mantle. It has the same composition as the lower mantle but behaves plastically and slowly flows. Partial melting within the asthenosphere generates *magma* (molten material), some of which rises to the surface because it is less dense than the rock from which it was derived. The upper mantle surrounds the asthenosphere. The solid upper mantle and the overlying crust constitute the **lithosphere,** which is broken into numerous individual pieces called **plates**

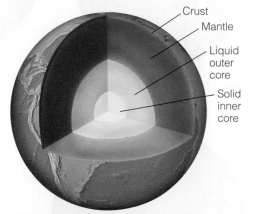

Crust
Mantle
Liquid outer core
Solid inner core

a Early Earth probably had a uniform composition and density throughout.

b The temperature of early Earth reached the melting point of iron and nickel, which, being denser than silicate minerals, settled to Earth's center. At the same time, the lighter silicates flowed upward to form the mantle and the crust.

c In this way, a differentiated Earth formed, consisting of a dense iron–nickel core, an iron-rich silicate mantle, and a silicate crust with continents and ocean basis.

• *Active* Figure 1.6 **Homogeneous Accretion Theory for the Formation of a Differentiated Earth** Visit the Geology Resource Center to view this and other active figures at www.cengage.com/sso.

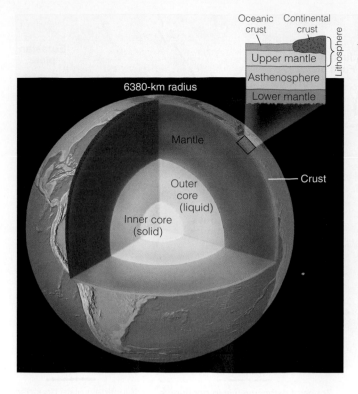

• Figure 1.7 **Cross Section of Earth Illustrating the Core, Mantle, and Crust** The enlarged portion shows the relationship between the lithosphere (composed of the continental crust, oceanic crust, and upper mantle) and the underlying asthenosphere and lower mantle.

silicon and aluminum. *Oceanic crust* is thin (5–10 km), denser than continental crust (3.0 g/cm³), and is composed of the dark igneous rocks *basalt* and *gabbro* (see Figures 2.11a and b).

The recognition that the lithosphere is divided into rigid plates that move over the asthenosphere forms the foundation of **plate tectonic theory** (• Figure 1.9). Zones of volcanic activity, earthquakes, or both mark most plate boundaries. Along these boundaries, plates diverge, converge, or slide sideways past each other (• Figure 1.10).

The acceptance of plate tectonic theory is recognized as a major milestone in the geologic sciences, comparable to the revolution that Darwin's theory of evolution caused in biology. Plate tectonic theory has provided a framework for interpreting the composition, structure, and internal processes of Earth on a global scale. It has led to the realization that the continents and ocean basins are part of a lithosphere-atmosphere-hydrosphere system that evolved together with Earth's interior (Table 1.2).

A revolutionary concept when it was proposed in the 1960s, plate tectonic theory has had far-reaching consequences in all fields of geology because it provides the basis for relating many seemingly unrelated phenomena. Besides being responsible for the major features of Earth's crust, plate movements also affect the formation and occurrence of Earth's natural resources, as well as influencing the distribution and evolution of the world's biota.

that move over the asthenosphere, partially as a result of underlying *convection cells* (• Figure 1.8). Interactions of these plates are responsible for such phenomena as earthquakes, volcanic eruptions, and the formation of mountain ranges and ocean basins.

The **crust,** Earth's outermost layer, consists of two types. *Continental crust* is thick (20–90 km), has an average density of 2.7 g/cm³, and contains considerable

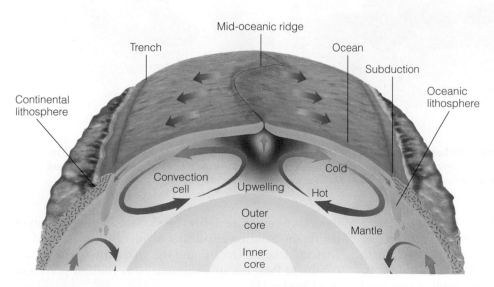

• *Active* Figure 1.8 **Movement of Earth's Plates** Earth's plates are thought to move partially as a result of underlying mantle convection cells in which warm material from deep within Earth rises toward the surface, cools, and then, upon losing heat, descends back into the interior, as shown in this diagrammatic cross section. Visit the Geology Resource Center to view this and other active figures at www.cengage.com/sso.

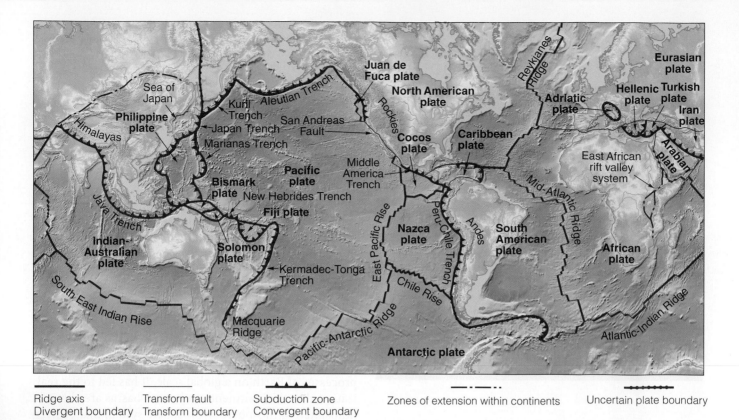

Ridge axis ▲▲▲▲ **Transform fault** **Subduction zone** ——·—··—· **Uncertain plate boundary**
Divergent boundary **Transform boundary** **Convergent boundary** **Zones of extension within continents**

● *Active* Figure 1.9 **Earth's Plates** Earth's lithosphere is divided into rigid plates of various sizes that move over the asthenosphere. Visit the Geology Resource Center to view this and other active figures at www.cengage.com/sso.

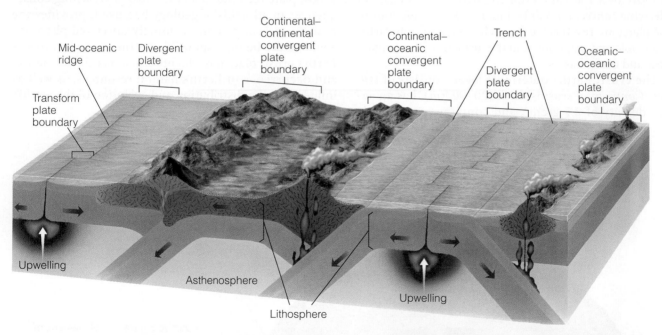

● *Active* Figure 1.10 **Relationship Between Lithosphere, Asthenosphere, and Plate Boundaries** An idealized block diagram illustrating the relationship between the lithosphere and the underlying asthenosphere and the three principal types of plate boundaries: divergent, convergent, and transform. Visit the Geology Resource Center to view this and other active figures at www.cengage.com/sso.

The impact of plate tectonic theory has been particularly notable in the interpretation of Earth's history. For example, the Appalachian Mountains in eastern North America and the mountain ranges of Greenland, Scotland, Norway, and Sweden are not the result of unrelated mountain-building episodes but, rather, are part of a larger mountain-building event that involved the closing of an ancient "Atlantic Ocean" and the formation of the supercontinent Pangaea approximately 251 million years ago (see Chapter 11).

TABLE 1.2	Plate Tectonics and Earth Systems

Solid Earth

Plate tectonics is driven by convection in the mantle and in turn drives mountain building and associated igneous and metamorphic activity.

Atmosphere

Arrangement of continents affects solar heating and cooling, and thus winds and weather systems. Rapid plate spreading and hot-spot activity may release volcanic carbon dioxide and affect global climate.

Hydrosphere

Continental arrangement affects ocean currents. Rate of spreading affects volume of mid-oceanic ridges and hence sea level. Placement of continents may contribute to onset of ice ages.

Biosphere

Movement of continents creates corridors or barriers to migration, the creation of ecological niches, and the transport of habitats into more or less favorable climates.

Extraterrestrial

Arrangement of continents affects free circulation of ocean tides and influences tidal slowing of Earth's rotation.

Source: Adapted by permission from Stephen Dutch, James S. Monroe, and Joseph Moran, Earth Science (Minneapolis/St. Paul: West Publishing Co., 1997).

Organic Evolution and the History of Life

Plate tectonic theory provides us with a model for understanding the internal workings of Earth and its effect on Earth's surface. The theory of **organic evolution** (whose central thesis is that all present-day organisms are related, and that they have descended with modifications from organisms that lived during the past) provides the conceptual framework for understanding the history of life. Together, the theories of plate tectonics and organic evolution have changed the way we view our planet, and we should not be surprised at the intimate association between them. Although the relationship between plate tectonic processes and the evolution of life is incredibly complex, paleontologic data provide indisputable evidence of the influence of plate movement on the distribution of organisms.

The publication in 1859 of Darwin's *On the Origin of Species by Means of Natural Selection* revolutionized biology and marked the beginning of modern evolutionary biology. With its publication, most naturalists recognized that evolution provided a unifying theory that explained an otherwise encyclopedic collection of biologic facts.

When Darwin proposed his theory of organic evolution, he cited a wealth of supporting evidence, including the way organisms are classified, embryology, comparative anatomy, the geographic distribution of organisms, and, to a limited extent, the fossil record. Furthermore, Darwin proposed that *natural selection,* which results in the survival to reproductive age of those organisms best adapted to their environment, is the mechanism that accounts for evolution.

Perhaps the most compelling evidence in favor of evolution can be found in the fossil record. Just as the geologic record allows geologists to interpret physical events and conditions in the geologic past, **fossils,** which are the remains or traces of once-living organisms, not only provide evidence that evolution has occurred but also demonstrate that Earth has a history extending beyond that recorded by humans. The succession of fossils in the rock record provides geologists with a means for dating rocks and allowed for a relative geologic time scale to be constructed in the 1800s.

Geologic Time and Uniformitarianism

An appreciation of the immensity of geologic time is central to understanding the evolution of Earth and its biota. Indeed, time is one of the main aspects that sets geology apart from the other sciences except astronomy. Most people have difficulty comprehending geologic time because they tend to think in terms of the human perspective—seconds, hours, days, and years. Ancient history is what occurred hundreds or even thousands of years ago. When geologists talk of ancient geologic history, however, they are referring to events that happened hundreds of millions, or even billions, of years ago. To a geologist, recent geologic events are those that occurred within the last million years or so.

It is also important to remember that Earth goes through cycles of much longer duration than the human perspective of time. Although they may have disastrous effects on the human species, global warming and cooling are part of a larger cycle that has resulted in numerous glacial advances and retreats during the past 1.8 million years. Because of their geologic perspective on time and how the various Earth subsystems and cycles are interrelated, geologists can make valuable contributions to many of the current environmental debates, such as those involving global warming and sea level changes, both topics of which are discussed in subsequent chapters.

The **geologic time scale** subdivides geologic time into a hierarchy of increasingly shorter time intervals; each time subdivision has a specific name. The geologic time scale resulted from the work of many 19th-century geologists who pieced together information from numerous rock exposures and constructed a chronology based on changes in Earth's biota through time. Subsequently, with the discovery of radioactivity in 1895 and the development of various radiometric dating techniques, geologists have been able to assign ages (also known as absolute ages) in years to the subdivisions of the geologic time scale (• Figure 1.11).

One of the cornerstones of geology is the **principle of uniformitarianism,** which is based on the premise that present-day processes have operated throughout geologic time. Therefore, in order to understand and interpret geologic events from evidence preserved in rocks, we must first understand present-day processes and their results. In fact, uniformitarianism fits in completely with the system approach we are following for the study of Earth.

Uniformitarianism is a powerful principle that allows us to use present-day processes as the basis for interpreting the past and for predicting potential future events. We should keep in mind, however, that uniformitarianism does not exclude sudden or catastrophic events such as volcanic eruptions, earthquakes, tsunami, landslides, or floods. These are processes that shape our modern world, and some geologists view Earth history as a series of such short-term or punctuated events. This view is certainly in keeping with the modern principle of uniformitarianism.

Furthermore, uniformitarianism does not require that the rates and intensities of geologic processes be constant through time. We know that volcanic activity was more intense in North America 5 to 10 million years ago than it is today and that glaciation has been more prevalent during the last several million years than in the previous 300 million years.

What uniformitarianism means is that even though the rates and intensities of geologic processes have varied during the past, the physical and chemical laws of nature have remained the same. Although Earth is in a dynamic state of change and has been ever since it formed, the processes that shaped it during the past are the same ones operating today.

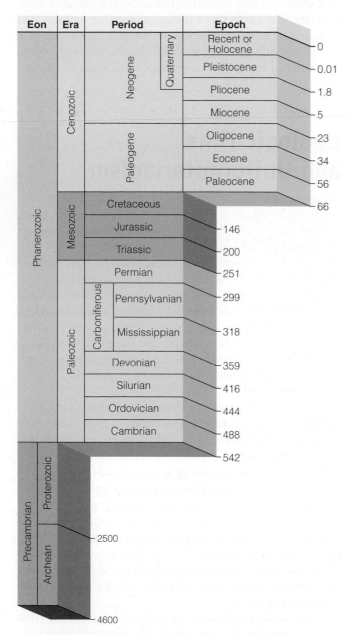

• **Figure 1.11 The Geologic Time Scale** The numbers to the right of the columns are the ages in millions of years before the present. Dates are from Gradstein, F., J. Ogg, and A. Smith. A Geologic Time Scale 2004 (Cambridge, UK: Cambridge University Press, 2005), Figure 1.2.

How Does the Study of Historical Geology Benefit Us?

The most meaningful lesson to learn from the study of historical geology is that Earth is an extremely complex planet in which interactions are taking place between its various subsystems and have been for the past 4.6 billion years. If we want to ensure the survival of the human species, we must understand how the various subsystems work and interact with each other and, more importantly, how our actions affect the delicate balance between these systems. We can do this, in part, by studying what has happened in the past, particularly on the global scale, and use that information to try to determine how our actions might affect the delicate balance between Earth's various subsystems in the future.

The study of geology goes beyond learning numerous facts about Earth. Moreover, we don't just study geology—we *live* it. Geology is an integral part of our lives. Our standard of living depends directly on our consumption of natural resources, resources that formed millions and billions of years ago. However, the way we consume natural resources and interact with the environment, as individuals and as a society, also determines our ability to pass on this standard of living to the next generation.

As you study the various topics and historical accounts covered in this book, keep in mind the three major themes (plate tectonics, geologic time, and organic evolution) discussed in this chapter and how, like the parts of a system, they are interrelated and responsible for the 4.6-billion-year history of Earth. View each chapter's topic in the context of how it fits in the whole Earth system, and remember that Earth's history is a continuum and the result of interaction between its various subsystems. In fact, the Epilogue addresses the issue of Earth's changing history in terms of the interaction of its various subsystems and cycles and the impact of humans as it relates to such issues as global warming both from the human and geologic time-frame perspective.

Historical geology is not a dry history of Earth but a vibrant, dynamic science in which what we see today is based on what went on before. As Stephen Jay Gould states in his book *Wonderful Life: The Burgess Shale and the Nature of History* (1989), "Play the tape of life again starting with the Burgess Shale, and a different set of survivors—not including vertebrates this time—would grace our planet today."

SUMMARY

- Scientists view Earth as a system of interconnected components that interact and affect each other. The principal subsystems of Earth are the atmosphere, hydrosphere, biosphere, lithosphere, mantle, and core. Earth is considered a dynamic planet that continually changes because of the interaction among its various subsystems and cycles.

- Geology, the study of Earth, is divided into two broad areas: Physical geology is the study of Earth materials, as well as the processes that operate within Earth and on its surface; historical geology examines the origin and evolution of Earth, its continents, oceans, atmosphere, and life.

- The scientific method is an orderly, logical approach that involves gathering and analyzing facts about a particular phenomenon, formulating hypotheses to explain the phenomenon, testing the hypotheses, and finally proposing a theory. A theory is a testable explanation for some natural phenomenon that has a large body of supporting evidence. Both the theory of organic evolution and plate tectonic theory are theories that revolutionized biology and geology, respectively.

- The universe began with the Big Bang approximately 14 billion years ago. Astronomers have deduced this age by observing that celestial objects are moving away from each other in an ever-expanding universe. Furthermore, the universe has a pervasive background radiation of 2.7 K above absolute zero (2.7 K = −270.3°C), which is thought to be the faint afterglow of the Big Bang.

- About 4.6 billion years ago, our solar system formed from a rotating cloud of interstellar matter. As this cloud condensed, it eventually collapsed under the influence of gravity and flattened into a counterclockwise rotating disk. Within this rotating disk, the Sun, planets, and moons formed from the turbulent eddies of nebular gases and solids.

- Earth formed from a swirling eddy of nebular material 4.6 billion years ago, accreting as a solid body and soon thereafter differentiated into a layered planet during a period of internal heating.

- Earth is differentiated into layers. The outermost layer is the crust, which is divided into continental and oceanic portions. The crust and underlying solid part of the upper mantle, also known as the lithosphere, overlie the asthenosphere, a zone that behaves plastically and flows slowly. The asthenosphere is underlain by the solid lower mantle. Earth's core consists of an outer liquid portion and an inner solid portion.

- The lithosphere is broken into a series of plates that diverge, converge, and slide sideways past one another.

- Plate tectonic theory provides a unifying explanation for many geological features and events. The interaction between plates is responsible for volcanic eruptions, earthquakes, the formation of mountain ranges and ocean basins, and the recycling of rock materials.

- The central thesis of the theory of organic evolution is that all living organisms are related and evolved (descended with modifications) from organisms that existed in the past.

- Time sets geology apart from the other sciences except astronomy, and an appreciation of the immensity of geologic time is central to understanding Earth's evolution. The geologic time scale is the calendar geologists use to date past events.

- The principle of uniformitarianism is basic to the interpretation of Earth history. This principle holds that the laws of nature have been constant through time and that the same processes operating today have operated in the past, although not necessarily at the same rates.

- Geology is an integral part of our lives. Our standard of living depends directly on our consumption of natural resources, resources that formed millions and billions of years ago. Furthermore, an appreciation of Earth history and the relationship between that history and the interaction of Earth's various subsystems and cycles is critical to an understanding of how humans are affecting this history and our role in such current issues as global warming.

IMPORTANT TERMS

asthenosphere, p. 10
Big Bang, p. 5
core, p. 10
crust, p. 11
fossil, p. 13
geologic time scale, p. 13
geology, p. 4

hypothesis, p. 4
Jovian planets, p. 6
lithosphere, p. 10
mantle, p. 10
organic evolution, p. 13
plate, p. 10
plate tectonic theory, p. 11

principle of uniformitarianism, p. 14
scientific method, p. 4
solar nebula theory, p. 6
system, p. 3
terrestrial planets, p. 6
theory, p. 4

REVIEW QUESTIONS

1. The change in frequency of a sound wave caused by movement of its source relative to an observer is know as the
   a. _____ Curie point; b. _____ Hubble shift; c. _____ Doppler effect; d. _____ Quasar force; e. _____ none of the previous answers.

2. The premise that present-day processes have operated throughout geologic time is the principle of
   a. _____ organic evolution; b. _____ plate tectonics; c. _____ uniformitarianism; d. _____ geologic time; e. _____ scientific deduction.

3. The study of the origin and evolution of Earth is
   a. _____ astronomy; b. _____ historical geology; c. _____ astrobiology; d. _____ physical geology; e. _____ paleontology.

4. The concentric layer that comprises most of Earth's volume is the
   a. _____ inner core; b. _____ outer core; c. _____ mantle; d. _____ asthenosphere; e. _____ crust.

5. Plates are composed of
   a. _____ the crust and upper mantle; b. _____ the asthenosphere and upper mantle; c. _____ the crust and asthenosphere; d. _____ continental and oceanic crust only; e. _____ the core and mantle.

6. A combination of related parts interacting in an organized fashion is
   a. _____ a cycle; b. _____ a theory; c. _____ uniformitarianism; d. _____ a hypothesis; e. _____ a system.

7. The movement of plates is thought to result from
   a. _____ density differences between the inner and outer core; b. _____ rotation of the mantle around the core; c. _____ gravitational forces; d. _____ the Coriolis effect; e. _____ convection cells.

8. Which of the following statements about a scientific theory is *not* true?
   a. _____ It is an explanation for some natural phenomenon; b. _____ It is a conjecture or guess; c. _____ It has a large body of supporting evidence; d. _____ It is testable. e. _____ Predictive statements can be derived from it.

9. What two observations led scientists to conclude that the Big Bang occurred approximately 14 billion years ago?
   a. _____ A steady-state universe and background radiation of 2.7 K above absolute zero; b. _____ A steady-state universe and opaque background radiation; c. _____ An expanding universe and opaque background radiation; d. _____ An expanding universe and background radiation of 2.7 K above absolute zero. e. _____ A shrinking universe and opaque background radiation.

10. That all living organisms are the descendants of different life-forms that existed in the past is the central claim of
    a. _____ the principle of fossil succession; b. _____ the principle of uniformitarianism; c. _____ plate tectonics; d. _____ organic evolution; e. _____ none of the previous answers.

11. Why is viewing Earth as a system a good way to study it? Are humans a part of the Earth system? If so, what role, if any, do we play in Earth's evolution?

12. Why is Earth considered a dynamic planet? What are its three concentric layers and their characteristics?

13. Explain how the principle of uniformitarianism allows for catastrophic events.

14. Why is plate tectonic theory so important to geology? How does it fit into a systems approach to the study of Earth?

15. What is the Big Bang? What evidence do we have that the universe began approximately 14 billion years ago?

16. How does the solar nebula theory account for the formation of our solar system, its features, and evolutionary history?

17. Discuss why plate tectonic theory is a unifying theory of geology.

18. Why is it important to have a basic knowledge and understanding of historical geology?

APPLY YOUR KNOWLEDGE

1. Describe how you would use the scientific method to formulate a hypothesis explaining the similarity of mountain ranges on the east coast of North America and those in England, Scotland, and the Scandinavian countries. How would you test your hypothesis?

2. Discuss why an accurate geologic time scale is particularly important for geologists in examining global temperature changes during the past and how an understanding of geologic time is crucial to the current debate on global warming and its consequences.

3. An important environmental issue facing the world today is global warming. How can this problem be approached from a global systems perspective? What are the possible consequences of global warming, and can we really do anything about it? Are there ways to tell if global warming occurred in the geologic past?

Sue Monroe

CHAPTER 2

MINERALS AND ROCKS

Mount Whitney (the highest peak on the right) in California and the adjacent peaks are made up of granite which in turn is composed of the minerals quartz, potassium feldspars, plagioclase feldspars, and small amounts of one or two other minerals. The rocks in this view make up the Sierra Nevada batholith, a huge body of rock that cooled from magma far below Earth's surface. Subsequent uplift of the Sierra Nevada and deep erosion has exposed these rocks at the surface.

[OUTLINE]

Introduction

Matter and Its Composition
   Elements and Atoms
   Bonding and Compounds

Minerals—The Building Blocks of Rocks
   How Many Minerals Are There?
   Rock-Forming Minerals and the Rock Cycle

Igneous Rocks
   Texture and Composition
   Classifying Igneous Rocks

Sedimentary Rocks
   Sediment Transport, Deposition, and Lithification
   Classification of Sedimentary Rocks

Metamorphic Rocks
   What Causes Metamorphism?
   Metamorphic Rock Classification

Plate Tectonics and the Rock Cycle

Summary

At the end of this chapter, you will have learned that

- Chemical elements are composed of atoms, all of the same kind, whereas compounds form when different atoms bond together. Most minerals are compounds, which are characterized as naturally occurring, inorganic, crystalline solids.

- Of the 3500 or so minerals known, only a few, perhaps two dozen, are common in rocks, but many others are found in small quantities in rocks and some are important natural resources.

- Cooling and crystallization of magma or lava and the consolidation of pyroclastic materials account for the origin of igneous rocks.

- Geologists use mineral content (composition) and textures to classify plutonic rocks (intrusive igneous rocks) and volcanic rocks (extrusive igneous rocks).

- Mechanical and chemical weathering of rocks yields sediment that is transported, deposited, and then lithified to form detrital and chemical sedimentary rocks.

- Texture and composition are the criteria geologists use to classify sedimentary rocks.

- Any type of rock may be altered by heat, pressure, fluids, or any combination of these, to form metamorphic rocks.

- One feature used to classify metamorphic rocks is foliation—that is, a platy or layered aspect, but some lack this feature and are said to be nonfoliated.

- The fact that Earth materials are continually recycled and that the three families of rocks are interrelated is summarized in the rock cycle.

Introduction

Ice probably does not come to mind when you hear the word mineral, and yet it is a mineral because it is a naturally occurring, inorganic, crystalline solid, meaning that its atoms of hydrogen and oxygen are arranged in a specific three-dimensional pattern, unlike the arrangement of atoms in liquids and gases. Furthermore, ice has a specific chemical composition (H_2O) and characteristic physical properties such as hardness and density. Thus, a **mineral** is a naturally occurring, inorganic crystalline solid, with a specified chemical composition, and distinctive physical properties (● Figure 2.1a, b). A **rock,** by contrast, is made up of one or more minerals (Figure 2.1c), just as a forest is made up of trees either of the same kind or different kinds.

Some minerals are beautiful and have been a source of fascination for thousands of years. As a matter of fact, several minerals and rocks have served as religious symbols or talismans, or have been worn, carried, applied externally, or ingested for their presumed mystical or curative powers. For example, diamond is supposed to ward off evil spirits, sickness, and floods, and relating gemstones to birth month gives them great appeal to many people.

Many minerals and rocks are important economically and account for complex economic and political ties between nations. The United States has no domestic production of manganese, a necessary element for the production of steel, nor does it produce any cobalt for use in magnets and corrosion- and wear-resistant alloys. Accordingly, the United States must import all the manganese and cobalt it uses as well as many other resources. Iron ore, on the other hand, is found in abundance in the Great Lakes region of the United States and Canada. Even some rather common minerals are important; quartz, the mineral that makes up most of the world's sand, is used to make glass, optical instruments, and sandpaper.

Rocks, too, find many uses. The phosphate used in fertilizers and animal feed supplements comes from phosphorous-rich rocks, and coal is an energy resource most of which is burned to generate electricity. Some rocks are crushed and used as aggregate in cement for roadbeds, foundations, and sidewalks, or they are cut and polished for countertops, mantelpieces, monuments, and tombstones. In fact, rock has been used for construction for thousands of years. Our primary interest in rocks, however, lies in the fact that they constitute the **geologic record,** which is the record of prehistoric physical and biologic events, a record that you will learn to decipher in this course.

Matter and Its Composition

Matter is anything that has mass and occupies space, so it exists as solids, liquids, gases, and plasma. The last is an ionized gas, as in neon lights and the Sun, but here we are interested mostly in solids, because by definition minerals are solids and rocks are composed of minerals. Liquids and gases are also important, though, because many features we see in the geologic record resulted from moving liquids (running water) and gases (wind).

Elements and Atoms
Matter consists of chemical **elements,** each of which is composed of tiny particles called **atoms,** the smallest units of matter that retain the characteristics of an element. Atoms have a compact nucleus made up of one or more *protons*—particles with a positive electrical charge—and electrically neutral *neutrons*. Negatively charged *electrons* rapidly orbit the nucleus at specific distances in one or more *electron shells* (● Figure 2.2).

• Figure 2.1 **Minerals and Rock**

a These beautiful emeralds are gemstones because they can be cut, polished, and used for decorative purposes. This specimen is on display at the Natural History Museum in Vienna, Austria.

b Quartz, a very common mineral, is an important part of many rocks. In fact, most of the world's sand is made up of quartz.

c This metamorphic rock, which is composed of several minerals, lies on the Lake Superior shoreline at Marquette, Michigan.

The number of protons in an atom's nucleus determines its **atomic number;** carbon has 6 protons, so its atomic number is 6 (Figure 2.2). An atom's **atomic mass number,** in contrast, is found by adding the number of protons and neutrons in the nucleus (electrons have negligible mass). However, the number of neutrons in the nucleus of an element might vary. Carbon atoms (with 6 protons) have 6, 7, or 8 neutrons, making three *isotopes,* or different forms, of carbon (• Figure 2.3). Some elements have only one isotope, but many have several.

Bonding and Compounds

The process whereby atoms join to other atoms is known as **bonding.** Should atoms of two or more elements bond, the resulting substance is a **compound.** Thus, gaseous oxygen is an element, whereas ice, made up of hydrogen and oxygen (H_2O), is a compound. Most minerals are compounds, but there are a few important exceptions.

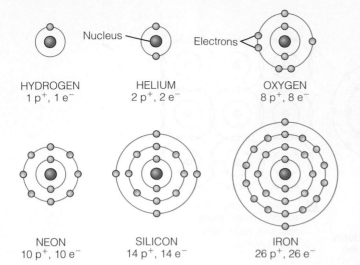

			Distribution of Electrons			
Element	Symbol	Atomic Number	First Shell	Second Shell	Third Shell	Fourth Shell
Hydrogen	H	1	1	—	—	—
Helium	He	2	2	—	—	—
Carbon	C	6	2	4	—	—
Oxygen	O	8	2	6	—	—
Neon	Ne	10	2	8	—	—
Sodium	Na	11	2	8	1	—
Magnesium	Mg	12	2	8	2	—
Aluminum	Al	13	2	8	3	—
Silicon	Si	14	2	8	4	—
Phosphorus	P	15	2	8	5	—
Sulfur	S	16	2	8	6	—
Chlorine	Cl	17	2	8	7	—
Potassium	K	19	2	8	8	1
Calcium	Ca	20	2	8	8	2
Iron	Fe	26	2	8	14	2

• **Figure 2.2 Shell Model for Atoms** The shell model for several atoms and their electron configurations. A circle represents the nucleus of each atom, but remember that atomic nuclei are made up of protons and neutrons as shown in Figure 2.3.

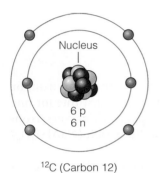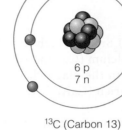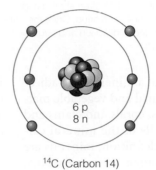

^{12}C (Carbon 12) ^{13}C (Carbon 13) ^{14}C (Carbon 14)

• **Figure 2.3 Isotopes of Carbon**
The element carbon has 6 protons (p) in its nucleus so its atomic number is 6, but its atomic mass number may be 12, 13, or 14 depending on how many neutrons (n) are in the nucleus.

Except for hydrogen with one proton and one electron, the innermost electron shell of an atom has no more than two electrons, and the outermost shell contains no more than eight; these outer ones are those involved in chemical bonding. *Ionic* and *covalent bonding* are the most important types in minerals, although some useful properties of certain minerals result from metallic bonding and van der Waals bonds or forces. A few elements, known as *noble gases,* have complete outer shells with eight electrons, so rarely react with other elements to form compounds.

One way for the noble gas configuration of eight outer electrons to be attained is by transfer of one or more electrons from one atom to another. A good example is sodium (Na) and chlorine (Cl); sodium has only one electron in its outer shell, whereas chlorine has seven. Sodium loses its outer electron, leaving the next shell with eight (• Figure 2.4a). But, now sodium has one fewer electron (negative charge) than it has protons (positive charge), so it is a positively charged *ion,* symbolized Na^{+1}.

The electron lost by sodium goes into chlorine's outer shell, which had seven to begin with, so now it has eight (Figure 2.4a). In this case, though, chlorine has one too many negative charges and is thus an ion symbolized Cl^{-1}.

An attractive force exists between the Na^{+1} and Cl^{-1} ions, so an *ionic bond* forms between them, yielding the mineral halite (NaCl).

Covalent bonds result when the electron shells of adjacent atoms overlap and they share electrons. A carbon atom in diamond shares all four of its outer electrons with a neighbor to produce the stable noble gas configuration (Figure 2.4b). Among the most common minerals, the silicates, silicon forms partly covalent and partly ionic bonds with oxygen.

Minerals—The Building Blocks of Rocks

A mineral's composition is shown by a chemical formula, a shorthand way of indicating how many atoms of different kinds it contains. For example, quartz (SiO_2) is made up of one silicon atom for every two oxygen atoms, whereas the formula for orthoclase is $KAlSi_3O_8$. A few minerals known as *native elements,* such as gold (Au) and diamond (C), consist of only one element and accordingly are not compounds.

• Figure 2.4 **Ionic and Covalent Bonding**

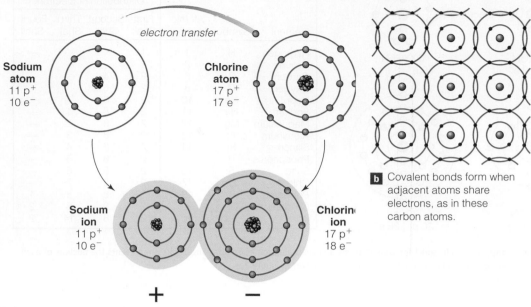

electron transfer

Sodium
atom
11 p⁺
10 e⁻

Chlorine
atom
17 p⁺
17 e⁻

Sodium
ion
11 p⁺
10 e⁻

Chlorine
ion
17 p⁺
18 e⁻

+   −

b Covalent bonds form when adjacent atoms share electrons, as in these carbon atoms.

a In ionic bonding, the electron in the outermost electron shell of sodium is transferred to the outermost electron shell of chlorine. Once the transfer has taken place, the positively charged sodium ion and negatively charged chlorine ion attract one another.

Recall our formal definition of a mineral. The adjective *inorganic* reminds us that animal and vegetable matter are not minerals. Nevertheless, corals and clams and some other organisms build shells of the mineral calcite ($CaCO_3$) or silica (SiO_2). By definition, minerals are **crystalline solids** in which their atoms are arranged in a specific three-dimensional framework. Ideally, minerals grow and form perfect crystals with planar surfaces (crystal faces), sharp corners, and straight edges (• Figure 2.5). In many cases, numerous minerals grow in proximity, as in cooling lava, and thus do not develop well-formed crystals.

Some minerals have very specific chemical compositions, but others have a range of compositions because one element can substitute for another if the ions of the two elements have the same electrical charge and are about the same size. Iron and magnesium meet these criteria and thus substitute for one another in olivine

$\{(Fe,Mg)_2SiO_4\}$, which may have magnesium, iron, or both. Calcium (Ca) and sodium (Na) substitute for one another in the plagioclase feldspars, which vary from calcium-rich ($CaAl_2Si_2O_8$) to sodium-rich ($NaAlSi_3O_8$) varieties.

Composition and structure control the characteristic physical properties of minerals. These properties are particularly useful for mineral identification (see Appendix C).

How Many Minerals Are There? Geologists recognize several mineral groups, each of which is composed of minerals sharing the same negatively charged ion or ion group (Table 2.1). More than 3500 minerals are known, but only about two dozen are particularly common, although many others are important resources.

• Figure 2.5 **A Variety of Mineral Crystal Shapes**

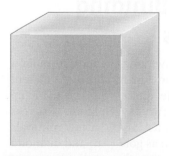

a Cubic crystals are typical of the minerals halite and galena.

b Pyritohedron crystals such as those of pyrite have 12 sides.

c Diamond has octahedral, or 8-sided, crystals.

d A prism terminated by a pyramid is found in quartz.

TABLE **2.1**

Mineral Group	Negatively Charged Ion or Ion Group	Examples	Composition
Carbonate	$(CO_3)^{-2}$	Calcite	$CaCO_3$
		Dolomite	$CaMg(CO_3)_2$
Halide	Cl^{-1}, F^{-1}	Halite	$NaCl$
		Fluorite	CaF_2
Native element	—	Gold	Au
		Silver	Ag
		Diamond	C
		Graphite	C
Phosphate	$(PO_4)^{-3}$	Apatite	$Ca_5(PO_4)_3$ (F, Cl)
Oxide	O^{-2}	Hematite	Fe_2O_3
		Magnetite	Fe_3O_4
Silicate	$(SiO_4)^{-4}$	Quartz	SiO_2
		Potassium feldspar	$KAlSi_3O_8$
		Olivine	$(Mg,Fe)_2SiO_4$
Sulfate	$(SO_4)^{-2}$	Anhydrite	$CaSO_4$
		Gypsum	$CaSO_4 \cdot 2H_2O$
Sulfide	S^{-2}	Galena	PbS
		Pyrite	FeS_2

Some of the Mineral Groups Recognized by Geologists

Silicate Minerals Given that oxygen (62.6%) and silicon (21.2%) account for nearly 84% of all atoms in Earth's crust, you might expect these elements to be common in minerals. And, indeed, they are. In fact, a combination of silicon and oxygen is called *silica,* and the minerals made up of silica are **silicates.** Silicates account for about one-third of all known minerals and make up perhaps 95% of Earth's crust.

All silicates are composed of a basic building block called the *silica tetrahedron,* consisting of one silicon atom surrounded by four oxygen atoms. These tetrahedra exist in minerals as isolated units bonded to other elements, or they may be arranged in single chains, double chains, sheets, or complex three-dimensional networks, thus accounting for the incredible diversity of silicate minerals.

Among the silicates, geologists recognize *ferromagnesian silicates,* which contain iron (Fe), magnesium (Mg), or both (• Figure 2.6a). They tend to be dark colored and denser than the *nonferromagnesian silicates,* which, of course, lack these elements (Figure 2.6b).

Other Mineral Groups Representatives of the several other mineral groups in Table 2.1 may be important resources, but most of them are not common constituents of rocks. For example, the oxides have oxygen combined with some other element as in hematite (Fe_2O_3) and magnetite (Fe_3O_4), which are iron ores. In the sulfates, an element is combined with the complex sulfate molecule (SO_4)$^{-2}$ as in gypsum ($CaSO_4 \cdot 2H_2O$), much of which is used for drywall or wallboard.

One group of minerals that deserves special attention is the **carbonates,** which include those with the carbonate ion (CO_3)$^{-2}$ as in the mineral calcite ($CaCO_3$). Many other carbonate minerals are known, but only dolomite [$CaMg(CO_3)_2$] is very common. Calcite and dolomite are important because they make up the sedimentary rocks limestone and dolostone, respectively, both of which are widespread at or near Earth's surface.

Rock-Forming Minerals and the Rock Cycle
Geologists have identified hundreds of minerals in rocks, but only a few are common enough to be designated as **rock-forming minerals**—that is, minerals

• **Figure 2.6** **Some of the Common Silicate Minerals** The ferromagnesian silicates tend to be darker colored and denser than the nonferromagnesian silicates.

Olivine / Augite / Hornblende / Biotite

Sue Monroe

a Ferromagnesian silicates

Quartz / Orthoclase / Plagioclase / Muscovite

Sue Monroe

b Nonferromagnesian silicates

that are essential for the identification and classification of rocks. As you might expect from our previous discussion, most rock-forming minerals are silicates (Figure 2.6), but the carbonate minerals calcite and dolomite are also important. In most cases, minerals present in small amounts, called *accessory minerals,* can be ignored.

The **rock cycle** is a pictorial representation of events leading to the origin, destruction and/or changes, and reformation of rocks as a consequence of Earth's internal and surface processes (• Figure 2.7). Furthermore it shows that the three major rock groups—igneous, sedimentary, and metamorphic—are interrelated; that is, any rock type can be derived from the others. Figure 2.7 shows that the ideal cycle involves those events depicted on the circle leading from magma to igneous rocks and so on. Notice also that the circle has several internal arrows indicating interruptions in the cycle.

Igneous Rocks

The ultimate source of all **igneous rocks** is **magma,** molten rock below Earth's surface. The term **lava** is used to describe magma that reaches the surface. Igneous rocks also result from magma that erupts explosively thereby forming particulate matter known as **pyroclastic materials,** such as volcanic ash, which become consolidated. Geologists characterize the igneous rocks that formed at the surface as **volcanic rocks** or **extrusive igneous rocks,** and those that formed below the surface as **plutonic rocks** or **intrusive igneous rocks.** The plutonic or intrusive rocks are found in several types of *plutons* (intrusive igneous bodies) shown in • Figure 2.8.

Texture and Composition
Several igneous rock textures tell us something about how the rocks formed in the first place. For instance, rapid cooling in a lava flow

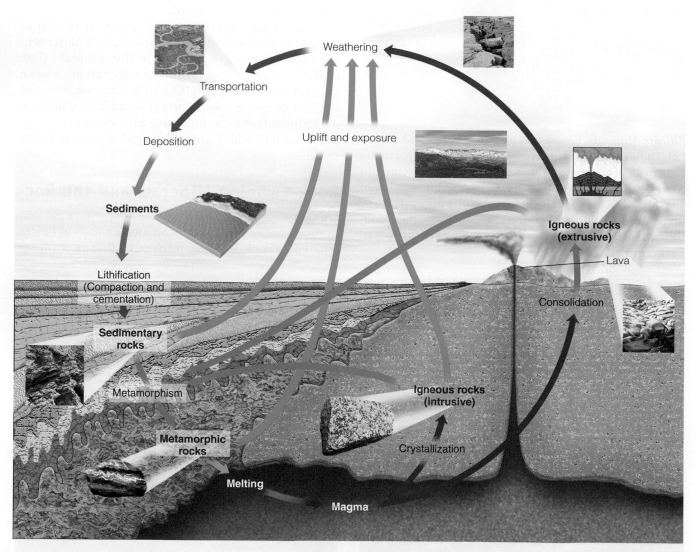

• **Figure 2.7 The Rock Cycle** This cycle shows the interrelationships among Earth's internal and surface processes and how the three major rock groups are related. An ideal cycle includes the events on the outer margin of the cycle, but interruptions indicated by internal arrows are common.

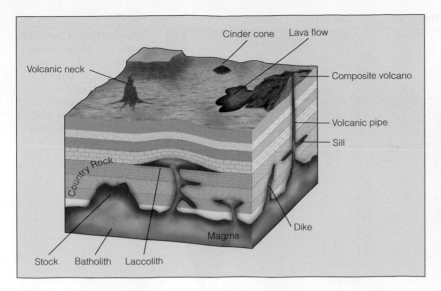

• **Figure 2.8 Block Diagram Showing Plutons or Intrusive Bodies of Igneous Rock** Notice that some plutons cut across the layering in the intruded rock (country rock) and are said to be discordant, but others are parallel to the layering and are concordant.

textures are usually sufficient to determine whether an igneous rock is volcanic or plutonic.

Some igneous rocks, though, have a combination of markedly different-sized minerals—a so-called *porphyritic texture;* the large minerals are *phenocrysts,* whereas the smaller ones constitute the rock's *groundmass* (Figure 2.9c). A porphyritic texture might indicate a two-stage cooling history in which magma began cooling below the surface and then was expelled onto the surface where cooling continued. The resulting igneous rocks are characterized as *porphyry.*

A *glassy texture* results from cooling so rapidly that the atoms in lava have too little time to form the three-dimensional framework of minerals. As a result, the natural glass *obsidian* forms (Figure 2.9d). Cooling lava might have a large content of trapped water vapor and other gases that form small holes or cavities called *vesicles;* rocks with numerous vesicles are *vesicular* (Figure 2.9e). And finally, a *pyroclastic* or *fragmental texture* characterizes igneous rocks composed of pyroclastic materials (Figure 2.9f).

With few exceptions, the primary constituent of magma is silica, but the silica content varies enough for us

results in a fine-grained or *aphanitic* texture, in which individual minerals are too small to see without magnification (• Figure 2.9a). In contrast, a coarse-grained or *phaneritic texture* is the outcome of comparatively slow cooling that takes place in plutons (Figure 2.9b). Thus, these two

• **Figure 2.9 Igneous Rock Textures** Texture is one criterion used to classify igneous rocks and it tells something about the history of these rocks.

a Rapid cooling as in a lava flow results in many small minerals and an aphanitic or fine-grained texture.

b Slower cooling in plutons yields a phaneritic or coarse grained texture.

c A porphyritic texture indicates a complex cooling history.

d Obsidian has a glassy texture because magma cooled too quickly for mineral crystals to form.

e Gases expand in cooling lava and yield a vesicular texture.

f Microscopic view of a fragmental texture. The colorless, angular objects are pieces of volcanic glass measuring up to 2 mm.

Sue Monroe

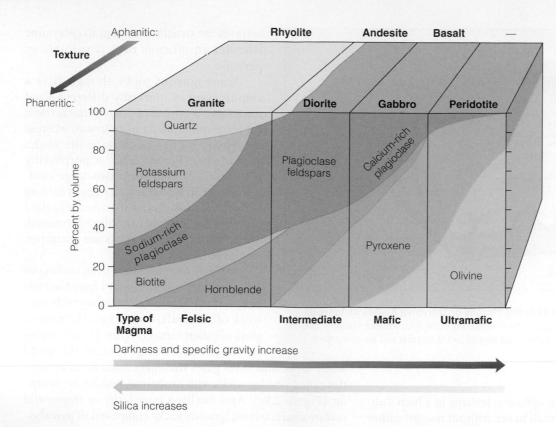

• **Figure 2.10**
Classification of Igneous Rocks This diagram shows the proportions of the main minerals and the textures of common igneous rocks. For example, an aphanitic (fine-grained) rock of mostly calcium-rich plagioclase and pyroxene is basalt.

to recognize magmas characterized as *felsic* (>65% silica), *intermediate* (53–65% silica), *mafic* (45–52% silica), and *ultramatic* (<45% silica).Felsic magma also contains considerable sodium, potassium, and aluminum, but little calcium, iron, and magnesium. In contrast, mafic magma has proportionately more calcium, iron, and magnesium. Intermediate magma, of course, has a composition between those of felsic and mafic magmas.

Classifying Igneous Rocks

Texture and composition are the criteria that geologists use to classify most igneous rocks. Notice in • Figure 2.10 that all of the rocks shown, except peridotite, are pairs; each member of a pair has the same composition but a different texture. Thus, basalt (aphanitic) and gabbro (phaneritic) have the same minerals, and because they contain a large proportion of ferromagnesian silicates they tend to be dark colored (• Figure 2.11a and b). Rhyolite (aphanitic) and granite (phaneritic) also have the same composition but differ in texture, and because they are made up largely of nonferromagnesian silicates, they are light colored (Figure 2.11e and f). We can infer from their textures that basalt, andesite, and rhyolite are volcanic rocks, whereas gabbro, diorite, and granite are plutonic. Notice also from Figure 2.10 that composition is related to the type of magma from which the rocks cooled.

For a few igneous rocks, texture is the only criterion for classification (• Figure 2.12a). Tuff is composed of volcanic ash, a designation for pyroclastic materials measuring less than 2 mm (Figure 2.12b). Some ash deposits are so hot when they form that the particles fuse together, forming *welded tuff*. Consolidation of larger pyroclastic materials, such as *lapilli* (2–64 mm) and blocks and bombs* (>64 mm), yields *volcanic breccia*. *Obsidian* and *pumice* are varieties of volcanic glass. The former looks like red, brown, or black glass (Figure 2.9d), whereas the latter has a frothy appearance, because it has numerous vesicles (Figure 2.12c).

Sedimentary Rocks

Notice in the rock cycle (Figure 2.7) that weathering processes disintegrate or decompose rocks at or near the surface. Accordingly, rocks are broken into smaller particles of gravel (>2 mm), sand (1/16–2 mm), silt (1/256–1/16 mm), and clay (<1256 mm), and some minerals may be dissolved—that is, taken into solution. In either case, these solid or dissolved materials are commonly transported elsewhere and deposited as *sediment*, perhaps as sand on a beach or as minerals extracted from solution by inorganic chemical processes of by the activities of

---

*Blocks are angular, whereas bombs are smooth and commonly shaped like a teardrop.

a Basalt

c Andesite

e Rhyolite

b Gabbro

d Diorite

f Granite

Sue Monroe

• Figure 2.12 **The Igneous Rocks in this Chart are Classified Mainly by their Textures**

Composition	Felsic ⟷ Mafic	
Texture — Vesicular	Pumice	Scoria
Glassy	Obsidian	
Pyroclastic or Fragmental	⟷ Volcanic Breccia ⟷	
	Tuff/welded tuff	

a Vesicular, glassy, and pyroclastic are textures found in some igneous rocks.

James S. Monroe

b This outcrop (rock exposure) in Colorado has a basalt lava flow at the top underlain by tuff (composed of volcanic ash), and volcanic breccia composed of angular fragments of volcanic rock.

Sue Monroe

c Pumice is glassy and vesicular; its density is so low that it floats.

organisms. Any sedimentary deposit may be lithified—that is, transformed into **sedimentary rock,** which by definition is any rock made up of sediment.

Sediment Transport, Deposition, and Lithification

Sediment transport involves removing sediment from its source area and carrying it elsewhere. Running water is the most effective method of transport, but glaciers, wind, and waves are also important in some areas. In any case, transported sediment must eventually be deposited as in a stream channel, on a beach, or on the seafloor. Sediment also accumulates as minerals that precipitate from solution or that organisms extract from solution. In all cases, sediment is an unconsolidated aggregate of solid particles, such as sand or mud.

Lithification is the geologic phenomenon of converting sediment into sedimentary rock. As sediment accumulates, *compaction*, resulting from pressure exerted by the weight of overlying sediments, reduces the amount of pore space (the void space between particles) and thus the volume of the deposit (• Figure 2.13). Compaction alone is usually sufficient for lithification of mud (particles measuring less than $1/16$ mm). But for gravel and sand, *cementation* is necessary, which involves the precipitation of minerals within pores that effectively binds the sediment together (Figure 2.13). The most common cements are calcium carbonate ($CaCO_3$), silica (SiO_2), and, in a distant third place, iron oxides and hydroxides such as hematite (Fe_2O_3) and limonite [$FeO(OH)$], respectively.

Classification of Sedimentary Rocks

Detrital sedimentary rocks are made up of detritus—that is, the solid particles such as gravel, sand, and mud derived from preexisting rocks. In contrast, **chemical sedimentary rocks** are made up of minerals derived from materials in solution and extracted by either inorganic chemical processes or by the activities of organisms. The subcategory called *biochemical sedimentary rocks* is so named because of the importance of organisms.

Detrital Sedimentary Rocks All detrital sedimentary rocks are composed of fragments or particles, known as *clasts,* derived from any preexisting rock, and defined primarily by the size of their constituent clasts (Figure 2.13). *Conglomerate and sedimentary breccia* are both made up of gravel (detrital particles measuring more than 2 mm). The only difference between them is that conglomerate (• Figure 2.14a) has rounded clasts, as opposed to the

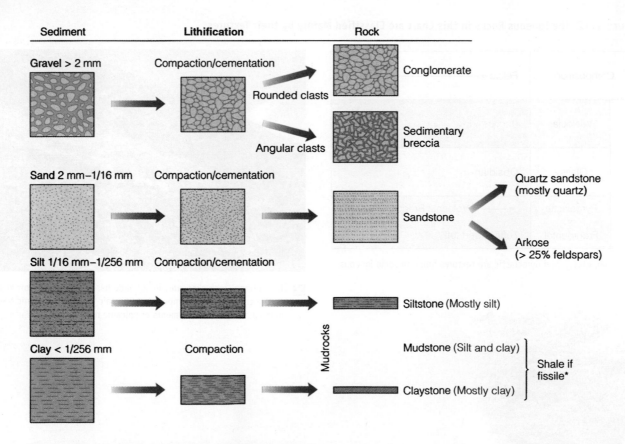

*Fissile refers to rocks that split along closely spaced planes.

• **Figure 2.13** **Lithification and Classification of Detrital Sedimentary Rocks** Notice that little compaction takes place in gravel and sand, that two types of sandstone are shown, and that mudrock is a collective term for detrital sedimentary rocks made up of silt and clay.

• Figure 2.14 **Detrital Sedimentary Rocks**

a A layer of conglomerate overlying sandstone. The largest clast measures about 30 cm across.

b Sedimentary breccia in Death Valley, California. Notice the angular gravel-sized particles. The largest clast is about 12 cm across.

c Exposure of shale in Tennessee.

angular ones in sedimentary breccia (Figure 2.14b). Particle rounding results from wear and abrasion during transport.

The term *sandstone* applies to any detrital sedimentary rock composed of sand (particles measuring 1/16–2mm) (Figure 2.13). *Quartz sandstone* is the most common, and, as its name implies, is made up mostly of quartz. Another fairly common type of sandstone is *arkose*, which in addition to quartz has at least 25% feldspar minerals (Figure 2.13).

All detrital sedimentary rocks composed of particles measuring less than 1/16 mm are collectively called *mudrocks* (Figure 2.13). Among the mudrocks we distinguish between *siltstone* composed of silt-sized particles, *mudstone* with a mixture of silt- and clay-sized particles, and *claystone,* which is composed mostly of particles measuring less than 1/256 mm. However, some mudstones and claystones are further designated as *shale* if they are fissile—meaning that they break along closely spaced parallel planes (Figures 2.13 and 2.14c).

Chemical Sedimentary Rocks You already know that chemical sedimentary rocks result when minerals are extracted from solution by inorganic chemical processes or by organisms. Some of the resulting rocks have a *crystalline texture*—that is, they are made up of an interlocking mosaic of mineral crystals. But others have a clastic texture—an accumulation of broken shells on a beach if lithified becomes a type of limestone.

Limestone and *dolostone* are the most common rocks in this category (• Figure 2.15), and geologists refer to both as **carbonate rocks** because they are composed of carbonate minerals: calcite and dolomite, respectively (Table 2.2). Actually, most dolomite was originally limestone that was altered when magnesium replaced some of the calcium in calcite. Furthermore, most limestone is also biochemical, because organisms are so important in its origin.

The **evaporites** such as *rock salt* and *rock gypsum* form by inorganic chemical precipitation of minerals from solutions made concentrated by evaporation (Table 2.2, • Figures 2.16a and b). If you were to evaporate seawater, for instance, you would end up with a small amount of rock gypsum and considerably more rock salt. Several other evaporites would also form but only in very small amounts.

Chert, a dense, hard rock, consists of silica, some of which forms as oval to spherical masses within other rocks by inorganic chemical precipitation. Other chert consists of layers of microscopic shells of silica-secreting organisms and is thus a biochemical sedimentary rock (Table 2.2, Figure 2.16c).

A biochemical sedimentary rock of great economic importance is *coal,* which consists of partially altered, compressed remains of land plants. Coal forms in oxygen-deficient swamps and bogs where the decomposition process is interrupted and the accumulating vegetation alters first to *peat* and if buried and compressed, it changes to coal (Table 2.2, Figure 2.16d).

• **Figure 2.15** **Chemical Sedimentary Rocks**

a Limestone with numerous fossil shells is called fossiliferous limestone.

c Coquina is limestone made up entirely of broken shells. Compare with the fossiliferous limestone in **a** .

b This oolitic limestone is made up partly of ooids (see inset), which are rather spherical grains of calcium carbonate.

Metamorphic Rocks

All sedimentary and igneous rocks form under their own specific conditions of temperature, pressure, and fluid activity. However, should they be subjected to different conditions, they may be altered in the solid state—that is, they become **metamorphic rocks.** Although sedimentary rocks followed by igneous rocks are most susceptible to

---

TABLE **2.2**	Classification of Chemical and Biochemical Sedimentary Rocks

Chemical Sedimentary Rocks		
Texture	**Composition**	**Rock Name**
Crystalline	Calcite ($CaCO_3$)	Limestone ⎤
Crystalline	Dolomite [$CaMg(CO_3)_2$]	Dolostone ⎦ —— Carbonates
Crystalline	Gypsum ($CaSO_4 \cdot 2H_2O$)	Rock gypsum ⎤
Crystalline	Halite (NaCl)	Rock salt ⎦ —— Evaporites

Biochemical Sedimentary Rocks		
Texture	**Composition**	**Rock Name**
Clastic	Calcium carbonate ($CaCO_3$) shells	Limestone (various types such as chalk and coquina)
Usually Crystalline	Altered microscopic shells of silicon dioxide (SiO_2)	Chert
—	Mostly carbon from altered plant remains	Coal

• Figure 2.16 **Evaporites, Chert, and Coal**

a This cylindrical core of rock salt was taken from an oil well in Michigan.

b Bedded chert exposed in Marin County, California. Most of the layers are about 5 cm thick.

c Rock gypsum. When deeply buried, gypsum ($CaSO_4 \cdot 2H_2O$) loses its water and is converted to anhydrite ($CaSO_4$).

d Bituminous coal is the most common type of coal used for fuel.

metamorphism, any type of rock, including metamorphic rocks, can be changed (metamorphosed) under the right conditions. Metamorphic changes may be compositional (new minerals form) or textural (minerals become aligned) or both (• Figure 2.17). Some of these changes are minor, so the parent rock is easily recognized, but changes may be so great that it can be very difficult to identify the parent rock.

• Figure 2.17 **Foliated Texture**

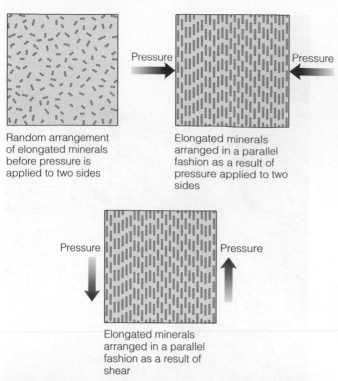

Random arrangement of elongated minerals before pressure is applied to two sides

Pressure → ← Pressure

Elongated minerals arranged in a parallel fashion as a result of pressure applied to two sides

Pressure ↓ ↑ Pressure

Elongated minerals arranged in a parallel fashion as a result of shear

a When rocks are subjected to differential pressure, the mineral grains are typically arranged in a parallel fashion, producing a foliated texture.

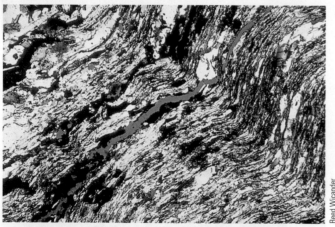

Reed Wicander

b Photomicrograph of a metamorphic rock with a foliated texture showing the parallel arrangement of mineral grains.

What Causes Metamorphism?

Heat, pressure, and chemical fluids—the agents responsible for metamorphism—may act singly or in any combination, and the time during which they are effective varies considerably. For example, a lava flow may bake the underlying rocks for a short time but otherwise has little effect, whereas the rocks adjacent to plutons, especially batholiths (Figure 2.8), may be altered over a long period and for a great distance.

Heat is important because it increases the rates of chemical reactions that may yield minerals different from those in the parent rock. Sources of heat include plutons,

especially the larger ones such as stocks and batholiths, and deep burial, because Earth's temperature increases with depth. Deep burial also subjects rocks to *lithostatic pressure* resulting from the weight of the overlying rocks. Under these conditions minerals in a rock become more closely packed and may *recrystallize*—that is, form smaller and denser minerals. Lithostatic pressure operates with the same intensity in all directions, but *differential pressure* exerts force more intensely from one direction, as occurs at convergent plate boundaries.

In any region where metamorphism takes place, water is present in varying amounts and may contain ions in solution that enhance metamorphism by increasing the rate of chemical reactions. Accordingly, *fluid activity* is also an important metamorphic agent.

Contact metamorphism takes place when heat and chemical fluids from an igneous body alter adjacent rocks. The rocks in contact with a batholith may be heated to nearly 1000°C, and of course fluids from the magma also bring about changes. The degree of metamorphism decreases with increasing distance from a pluton until the surrounding rocks are unaffected.

Most metamorphic rocks result from **regional metamorphism,** which takes place over large but elongated areas as the result of tremendous pressure, elevated temperatures, and fluid activity. This kind of metamorphism is most obvious along convergent plate boundaries where the rocks are intensely deformed during convergence and subduction. It also takes place at divergent plate boundaries, though usually at shallower depths, and here only high temperature and fluid activity are important.

Dynamic metamorphism is much more restricted in its extent being confined to zones adjacent to faults (fractures along which rocks have moved) where high levels of differential pressure develop.

Metamorphic Rock Classification

Many metamorphic rocks, especially those subjected to intense differential pressure, have their platy and elongate minerals aligned in a parallel fashion, giving them a *foliated texture* (Figure 2.17). In contrast, some metamorphic rocks do not develop any discernable orientation of their minerals. Instead, they consist of a mosaic of roughly equidimensional minerals and have a *nonfoliated texture.*

Foliated Metamorphic Rocks Slate, a very fine-grained rock, results from low-grade metamorphism of mudrocks or, more rarely, volcanic ash (Table 2.3, • Figure 2.18a). *Phyllite* is coarser-grained than slate, but the minerals are still too small to see without magnification. Actually, the change from some kind of mudrock (shale perhaps) to slate and then to phyllite is part of a continuum, and so is the origin of *schist*. In schist, though, the elongate and platy minerals are clearly visible and they impart a *schistosity* or *schistose foliation* to the rock (Table 2.3, Figure 2.18b).

TABLE **2.3** Classification of Common Metamorphic Rocks

Texture	Metamorphic Rock	Typical Minerals	Metamorphic Grade	Characteristics of Rocks	Parent Rock
Foliated	Slate	Clays, micas, chlorite	Low	Fine-grained, splits easily into flat pieces	Mudrocks, volcanic ash
	Phyllite	Fine-grained quartz, micas, chlorite	Low to medium	Fine-grained, glossy or lustrous sheen	Mudrocks
	Schist	Micas, chlorite, quartz, talc, hornblende, garnet, staurolite, graphite	Low to high	Distinct foliation, minerals visible	Mudrocks, carbonates, mafic igneous rocks
	Gneiss	Quartz, feldspars, hornblende, micas	High	Segregated light and dark bands visible	Mudrocks, sandstones, felsic igneous rocks
	Amphibolite	Hornblende, plagioclase	Medium to high	Dark-colored, weakly foliated	Mafic igneous rocks
	Migmatite	Quartz, feldspars, hornblende, micas	High	Streaks or lenses of granite intermixed with gneiss	Felsic igneous rocks mixed with metamorphic rocks
Nonfoliated	Marble	Calcite, dolomite	Low to high	Interlocking grains of calcite or dolomite, reacts with HCl	Limestone or dolostone
	Quartzite	Quartz	Medium to high	Interlocking quartz grains, hard, dense	Quartz sandstone
	Greenstone	Chlorite, epidote, hornblende	Low to high	Fine-grained, green	Mafic igneous rocks
	Hornfels	Micas, garnet, andalusite, cordierite, quartz	Low to medium	Fine-grained, equidimensional grains, hard, dense	Mudrocks
	Anthracite	Carbon	High	Black, lustrous, subconchoidal fracture	Coal

Gneiss, with its alternating dark and light bands of minerals, is one of the more attractive metamorphic rocks (Table 2.3, Figures 2.1c, 2.18c). Quartz and feldspars are most common in the light-colored bands, and biotite and hornblende are found in the dark-colored bands. Most gneiss probably forms from regional metamorphism of clay-rich sedimentary rocks, but metamorphism of granite and other metamorphic rocks also yields gneiss.

In some areas of regional metamorphism, exposures of "mixed rocks" having both igneous and metamorphic characteristics are present. These rocks, called *migmatites*, consist of streaks or lenses of granite intermixed with ferromagnesian mineral-rich metamorphic rocks (Figure 2.18d). Where do you think these rocks would be in the rock cycle?

Nonfoliated Metamorphic Rocks Nonfoliated metamorphic rocks may form by either contact or regional metamorphism, but in all cases they lack the platy and elongate minerals found in foliated rocks. *Marble*, a well-known metamorphic rock composed of calcite or dolomite, results from metamorphism of limestone or dolostone (Table 2.3, • Figure 2.19a). Some marble is attractive and has been a favorite with sculptors and for building material for centuries. Metamorphism of quartz sandstone yields *quartzite*, a hard, compact rock (Table 2.3, Figure 2.19b).

The name *greenstone* applies to any compact, dark green, altered mafic igneous rock that formed under metamorphic conditions. Minerals such as chlorite and epidote give it its green color. Several varieties of *hornfels* are known, most of which formed when clay-rich sedimentary rocks were altered during contact metamorphism.

Anthracite is a black, lustrous, hard coal with mostly carbon and a low amount of volatile matter. It forms from other types of coal.

• Figure 2.18 **Foliated Metamorphic Rocks**

a Slate

b Schist

c Gneiss

d Migmatite shows features of both metamorphic and igneous rocks

a Marble

b Quartzite

• Figure 2.19 **Nonfoliated Metamorphic Rocks** What were these rocks before they were metamorphosed?

Plate Tectonics and the Rock Cycle

Notice in the rock cycle (Figure 2.7) that interactions among Earth systems are responsible for the origin and alternation of rocks. The atmosphere, hydrosphere, and biosphere acting on Earth materials account for weathering, transport, and deposition, whereas Earth's internal heat is responsible for melting and contributes to metamorphism. Plate tectonics also plays an important role in recycling Earth materials.

Sediment along continental margins may become lithified and be incorporated into a moving plate along with underlying oceanic crust. Where plates collide at convergent plate boundaries, heat and pressure commonly lead to metamorphism, igneous activity, and the origin of new rocks. In addition, some of the rocks in a subducted plate are deformed and incorporated into an evolving mountain system that in turn is weathered and eroded, yielding sediment to begin yet another cycle.

SUMMARY

- Elements are composed of atoms, which have a nucleus of protons and neutrons around which electrons orbit in electron shells.
- The number of protons in a nucleus determines the atomic number of an element; the atomic mass number is the number of protons plus neutrons in the nucleus.
- Atoms bond by transferring electrons from one atom to another (ionic bond) or by sharing electrons (covalent bond). Most minerals are compounds of two or more elements bonded together.
- By far the most common minerals are the silicates (composed of at least silicon and oxygen), but carbonate minerals (containing the CO_3 ion) are prevalent in some rocks.
- The two broad groups of igneous rocks, plutonic (or intrusive) and volcanic (or extrusive) are classified by composition and texture. However, for a few volcanic rocks texture is the main consideration.
- Sedimentary rocks are also grouped into two broad categories: detrital (made up of solid particles of preexisting rocks) and chemical/biochemical (composed of minerals derived by inorganic chemical processes or the activities of organisms).
- Lithification involving compaction and cementation is the process whereby sediment is transformed into sedimentary rock.
- Metamorphic rocks result from compositional and/or textural transformation of other rocks by heat, pressure, and fluid activity. Most metamorphism is regional, occurring deep within the crust over large areas, but some, called contact metamorphism, take place adjacent to hot igneous rocks.
- Metamorphism imparts a foliated texture to many rocks (parallel alignment of minerals), but some rocks have a mosaic of equidimensional minerals and are nonfoliated.
- Metamorphic rocks are classified largely by their textures, but composition is a consideration for some.
- Plate tectonics, which is driven by Earth's internal heat coupled with surface processes such as weathering, erosion, and deposition, accounts for the recycling of Earth materials in the rock cycle.

IMPORTANT TERMS

atom, p. 19
atomic mass number, p. 20
atomic number, p. 20
bonding, p. 20
carbonate mineral, p. 23
carbonate rock, p. 29
chemical sedimentary rock, p. 28
compound, p. 20
contact metamorphism, p. 32
crystalline solid, p. 22
detrital sedimentary rock, p. 28

dynamic metamorphism, p. 32
element, p. 19
evaporite, p. 29
extrusive igneous rock, p. 24
geologic record, p. 19
igneous rock, p. 24
intrusive igneous rock, p. 24
lava, p. 24
lithification, p. 28
magma, p. 24
metamorphic rock, p. 31

mineral, p. 19
plutonic rock, p. 24
pyroclastic materials p. 24
regional metamorphism, p. 32
rock, p. 19
rock cycle, p. 24
rock-forming mineral, p. 23
sedimentary rock, p. 28
silicate, p. 23
volcanic rock, p. 24

REVIEW QUESTIONS

1. Any rock that has been altered from a previous state by heat, pressure, and chemical fluids is a _____ rock.
a. _____ plutonic; b. _____ metamorphic; c. _____ ferromagnesian; d. _____ sedimentary; e. _____ evaporite.

2. Although silicates are the most common minerals, the _____ are also important because they are the main components of limestone and dolostone.
a. _____ oxides; b. _____ sulfides; c. _____ carbonates; d. _____ hydroxides; e. _____ phosphates.

3. Lithification is the processes whereby
a. _____ lava cools and forms an aphanitic texture;
b. _____ atoms of two different elements join together;
c. _____ organic matter is converted into granite; d. _____ rocks are altered by heat at the margin of a pluton; e. _____ sediment is converted into sedimentary rock.

4. Which one of the following is a phaneritic igneous rock?
a. _____ gabbro; b. _____ sandstone; c. _____ coal; d. _____ phyllite; e. _____ dolostone.

5. If a naturally occurring, solid substance has all of is atoms arranged in a specific three-dimensional framework it is said to be
a. _____ covalently bonded; b. _____ crystalline; c. _____ porphyritic; d. _____ biochemical; e. _____ sedimentary.

6. If you were to encounter an igneous rock in which the minerals were clearly visible, you would be justified in concluding that the rock is
   a. _____ detrital; b. _____ foliated; c. _____ plutonic; d. _____ metamorphic; e. _____ extraterrestrial.
7. Most limestones have a large component of calcite that was extracted from seawater by
   a. _____ evaporation; b. _____ melting; c. _____ weathering; d. _____ organisms; e. _____ dynamic metamorphism.
8. An atom with 6 protons and 8 neutrons has an atomic mass number of
   a. _____ 14; b. _____ 2; c. _____ 6; d. _____ 8; e. _____ 48.
9. A rock made up of detrital particles measuring between 1/16 and 2 mm is
   a. _____ sandstone; b. _____ chert; c. _____ rhyolite; d. _____ schist; e. _____ migmatite.
10. The particles ejected by volcanoes during explosive eruptions are collectively called
    a. _____ biochemical constituents; b. _____ volcanic compounds; c. _____ carbonate minerals; d. _____ pyroclastic materials; e. _____ intrusive metamorphics.

11. How do plutonic rocks differ from volcanic rocks? Give one example of each.
12. What are atoms, elements, and compounds?
13. How is it possible for some chemical elements to substitute for one another in the mineral olivine?
14. Describe how the lithification of sand and mud take place.
15. Compare regional and contact metamorphism. Give an example of a rock that forms under each condition.
16. Why is ice a mineral but water vapor and liquid water are not minerals?
17. How do evaporites form, and which ones are common?
18. What are the criteria for identifying conglomerate and sedimentary breccia?
19. Cubic crystals of the minerals pyrite, halite, and galena are common, so how could you tell them apart? See Appendix C.
20. What are the two groups of silicate minerals, and how do they differ from one another?

APPLY YOUR KNOWLEDGE

1. Examine the two igneous rock specimens shown here. Can you tell which is volcanic and which is plutonic? If so, how? Specimen 1 is made up of 38% calcium-rich plagioclase, 10% hornblende, 40% pyroxene, and 12% olivine. Specimen 2 has 15% quartz, 55% potassium feldspar, 20% sodium-rich plagioclase, and 10% biotite and hornblende. Classify both rock specimens.

2. You know that minerals are made up of chemical elements and that rocks are composed of minerals, but despite your best efforts the students in your science class routinely mistake one for the other. Can you think of any analogies that might help them distinguish between minerals and rocks?

3. Examine the rocks in Figures 2.14a and b and notice the differences between the clasts. In which specimen do you think the clasts were transported the greatest distance? How do you know?

4. Use the rock cycle to explain how any one of the three main families of rocks can be derived from any one of the other families.

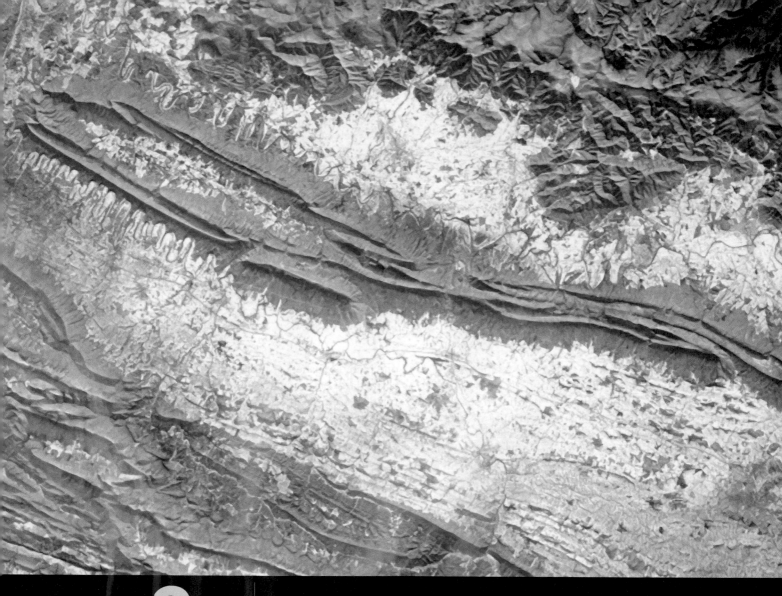

NASA Johnson Space Center—Earth Sciences and Image Analysis

CHAPTER 3

PLATE TECTONICS: A UNIFYING THEORY

Satellite image of the Appalachian Mountains in Virginia, USA. The Appalachians are the result of subduction (one plate plunging beneath another plate) along a convergent plate boundary. Beginning in the Middle Ordovician, the east coast of North America (Laurentia) and the west coast of Europe (Baltica) became active oceanic-continental convergent plate boundaries, resulting in mountain building activity along their respective margins. As convergence continued throughout the rest of the Paleozoic, the two continents eventually collided to form a large mountain range consisting of numerous large-scale folds (see Chapters 10 and 11). Following a period of erosion during the Mesozoic, the ancient Appalachians that formed during the Paleozoic were uplifted and exposed, resulting in the present landscape of ridges and valleys (see Chapters 14 and 16).

[OUTLINE]

Introduction

Early Ideas About Continental Drift

Alfred Wegener and the Continental Drift Hypothesis

Additional Support for Continental Drift

Paleomagnetism and Polar Wandering

Magnetic Reversals and Seafloor Spreading

Plate Tectonics and Plate Boundaries

Perspective *Tectonics of the Terrestrial Planets*

Divergent Boundaries

An Example of Ancient Rifting

Convergent Boundaries

Recognizing Ancient Convergent Boundaries

Transform Boundaries

Hot Spots and Mantle Plumes

How Are Plate Movement and Motion Determined?

The Driving Mechanism of Plate Tectonics

How Are Plate Tectonics and Mountain Building Related?

Terrane Tectonics

Plate Tectonics and the Distribution of Life

Plate Tectonics and the Distribution of Natural Resources

Summary

At the end of this chapter, you will have learned that

- Plate tectonics is the unifying theory of geology and has revolutionized geology.

- The hypothesis of continental drift was based on considerable geologic, paleontologic, and climatologic evidence.

- The hypothesis of seafloor spreading accounts for continental movement and the idea that thermal convection cells provide a mechanism for plate movement.

- The three types of plate boundaries are divergent, convergent, and transform. Along these boundaries new plates are formed, consumed, or slide past one another.

- Interaction along plate boundaries accounts for most of Earth's earthquake and volcanic activity.

- The rate of movement and motion of plates can be calculated in several ways.

- Some type of convective heat system is involved in plate movement.

- Plate movement affects the distribution of natural resources.

- Plate movement affects the distribution of the world's biota and has influenced evolution.

Introduction

Imagine it is the day after Christmas, December 26, 2004, and you are vacationing on a beautiful beach in Thailand. You look up from the book you're reading to see the sea suddenly retreat from the shoreline, exposing a vast expanse of seafloor that had moments before been underwater and teeming with exotic and colorful fish. It is hard to believe that within minutes of this unusual event, a powerful tsunami will sweep over your resort and everything in its path for several kilometers inland. Within hours, the coasts of Indonesia, Sri Lanka, India, Thailand, Somalia, Myanmar, Malaysia, and the Maldives will be inundated by the deadliest tsunami in history. More than 220,000 will die, and the region will incur billions of dollars in damage.

One year earlier, on December 26, 2003, violent shaking from an earthquake awakened hundreds of thousands of people in the Bam area of southeastern Iran. When the 6.6-magnitude earthquake was over, an estimated 43,000 people were dead, at least 30,000 were injured, and approximately 75,000 survivors were left homeless. The amount of destruction that this 6.6-magnitude earthquake caused is staggering. At least 85% of the structures in the Bam area were destroyed or damaged. Collapsed buildings were everywhere, streets were strewn with rubble, and all communications were knocked out.

Now go back another $12^1/_2$ years to June 15, 1991, when Mount Pinatubo in the Philippines erupted violently, discharging huge quantities of ash and gases into the atmosphere. Fortunately, in this case, warnings of an impending eruption were broadcast and heeded, resulting in the evacuation of 200,000 people from areas around the volcano. Unfortunately, the eruption still caused at least 364 deaths, not only from the eruption but also from ensuing mudflows.

What do these three tragic events and other equally destructive volcanic eruptions and earthquakes have in common? They are part of the dynamic interactions involving Earth's plates. When two plates come together, one plate is pushed or pulled under the other plate, triggering large earthquakes such as the ones that shook India in 2001, Iran in 2003, Pakistan in 2005, and Indonesia in 2006. If the conditions are right, earthquakes can produce a tsunami such as the one in 2004 or the 1998 Papua New Guinea tsunami that killed more than 2200 people.

As the descending plate moves downward and is assimilated into Earth's interior, magma is generated. Being less dense than the surrounding material, the magma rises toward the surface, where it may erupt as a volcano, as Mount Pinatubo did in 1991. It, therefore, should not be surprising that the distribution of volcanoes and earthquakes closely follows plate boundaries.

As we stated in Chapter 1, *plate tectonic theory* has had significant and far-reaching consequences in all fields of geology, because it provides the basis for relating many seemingly unrelated phenomena. The interactions between moving plates determine the location of continents, ocean basins, and mountain systems, all of which, in turn, affect atmospheric and oceanic circulation patterns that ultimately determine global climates (see Table 1.2). Plate movements have also profoundly influenced the geographic distribution, evolution, and extinction of plants and animals. Furthermore, the formation and distribution of many geologic resources, such as metal ores, are related to plate tectonic processes, so geologists incorporate plate tectonic theory into their prospecting efforts.

If you're like most people, you probably have only a vague notion of what plate tectonic theory is. Yet plate tectonics affects all of us. Volcanic eruptions, earthquakes, and tsunami are the result of interactions between plates. Global weather patterns and oceanic currents are caused, in part, by the configuration of the continents and ocean basins. The formation and distribution of many natural resources are related to plate movement, and thus have an impact on the economic well-being and political decisions of nations. It is therefore important to understand this unifying theory, not only because it impacts us as individuals and citizens of nation–states but also because it ties together many aspects of the geology you will be studying.

Early Ideas About Continental Drift

The idea that Earth's past geography was different from today is not new. The earliest maps showing the east coast of South America and the west coast of Africa probably provided people with the first evidence that continents may have once been joined together, then broke apart and moved to their present locations.

During the late 19th century, Austrian geologist Edward Suess noted the similarities between the Late Paleozoic plant fossils of India, Australia, South Africa, and South America, as well as evidence of glaciation in the rock sequences of these continents. The plant fossils comprise a unique flora that occurs in the coal layers just above the glacial deposits of these southern continents. This flora is very different from the contemporaneous coal swamp flora of the northern continents and is collectively known as the *Glossopteris* **flora,** after its most conspicuous genus (• Figure 3.1).

In his book, *The Face of the Earth*, published in 1885, Suess proposed the name *Gondwanaland* (or **Gondwana,** as we will use here) for a supercontinent composed of the aforementioned southern continents. Abundant fossils of the *Glossopteris* flora are found in coal seams in Gondwana, a province in India. Suess thought these southern continents were at one time connected by land bridges over which plants and animals migrated. Thus,

in his view, the similarities of fossils on these continents were due to the appearance and disappearance of the connecting land bridges.

The American geologist Frank Taylor published a pamphlet in 1910 presenting his own theory of continental drift. He explained the formation of mountain ranges as a result of the lateral movement of continents. He also envisioned the present-day continents as parts of larger polar continents that eventually broke apart and migrated toward the equator after Earth's rotation was supposedly slowed by gigantic tidal forces. According to Taylor, these tidal forces were generated when Earth captured the Moon approximately 100 million years ago.

Although we now know that Taylor's mechanism is incorrect, one of his most significant contributions was his suggestion that the Mid-Atlantic Ridge, discovered by the 1872–1876 British HMS *Challenger* expeditions, might mark the site along which an ancient continent broke apart to form the present-day Atlantic Ocean.

Alfred Wegener and the Continental Drift Hypothesis

Alfred Wegener, a German meteorologist (• Figure 3.2), is generally credited with developing the hypothesis of **continental drift.** In his monumental book, *The Origin of Continents and Oceans* (first published in 1915), Wegener proposed that all landmasses were originally united in a single supercontinent that he named **Pangaea,** from the Greek meaning "all land." Wegener portrayed his grand concept of continental movement in a series of maps showing the breakup of Pangaea and the movement of the various continents to their present-day locations.

• **Figure 3.1 Fossil** *Glossopteris* **Leaves** Plant fossils, such as these *Glossopteris* leaves from the Upper Permian Dunedoo Formation in Australia, are found on all five of the Gondwana continents. Their presence on continents with widely varying climates today is evidence that the continents were at one time connected. The distribution of the plants at that time was in the same climatic latitudinal belt.

• Figure 3.2 **Alfred Wegener** Alfred Wegener, a German meteorologist, proposed the continental drift hypothesis in 1912 based on a tremendous amount of geologic, paleontologic, and climatologic evidence. He is shown here waiting out the Arctic winter in an expedition hut in Greenland.

Wegener amassed a tremendous amount of geologic, paleontologic, and climatologic evidence in support of continental drift; however, initial reaction of scientists to his then-heretical ideas can best be described as mixed.

What evidence did Wegener use to support his hypothesis of continental drift? First, Wegener noted that the shorelines of continents fit together, forming a large supercontinent (• Figure 3.3), and that marine, nonmarine, and glacial rock sequences of Pennsylvanian to Jurassic age are almost identical for all five Gondwana continents, strongly indicating that they were joined together at one time (• Figure 3.4). Furthermore, mountain ranges and glacial deposits match up when continents are united into a single landmass (• Figure 3.5). And last, many of the same extinct plant and animal groups are found today on widely separated continents, indicating that the continents must have been close to each other at one time (• Figure 3.6). Wegener argued that this vast amount of evidence from a variety of sources surely indicated that the continents must have been close to each other at some time in the past.

• *Active* Figure 1.1 **Continental Fit** When continents are placed together based on their outlines, the best fit isn't along their present-day coastlines, but rather along the continental slope at a depth of about 2000 m, where erosion would be minimal. Visit the Geology Resource Center to view this and other active figures at www.cengage.com/sso.

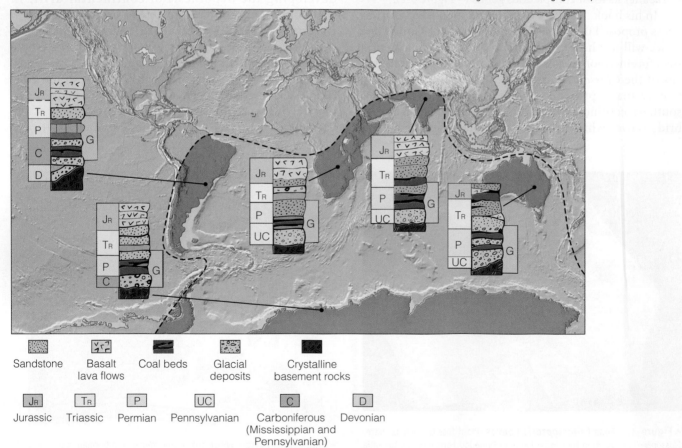

• Figure 3.4 **Similarity of Rock Sequences on the Gondwana Continents** Sequences of marine, nonmarine, and glacial rocks of Pennsylvanian (UC) to Jurassic (JR) age are nearly the same on all five Gondwana continents (South America, Africa, India, Australia, and Antarctica). These continents are widely separated today and have different environments and climates ranging from tropical to polar. Thus, the rocks forming on each continent are very different. When the continents were all joined together in the past, however, the environments of adjacent continents were similar and the rocks forming in those areas were similar. The range indicated by G in each column is the age range (Carboniferous-Permian) of the *Glossopteris* flora.

• Figure 3.5 **Glacial Evidence Indicating Continental Drift**

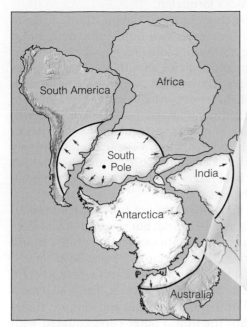

a When the Gondwana continents are placed together so that South Africa is located at the South Pole, the glacial movements indicated by striations (red arrows) found on rock outcrops on each continent make sense. In this situation, the glacier (white area) is located in a polar climate and has moved radially outward from its thick central area toward its periphery.

b Glacial striations (scratch marks) on an outcrop of Permian-age bedrock exposed at Hallet's Cove, Australia, indicate the general direction of glacial movement more than 200 million years ago. As a glacier moves over a continent's surface, it grinds and scratches the underlying rock. The scratch marks that are preserved on a rock's surface (glacial striations) thus provide evidence of the direction (red arrows) the glacier moved at that time.

Courtesy of Scott Katz

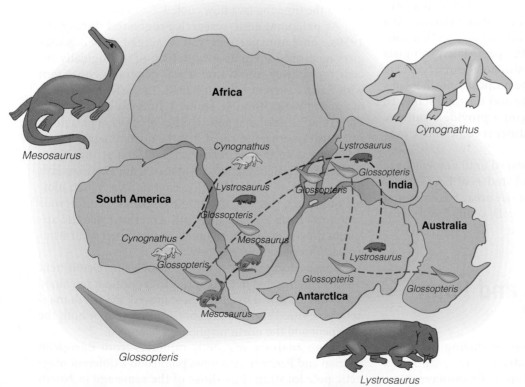

• Figure 3.6 **Fossil Evidence Supporting Continental Drift** Some of the animals and plants whose fossils are found today on the widely separated continents of South America, Africa, India, Australia, and Antarctica. During the Late Paleozoic Era, these continents were joined together to form Gondwana, the southern landmass of Pangaea. Plants of the *Glossopteris* flora are found on all five continents, which today have widely different climates; however, during the Pennsylvanian and Permian periods, they were all located in the same general climatic belt. *Mesosaurus* is a freshwater reptile whose fossils are found only in similar nonmarine Permian-age rocks in Brazil and South Africa. *Cynognathus* and *Lystrosaurus* are land reptiles that lived during the Early Triassic Period. Fossils of *Cynognathus* are found in South America and Africa, whereas fossils of *Lystrosaurus* have been recovered from Africa, India, and Antarctica. It is hard to imagine how a freshwater reptile and land-dwelling reptiles could have swum across the wide oceans that presently separate these continents. It is more logical to assume that the continents were once connected. Modified from E.H. Colbert, *Wandering Lands and Animals*, 1973, 72, Figure 31.

Additional Support for Continental Drift

With the publication of *The Origin of Continents and Oceans* and its subsequent four editions, some scientists began to take Wegener's unorthodox views seriously.

Alexander du Toit, a South African geologist, was one of Wegener's more ardent supporters. He further developed Wegener's arguments and gathered more geologic and paleontologic evidence in support of continental drift. In 1937, du Toit published *Our Wandering Continents*, in which he contrasted the glacial deposits of Gondwana with coal deposits of the same age found in the continents of the Northern Hemisphere. To resolve this apparent climatologic paradox, du Toit moved the Gondwana continents to the South Pole and brought the northern continents together such that the coal deposits were located at the equator. He named this northern landmass **Laurasia.** It consisted of present-day North America, Greenland, Europe, and Asia (except for India).

Du Toit also provided paleontologic support for continental drift by noting that fossils of the freshwater reptile *Mesosaurus* are found in Permian-age rocks in certain regions of Brazil and South Africa and nowhere else in the world (Figure 3.6). Because the physiologies of freshwater and marine animals are completely different, it is hard to imagine how a freshwater reptile could have swum across the Atlantic Ocean and found a freshwater environment nearly identical to its former habitat. Moreover, if *Mesosaurus* could have swum across the ocean, its fossil remains should be widely dispersed. It is more logical to assume that *Mesosaurus* lived in lakes in what are now adjacent areas of South America and Africa but were once united into a single continent.

Notwithstanding all of the empirical evidence presented by Wegener and later by du Toit and others, most geologists simply refused to entertain the idea that continents might have moved in the past. The geologists were not necessarily being obstinate about accepting new ideas; rather, they found the proposed mechanisms for continental drift inadequate and unconvincing. In part, this was because no one could provide a suitable mechanism to explain how continents could move over Earth's surface.

Interest in continental drift waned until new evidence from oceanographic research and studies of Earth's magnetic field showed that the present-day ocean basins are not as old as continents, but are geologically young features that resulted from the breakup of Pangea.

Paleomagnetism and Polar Wandering

Interest in continental drift revived during the 1950s as a result of evidence from paleomagnetic studies, a relatively new discipline at the time. **Paleomagnetism** is the remanent magnetism in ancient rocks recording the direction and intensity of Earth's magnetic poles at the time of the rock's formation.

Earth can be thought of as a giant dipole magnet (meaning that it possesses two unlike magnetic poles referred to as the north and south pole) in which the magnetic poles essentially coincide with the geographic poles (• Figure 3.7). This arrangement means that the strength of the magnetic field is not constant, but varies. Notice in Figure 3.7 that the lines of magnetic force around Earth parallel its surface only near the equator. As the lines of force approach the poles, they are oriented at increasingly larger angles with respect to the surface, and the strength of the magnetic field increases; it is strongest at the poles and weakest at the equator.

Experts on magnetism do not fully understand all aspects of Earth's magnetic field, but most agree that electrical currents resulting from convection in the liquid outer core generate it. Furthermore, it must be generated continuously or it would decay and Earth would have no magnetic field in as little as 20,000 years. The model most widely accepted now is that thermal and compositional convection within the liquid outer core, coupled with Earth's rotation, produce complex electrical currents or a *self-exciting dynamo* that, in turn, generates the magnetic field.

When magma cools, the magnetic iron-bearing minerals align themselves with Earth's magnetic field, recording both its direction and strength. The temperature at which iron-bearing minerals gain their magnetization is called the **Curie point.** As long as the rock is not subsequently heated above the Curie point, it will preserve that remanent magnetism. Thus, an ancient lava flow provides a record of the orientation and strength of Earth's magnetic field at the time the lava flow cooled.

As paleomagnetic research progressed during the 1950s, some unexpected results emerged. When geologists measured the paleomagnetism of geologically recent rocks, they found that it was generally consistent with Earth's current magnetic field. The paleomagnetism of ancient rocks, though, showed different orientations. For example, paleomagnetic studies of Silurian lava flows in North America indicated that the north magnetic pole was located in the western Pacific Ocean at that time, whereas the paleomagnetic evidence from Permian lava flows pointed to yet another location in Asia. When plotted on a map, the paleomagnetic readings of numerous lava flows from all ages in North America trace the apparent movement of the magnetic pole (called *polar wandering*) through time (• Figure 3.8).

This paleomagnetic evidence from a single continent could be interpreted in three ways: The continent remained fixed, and the north magnetic pole moved; the north magnetic pole stood still and the continent moved; or both the continent and the north magnetic pole moved.

Upon analysis, magnetic minerals from European Silurian and Permian lava flows pointed to a different magnetic pole location than those of the same age in North America (Figure 3.8). Furthermore, analysis of lava flows

• Figure 3.7 **Earth's Magnetic Field**

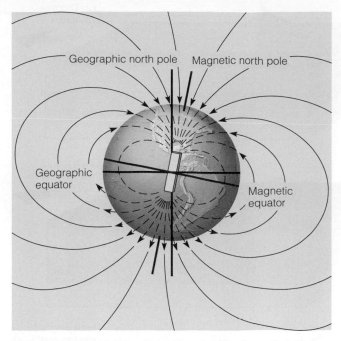

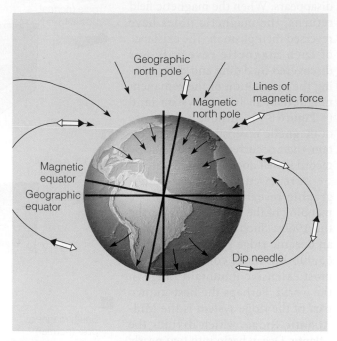

a Earth's magnetic field has lines of force just like those of a bar magnet.

b The strength of the magnetic field changes from the magnetic equator to the magnetic poles. This change in strength causes a dip needle (a magnetic needle that is balanced on the tip of a support so that it can freely move vertically) to be parallel to Earth's surface only at the magnetic equator, where the strength of the north and south magnetic poles are equally balanced. Its inclination or dip with respect to Earth's surface increases as it moves toward the magnetic poles, until it is at 90 degrees or perpendicular to Earth's surface at the magnetic poles.

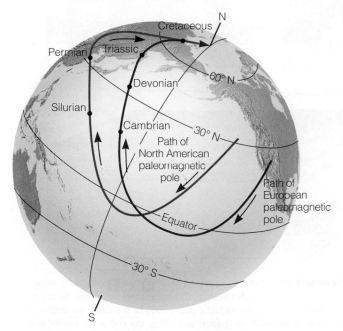

• Figure 3.8 **Polar Wandering** The apparent paths of polar wandering for North America and Europe. The apparent location of the north magnetic pole is shown for different periods on each continent's polar wandering path. If the continents have not moved through time, and because Earth has only one magnetic pole, the paleomagnetic readings for the same time in the past, taken on different continents, should all point to the same location. However, the north magnetic pole has different locations for the same time in the past when measured on different continents, indicating multiple north magnetic poles. The logical explanation for this dilemma is that the north magnetic pole has remained at the same approximate geographic location during the past, and the continents have moved.

from all continents indicated that each continent seemingly had its own series of magnetic poles. Does this really mean there were different north magnetic poles for each continent? That would be highly unlikely and difficult to reconcile with the theory accounting for Earth's magnetic field.

The best explanation for such data is that the magnetic poles have remained near their present locations at the geographic north and south poles and the continents have moved. When the continental margins are fit together so that the paleomagnetic data point to only one magnetic pole, we find, just as Wegener did, that the rock sequences and glacial deposits match, and that the fossil evidence is consistent with the reconstructed paleogeography.

Magnetic Reversals and Seafloor Spreading

Geologists refer to Earth's present magnetic field as being normal—that is, with the north and south magnetic poles located approximately at the north and south geographic poles, respectively. At various times in the geologic past, however, Earth's magnetic field has completely reversed, that is, the magnetic north and south poles reverse positions, so that the magnetic north pole becomes the magnetic south pole, and the magnetic south pole becomes the magnetic north pole. During such a reversal, the magnetic field weakens until it temporarily disappears. When the magnetic field returns, the magnetic poles have reversed their position. The existence of such **magnetic reversals** was discovered by dating and determining the orientation of the remanent magnetism in lava flows on land (• Figure 3.9). Although the cause of magnetic reversals is still uncertain, their occurrence in the geologic record is well documented.

A renewed interest in oceanographic research led to extensive mapping of the ocean basins during the 1960s. Such mapping revealed an oceanic ridge system more than 65,000 km long, constituting the most extensive mountain range in the world. Perhaps the best-known part of the ridge system is the Mid-Atlantic Ridge, which divides the Atlantic Ocean basin into two nearly equal parts (• Figure 3.10).

As a result of the oceanographic research conducted in the 1950s, Harry Hess of Princeton University proposed, in a 1962 landmark paper, the theory of **seafloor spreading** to account for continental movement. He suggested that continents do not move through oceanic crust as do ships plowing through sea ice, but rather that the continents and oceanic crust move together as a single unit. Thus, the theory of seafloor spreading answered a major objection of the opponents of continental drift—namely, how could continents move through oceanic crust? As we now know, continents move with the oceanic crust as part of a lithospheric system.

Hess further postulated that the seafloor separates at oceanic ridges, where new crust is formed by upwelling magma. As the magma cools, the newly formed oceanic crust moves laterally away from the ridge.

As a mechanism to drive this system, Hess revived the idea (first proposed in the late 1920s by the British geologist Arthur Holmes) of a heat transfer system—or **thermal convection cells**—within the mantle as a mechanism to move the plates. According to Hess, hot magma rises from the mantle, intrudes along fractures defining oceanic ridges, and thus forms new crust. Cold crust is sub-

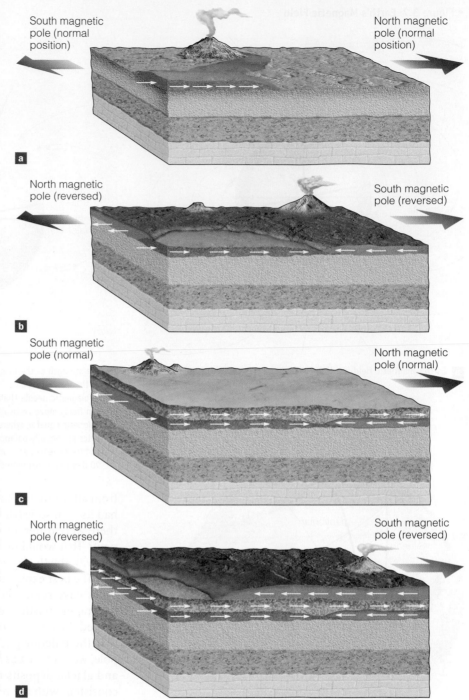

• **Figure 3.9** **Magnetic Reversals** During the time period shown (**a** and **d**), volcanic eruptions produced a succession of overlapping lava flows. At the time of these volcanic eruptions, Earth's magnetic field completely reversed—that is, the north magnetic pole moved to the south geographic pole, and the south magnetic pole moved to the north geographic pole. Thus, the end of the needle of a magnetic compass that today would point to the North Pole would point to the South Pole if the magnetic field should again reverse. We know that Earth's magnetic field has reversed numerous times in the past because when lava flows cool below the Curie point, magnetic minerals within the flow orient themselves parallel to the magnetic field at the time. They thus record whether the magnetic field was normal or reversed at that time. The white arrows in this diagram show the direction of the north magnetic pole for each individual lava flow, thus confirming that Earth's magnetic field has reversed during the past.

ducted back into the mantle at oceanic trenches, where it is heated and recycled, thus completing a thermal convection cell (see Figure 1.8).

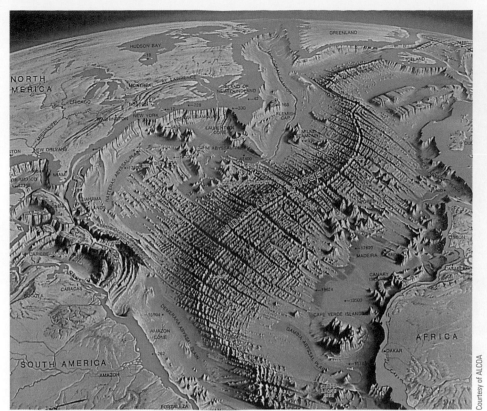

Courtesy of ALCOA

• Figure 3.10 **Topography of the Atlantic Ocean Basin** Artistic view of what the Atlantic Ocean basin would look like without water. The major feature is the Mid-Atlantic Ridge, an oceanic ridge system that is longer than 65,000 km and divides the Atlantic Ocean basin in half. It is along such oceanic ridges that the seafloor is separating and new oceanic crust is forming from magma upwelling from Earth's interior.

How could Hess's hypothesis be confirmed? Magnetic surveys of the oceanic crust revealed a pattern of striped **magnetic anomalies** (deviations from the average strength of Earth's present-day magnetic field) in the rocks that are both parallel to and symmetric around the oceanic ridges

(• Figure 3.11). A positive magnetic anomaly results when Earth's magnetic field at the time of oceanic crust formation along an oceanic ridge summit was the same as today, thus yielding a stronger than normal (positive) magnetic signal. A negative magnetic anomaly results

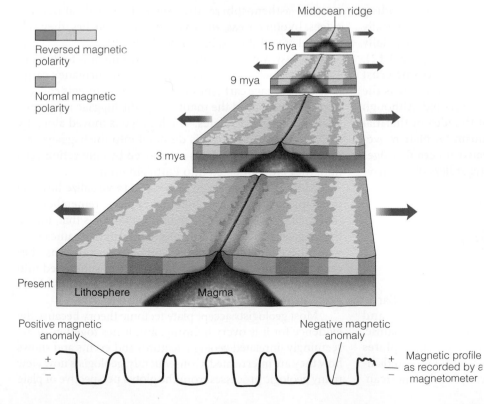

• Figure 3.11 **Magnetic Anomalies and Seafloor Spreading** The sequence of magnetic anomalies preserved within the oceanic crust is both parallel to and symmetric around oceanic ridges. Basaltic lava intruding into an oceanic ridge today and spreading laterally away from the ridge records Earth's current magnetic field or polarity (considered by convention to be normal). Basaltic intrusions 3, 9, and 15 million years ago record Earth's reversed magnetic field at those times. This schematic diagram shows how the solidified basalt moves away from the oceanic ridge (or spreading ridge), carrying with it the magnetic anomalies that are preserved in the oceanic crust. Magnetic anomalies are magnetic readings that are either higher (positive magnetic anomalies) or lower (negative magnetic anomalies) than Earth's current magnetic field strength. The magnetic anomalies are recorded by a magnetometer, which measures the strength of the magnetic field. Modified from Kious and Tiling, USGS and Hyndman & Hyndman, *Natural Hazards and Disasters,* Brooks/Cole, 2006, p. 15, Fig. 2.6b.

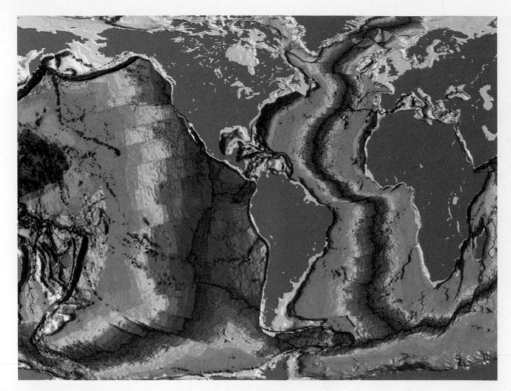

• Figure 3.12 **Age of the World's Ocean Basins** The age of the world's ocean basins has been determined from magnetic anomalies preserved in oceanic crust. The red colors adjacent to the oceanic ridges are the youngest oceanic crust. Moving laterally away from the ridges, the red colors grade to yellow at 48 million years ago, to green at 68 million years ago, and to dark blue some 155 million years ago. The darkest blue color is adjacent to the continental margins and is just somewhat less than 180 million years old. Magnetic anomalies demonstrate that the youngest oceanic crust is adjacent to the spreading ridges and that its age increases away from the ridge axis.

when Earth's magnetic field at the time of oceanic crust formation was reversed, therefore yielding a weaker than normal (negative) magnetic signal.

Thus, as new oceanic crust forms at oceanic ridge summits and records Earth's magnetic field at the time, the previously formed crust moves laterally away from the ridge. These magnetic stripes, therefore, represent times of normal and reversed polarity at oceanic ridges (where up-welling magma forms new oceanic crust), and conclusively confirm Hess's theory of seafloor spreading.

One of the consequences of the seafloor spreading theory is its confirmation that ocean basins are geologically young features whose openings and closings are partially responsible for continental movement (• Figure 3.12). Radiometric dating reveals that the oldest oceanic crust is somewhat less than 180 million years old, whereas the oldest continental crust is 3.96 billion years old. Although geologists do not universally accept the idea of thermal convection cells as a driving mechanism for plate movement, most accept that plates are created at oceanic ridges and destroyed at deep-sea trenches, regardless of the driving mechanism involved.

Plate Tectonics and Plate Boundaries

Plate tectonic theory is based on a simple model of Earth. The rigid lithosphere, composed of both oceanic and continental crust as well as the underlying upper mantle, consists of numerous variable-sized pieces called **plates** (• Figure 3.13). There are seven major plates (Eurasian, Indian-Australian, Antarctic, North American, South American,

Pacific, and African), and numerous smaller ones ranging from only a few tens to several hundreds of kilometers in width (for example, the Nazca plate). Plates also vary in thickness; those composed of upper mantle and continental crust are as much as 250 km thick, whereas those of upper mantle and oceanic crust are up to 100 km thick.

The lithosphere overlies the hotter and weaker semi-plastic asthenosphere. It is thought that movement resulting from some type of heat-transfer system within the asthenosphere causes the overlying plates to move. As plates move over the asthenosphere, they separate, mostly at oceanic ridges. In other areas, such as at oceanic trenches, they collide and are subducted back into the mantle.

An easy way to visualize plate movement is to think of a conveyor belt moving luggage from an airplane's cargo hold to a baggage cart. The conveyor belt represents convection currents within the mantle, and the luggage represents Earth's lithospheric plates. The luggage is moved along by the conveyor belt until it is dumped into the baggage cart in the same way that plates are moved by convection cells until they are subducted into Earth's interior.

Although this analogy allows you to visualize how the mechanism of plate movement takes place, remember that this analogy is limited. The major limitation is that, unlike the luggage, plates consist of continental and oceanic crust, which have different densities, and only oceanic crust, because it is denser than continental crust, is subducted into Earth's interior. Nonetheless, this analogy does provide an easy way to visualize plate movement.

Most geologists accept plate tectonic theory because the evidence for it is overwhelming, and it ties together many seemingly unrelated geologic features and events and shows how they are interrelated. Consequently, geologists now view many geologic processes from the global perspective of plate

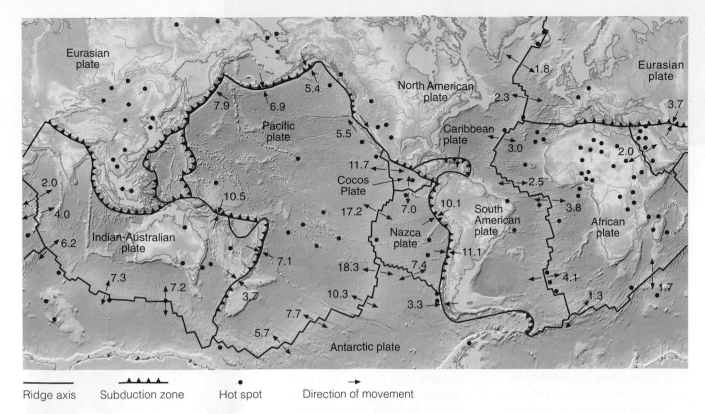

Ridge axis   Subduction zone   Hot spot   Direction of movement

• **Figure 3.13 Earth's Plates** A world map showing Earth's plates, their boundaries, their relative motion and average rates of movement in centimeters per year, and hot spots. Data from J. B. Minster and T. H. Jordan "Present Day Plate Motions," *Journal of Geophysical Research, 83* (1978) 5331–51. Copyright © 1978 American Geophysical Union. Modified by permission of American Geophysical Union.

tectonic theory in which interactions along plate margins are responsible for such phenomena as mountain building, earthquake activity, and volcanism. Furthermore, because all the inner planets have had a similar origin and early history, geologists are interested in determining whether plate tectonics is unique to Earth or whether it operates in the same way on other planets (see Perspective).

Because it appears that plate tectonics has operated since at least the Proterozoic Eon (see Chapter 9), it is important that we understand how plates move and interact with each other and how ancient plate boundaries are recognized. After all, the movement of plates has profoundly affected the geologic and biologic history of this planet.

Geologists recognize three major types of plate boundaries: *divergent, convergent,* and *transform* (Table 3.1). Along these boundaries, new plates are formed, are consumed, or slide laterally past one another. To understand the implications of plate interactions in terms of how they have affected Earth's history, geologists must study present-day plate boundaries.

TABLE **3.1**	**Types of Plate Boundaries**		
Type	**Example**	**Landforms**	**Volcanism**
Divergent			
Oceanic	Mid-Atlantic Ridge	Mid-oceanic ridge with axial rift valley	Basalt
Continental	East African Rift Valley	Rift Valley	Basalt and rhyolite, no andesite
Convergent			
Oceanic–oceanic	Aleutian Islands	Volcanic island arc, offshore oceanic trench	Andesite
Oceanic–continental	Andes	Offshore oceanic trench, volcanic mountain chain, mountain belt	Andesite
Continental–continental	Himalayas	Mountain belt	Minor
Transform	San Andreas fault	Fault valley	Minor

Tectonics of the Terrestrial Planets

The four inner, or terrestrial, planets— Mercury, Venus, Earth, and Mars—all had a similar early history involving accretion, differentiation into a metallic core and silicate mantle and crust, and formation of an early atmosphere by outgassing. Their early history was also marked by widespread volcanism and meteorite impacts, both of which helped modify their surfaces. Whereas three terrestrial planets as well as some of the Jovian moons display internal activity, Earth appears to be unique in that its surface is broken into a series of plates.

Images of Mercury sent back by *Mariner 10* show a heavily cratered surface (Figure 1) with the largest impact basins filled with what appear to be lava flows similar to the lava plains on Earth's Moon.

Another feature of Mercury's surface is a large number of lobate scarps that cut through the lava plains and impact basins. These scarps have been interpreted as thrust faults (a fracture along which one block of rocks has been pushed over an adjacent block of rocks) that resulted from cooling and compression during Mercury's early history.

Of all the planets, Venus (Figure 2) is the most similar in size and mass to Earth, but it differs in most other respects. Whereas Earth is dominated by plate tectonics, Venus shows no evidence of current or geologically-recent plate tectonic activity, although there is still debate as to whether there was a period of active plate tectonics early

in its history. Instead, volcanism is thought to be the dominant force in the evolution of the Venusian surface (Figure 3). Even though no active volcanism has been observed on Venus, much of its surface has

Figure 3 Volcano Sapas Mons of Venus, contains two lava-filled craters and is flanked by lava flows, attesting to the volcanic activity that was once common on Venus.

JPL/ NASA

Figure 1 A color-enhanced photomosaic of Mercury shows its heavily cratered surface, which has changed very little since its early history.

JPL/ NASA

Figure 2 A color-enhanced photomosaic of Venus based on radar images beamed back to Earth by the *Magellan* spacecraft. This image shows impact craters and volcanic features characteristic of the planet.

been shaped by volcanic activity as indicated by the numerous giant volcanoes and volcanic pancake domes. These structures are thought to be the product of rising convection currents of magma.

Mars (Figure 4), popularly known as the Red Planet, has numerous features that indicate an extensive early period of volcanism. These include Olympus Mons (Figure 5), the solar system's largest volcano, lava flows, and uplifted regions thought to have resulted from mantle convection. In addition to volcanic features, Mars displays abundant evidence of tensional tectonics, including numerous faults and large fault-produced valley structures. It is currently hypothesized that Mars may have experienced a period of plate tectonics during its very early history, which was followed by the initiation of superplumes and volcanism.

Although not a terrestrial planet, Io (Figure 6), the innermost of Jupiter's Galilean moons, must be mentioned. Images from the *Voyager* and *Galileo* spacecrafts show that Io has no impact craters. In fact, more than a hundred active volcanoes are visible on the moon's surface, and the sulfurous gas and ash erupted by these volcanoes bury any newly formed meteorite impact craters. Because of its proximity to Jupiter, the heat source of Io is probably tidal heating, in which the resulting friction is enough to at least partially melt Io's interior and drive its volcanoes.

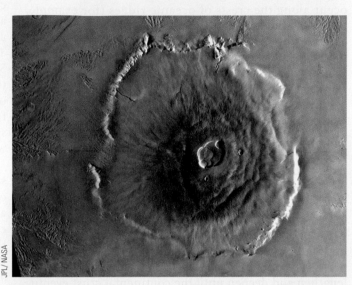

Figure 5 A vertical view of Olympus Mons, a volcano on Mars, and the largest volcano in our solar system. The edge of the Olympus Mons crater is marked by a cliff several kilometers high, rather than a moat as in Mauna Loa, Earth's largest volcano.

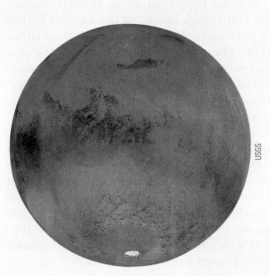

Figure 4 A photomosaic of Mars shows a variety of geologic structures, including the southern polar ice cap.

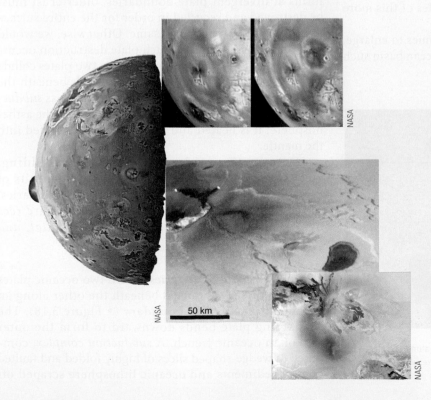

50 km

Figure 6 Volcanic features of Io, the innermost moon of Jupiter. As shown in these digitally enhanced color images, Io is a very volcanically active moon.

Divergent Boundaries

Divergent plate boundaries, or *spreading ridges,* occur where plates are separating and new oceanic lithosphere is forming. Divergent boundaries are places where the crust is extended, thinned, and fractured as magma—derived from the partial melting of the mantle—rises to the surface. The magma is almost entirely basaltic and intrudes into vertical fractures to form dikes and pillow lava flows (• Figure 3.14). As successive injections of magma cool and solidify, they form new oceanic crust and record the intensity and orientation of Earth's magnetic field (Figure 3.11). Divergent boundaries most commonly occur along the crests of oceanic ridges (for example, the Mid-Atlantic Ridge). Oceanic ridges are thus characterized by rugged topography with high relief resulting from displacement of rocks along large fractures, shallow-depth earthquakes, high heat flow, and basaltic flows or pillow lavas.

Divergent boundaries are also present under continents during the early stages of continental breakup. When magma wells up beneath a continent, the crust is initially elevated, stretched, and thinned, producing fractures, faults, rift valleys, and volcanic activity (• Figure 3.15a). As magma intrudes into faults and fractures, it solidifies or flows out onto the surface as lava flows; the latter often covering the rift valley floor (Figure 3.15b). The East African Rift Valley is an excellent example of continental breakup at this stage (• Figure 3.16a).

As spreading proceeds, some rift valleys continue to lengthen and deepen until the continental crust eventually breaks and a narrow linear sea is formed, separating two continental blocks (Figure 3.15c). The Red Sea, separating the Arabian Peninsula from Africa (Figure 3.16b), and the Gulf of California, which separates Baja California from mainland Mexico, are good examples of this more advanced stage of rifting.

As a newly created narrow sea continues to enlarge, it may eventually become an expansive ocean basin such as the Atlantic Ocean basin is today, separating North and South America from Europe and Africa by thousands of kilometers (Figure 3.15d). The Mid-Atlantic Ridge is the boundary between these diverging plates (Figure 3.10); the American plates are moving westward, and the Eurasian and African plates are moving eastward.

An Example of Ancient Rifting

What features in the geologic record can geologists use to recognize ancient rifting? Associated with regions of continental rifting are faults, dikes, sills, lava flows, and thick sedimentary sequences within rift valleys, all features that are preserved in the geologic record. The Triassic fault-block basins of the eastern United States are a good example of ancient continental rifting (• Figure 3.17a). These fault-block basins mark the zone of rifting that occurred when North America split apart from Africa (see Chapter 14). The basins contain thousands of meters of continental sediment and are riddled with dikes and sills Figure 3.17b.

Pillow lavas (Figure 3.14), in association with deep-sea sediment, are also evidence of ancient rifting. The presence of pillow lavas marks the formation of a spreading ridge in a narrow linear sea. A narrow linear sea forms when the continental crust in the rift valley finally breaks apart, and the area is flooded with seawater. Magma, intruding into the sea along this newly formed spreading ridge, solidifies as pillow lavas, which are preserved in the geologic record, along with the sediment that was deposited on them.

Convergent Boundaries

Whereas new crust forms at divergent plate boundaries, older crust must be destroyed and recycled in order for the entire surface area of Earth to remain the same. Otherwise, we would have an expanding Earth. Such plate destruction occurs at **convergent plate boundaries,** where two plates collide and the leading edge of one plate descends beneath the margin of the other plate by a process known as *subduction.* As the subducting plate moves down into the asthenosphere, it is heated and eventually incorporated into the mantle.

Deformation, volcanism, mountain building, metamorphism, earthquake activity, and deposits of valuable minerals characterize convergent boundaries. Three types of convergent plate boundaries are recognized: *oceanic–oceanic, oceanic–continental, and continental–continental.*

Oceanic–Oceanic Boundaries When two oceanic plates converge, one is subducted beneath the other along an **oceanic–oceanic plate boundary** (• Figure 3.18). The subducting plate bends downward to form the outer wall of an oceanic trench. A *subduction complex,* composed of wedge-shaped slices of highly folded and faulted marine sediments and oceanic lithosphere scraped off

• Figure 3.14 **Pillow Lavas** Pillow lavas forming along the Mid-Atlantic Ridge. Their distinctive bulbous shape is the result of underwater eruption.

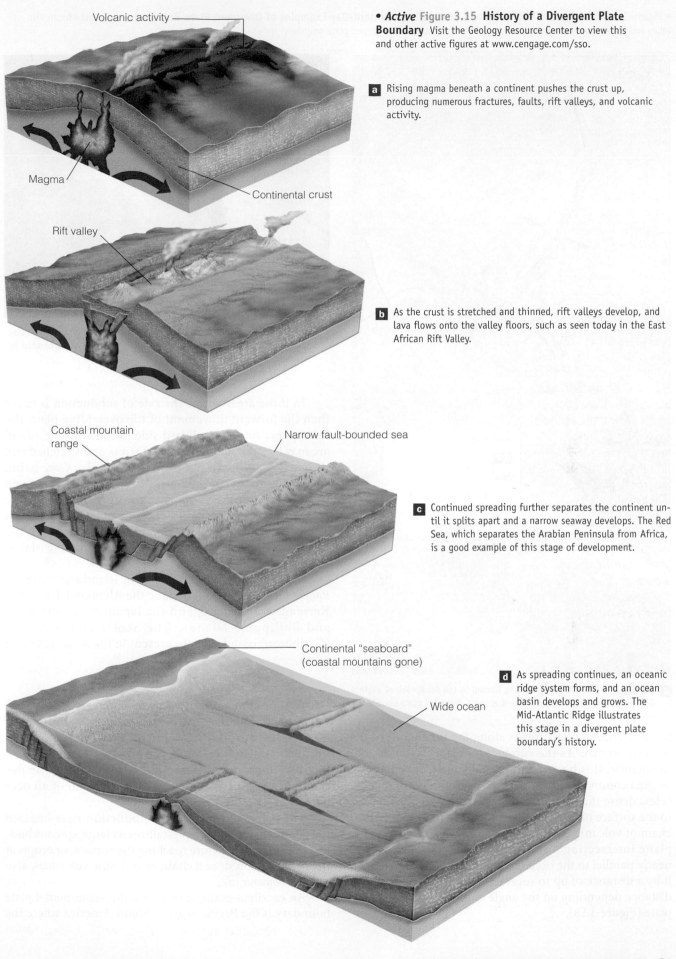

Volcanic activity

Magma

Continental crust

• *Active* Figure 3.15 **History of a Divergent Plate Boundary** Visit the Geology Resource Center to view this and other active figures at www.cengage.com/sso.

a Rising magma beneath a continent pushes the crust up, producing numerous fractures, faults, rift valleys, and volcanic activity.

Rift valley

b As the crust is stretched and thinned, rift valleys develop, and lava flows onto the valley floors, such as seen today in the East African Rift Valley.

Coastal mountain range

Narrow fault-bounded sea

c Continued spreading further separates the continent until it splits apart and a narrow seaway develops. The Red Sea, which separates the Arabian Peninsula from Africa, is a good example of this stage of development.

Continental "seaboard" (coastal mountains gone)

Wide ocean

d As spreading continues, an oceanic ridge system forms, and an ocean basin develops and grows. The Mid-Atlantic Ridge illustrates this stage in a divergent plate boundary's history.

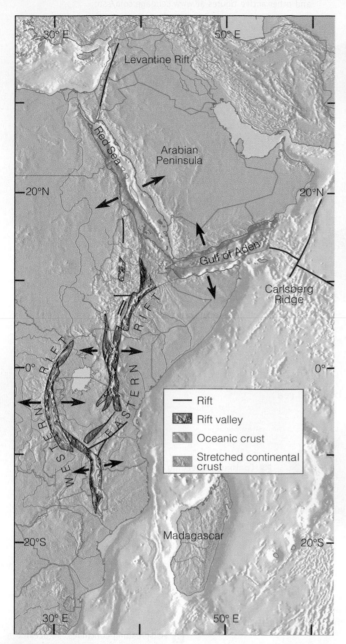

Legend:
- — Rift
- ▨ Rift valley
- ▨ Oceanic crust
- ▨ Stretched continental crust

a The East African Rift Valley is being formed by the separation of eastern Africa from the rest of the continent along a divergent plate boundary.

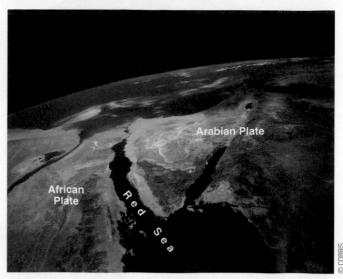

b The Red Sea represents a more advanced stage of rifting, in which two continental blocks (Africa and the Arabian Peninsula) are separated by a narrow sea.

the descending plate, forms along the inner wall of the oceanic trench. As the subducting plate descends into the mantle, it is heated and partially melted, generating magma commonly of andesitic composition. This magma is less dense than the surrounding mantle rocks and rises to the surface of the nonsubducted plate to form a curved chain of volcanic islands called a *volcanic island arc* (any plane intersecting a sphere makes an arc). This arc is nearly parallel to the oceanic trench and is separated from it by a distance of up to several hundred kilometers—the distance depending on the angle of dip of the subducting plate (Figure 3.18).

In those areas where the rate of subduction is faster than the forward movement of the overriding plate, the lithosphere on the landward side of the volcanic island arc may be subjected to tensional stress and stretched and thinned, resulting in the formation of a *back-arc basin*. This back-arc basin may grow by spreading if magma breaks through the thin crust and forms new oceanic crust (Figure 3.18). A good example of a back-arc basin associated with an oceanic–oceanic plate boundary is the Sea of Japan between the Asian continent and the islands of Japan.

Most present-day active volcanic island arcs are in the Pacific Ocean basin and include the Aleutian Islands, the Kermadec–Tonga arc, and the Japanese (Figure 3.18) and Philippine Islands. The Scotia and Antillean (Caribbean) island arcs are present in the Atlantic Ocean basin.

Oceanic–Continental Boundaries When an oceanic and a continental plate converge, the denser oceanic plate is subducted under the continental plate along an **oceanic–continental plate boundary** (• Figure 3.19). Just as at oceanic–oceanic plate boundaries, the descending oceanic plate forms the outer wall of an oceanic trench.

The magma generated by subduction rises beneath the continent and either crystallizes as large igneous bodies (called *plutons*) before reaching the surface, or erupts at the surface to produce a chain of andesitic volcanoes, also called a *volcanic arc*.

An excellent example of an oceanic–continental plate boundary is the Pacific coast of South America where the

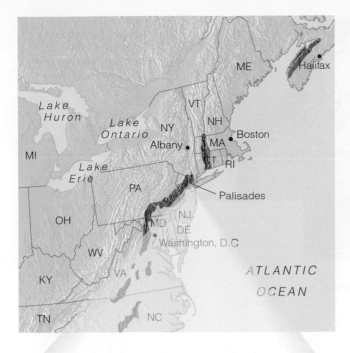

• **Figure 3.17** **Triassic Fault-Block Basins of Eastern North America—An Example of Ancient Rifting**

a Triassic fault-block basin deposits appear in numerous locations throughout eastern North America. These fault-block basins are good examples of ancient continental rifting, and during the Triassic Period they looked like today's fault-block basins (rift valleys) of the East African Rift Valley.

Courtesy of John M. Faivre

b Palisades of the Hudson River. This sill (tabular-shaped concordant igneous intrusion) was one of many that were intruded into the fault-block basin sediments during the Late Triassic rifting that marked the separation of North America from Africa.

oceanic Nazca plate is currently being subducted under South America (Figure 3.19). The Peru-Chile Trench marks the site of subduction, and the Andes Mountains are the resulting volcanic mountain chain on the nonsubducting plate. This particular example demonstrates the effect that plate tectonics has on our lives. For instance, earthquakes are commonly associated with subduction zones, and the western side of South America is the site of frequent and devastating earthquakes.

Continental–Continental Boundaries   Two continents approaching each other are initially separated by an ocean floor that is being subducted under one continent. The edge of that continent displays the features characteristic of oceanic–continental convergence. As the ocean floor continues to be subducted, the two continents come closer together until they eventually collide. Because continental lithosphere, which consists of continental crust and the upper mantle, is less dense than oceanic lithosphere (oceanic crust and upper mantle), it cannot sink into the asthenosphere. Although one continent may partly slide under the other, it cannot be pulled or pushed down into a subduction zone (• Figure 3.20).

When two continents collide, they are welded together along a zone marking the former site of subduction. At this **continental–continental plate boundary,** an interior mountain belt is formed consisting of deformed sediments and sedimentary rocks, igneous intrusions, metamorphic rocks, and fragments of oceanic crust. In addition, the entire region is subjected to numerous earthquakes. The Himalayas in central Asia, the world's youngest and highest mountain system, resulted from the collision between

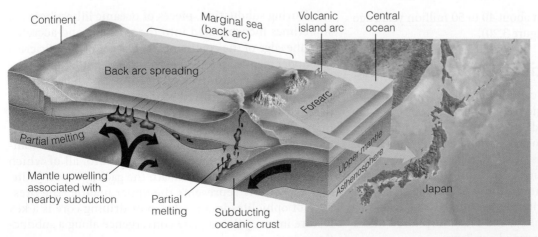

• *Active* **Figure 3.18**
Oceanic–Oceanic Convergent Plate Boundary An oceanic trench forms where one oceanic plate is subducted beneath another. On the nonsubducted plate, a volcanic island arc forms from the rising magma generated from the subducting plate. The Japanese Islands are a volcanic island arc resulting from the subduction of one oceanic plate beneath another oceanic plate. Visit the Geology Resource Center to view this and other active figures at www.cengage.com/sso.

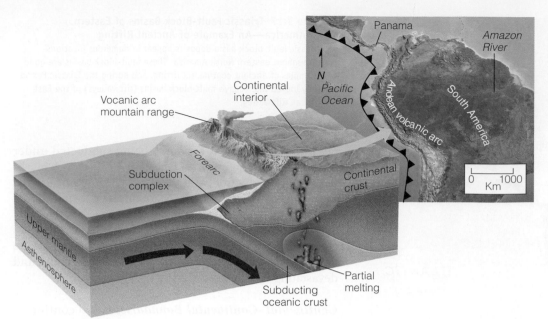

• **Active** Figure 3.19
Oceanic–Continental Convergent Plate Boundary When an oceanic plate is subducted beneath a continental plate, an andesitic volcanic mountain range is formed on the continental plate as a result of rising magma. The Andes Mountains in South America are one of the best examples of continuing mountain building at an oceanic–continental convergent plate boundary. Visit the Geology Resource Center to view this and other active figures at www.cengage.com/sso.

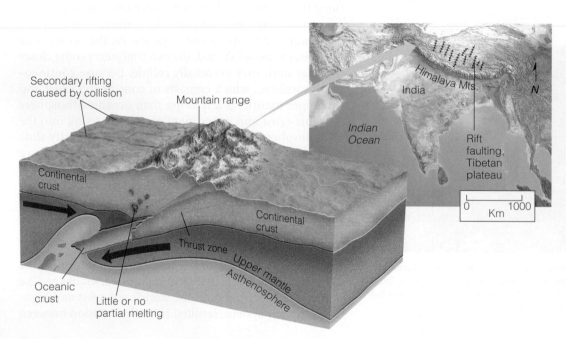

• **Active** Figure 3.20
Continental–Continental Convergent Plate Boundary When two continental plates converge, neither is subducted because of their great thickness and low and equal densities. As the two continental plates collide, a mountain range is formed in the interior of a new and larger continent. The Himalayas in central Asia resulted from the collision between India and Asia approximately 40 to 50 million years ago. Visit the Geology Resource Center to view this and other active figures at www.cengage.com/sso.

India and Asia that began about 40 to 50 million years ago and is still continuing (Figure 3.20).

Recognizing Ancient Convergent Plate Boundaries

How can former subduction zones be recognized in the geologic record? Igneous rocks provide one clue to ancient subduction zones. The magma that erupted at the surface, forming island arc volcanoes and continental volcanoes, is of andesitic composition. Another clue is the zone of intensely deformed rocks between the deep-sea trench where subduction took place and the area of igneous activity. Here, sediments and submarine rocks are folded, faulted, and metamorphosed into a chaotic mixture of rocks termed a *mélange*.

During subduction, pieces of oceanic lithosphere are sometimes incorporated into the mélange and accreted onto the edge of the continent. Such slices of oceanic crust and upper mantle are called *ophiolites* (• Figure 3.21). They consist of a layer of deep-sea sediments that includes graywackes (poorly sorted sandstones containing abundant feldspars and rock fragments, usually in a clay-rich matrix), black shales, and cherts. These deep-sea sediments are underlain by pillow lavas, a sheeted dike complex, massive gabbro, and layered gabbro, all of which form the oceanic crust. Beneath the gabbro is peridotite, which probably represents the upper mantle. The presence of ophiolites in an outcrop or drilling core is a key feature in recognizing plate convergence along a subduction zone.

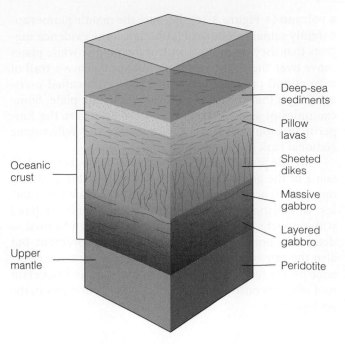

Deep-sea sediments

Pillow lavas

Sheeted dikes

Massive gabbro

Layered gabbro

Peridotite

Oceanic crust

Upper mantle

• **Figure 3.21 Ophiolites** Ophiolites are sequences of rock on land consisting of deep-sea sediments, oceanic crust, and upper mantle. Ophiolites are one feature used to recognize ancient convergent plate boundaries.

Elongate belts of folded and faulted marine sedimentary rocks, andesites, and ophiolites are found in the Appalachians, Alps, Himalayas, and Andes mountains. The combination of such features is significant evidence that these mountain ranges resulted from deformation along convergent plate boundaries.

Transform Boundaries The third type of plate boundary is a **transform plate boundary.** These mostly occur along fractures in the seafloor, known as *transform*

faults, where plates slide laterally past one another, roughly parallel to the direction of plate movement. Although lithosphere is neither created nor destroyed along a transform boundary, the movement between plates results in a zone of intensely shattered rock and numerous shallow-depth earthquakes.

Transform faults "transform" or change one type of motion between plates into another type of motion. Most commonly, transform faults connect two oceanic ridge segments (• Figure 3.22), but they can also connect ridges to trenches, and trenches to trenches. Although the majority of transform faults are in oceanic crust and are marked by distinct fracture zones, they may also extend into continents.

One of the best-known transform faults is the San Andreas Fault in California. It separates the Pacific plate from the North American plate and connects spreading ridges in the Gulf of California with the Juan de Fuca and Pacific plates off the coast of northern California (• Figure 3.23). Many of the earthquakes that affect California are the result of movement along this fault (see Chapter 16).

Unfortunately, transform faults generally do not leave any characteristic or diagnostic features except for the obvious displacement of the rocks with which they are associated. This displacement is usually large, on the order of tens to hundreds of kilometers. Such large displacements in ancient rocks can sometimes be related to transform fault systems.

Hot Spots and Mantle Plumes

Before leaving the topic of plate boundaries, we should briefly mention an intraplate feature found beneath both oceanic and continental plates. A **hot spot** is the location on Earth's surface where a stationary column of magma, originating deep within the mantle, has slowly risen to the surface (*mantle plume*) and formed

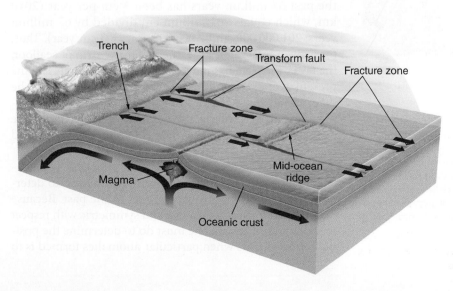

Trench

Fracture zone

Transform fault

Fracture zone

Magma

Mid-ocean ridge

Oceanic crust

• *Active* **Figure 3.22 Transform Plate Boundaries** Horizontal movement between plates occurs along transform faults. Extensions of transform faults on the seafloor form fracture zones. Most transform faults connect two oceanic ridge segments. Note that relative motion between the plates only occurs between the two ridges. Visit the Geology Resource Center to view this and other active figures at www.cengage.com/sso.

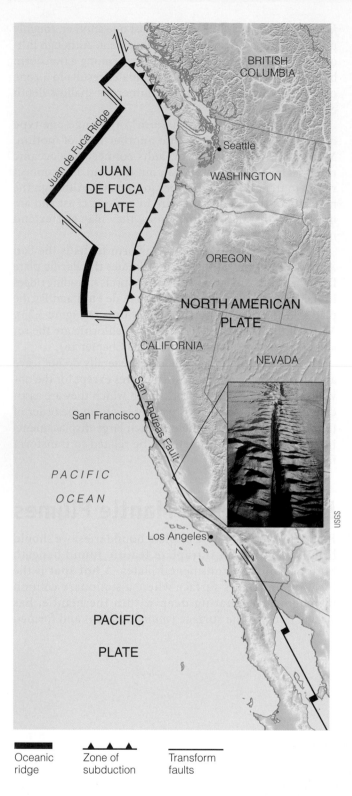

Oceanic ridge (symbol) **Zone of subduction** (symbol) **Transform faults** (symbol)

• **Figure 3.23** **The San Andreas Fault—A Transform Plate Boundary** The San Andreas Fault is a transform fault separating the Pacific plate from the North American plate. It connects the spreading ridges in the Gulf of California with the Juan de Fuca and Pacific plates off the coast of northern California. Movement along the San Andreas Fault has caused numerous earthquakes. The insert photograph shows a segment of the San Andreas Fault as it cuts through the Carrizo Plain, California.

a volcano (• Figure 3.24). Because the mantle plumes apparently remain stationary (although some evidence suggests that they might not) within the mantle while plates move over them, the resulting hot spots leave a trail of extinct and progressively older volcanoes called *aseismic ridges* that record the movement of the plate. Some examples of aseismic ridges and hot spots are the Emperor Seamount–Hawaiian Island chain and Yellowstone National Park in Wyoming.

Mantle plumes and hot spots help geologists explain some of the geologic activity occurring within plates as opposed to activity occurring at or near plate boundaries. In addition, if mantle plumes are essentially fixed with respect to Earth's rotational axis, they can be used to determine not only the direction of plate movement, but also the rate of movement. They can also provide reference points for determining paleolatitude, an important tool when reconstructing the location of continents in the geologic past.

How Are Plate Movement and Motion Determined?

How fast and in what direction are Earth's plates moving? Do they all move at the same rate? Rates of plate movement can be calculated in several ways. The least accurate method is to determine the age of the sediments immediately above any portion of the oceanic crust and then divide the distance from the spreading ridge by that age. Such calculations give an average rate of movement.

A more accurate method of determining both the average rate of movement and relative motion is by dating the magnetic anomalies in the crust of the seafloor (Figure 3.12). The distance from an oceanic ridge axis to any magnetic anomaly indicates the width of new seafloor that formed during that time interval. For example, if the distance between the present-day Mid-Atlantic Ridge and anomaly 31 is 2010 km, and anomaly 31 formed 67 million years ago (• Figure 3.25), then the average rate of movement during the past 67 million years has been 3 cm per year (2010 km, which equals 201 million cm divided by 67 million years; 201,000,000 cm/67,000,000 years = 3 cm/year). Thus, for a given interval of time, the wider the strip of seafloor, the faster the plate has moved. In this way, not only can the present average rate of movement and relative motion be determined (Figure 3.13), but the average rate of movement during the past can also be calculated by dividing the distance between anomalies by the amount of time elapsed between anomalies.

Geologists use magnetic anomalies not only to calculate the average rate of plate movement, but also to determine plate positions at various times in the past. Because magnetic anomalies are parallel and symmetric with respect to spreading ridges, all one must do to determine the position of continents when particular anomalies formed is to

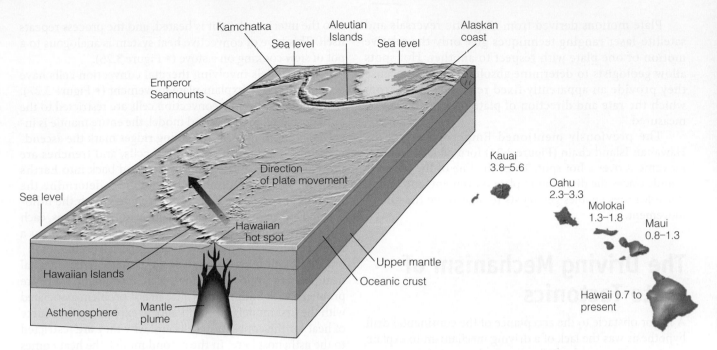

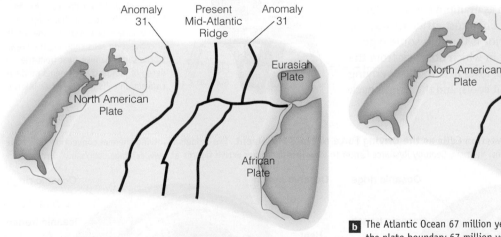

• *Active* Figure 3.24 **Hot Spots** A hot spot is the location where a stationary mantle plume has risen to the surface and formed a volcano. The Emperor Seamount–Hawaiian Island chain formed as a result of the Pacific plate moving over a mantle plume, and the line of volcanic islands in this chain traces the direction of plate movement. The island of Hawaii and Loihi Seamount are the only current hot spots of this island chain. The numbers indicate the age of the islands in millions of years. Visit the Geology Resource Center to view this and other active figures at www.cengage.com/sso.

• Figure 3.25 **Reconstructing Plate Positions Using Magnetic Anomalies**

a The present North Atlantic, showing the Mid-Atlantic Ridge and magnetic anomaly 31, which formed 67 million years ago.

b The Atlantic Ocean 67 million years ago. Anomaly 31 marks the plate boundary 67 million years ago. By moving the anomalies back together, along with the plates they are on, we can reconstruct the former positions of the continents.

move the anomalies back to the spreading ridge, which will also move the continents with them (Figure 3.25). Unfortunately, subduction destroys oceanic crust and the magnetic record it carries. Thus, we have an excellent record of plate movements since the breakup of Pangaea, but not as good an understanding of plate movement before that time.

The average rate of movement as well as the relative motion between any two plates can also be determined by satellite-laser ranging techniques. Laser beams from a station on one plate are bounced off a satellite (in geosynchronous orbit) and returned to a station on a different plate. As the plates move away from each other, the laser beam takes more time to go from the sending station to the stationary satellite and back to the receiving station. This difference in elapsed time is used to calculate the rate of movement and the relative motion between plates.

Plate motions derived from magnetic reversals and satellite-laser ranging techniques give only the relative motion of one plate with respect to another. Hot spots allow geologists to determine absolute motion because they provide an apparently fixed reference point from which the rate and direction of plate movement can be measured.

The previously mentioned Emperor Seamount-Hawaiian Island chain (Figure 3.24) formed as a result of movement over a hot spot. Thus, the line of the volcanic islands traces the direction of plate movement, and dating the volcanoes enables geologists to determine the rate of movement.

The Driving Mechanism of Plate Tectonics

A major obstacle to the acceptance of the continental drift hypothesis was the lack of a driving mechanism to explain continental movement. When it was shown that continents and ocean floors moved together, not separately, and that new crust formed at spreading ridges by rising magma, most geologists accepted some type of convective heat system (convection cells) as the basic process responsible for plate motion. The question, however, remains: What exactly drives the plates?

Most of the heat from Earth's interior results from the decay of radioactive elements, such as uranium (see Chapter 4), in the core and lower mantle. The most efficient way for this heat to escape Earth's interior is through some type of slow convection of mantle rock in which hot rock from the interior rises toward the surface, loses heat to the overlying lithosphere, becomes denser as it cools, and then sinks back

into the interior where it is heated, and the process repeats itself. This type of convective heat system is analogous to a pot of stew cooking on a stove (• Figure 3.26).

Two models involving thermal convection cells have been proposed to explain plate movement (• Figure 3.27). In one model, thermal convection cells are restricted to the asthenosphere; in the second model, the entire mantle is involved. In both models, spreading ridges mark the ascending limbs of adjacent convection cells, and trenches are present where convection cells descend back into Earth's interior. The convection cells, therefore, determine the location of spreading ridges and trenches, with the lithosphere lying above the thermal convection cells. Thus, each plate corresponds to a single convection cell and moves as a result of the convective movement of the cell itself.

Although most geologists agree that Earth's internal heat plays an important role in plate movement, there are problems with both models. The major problem associated with the first model is the difficulty in explaining the source of heat for the convection cells and why they are restricted to the asthenosphere. In the second model, the heat comes

• Figure 3.26 **Convection in a Pot of Stew** Heat from the stove is applied to the base of the stew pot causing the stew to heat up. As heat rises through the stew, pieces of the stew are carried to the surface, where the heat is dissipated, the pieces of stew cool, and then sink back to the bottom of the pot. The bubbling seen at the surface of the stew is the result of convection cells churning the stew. In the same manner, heat from the decay of radioactive elements produce convection cells within Earth's interior.

• *Active* Figure 3.27 **Thermal Convection Cells as the Driving Force of Plate Movement** Two models involving thermal convection cells have been proposed to explain plate movement. Visit the Geology Resource Center to view this and other active figures at www.cengage.com/sso.

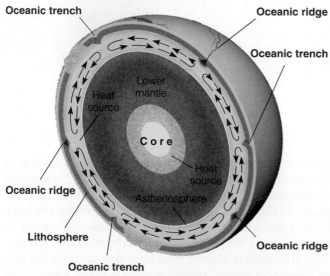

a In the first model, thermal convection cells are restricted to the asthenosphere.

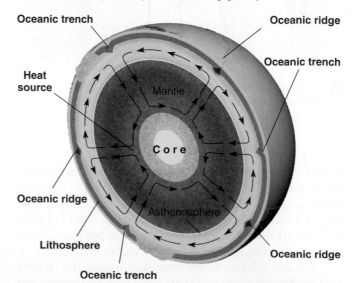

b In the second model, thermal convection cells involve the entire mantle.

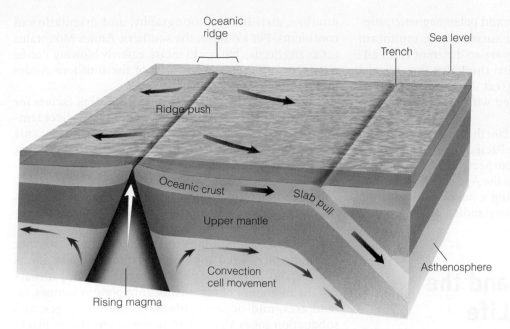

Oceanic ridge

Trench   Sea level

Ridge push

Oceanic crust

Slab pull

Upper mantle

Convection cell movement

Rising magma

Asthenosphere

• Figure 3.28 **Plate Movement Resulting from Gravity-Driven Mechanisms** Plate movement is also thought to occur, at least partially, from gravity-driven "slab-pull" or "ridge-push" mechanisms. In slab-pull, the edge of the subducting plate descends into the interior, and the rest of the plate is pulled downward. In ridge-push, rising magma pushes the oceanic ridges higher than the rest of the oceanic crust. Gravity thus pushes the oceanic lithosphere away from the ridges and toward the trenches.

from the outer core, but it is still not known how heat is transferred from the outer core to the mantle. Nor is it clear how convection can involve both the lower mantle and the asthenosphere.

In addition to some type of thermal convection system driving plate movement, some geologists think plate movement occurs because of a mechanism involving "slab-pull" or "ridge-push," both of which are gravity driven but still dependent on thermal differences within Earth (• Figure 3.28). In slab-pull, the subducting cold slab of lithosphere, being denser than the surrounding warmer asthenosphere, pulls the rest of the plate along as it descends into the asthenosphere. As the lithosphere moves downward, there is a corresponding upward flow back into the spreading ridge.

Operating in conjunction with slab-pull is the ridge-push mechanism. As a result of rising magma, the oceanic ridges are higher than the surrounding oceanic crust. It is thought that gravity pushes the oceanic lithosphere away from the higher spreading ridges and toward the trenches.

Currently, geologists are fairly certain that some type of convective system is involved in plate movement, but the extent to which other mechanisms, such as slab-pull and ridge-push, are involved is still unresolved. However, the fact that plates have moved in the past and are still moving today has been proven beyond a doubt. And although a comprehensive theory of plate movement has not yet been developed, more and more of the pieces are falling into place as geologists learn more about Earth's interior.

How Are Plate Tectonics and Mountain Building Related?

What role does plate tectonics play in mountain building? An **orogeny** is an episode of intense rock

deformation or mountain building. Orogenies are the consequence of compressive forces related to plate movement. As one plate is subducted under another plate, sedimentary and volcanic rocks are folded and faulted along the plate margin, while the more deeply buried rocks are subjected to regional metamorphism. Magma generated within the mantle either rises to the surface to erupt as andesitic volcanoes or cools and crystallizes beneath the surface, forming intrusive igneous bodies. Typically, most orogenies occur along either oceanic–continental or continental–continental plate boundaries.

As we discussed earlier in this chapter, ophiolites are evidence of ancient convergent plate boundaries. The slivers of ophiolites found in the interiors of such mountain ranges as the Alps, Himalayas, and Urals mark the sites of former subduction zones. The relationship between mountain building and the opening and closing of ocean basins is called the *Wilson cycle* in honor of the Canadian geologist J. Tuzo Wilson, who first suggested that an ancient ocean had closed to form the Appalachian Mountains and then reopened and widened to form the present Atlantic Ocean.

According to some geologists, much of the geology of continents can be described in terms of a succession of Wilson cycles. We will see in Chapter 14, however, that new evidence concerning the movement of microplates and their accretion at the margin of continents must also be considered when dealing with the tectonic history of a continent.

Terrane Tectonics During the 1970s and 1980s, geologists discovered that parts of many mountain systems are composed of small, accreted lithospheric blocks that clearly originated elsewhere. These **terranes,** as they are called, differ completely in their fossil content, stratigraphy, structural trends, and paleomagnetic properties from the rocks of the surrounding mountain system. In fact, these terranes are so different from adjacent rocks that most geologists think that they formed elsewhere and were carried great distances as parts of other plates until they collided with other terranes or continents.

To date, most terranes in North America have been identified in mountains of the Pacific coast region, but a number of such terranes are suspected to be present in other mountain systems, such as the Appalachians. Terrane tectonics is, however, providing a new way of viewing Earth and gaining a better understanding of the geologic history of the continents.

Plate Tectonics and the Distribution of Life

Plate tectonic theory is as revolutionary and far-reaching in its implications for geology as the theory of evolution was for biology when it was proposed. Interestingly, it was the fossil evidence that convinced Wegener, Suess, and du Toit, as well as many other geologists, of the correctness of continental drift. Together, the theories of plate tectonics and evolution have changed the way we view our planet, and we should not be surprised at the intimate association between them. Although the relationship between plate tectonic processes and the evolution of life is incredibly complex, paleontologic data provide convincing evidence of the influence of plate movement on the distribution of organisms.

The present distribution of plants and animals is not random but is controlled largely by climate and geographic barriers. The world's biotas occupy *biotic provinces*, which are regions characterized by a distinctive assemblage of plants and animals. Organisms within a province have similar ecologic requirements, and the boundaries separating provinces are, therefore, natural ecologic breaks. Climatic or geographic barriers are the most common province boundaries, and plate movement largely controls these.

Because adjacent provinces usually have less than 20% of their species in common, global diversity is a direct reflection of the number of provinces; the more provinces there are, the greater the global diversity. When continents break up, for example, the opportunity for new provinces to form increases, with a resultant increase in diversity. Just the opposite occurs when continents come together. Plate tectonics thus plays an important role in the distribution of organisms and their evolutionary history.

The complex interaction between wind and ocean currents has a strong influence on the world's climates. Wind and ocean currents are strongly influenced by the number, distribution, topography, and orientation of continents. For example, the southern Andes Mountains act as an effective barrier to moist, easterly blowing Pacific winds, resulting in a desert east of the southern Andes that is virtually uninhabitable.

Temperature is one of the major limiting factors for organisms, and province boundaries often reflect temperature barriers. Because atmospheric and oceanic temperatures decrease from the equator to the poles, most species exhibit a strong climatic zonation. This biotic zonation parallels the world's latitudinal atmospheric and oceanic circulation patterns. Changes in climate, thus, have a profound effect on the distribution and evolution of organisms.

The distribution of continents and ocean basins not only influences wind and ocean currents but also affects provinciality by creating physical barriers to, or pathways for, the migration of organisms. Intraplate volcanoes, island arcs, mid-oceanic ridges, mountain ranges, and subduction zones all result from the interaction of plates, and their orientation and distribution strongly influence the number of provinces and hence total global diversity. Thus, provinciality and diversity will be highest when there are numerous small continents spread across many zones of latitude.

When a geographic barrier separates a once-uniform fauna, species may undergo divergence. If conditions on opposite sides of the barrier are sufficiently different, then species must adapt to the new conditions, migrate, or become extinct. Adaptation to the new environment by various species may involve enough change that new species eventually evolve.

The marine invertebrates found on opposite sides of the Isthmus of Panama provide an excellent example of divergent evolution (see Chapter 7) caused by the formation of a geographic barrier. Prior to the rise of this land connection between North and South America, a homogeneous population of bottom-dwelling invertebrates inhabited the shallow seas of the area. After the rise of the Isthmus of Panama by subduction of the Pacific plate approximately 5 million years ago, the original population was divided into two populations—one in the Caribbean Sea, and the other in the Pacific Ocean. In response to the changing environment, new species evolved on opposite sides of the isthmus (● Figure 3.29).

The formation of the Isthmus of Panama also influenced the evolution of the North and South American mammalian faunas (see Chapter 18). During most of the Cenozoic Era, South America was an island continent, and its mammalian fauna evolved in isolation from the rest of the world's faunas. When North and South America were connected by the Isthmus of Panama, migrants from North America replaced most of the indigenous South American mammals. Surprisingly, only a few South American mammal groups migrated northward.

• Figure 3.29 **Plate Tectonics and the Distribution of Organisms**

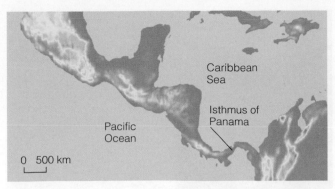

a The Isthmus of Panama forms a barrier that divides a once-uniform fauna of molluscs.

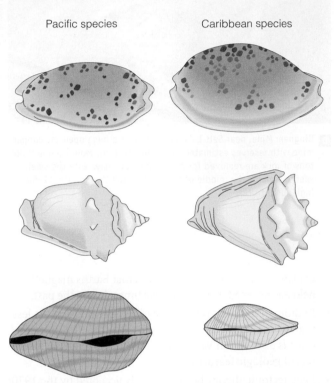

Pacific species   Caribbean species

b Divergent evolution of gastropod and bivalve species occurred as a result of the formation of the Isthmus of Panama. Each pair belongs to the same genus but is a different species.

Plate Tectonics and the Distribution of Natural Resources

In addition to being responsible for the major features of Earth's crust and influencing the distribution and evolution of the world's biota, plate movement also affects the formation and distribution of some natural resources. Consequently, geologists are using plate tectonic theory in their search for petroleum and mineral deposits and in explaining the occurrence of these natural resources.

Although large concentrations of petroleum occur in many areas of the world, more than 50% of all proven reserves are in the Persian Gulf region. The reason for this is paleogeography and plate movement. Elsewhere in the world, plate tectonics is also responsible for concentrations of petroleum. The formation of the Appalachians, for example, resulted from the compressive forces generated along a convergent plate boundary and provided the structural traps necessary for petroleum to accumulate (see Chapters 10 and 11).

Many metallic mineral deposits such as copper, gold, lead, silver, tin, and zinc are related to igneous and associated hydrothermal (hot water) activity. So, it is not surprising that a close relationship exists between plate boundaries and the occurrence of these valuable deposits.

The magma generated by partial melting of a subducting plate rises toward the surface, and as it cools, it precipitates and concentrates various metallic ores. Many of the world's major metallic ore deposits, such as the porphyry copper deposits of western North and South America (• Figure 3.30a), are associated with convergent plate boundaries. The majority of the copper deposits in the Andes and southwestern United States were formed less than 60 million years ago when oceanic plates were subducted under the North and South American plates. The rising magma and associated hydrothermal fluids carried minute amounts of copper, which were originally widely disseminated but eventually became concentrated in the cracks and fractures of the surrounding andesites. These low-grade copper deposits contain from 0.2 to 2% copper and are extracted from large open-pit mines (Figure 3.30b).

Divergent plate boundaries also yield valuable ore deposits. The island of Cyprus in the Mediterranean is rich in copper and has been supplying all or part of the world's needs for the past 3000 years. The concentration of copper on Cyprus formed as a result of precipitation adjacent to hydrothermal vents along a divergent plate boundary. This deposit was brought to the surface when the copper-rich seafloor collided with the European plate, warping the seafloor and forming Cyprus.

Studies indicate that minerals containing such metals as copper, gold, iron, lead, silver, and zinc are currently forming in the Red Sea. The Red Sea is opening as a result of plate divergence and represents the earliest stage in the growth of an ocean basin (Figures 3.15c and 3.16a).

It is becoming increasingly clear that if we are to keep up with the continuing demands of a global industrialized society, the application of plate tectonic theory to the origin and distribution of mineral resources is essential. We will discuss natural resources and their formation as part of the geologic history of Earth in later chapters in this book.

Figure 3.30 Copper Deposits and Convergent Plate Boundaries

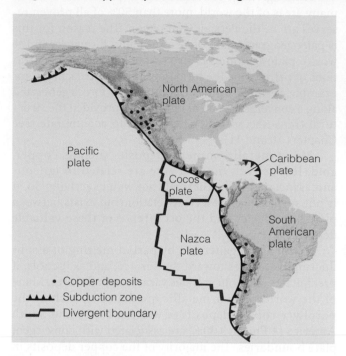

a Valuable copper deposits are located along the west coasts of North and South America in association with convergent plate boundaries. The rising magma and associated hydrothermal activity resulting from subduction carried small amounts of copper, which became trapped and concentrated in the surrounding rocks through time.

Copyright and photograph by Dr. Parvinder S. Sethi

b Bingham Mine, near Salt Lake City, Utah, is a huge open-pit copper mine with reserves estimated at 1.7 billion tons. More than 400,000 tons of rock are removed for processing each day. Note the small specks towards the middle of the photograph that are the 12-foot high dump trucks!

SUMMARY

- The concept of continental movement is not new. The earliest maps showing the similarity between the east coast of South America and the west coast of Africa provided the first evidence that continents may once have been united and subsequently separated from each other.

- Alfred Wegener is generally credited with developing the hypothesis of continental drift. He provided abundant geologic and paleontologic evidence to show that the continents were once united in one supercontinent, which he named Pangaea. Unfortunately, Wegener could not explain how the continents moved, and most geologists ignored his ideas.

- The hypothesis of continental drift was revived during the 1950s when paleomagnetic studies of rocks indicated the presence of multiple magnetic north poles instead of just one as there is today. This paradox was resolved by constructing a map in which the continents could be moved into different positions such that the paleomagnetic data would then be consistent with a single magnetic north pole.

- Seafloor spreading was confirmed by the discovery of magnetic anomalies in the ocean crust that were both parallel to and symmetric around the ocean ridges, indicating that new oceanic crust must have formed as the seafloor was spreading. The pattern of oceanic magnetic anomalies matched the pattern of magnetic reversals already known from

continental lava flows, and showed that Earth's magnetic field has reversed itself numerous times during the past.

- Radiometric dating reveals the oldest oceanic crust is less than 180 million years old, whereas the oldest continental crust is 3.96 billion years old. Clearly, the ocean basins are recent geologic features.

- Plate tectonic theory became widely accepted by the 1970s because the evidence overwhelmingly supports it and because it provides geologists with a powerful theory for explaining such phenomena as volcanism, earthquake activity, mountain building, global climatic changes, the distribution of the world's biota, and the distribution of some mineral resources.

- Geologists recognize three types of plate boundaries: divergent boundaries, where plates move away from each other; convergent boundaries, where two plates collide; and transform boundaries, where two plates slide past each other.

- Ancient plate boundaries can be recognized by their associated rock assemblages and geologic structures. For divergent boundaries, these may include rift valleys with thick sedimentary sequences and numerous dikes and sills. For convergent boundaries, ophiolites and andesitic rocks are two characteristic features. Transform faults generally do not leave any characteristic or diagnostic features in the rock record.

- Although a comprehensive theory of plate movement has yet to be developed, geologists think that some type of convective heat system is the major driving force.
- The average rate of movement and relative motion of plates can be calculated in several ways. The results of these different methods all agree and indicate that the plates move at different average velocities.
- The absolute motion of plates can be determined by the movement of plates over mantle plumes. A mantle plume is an apparently stationary column of magma that rises to the surface where it becomes a hot spot and forms a volcano.

- Geologists now realize that plates can grow when terranes collide with the margins of continents.
- The relationship between plate tectonic processes and the evolution of life is complex. The distribution of plants and animals is not random, but is controlled mostly by climate and geographic barriers, which are influenced, to a great extent, by the movement of plates.
- A close relationship exists between the formation of some mineral deposits and petroleum, and plate boundaries. Furthermore, the formation and distribution of some natural resources are related to plate movements.

IMPORTANT TERMS

continental–continental plate boundary, p. 53
continental drift, p. 39
convergent plate boundary, p. 50
Curie point, p. 42
divergent plate boundary, p. 50
Glossopteris flora, p. 39
Gondwana, p. 39
hot spot, p. 55

Laurasia, p. 42
magnetic anomaly, p. 45
magnetic reversal, p. 44
oceanic–continental plate boundary, p. 52
oceanic–oceanic plate boundary, p. 50
orogeny, p. 59
paleomagnetism, p. 42

Pangaea, p. 39
plate, p. 46
plate tectonic theory, p. 46
seafloor spreading, p. 44
terrane, p. 59
thermal convection cell, p. 44
transform fault, p. 55
transform plate boundary, p. 55

REVIEW QUESTIONS

1. The man credited with developing the continental drift hypothesis is
   a. _____ Wilson; b. _____ Hess; c. _____ Vine; d. _____ Wegener; e. _____ du Toit.

2. The southern part of Pangaea, consisting of South America, Africa, India, Australia, and Antarctica, is called
   a. _____ Gondwana; b. _____ Laurentia; c. _____ Atlantis; d. _____ Laurasia; e. _____ Pacifica.

3. The Hawaiian Island chain and Yellowstone National Park, Wyoming, are examples of
   a. _____ oceanic–oceanic plate boundaries; b. _____ hot spots; c. _____ divergent plate boundaries; d. _____ transform plate boundaries; e. _____ oceanic–continental plate boundaries.

4. Hot spots and aseismic ridges can be used to determine the
   a. _____ location of divergent plate boundaries; b. _____ absolute motion of plates; c. _____ location of magnetic anomalies in the oceanic crust; d. _____ relative motion of plates; e. _____ location of convergent plate boundaries.

5. Magnetic surveys of the ocean basins indicate that
   a. _____ the oceanic crust is youngest adjacent to mid-oceanic ridges; b. _____ the oceanic crust is oldest adjacent to mid-oceanic ridges; c. _____ the oceanic crust is youngest adjacent to the continents; d. _____ the oceanic crust is the same age everywhere; e. _____ answers b and c.

6. The driving mechanism of plate movement is thought to be
   a. _____ isostasy; b. _____ Earth's rotation; c. _____ thermal convection cells; d. _____ magnetism; e. _____ polar wandering.

7. Convergent plate boundaries are areas where
   a. _____ new continental lithosphere is forming; b. _____ new oceanic lithosphere is forming; c. _____ two plates come together; d. _____ two plates slide past each other; e. _____ two plates move away from each other.

8. The most common biotic province boundaries are
   a. _____ geographic barriers; b. _____ biologic barriers; c. _____ climatic barriers; d. _____ answers a and b; e. _____ answers a and c.

9. The Andes Mountains are a good example of what type of plate boundary?
   a. _____ continental–continental; b. _____ oceanic–oceanic; c. _____ oceanic–continental; d. _____ divergent; e. _____ transform.

10. Iron-bearing minerals in magma gain their magnetism and align themselves with the magnetic field when they cool through the
    a. _____ Curie point; b. _____ magnetic anomaly point; c. _____ thermal convection point; d. _____ hot spot point; e. _____ isostatic point.

11. Explain why global diversity increases with an increase in the number of biotic provinces. How does plate movement affect the number of biotic provinces?

12. What evidence convinced Wegener and others that continents must have moved in the past and at one time formed a supercontinent?

13. Why was the continental drift hypothesis proposed by Wegener, rejected by so many geologists for so long?

14. How did the theory of seafloor spreading, proposed by Harry Hess in 1962, overcome the objections of those opposed to continental drift?

15. Why is some type of thermal convection system thought to be the major force driving plate movement? How have "slab-pull" and "ridge-push," both mainly gravity driven, modified a purely thermal convection model for plate movement?

16. What can hot spots tell us about the absolute direction of plate movement?

17. Explain why such natural disasters as volcanic eruptions and earthquakes are associated with divergent and convergent plate boundaries.

18. In addition to the volcanic eruptions and earthquakes associated with convergent and divergent plate boundaries, why are these boundaries also important to the formation and accumulation of various metallic ore deposits?

19. Based on your knowledge of biology and the distribution of organisms throughout the world, how do you think plate tectonics has affected this distribution both on land and in the oceans?

20. Plate tectonic theory builds on the continental drift hypothesis and the theory of seafloor spreading. As such, it is a unifying theory of geology. Explain why it is a unifying theory.

APPLY YOUR KNOWLEDGE

1. Using the age for each of the Hawaiian Islands in Figure 3.24 and an atlas in which you can measure the distance between islands, calculate the average rate of movement per year for the Pacific plate since each island formed. Is the average rate of movement the same for each island? Would you expect it to be? Explain why it may not be.

2. If the movement along the San Andreas Fault, which separates the Pacific plate from the North American plate, averages 5.5 cm per year, how long will it take before Los Angeles is opposite San Francisco?

3. You've been selected to be part of the first astronaut team to go to Mars. While your two fellow crew members descend to the Martian surface, you'll be staying in the command module and circling the Red Planet. As part of the geologic investigation of Mars, one of the crew members will be mapping the geology around the landing site and deciphering the geologic history of the area. Your job will be to observe and photograph the planet's surface and try to determine whether Mars had an active plate tectonic regime in the past and whether there is current plate movement. What features would you look for, and what evidence might reveal current or previous plate activity?

Copyright and photograph by Dr. Parvinder S. Sethi

CHAPTER 4

GEOLOGIC TIME: CONCEPTS AND PRINCIPLES

The Grand Canyon, Arizona. Major John Wesley Powell led two expeditions down the Colorado River and through the canyon in 1869 and 1871. He was struck by the seemingly limitless time represented by the rocks exposed in the canyon walls and by the recognition that these rock layers, like the pages in a book, contain the geologic history of this region.

[OUTLINE]

Introduction

How is Geologic Time Measured?

Early Concepts of Geologic Time and Earth's Age

Relative Dating Methods

Establishment of Geology as a Science— The Triumph of Uniformitarianism over Neptunism and Catastrophism

   Neptunism and Catastrophism

   Uniformitarianism

   Modern View of Uniformitarianism

Lord Kelvin and a Crisis in Geology

Absolute Dating Methods

   Atoms, Elements, and Isotopes

   Radioactive Decay and Half-Lives

   Long-Lived Radioactive Isotope Pairs

   Fission-Track Dating

   Radiocarbon and Tree-Ring Dating Methods

Geologic Time and Climate Change

   Perspective *Denver's Weather— 280 Million Years Ago!*

Summary

- The concept of geologic time and its measurement have changed through human history.

- The fundamental principles of relative dating provide a means to interpret geologic history.

- The principle of uniformitarianism is fundamental to geology and prevailed over the concepts of neptunism and catastrophism because it provides a better explanation for observed geologic phenomena.

- The discovery of radioactivity provided geologists with a clock that could determine the ages of ancient rocks and validate that Earth was very old.

- Absolute dating methods are used to date geologic events in terms of years before present.

- The most accurate radiometric dates are obtained from igneous rocks.

- Geologic time is an important element in the study of climate change.

Introduction

In 1869, Major John Wesley Powell, a Civil War veteran who had lost his right arm in the battle of Shiloh, led a group of hardy explorers down the uncharted Colorado River through the Grand Canyon. With no maps or other information, Powell and his group ran the many rapids of the Colorado River in fragile wooden boats, hastily recording what they saw. Powell wrote in his diary that "all about me are interesting geologic records. The book is open and I read as I run."

From this initial reconnaissance, Powell led a second expedition down the Colorado River in 1871. This second trip included a photographer, a surveyor, and three topographers. Members of the expedition made detailed topographic and geologic maps of the Grand Canyon area as well as the first photographic record of the region.

Probably no one has contributed as much to the understanding of the Grand Canyon as Major Powell. In recognition of his contributions, the Powell Memorial was erected on the South Rim of the Grand Canyon in 1969 to commemorate the 100th anniversary of this history-making first expedition.

Most tourists today, like Powell and his fellow explorers in 1869, are astonished by the seemingly limitless time represented by the rocks exposed in the walls of the Grand Canyon. For most visitors, viewing a 1.5-kilometer-deep cut into Earth's crust is the only encounter they'll ever have with the magnitude of geologic time. When standing on the rim and looking down into the Grand Canyon, we are really looking far back in time, all the way back to the early history of our planet. In fact, more than 1 billion years of history are preserved in the rocks of the Grand Canyon, and reading what is preserved in those rocks is, just as Powell noted in his diary more than 100 years ago, like reading the pages in a history book.

In reading this "book"—the rock layers of the Grand Canyon—we learn that this area underwent episodes of mountain building as well as periods of advancing and retreating shallow seas. How do we know this? The answer lies in applying the principles of relative dating to the rocks we see exposed in the canyon walls, and also in recognizing that present-day processes have operated throughout Earth history. In this chapter, you will learn what those principles are and how they can be used to determine geologic history.

We begin this chapter by asking the question "What is time?" We seem obsessed with time, and we organize our lives around it. Yet most of us feel we don't have enough of it—we are always running "behind" or "out of time." Whereas physicists deal with extremely short intervals of time, and geologists deal with incredibly long periods of time, most of us tend to view time from the perspective of our own existence; that is, we partition our lives into seconds, hours, days, weeks, months, and years. Ancient history is what occurred hundreds or even thousands of years ago. Yet when geologists talk of ancient geologic history, they are referring to events that happened millions or even billions of years ago!

Vast periods of time set geology apart from most of the other sciences, and an appreciation of the immensity of geologic time is fundamental to understanding the physical and biological history of our planet. In fact, understanding and accepting the magnitude of geologic time are major contributions geology has made to the sciences.

How Is Geologic Time Measured?

In some respects, time is defined by the methods used to measure it. Geologists use two different frames of reference when discussing geologic time. **Relative dating** is placing geologic events in a sequential order as determined from their positions in the geologic record. Relative dating will not tell us how long ago a particular event took place; only that one event preceded another. A useful analogy for relative dating is a television guide that does not list the times programs are shown. You cannot tell what time a particular program will be shown, but by watching a few shows and checking the guide, you can determine whether you have missed the show or how many shows are scheduled before the one you want to see.

The various principles used to determine relative dating were discovered hundreds of years ago, and since then they have been used to construct the *relative geologic time scale* (• Figure 4.1). Furthermore, these principles are still widely used by geologists today, especially in reconstructing the geologic history of the terrestrial planets and their moons.

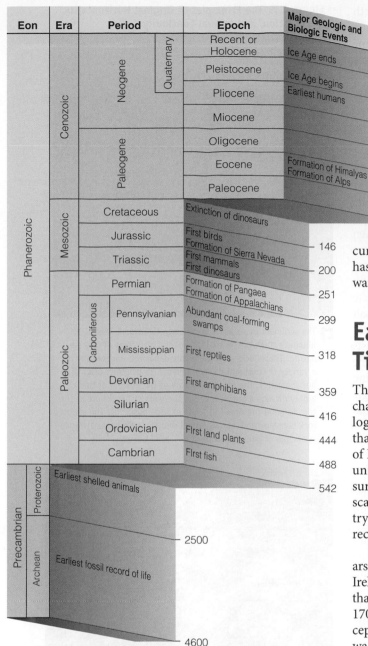

Eon	Era	Period		Epoch	Major Geologic and Biologic Events	Millions of Years Ago
Phanerozoic	Cenozoic	Neogene	Quaternary	Recent or Holocene	Ice Age ends	0
				Pleistocene	Ice Age begins	0.01
				Pliocene	Earliest humans	1.8
				Miocene		5
		Paleogene		Oligocene		23
				Eocene	Formation of Himalyas Formation of Alps	34
				Paleocene		56
	Mesozoic	Cretaceous			Extinction of dinosaurs	66
		Jurassic			First birds Formation of Sierra Nevada	146
		Triassic			First mammals First dinosaurs	200
	Paleozoic	Permian			Formation of Pangaea Formation of Appalachians	251
		Carboniferous	Pennsylvanian		Abundant coal-forming swamps	299
			Mississippian		First reptiles	318
		Devonian			First amphibians	359
		Silurian				416
		Ordovician			First land plants	444
		Cambrian			First fish	488
						542
Precambrian	Proterozoic				Earliest shelled animals	
						2500
	Archean				Earliest fossil record of life	
						4600

• **Figure 4.1 The Geologic Time Scale** Some of the major geologic and biologic events are indicated for the various eras, periods, and epochs. Dates are from Gradstein, F., Ogg, J. and Smith, A., *A Geologic Timescale 2004* (Cambridge, UK: Cambridge University Press, 2005), Figure 1.2.

Advances and refinements in absolute dating techniques during the 20th century have changed the way we view Earth in terms of when events occurred in the past and the rates of geologic change through time. The ability to accurately determine past climatic changes and their causes has important implications for the current debate on global warming and its effects on humans (see the Epilogue).

Early Concepts of Geologic Time and Earth's Age

The concept of geologic time and its measurement have changed throughout human history. Early Christian theologians were largely responsible for formulating the idea that time is linear rather than circular. When St. Augustine of Hippo (A.D. 354–430) stated that the Crucifixion was a unique event from which all other events could be measured, he helped establish the idea of the B.C. and A.D. time scale. This prompted many religious scholars and clerics to try to establish the date of creation by analyzing historical records and the genealogies found in Scripture.

One of the most influential and famous Christian scholars was James Ussher (1581–1665), Archbishop of Armagh, Ireland, who, based upon Old Testament genealogy, asserted that God created Earth on Sunday, October 23, 4004 B.C. In 1701, an authorized version of the Bible made this date accepted Church doctrine. For nearly a century thereafter, it was considered heresy to assume that Earth and all its features were more than about 6000 years old. Thus, the idea of a very young Earth provided the basis for most Western chronologies of Earth history prior to the 18th century.

During the 18th and 19th centuries, several attempts were made to determine Earth's age on the basis of scientific evidence rather than revelation. The French zoologist Georges Louis de Buffon (1707–1788) assumed Earth gradually cooled to its present condition from a molten beginning. To simulate this history, he melted iron balls of various diameters and allowed them to cool to the surrounding temperature. By extrapolating their cooling rate to a ball the size of Earth, he determined that Earth was at least 75,000 years old. Although this age was much older than that derived from Scripture, it was still vastly younger than we now know our planet to be.

Other scholars were equally ingenious in attempting to calculate Earth's age. For example, if deposition rates could be determined for various sediments, geologists reasoned that they could calculate how long it would take to deposit

Absolute dating provides specific dates for rock units or events expressed in years before the present. In our analogy of the television guide, the time when the programs are actually shown would be the absolute dates. In this way, you not only can determine whether you have missed a show (relative dating), but also know how long it will be until a show you want to see will be shown (absolute dating).

Radiometric dating is the most common method of obtaining absolute ages. Dates are calculated from the natural rates of decay of various radioactive elements present in trace amounts in some rocks. It was not until the discovery of radioactivity near the end of the 19th century that absolute ages could be accurately applied to the relative geologic time scale. Today, the geologic time scale is really a dual scale—a relative scale based on rock sequences with radiometric dates expressed as years before the present (Figure 4.1).

any rock layer. They could then extrapolate how old Earth was from the total thickness of sedimentary rock in its crust. Rates of deposition vary, however, even for the same type of rock. Furthermore, it is impossible to estimate how much rock has been removed by erosion, or how much a rock sequence has been reduced by compaction. As a result of these variables, estimates of Earth's age ranged from younger than 1 million years to older than 2 billion years.

Another attempt to determine Earth's age involved ocean salinity. Scholars assumed that Earth's ocean waters were originally fresh and that their present salinity was the result of dissolved salt being carried into the ocean basins by streams. Knowing the volume of ocean water and its salinity, John Joly, a 19th-century Irish geologist, measured the amount of salt currently in the world's streams. He then calculated that it would have taken at least 90 million years for the oceans to reach their present salinity level. This was still much younger than the now accepted age of 4.6 billion years for Earth, mainly because Joly had no way to calculate either how much salt had been recycled or the amount of salt stored in continental salt deposits and seafloor clay deposits.

Besides trying to determine Earth's age, the naturalists of the 18th and 19th centuries were formulating some of the fundamental geologic principles that are used in deciphering Earth history. From the evidence preserved in the geologic record, it was clear to them that Earth is very old and that geologic processes have operated over long periods of time.

Relative Dating Methods

Before the development of radiometric dating techniques, geologists had no reliable means of absolute dating and therefore depended solely on relative dating methods. Relative dating places events in sequential order but does not tell us how long ago an event took place. Although the principles of relative dating may now seem self-evident, their discovery was an important scientific achievement because they provided geologists with a means to interpret geologic history and develop a relative geologic time scale.

Six fundamental geologic principles are used in relative dating: superposition, original horizontality, lateral continuity, cross-cutting relationships (all discussed in this chapter), and inclusions and fossil succession (discussed in Chapter 5).

The 17th century was an important time in the development of geology as a science because of the widely circulated writings of the Danish anatomist Nicolas Steno (1638–1686). Steno observed that when streams flood, they spread out across their floodplains and deposit layers of sediment that bury organisms dwelling on the floodplain. Subsequent floods produce new layers of sediments that are deposited or superposed over previous deposits. When lithified, these layers of sediment become sedimentary rock.

Thus, in an undisturbed succession of sedimentary rock layers, the oldest layer is at the bottom and the youngest layer is at the top. This **principle of superposition** is the basis for relative-age determinations of strata and their contained fossils (• Figure 4.2a and chapter opening photograph).

• Figure 4.2 **The Principles of Original Horizontality, Superposition, and Lateral Continuity**

Copyright and photograph by Dr. Parvinder S. Sethi

a The sedimentary rocks of Bryce Canyon National Park, Utah, illustrate three of the six fundamental principles of relative dating. These rocks were originally deposited horizontally in a variety of continental environments (principle of original horizontality). The oldest rocks are at the bottom of this highly dissected landscape, and the youngest rocks are at the top, forming the rims (principle of superposition). The exposed rock layers extend laterally in all directions for some distance (principle of lateral continuity).

Reed Wicander

b These shales and limestones of the Postolonnec Formation, at Postolonnec Beach, Crozon Peninsula, France, were originally deposited horizontally but have been significantly tilted since their formation.

• Figure 4.3 **The Principle of Cross-Cutting Relationships**

Sheldon Judson

a A dark gabbro dike cuts across granite in Acadia National Park, Maine. The dike is younger than the granite it intrudes.

Reed Wicander

b A small fault (arrows show direction of movement) cuts across, and thus displaces, tilted sedimentary beds along Templin Highway in Castaic, California. The fault is therefore younger than the youngest beds that are displaced.

Establishment of Geology as a Science—The Triumph of Uniformitarianism over Neptunism and Catastrophism

Steno's principles were significant contributions to early geologic thought, but the prevailing concepts of Earth history continued to be those that could be easily reconciled with a literal interpretation of Scripture. Two of these ideas, *neptunism* and *catastrophism,* were particularly appealing and accepted by many naturalists. In the final analysis, however, another concept, uniformitarianism, became the underlying philosophy of geology because it provided a better explanation for observed geologic phenomena than either neptunism or catastrophism.

Neptunism and Catastrophism The concept of **neptunism** was proposed in 1787 by a German professor of mineralogy, Abraham Gottlob Werner (1749–1817). Although Werner was an excellent mineralogist, he is best remembered for his incorrect interpretation of Earth history. He thought that all rocks, including granite and basalt, were precipitated in an orderly sequence from a primeval, worldwide ocean (Table 4.1).

Werner's subdivision of Earth's crust by supposed relative age attracted a large following in the late 1700s and became almost universally accepted as the standard geologic column. Two factors account for this. First, Werner's charismatic personality, enthusiasm for geology, and captivating lectures popularized the concept. Second, neptunism included a worldwide ocean that easily conformed to the biblical deluge.

Steno also observed that, because sedimentary particles settle from water under the influence of gravity, sediment is deposited in essentially horizontal layers, thus illustrating the **principle of original horizontality** (Figure 4.2a and chapter opening photograph). Therefore, a sequence of sedimentary rock layers that is steeply inclined from the horizontal must have been tilted after deposition and lithification (Figure 4.2b).

Steno's third principle, the **principle of lateral continuity,** states that sediment extends laterally in all directions until it thins and pinches out or terminates against the edge of the depositional basin (Figure 4.2a).

The Scottish geologist James Hutton (1726–1797) is considered by many to be the founder of modern geology. His detailed studies and observation of rock exposures and present-day geologic processes were instrumental in establishing the *principle of uniformitarianism* (see Chapter 1). Furthermore, Hutton is also credited with discovering the **principle of cross-cutting relationships**, whereby he recognized that an igneous intrusion or a fault must be younger than the rocks it intrudes or displaces (• Figure 4.3).

TABLE **4.1**	Werner's Subdivision of the Rocks of Earth's Crust
Alluvial rocks	Unconsolidated sediments
Secondary rocks	Various fossiliferous, detrital, and chemical rocks such as sandstones, limestones, and coal, as well as basalts
Transitional rocks	Chemical and detrital rocks including fossiliferous rocks
Primative rocks	Oldest rocks of Earth (igneous and metamorphic

Despite Werner's personality and arguments, his neptunian theory failed to explain what happened to the tremendous amount of water that once covered Earth. An even greater problem was Werner's insistence that all igneous rocks were precipitated from seawater. It was Werner's failure to recognize the igneous origin of basalt that finally led to the downfall of neptunism.

From the late 18th century to the mid-19th century, the concept of **catastrophism,** proposed by the French zoologist Baron Georges Cuvier (1769–1832), dominated European geologic thinking. Cuvier explained the physical and biologic history of Earth as resulting from a series of sudden widespread catastrophes. Each catastrophe accounted for significant and rapid changes in Earth, including exterminating existing life in the affected area. Following a catastrophe, new organisms were either created or migrated in from elsewhere.

According to Cuvier, six major catastrophes had occurred in the past. These conveniently corresponded to the six days of biblical creation. Furthermore, the last catastrophe was taken to be the biblical deluge, so catastrophism had wide appeal, especially among theologians.

Eventually, both neptunism and catastrophism were abandoned as untenable hypotheses because their basic assumptions could not be supported by field evidence. The simplistic sequence of rocks predicted by neptunism (Table 4.1) was contradicted by field observations from many different areas of the world. Moreover, basalt was shown to be of igneous origin, and subsequent discoveries of volcanic rocks interbedded with secondary and primitive deposits proved that volcanic activity had occurred throughout Earth history. As more field observations from widely separated areas were made, naturalists realized that far more than six catastrophes were needed to account for Earth history. With the demise of neptunism and catastrophism, the principle of uniformitarianism, advocated by James Hutton and Charles Lyell, became the guiding philosophy of geology.

Uniformitarianism
Hutton observed the processes of wave action, erosion by running water, and sediment transport, and concluded that, given enough time, these processes could account for the geologic features in his native Scotland. He reasoned that "the past history of our globe must be explained by what can be seen to be happening now." This assumption that present-day processes have operated throughout geologic time was the basis for the **principle of uniformitarianism.**

Although Hutton developed a comprehensive theory of uniformitarian geology, it was Charles Lyell (1797–1875) who became the principal advocate and interpreter of uniformitarianism. William Whewell, however, coined the term itself, in 1832.

Hutton viewed Earth history as cyclical—that is, continents are worn down by erosion, the eroded sediment is deposited in the sea, and uplift of the seafloor creates new continents, thus completing a cycle. He thought the mechanism for uplift was thermal expansion from Earth's hot interior. Hutton's field observations, and experiments performed by his contemporaries involving the melting of basalt samples, convinced him that igneous rocks were the result of cooling magma. This interpretation of the origin of igneous rocks, called *plutonism,* eventually displaced the neptunian view that igneous rocks precipitated from seawater.

Hutton also recognized the importance of unconformities in his cyclical view of Earth history. At Siccar Point, Scotland, he observed steeply inclined metamorphic rocks that had been eroded and covered by flat-lying younger rocks (• Figure 4.4). It was clear to him that severe upheavals had tilted the lower rocks and formed mountains. These were then worn away and covered by younger, flat-lying rocks. The erosion surface meant there was a gap in the geologic record, and the rocks above and below this surface provided evidence that both mountain building and erosion had occurred. Although Hutton did not use the word *unconformity,* he was the first to understand and explain the significance of such gaps in the geologic record.

Hutton was also instrumental in establishing the concept that geologic processes had vast amounts of time in which to operate. Because Hutton relied on known processes to account for Earth history, he concluded that Earth must be very old. However, he estimated neither how old Earth

• *Active* **Figure 4.4 Angular Unconformity at Siccar Point, Scotland** It was at this location in 1788 that James Hutton first realized the significance of unconformities in interpreting Earth history. Visit the Geology Resource Center to view this and other active figures at www.cengage.com/sso.

was, nor how long it took to complete a cycle of erosion, deposition, and uplift. He merely allowed that "we find no vestige of a beginning, and no prospect of an end," which was in keeping with a cyclical view of Earth history.

Unfortunately, Hutton was not a particularly good writer, and so his ideas were not widely disseminated or accepted. In fact, neptunism and catastrophism continued to be the dominant geologic concepts well into the 1800s. In 1830, however, Charles Lyell published a landmark book, *Principles of Geology,* in which he championed Hutton's concept of uniformitarianism.

Instead of relying on catastrophic events to explain various Earth features, Lyell recognized that imperceptible changes brought about by present-day processes could, over long periods of time, have tremendous cumulative effects. Not only did Lyell effectively reintroduce and establish the concept of unlimited geologic time, but he also discredited catastrophism as a viable explanation of geologic phenomena. Through his writings, Lyell firmly established uniformitarianism as the guiding principle of geology. Furthermore, the recognition of virtually limitless amounts of time was also necessary for, and instrumental in, the acceptance of Darwin's 1859 theory of evolution (see Chapter 7).

Perhaps because uniformitarianism is such a general concept, scientists have interpreted it in different ways. Lyell's concept of uniformitarianism embodied the idea of a steady-state Earth in which present-day processes have operated at the same rate in the past as they do today. For example, the frequency of earthquakes and volcanic eruptions for any given period of time in the past must be the same as it is today. Or, if the climate in one part of the world became warmer, another area would have to become cooler so that overall the climate remains the same. By such reasoning, Lyell claimed that conditions for Earth as a whole, had been constant and unchanging through time.

Modern View of Uniformitarianism
Geologists today assume that the principles, or laws, of nature are constant but that rates and intensities of change have varied through time. For example, volcanic activity was more intense in North America during the Miocene Epoch than today, whereas glaciation has been more prevalent during the last 1.8 million years than in the previous 300 million years. Because rates and intensities of geologic processes have varied through time, some geologists prefer to use the term *actualism* rather than *uniformitarianism* to remove the idea of "uniformity" from the concept. Most geologists, though, still use the term *uniformitarianism* because it indicates that, even though rates and intensities of change have varied in the past, laws of nature have remained the same.

Uniformitarianism is a powerful concept that allows us, through analogy and inductive reasoning, to use present-day processes as the basis for interpreting the past and for predicting potential future events. It does not eliminate occasional, sudden, short-term events such as volcanic eruptions, earthquakes, floods, or even meteorite impacts as forces that shape our modern world. In fact, some geologists view Earth history as a series of such short-term, or punctuated, events, and this view is certainly in keeping with the modern principle of uniformitarianism.

Earth is in a state of dynamic change and has been since it formed. Although rates of change may have varied in the past, natural laws governing the processes have not.

Lord Kelvin and a Crisis in Geology

Lord Kelvin (1824–1907), a highly respected English physicist, claimed in a paper written in 1866 to have destroyed the uniformitarian foundation on which Huttonian–Lyellian geology was based. Kelvin did not accept Lyell's strict uniformitarianism, in which chemical reactions in Earth's interior were supposed to continually produce heat, allowing for a steady-state Earth. Kelvin rejected this idea as perpetual motion—impossible according to the known laws of physics—and accepted instead the assumption that Earth was originally molten.

Kelvin knew from deep mines in Europe that Earth's temperature increases with depth, and he reasoned that Earth is losing heat from its interior. By knowing the size of Earth, the melting temperature of rocks, and the rate of heat loss, Kelvin calculated the age at which Earth was entirely molten. From these calculations, he concluded that Earth could not be older than 400 million years or younger than 20 million years. This wide discrepancy in age reflected uncertainties in average temperature increases with depth and the various melting points of Earth's constituent materials.

After establishing that Earth was very old and that present-day processes operating over long periods of time account for geologic features, geologists were in a quandary. If they accepted Kelvin's dates, they would have to abandon the concept of seemingly limitless time that was the underpinning of uniformitarian geology and one of the foundations of Darwinian evolution, and squeeze events into a shorter time frame. Some geologists objected to such a young age for Earth, but their objections were seemingly based more on faith than on hard facts. Kelvin's quantitative measurements and arguments seemed flawless and unassailable to scientists at the time.

Kelvin's reasoning and calculations were sound, but his basic premises were false, thereby invalidating his conclusions. Kelvin was unaware that Earth has an internal heat source—radioactivity—that has allowed it to maintain a fairly constant temperature through time.* His 40-year

---

*Actually, Earth's temperature has decreased through time because the original amount of radioactive materials has been decreasing and therefore is not supplying as much heat. However, the temperature is decreasing at a rate considerably slower than would be required to lend any credence to Kelvin's calculations (see p. 165, Chapter 8).

campaign for a young Earth ended with the discovery of radioactivity near the end of the 19th century and the insight in 1905 that natural radioactive decay can be used in many cases to date the age of formation of rocks. His "unassailable calculations" were no longer valid, and his proof for a geologically young Earth collapsed. Kelvin's theory, like neptunism, catastrophism, and a worldwide flood, thus became another interesting footnote in the history of geology.

Although the discovery of radioactivity destroyed Kelvin's arguments, it provided geologists with a clock that could determine the ages of ancient rocks and validate what geologists had long thought—namely, that Earth was indeed very old!

Absolute Dating Methods

Although most of the isotopes of the 92 naturally occurring elements are stable, some are radioactive and spontaneously decay to other, more stable isotopes of elements, releasing energy in the process. The discovery in 1903 by Pierre and Marie Curie that radioactive decay produces heat meant that geologists finally had a mechanism for explaining Earth's internal heat that did not rely on residual cooling from a molten origin. Furthermore, geologists now had a powerful tool to date geologic events accurately and to verify the long time periods postulated by Hutton and Lyell.

Atoms, Elements, and Isotopes
As we discussed in Chapter 2, all matter is made up of chemical *elements*, each composed of extremely small particles called *atoms*. The *nucleus* of an atom is composed of *protons* (positively charged particles) and *neutrons* (neutral particles) with *electrons* (negatively charged particles) encircling it (see Figure 2.2). The number of protons defines an element's *atomic number* and helps determine its properties and characteristics.

The combined number of protons and neutrons in an atom is its *atomic mass number*. However, not all atoms of the same element have the same number of neutrons in their nuclei. These variable forms of the same element are called *isotopes* (see Figure 2.3). Most isotopes are stable, but some are unstable and spontaneously decay to a more stable form. It is the decay rate of unstable isotopes that geologists measure to determine the absolute age of rocks.

Radioactive Decay and Half-Lives
Radioactive decay is the process whereby an unstable atomic nucleus is spontaneously transformed into an atomic nucleus of a different element. Scientists recognize three types of radioactive decay, all of which result in a change of

atomic structure (• Figure 4.5). In *alpha decay*, two protons and two neutrons are emitted from the nucleus, resulting in the loss of two atomic numbers and four atomic mass numbers. In *beta decay*, a fast-moving electron is emitted from a neutron in the nucleus, changing that neutron to a proton and consequently increasing the atomic number by one, with no resultant atomic mass number change. *Electron capture decay* is when a proton captures an electron from an electron shell and thereby converts to a neutron, resulting in a loss of one atomic number but not changing the atomic mass number.

Some elements undergo only one decay step in the conversion from an unstable form to a stable form. For example, rubidium 87 decays to strontium 87 by a single beta emission, and potassium 40 decays to argon 40 by a single electron capture. Other radioactive elements

• Figure 4.5 **Three Types of Radioactive Decay**

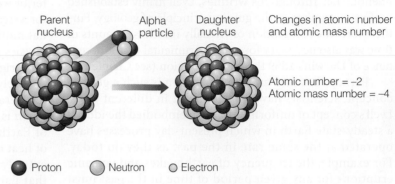

Changes in atomic number and atomic mass number

Atomic number = –2
Atomic mass number = –4

● Proton   ◯ Neutron   ◌ Electron

a Alpha decay, in which an unstable parent nucleus emits 2 protons and 2 neutrons.

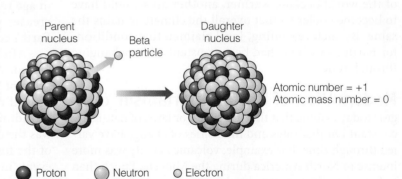

Atomic number = +1
Atomic mass number = 0

● Proton   ◯ Neutron   ◌ Electron

b Beta decay, in which an electron is emitted from a neutron in the nucleus.

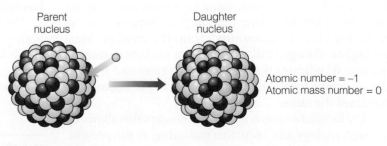

Atomic number = –1
Atomic mass number = 0

● Proton   ◯ Neutron   ◌ Electron

c Electron capture, in which a proton captures an electron and is thereby converted to a neutron.

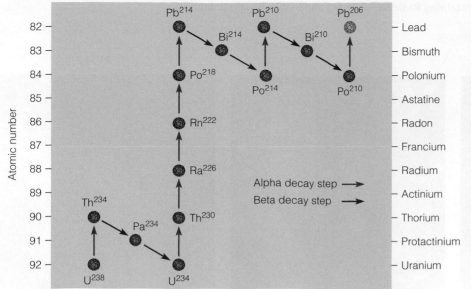

For example, an element with *1,000,000* parent atoms will have *500,000* parent atoms and *500,000* daughter atoms after one half-life. After two half-lives, it will have *250,000* parent atoms (one-half of the previous parent atoms, which is equivalent to one-fourth of the original parent atoms) and *750,000* daughter atoms. After three half-lives, it will have *125,000* parent atoms (one-half of the previous parent atoms, or one-eighth of the original parent atoms) and *875,000* daughter atoms, and so on, until the number of parent atoms remaining is so few that they cannot be accurately measured by present-day instruments.

undergo several decay steps. Uranium 235 decays to lead 207 by seven alpha steps and six beta steps, whereas uranium 238 decays to lead 206 by eight alpha and six beta steps (• Figure 4.6).

When we discuss decay rates, it is convenient to refer to them in terms of half-lives. The **half-life** of a radioactive element is the time it takes for one-half of the atoms of the original unstable *parent element* to decay to atoms of a new, stable *daughter element*. The half-life of a given radioactive element is constant and can be precisely measured. Half-lives of various radioactive elements range from less than a billionth of a second to 106 billion years.

Radioactive decay occurs at a geometric rate rather than a linear rate. Therefore, a graph of the decay rate produces a curve rather than a straight line (• Figure 4.7).

By measuring the parent–daughter ratio and knowing the half-life of the parent (which has been determined in the laboratory), geologists can calculate the age of a sample that contains the radioactive element. The parent–daughter ratio is usually determined by a *mass spectrometer,* an instrument that measures the proportions of elements of different masses.

Sources of Uncertainty The most accurate radiometric dates are obtained from igneous rocks. As magma cools and begins to crystallize, radioactive parent atoms are

• *Active* Figure 4.7 **Uniform, Linear Change Compared to Geometric Radioactive Decay** Visit the Geology Resource Center to view this and other active figures at www.cengage.com/sso.

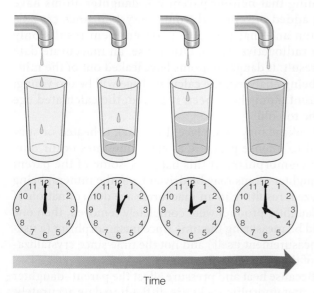

a Uniform, linear change is characteristic of many familiar processes. In this example, water is being added to a glass at a constant rate.

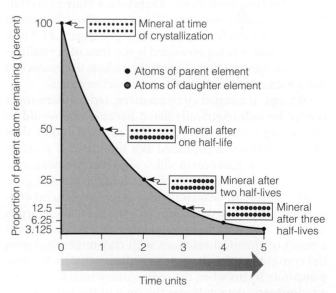

b A geometric radioactive decay curve, in which each time unit represents one half-life, and each half-life is the time it takes for half of the parent element to decay to the daughter element.

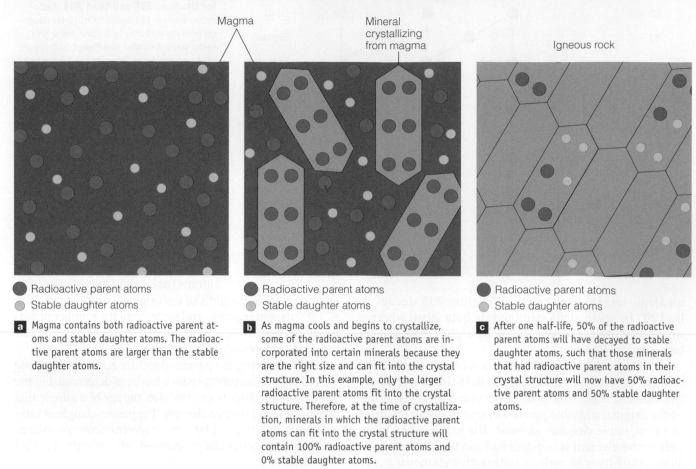

Magma   Mineral crystallizing from magma   Igneous rock

● Radioactive parent atoms
○ Stable daughter atoms

a Magma contains both radioactive parent atoms and stable daughter atoms. The radioactive parent atoms are larger than the stable daughter atoms.

● Radioactive parent atoms
○ Stable daughter atoms

b As magma cools and begins to crystallize, some of the radioactive parent atoms are incorporated into certain minerals because they are the right size and can fit into the crystal structure. In this example, only the larger radioactive parent atoms fit into the crystal structure. Therefore, at the time of crystallization, minerals in which the radioactive parent atoms can fit into the crystal structure will contain 100% radioactive parent atoms and 0% stable daughter atoms.

● Radioactive parent atoms
○ Stable daughter atoms

c After one half-life, 50% of the radioactive parent atoms will have decayed to stable daughter atoms, such that those minerals that had radioactive parent atoms in their crystal structure will now have 50% radioactive parent atoms and 50% stable daughter atoms.

separated from previously formed daughter atoms. Because they are the right size, some radioactive parent atoms are incorporated into the crystal structure of certain minerals. The stable daughter atoms, though, are a different size from the radioactive parent atoms and consequently cannot fit into the crystal structure of the same mineral as the parent atoms. Therefore, a mineral crystallizing in a cooling magma will contain radioactive parent atoms but no stable daughter atoms (• Figure 4.8). Thus, the time that is being measured is the time of crystallization of the mineral that contains the radioactive atoms and not the time of formation of the radioactive atoms.

Except in unusual circumstances, sedimentary rocks cannot be radiometrically dated because one would be measuring the age of a particular mineral rather than the time that it was deposited as a sedimentary particle. One of the few instances in which radiometric dates can be obtained on sedimentary rocks is when the mineral glauconite is present. Glauconite is a greenish mineral containing radioactive potassium 40, which decays to argon 40 (Table 4.2). It forms in certain marine environments as a result of chemical reactions with clay minerals during the conversion of sediments to sedimentary rock. Thus, glauconite forms when the sedimentary rock forms, and a radiometric date indicates the time of the sedimentary rock's origin. Being a gas, however, the daughter product,

argon, can easily escape from a mineral. Therefore, any date obtained from glauconite, or any other mineral containing the potassium 40 and argon 40 pair, must be considered a minimum age.

To obtain accurate radiometric dates, geologists must be sure that they are dealing with a *closed system*, meaning that neither parent nor daughter atoms have been added or removed from the system since crystallization and that the ratio between them results only from radioactive decay. Otherwise, an inaccurate date will result. If daughter atoms have leaked out of the mineral being analyzed, the calculated age will be too young; if parent atoms have been removed, the calculated age will be too old.

Leakage may take place if the rock is heated or subjected to intense pressure as can sometimes occur during metamorphism. If this happens, some of the parent or daughter atoms may be driven from the mineral being analyzed, resulting in an inaccurate age determination. If the daughter product were completely removed, then one would be measuring the time since metamorphism (a useful measurement itself) and not the time since crystallization of the mineral (• Figure 4.9).

Because heat and pressure affect the parent–daughter ratio, metamorphic rocks are difficult to date accurately. Remember that although the resulting parent–daughter

Isotopes		Half-Life of Parent (Years)	Effective Dating Range (Years)	Minerals and Rocks That Can Be Dated	
Parent	Daughter				
Uranium 238	Lead 206	4.5 billion	10 million to 4.6 billion	Zircon Uraninite	
Uranium 235	Lead 207	704 million			
Thorium 232	Lead 208	14 billion			
Rubidium 87	Strontium 87	48.8 billion	10 million to 4.6 billion	Muscovite Biotite Potassium feldspar Whole metamorphic or igneous rock	
Potassium 40	Argon 40	1.3 billion	100,000 to 4.6 billion	Glauconite Muscovite Biotite	Hornblende Whole volcanic rock

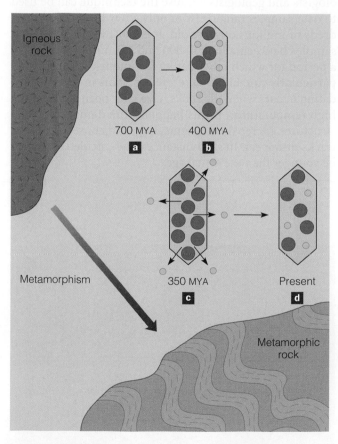

• Figure 4.9 **Effects of Metamorphism on Radiometric Dating** The effect of metamorphism in driving out daughter atoms from a mineral that crystallized 700 million years ago (MYA). The mineral is shown immediately after crystallization **a**, then at 400 MYA **b**, when some of the parent atoms had decayed to daughter atoms. Metamorphism at 350 MYA **c** drives the daughter atoms out of the mineral into the surrounding rock. If the rock has remained a closed chemical system throughout its history, dating the mineral today **d**, yields the time of metamorphism, whereas dating the whole rock provides the time of its crystallization, 700 MYA.

ratio of the sample being analyzed may have been affected by heat, the decay rate of the parent element remains constant, regardless of any physical or chemical changes.

To obtain an accurate radiometric date, geologists must make sure that the sample is fresh and unweathered and that it has not been subjected to high temperature or intense pressures after crystallization. Furthermore, it is sometimes possible to cross-check the radiometric date obtained by measuring the parent–daughter ratio of two different radioactive elements in the same mineral.

For example, naturally occurring uranium consists of both uranium 235 and uranium 238 isotopes. Through various decay steps, uranium 235 decays to lead 207, whereas uranium 238 decays to lead 206 (Figure 4.6). If the minerals that contain both uranium isotopes have remained closed systems, the ages obtained from each parent–daughter ratio should agree closely. If they do, they are said to be *concordant,* thus reflecting the time of crystallization of the magma. If the ages do not closely agree, then they are said to be *discordant,* and other samples must be taken and ratios measured to see which, if either, date is correct.

Recent advances and the development of new techniques and instruments for measuring various isotope ratios have enabled geologists to analyze not only increasingly smaller samples but with a greater precision than ever before. Presently, the measurement error for many radiometric dates is typically less than 0.5% of the age and, in some cases, is even better than 0.1%. Thus, for a rock 540 million years old (near the beginning of the Cambrian Period), the possible error could range from nearly 2.7 million years to as low as less than 540,000 years.

Long-Lived Radioactive Isotope Pairs

Table 4.2 shows the five common, long-lived parent–daughter isotope pairs used in radiometric dating. There are, however, several other long-lived parent–daughter isotope pairs that are also used for dating. Long-lived pairs have half-lives of millions or billions of years. All of these pairs were present when Earth formed and are still present in measurable quantities. Other shorter-lived radioactive isotope pairs have decayed to the point that only small quantities near the limit of detection remain.

The most commonly used isotope pairs are the uranium–lead and thorium–lead series, which are used principally to date ancient igneous intrusives, lunar samples, and some meteorites. The rubidium–strontium pair is also used for very old samples and has been effective in dating the oldest rocks on Earth as well as meteorites.

The potassium–argon method is typically used for dating fine-grained volcanic rocks from which individual crystals cannot be separated; hence, the whole rock is analyzed. Because argon is a gas, great care must be taken to ensure that the sample has not been subjected to heat, which would allow argon to escape; such a sample would yield an age that is too young.

Other long- and short-lived radioactive isotope pairs exist, but they are rather rare and used only in special situations. Three of these isotope pairs bear mentioning. They are argon–argon, samarium–neodymium, and thorium–uranium.

In argon–argon dating, the ratio of argon 30 to argon 40 is used. Just as with the potassium–argon method, fine-grained volcanic rocks are the most common rocks used in this dating method, but it can also be effectively used with metamorphic rocks. Developed in the 1960s, and further refined during the early 1980s, the argon–argon method has a reliable dating range of 2000 to 4.6 billion years.

The samarium–neodymium method uses the ratio of samarium 147 to neodymium 143, both of which are rare earth elements. Samarium 147 decays to neodymium 143 by a single alpha decay step, and has a half-life of 106 billion years. It is thus very useful in dating extremely old igneous and metamorphic rocks, as well as meteorites. The samarium–neodymium method has been helpful in determining the composition and evolution of Earth's mantle.

The uranium–thorium method is frequently used as a cross-check for carbon 14 dates, and for geologically young samples. In the uranium–thorium method, an age is calculated by determining the degree to which equilibrium has been established between the radioactive parent isotope uranium 234 and the radioactive isotope thorium 230. This technique is based on the fact that uranium is very slightly soluble in natural waters at or near Earth's surface, whereas thorium is not. Thus, carbonate minerals precipitating from surface or near-surface water, would typically not contain thorium. However, the trace amounts of uranium in the precipitated carbonates would, as part of the radioactive decay process, produce radioactive thorium 230, and eventually achieve equilibrium with the uranium 234. What is being measured then, is the degree to which equilibrium has been restored between thorium 230 and uranium 234. Because thorium 230 has a half-life of 75,000 years, the upper limit using this method is approximately 500,000 years.

Fission-Track Dating The emission of atomic particles resulting from the spontaneous decay of uranium within a mineral damages its crystal structure. The damage appears as microscopic linear tracks that are visible only after the mineral has been etched with hydrofluoric acid, an acid so powerful that its vapors can destroy one's sense of smell without careful handling. The age of the sample is determined from the number of fission tracks present and the amount of uranium the sample contains: the older the sample, the greater the number of tracks (• Figure 4.10).

Fission-track dating is of particular interest to archaeologists and geologists because the technique can be used to date samples ranging from only a few hundred to hundreds of millions of years old. It is most useful for dating samples between about 40,000 and 1.5 million years ago, a period for which other dating techniques are not always particularly suitable. One of the problems in fission-track dating occurs when the rocks have later been subjected to high temperatures. If this happens, the damaged crystal structures are repaired by annealing, and consequently the tracks disappear. In such instances, the calculated age will be younger than the actual age.

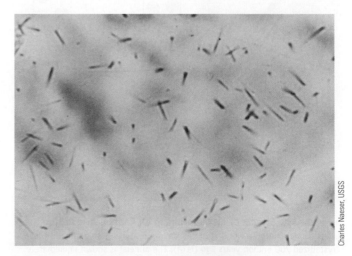

• Figure 4.10 **Fission-Track Dating** Each fission track (about 16 microns [= 16/1000 mm] long) in this apatite crystal is the result of the radioactive decay of a uranium atom. The apatite crystal, which has been etched with hydrofluoric acid to make the fission tracks visible, comes from one of the dikes at Shiprock, New Mexico, and has a calculated age of 27 million years.

Charles Naeser, USGS

Radiocarbon and Tree-Ring Dating
Methods Carbon is an important element in nature and is one of the basic elements found in all forms of life. It has three isotopes; two of these, carbon 12 and 13, are stable, whereas carbon 14 is radioactive (see Figure 2.3). Carbon 14 has a half-life of 5730 years plus or minus 30 years. The **carbon-14 dating** technique is based on the ratio of carbon 14 to carbon 12 and is generally used to date once-living material.

The short half-life of carbon 14 makes this dating technique practical only for specimens younger than about 70,000 years. Consequently, the carbon-14 dating method is especially useful in archaeology and has greatly helped unravel the events of the latter portion of the Pleistocene Epoch. For example, carbon-14 dates of maize from the Tehuacan Valley of Mexico have forced archaeologists to rethink their ideas of where the first center for maize domestication in Mesoamerica arose. Carbon-14 dating is also helping to answer the question of when humans began populating North America.

Carbon 14 is constantly formed in the upper atmosphere when cosmic rays, which are high-energy particles (mostly protons), strike the atoms of upper-atmospheric gases, splitting their nuclei into protons and neutrons. When a neutron strikes the nucleus of a nitrogen atom (atomic number 7, atomic mass number 14), it may be absorbed into the nucleus and a proton emitted. Thus, the atomic number of the atom decreases by one, whereas the atomic mass number stays the same. Because the atomic number has changed, a new element, carbon 14 (atomic number 6, atomic mass number 14), is formed. The newly formed carbon 14 is rapidly assimilated into the carbon cycle and, along with carbon 12 and 13, is absorbed in a nearly constant ratio by all living organisms (• Figure 4.11). When an organism dies, however, carbon 14 is not replenished, and the ratio of carbon 14 to carbon 12 decreases as carbon 14 decays back to nitrogen by a single beta-decay step (Figure 4.5).

Currently, the ratio of carbon 14 to carbon 12 is remarkably constant in both the atmosphere and in living organisms. There is good evidence, however, that the production of carbon 14, and thus the ratio of carbon 14 to carbon 12, has varied somewhat during the past several thousand years. This was determined by comparing ages established by carbon-14 dating of wood samples with ages established by counting annual tree rings in the same samples. As a result, carbon-14 ages have been corrected to reflect such variations in the past.

Tree-ring dating is another useful method for dating geologically recent events. The age of a tree can be determined by counting the growth rings in the lower part of the trunk. Each ring represents one year's growth, and the pattern of wide and narrow rings can be compared among trees to establish the exact year in which the rings were formed. The procedure of matching ring patterns from numerous trees and wood fragments in a given area is called *cross-dating*.

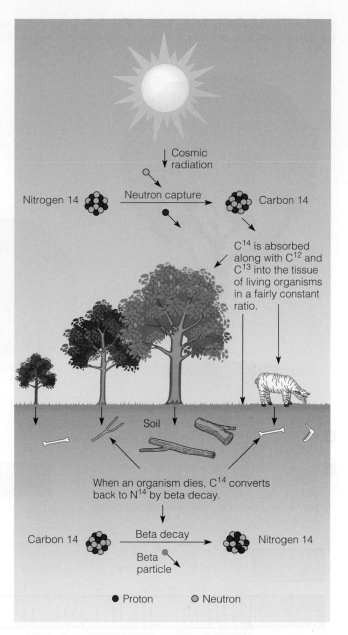

• *Active* Figure 4.11 **Carbon-14 Dating Method** The carbon cycle involves the formation of carbon 14 in the upper atmosphere, its dispersal and incorporation into the tissue of all living organisms, and its decay back to nitrogen 14 by beta decay. Visit the Geology Resource Center to view this and other active figures at www.cengage.com/sso.

By correlating distinctive tree-ring sequences from living to nearby dead trees, a time scale can be constructed that extends back to about 14,000 years ago (• Figure 4.12). By matching ring patterns to the composite ring scale, wood samples whose ages are not known can be accurately dated.

The applicability of tree-ring dating is somewhat limited because it can be used only where continuous tree records are found. It is therefore most useful in arid regions, particularly the southwestern United States, where trees live a very long time.

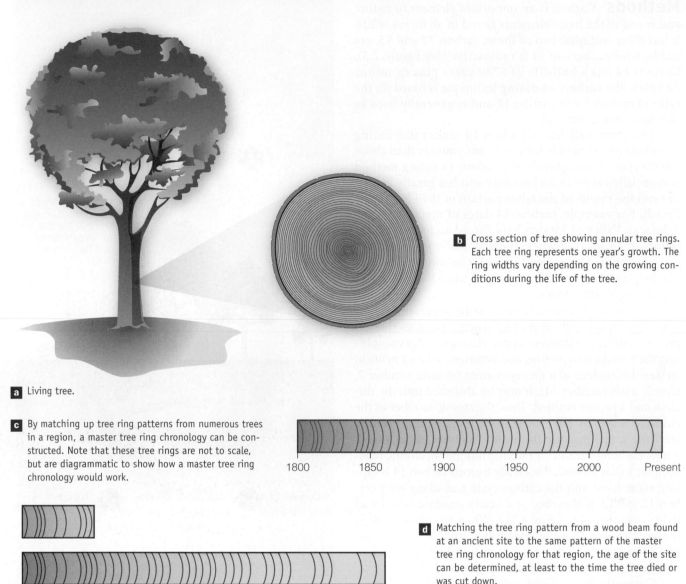

• Figure 4.12 Tree-Ring Dating Method In the cross-dating method, tree-ring patterns from different woods are compared to establish a ring-width chronology backward in time.

b Cross section of tree showing annular tree rings. Each tree ring represents one year's growth. The ring widths vary depending on the growing conditions during the life of the tree.

a Living tree.

c By matching up tree ring patterns from numerous trees in a region, a master tree ring chronology can be constructed. Note that these tree rings are not to scale, but are diagrammatic to show how a master tree ring chronology would work.

1800   1850   1900   1950   2000   Present

d Matching the tree ring pattern from a wood beam found at an ancient site to the same pattern of the master tree ring chronology for that region, the age of the site can be determined, at least to the time the tree died or was cut down.

1800   1850   1900   1950   2000   Present

Geologic Time and Climate Change

Given the debate concerning global warming and its possible implications, it is extremely important to be able to reconstruct past climatic regimes as accurately as possible (see Perspective). To model how Earth's climate system has responded to changes in the past and use that information for simulations of future climatic scenarios, geologists must have a geologic calendar that is as precise as possible.

New dating techniques with greater precision are providing geologists with more accurate dates for when and how long ago past climate changes occurred. The ability to accurately determine when past climate changes occurred helps geologists correlate these changes with regional and global geologic events to see whether there are any possible connections.

One interesting method that is becoming more common in reconstructing past climates is to analyze stalagmites from caves. Stalagmites are icicle-shaped structures rising from a cave floor and formed of calcium carbonate precipitated from evaporating water. A stalagmite, therefore, records a layered history, because each newly precipitated layer of calcium carbonate is younger than the previously precipitated layer (• Figure 4.13). Thus, a stalagmite's layers are oldest in the center at its base and are progressively younger as they move outward (principle of superposition).

Using techniques based on ratios of uranium 234 to thorium 230, geologists can achieve very precise radiometric dates on individual layers of a stalagmite. This technique

• Figure 4.13 **Stalagmites and Climate Change**

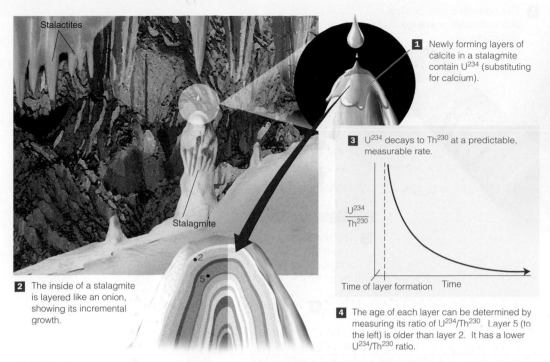

1 Newly forming layers of calcite in a stalagmite contain U^{234} (substituting for calcium).

3 U^{234} decays to Th^{230} at a predictable, measurable rate.

$$\frac{U^{234}}{Th^{230}}$$

Time of layer formation   Time

2 The inside of a stalagmite is layered like an onion, showing its incremental growth.

4 The age of each layer can be determined by measuring its ratio of U^{234}/Th^{230}. Layer 5 (to the left) is older than layer 2. It has a lower U^{234}/Th^{230} ratio.

a Stalagmites are icicle-shaped structures rising from the floor of a cave, and formed by the precipitation of calcium carbonate from evaporating water. A stalagmite is thus layered, with the oldest layer in the center and the youngest layers on the outside. Uranium 234 frequently substitutes for the calcium ion in the calcium carbonate of the stalagmite. Uranium 234 decays to thorium 230 at a predictable and measurable rate. Therefore, the age of each layer of the stalagmite can be dated by measuring the ratio of uranium 234 to thorium 230.

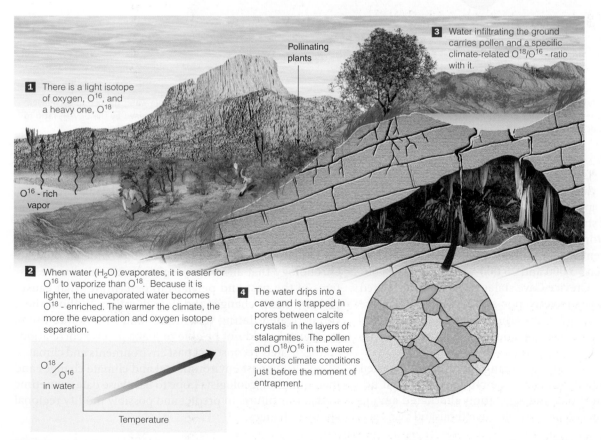

Pollinating plants

3 Water infiltrating the ground carries pollen and a specific climate-related O^{18}/O^{16} - ratio with it.

1 There is a light isotope of oxygen, O^{16}, and a heavy one, O^{18}.

O^{16} - rich vapor

2 When water (H$_2$O) evaporates, it is easier for O^{16} to vaporize than O^{18}. Because it is lighter, the unevaporated water becomes O^{18} - enriched. The warmer the climate, the more the evaporation and oxygen isotope separation.

$$\frac{O^{18}}{O^{16}}$$
in water

Temperature

4 The water drips into a cave and is trapped in pores between calcite crystals in the layers of stalagmites. The pollen and O^{18}/O^{16} in the water records climate conditions just before the moment of entrapment.

b There are two isotopes of oxygen, a light one, oxygen 16, and a heavy one, oxygen 18. Because the oxygen 16 isotope is lighter than the oxygen 18 isotope, it vaporizes more readily than the oxygen 18 isotope when water evaporates. Therefore, as the climate becomes warmer, evaporation increases, and the O^{18}/O^{16} ratio becomes higher in the remaining water. Water in the form of rain or snow percolates into the ground and becomes trapped in the pores between the calcite forming the stalagmites.

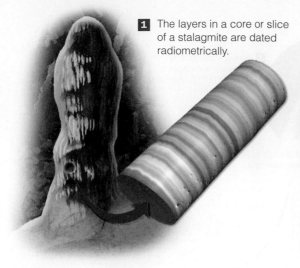

1 The layers in a core or slice of a stalagmite are dated radiometrically.

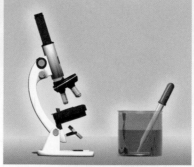

2 The pore water is analyzed in each layer for O^{18}/O^{16} and species of plants (from the pollen).

3 A record of climatic change is put together for the area of the caves.

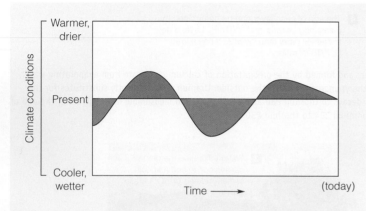

c The layers of a stalagmite can be dated by measuring the U^{234}/Th^{230} ratio, and the O^{18}/O^{16} ratio determined for the pore water trapped in each layer. Thus, a detailed record of climatic change for the area can be determined by correlating the climate of the area as determined by the O^{18}/O^{16} ratio to the time period determined by the U^{234}/Th^{230} ratio.

enables geologists to determine the age of materials much older than they can date by the carbon-14 method, and it is reliable back to approximately 500,000 years.

A study of stalagmites from Crevice Cave in Missouri revealed a history of climatic and vegetation change in the midcontinent region of the United States during the interval between 75,000 and 25,000 years ago. Dates obtained from the Crevice Cave stalagmites were correlated with major changes in vegetation and average temperature fluctuations, obtained from carbon 13 and oxygen 18 isotope profiles, to reconstruct a detailed picture of climate changes during this time period.

It was determined that during the interval between 75,000 and 55,000 years ago, the climate oscillated between warm and cold, and vegetation alternated among forest, savannah, and prairie. Fifty-five thousand years ago

the climate cooled, and there was a sudden change from grasslands to forest, which persisted until 25,000 years ago. This corresponds to the time when global ice sheets began building and advancing.

Thus, precise dating techniques in stalagmite studies using uranium 234 and thorium 230 provide an accurate chronology that allows geologists to model climate systems of the past and perhaps to determine what causes global climatic changes and their duration. Without these sophisticated dating techniques and others like them, geologists would not be able to make precise correlations and accurately reconstruct past environments and climates. By analyzing past environmental and climate changes and their duration, geologists hope to use these data, sometime in the near future, to predict and possibly modify regional climatic changes.

Denver's Weather—280 Million Years Ago!

With all the concern about global climate change, it might be worthwhile to step back a bit and look at climate change from a geologic perspective. We're all aware that some years are hotter than others and in some years we have more rain, but generally things tend to average out over time. We know that it will be hot in the summer in Arizona, and it will be very cold in Minnesota in the winter. We also know that scientists, politicians, and concerned people everywhere are debating whether humans are partly responsible for the global warming that Earth seems to be experiencing.

What about long-term climate change? We know that Earth has experienced periods of glaciation in the past—for instance, during the Proterozoic, the end of the Ordovician Period, and most recently during the Pleistocene Epoch. Earth has also undergone large-scale periods of aridity, such as during the end of the Permian and beginning of the Triassic periods. Such long-term climatic changes are probably the result of slow geographic changes related to plate tectonic activity. Not only are continents carried into higher and lower latitudes, but also their movement affects ocean circulation and atmospheric circulation patterns, which in turn affect climate, and result in climate changes.

Even though we can't physically travel back in time, geologists can reconstruct what the climate was like in the past. The distribution of plants and animals is controlled, in part, by climate. Plants are particularly sensitive to climate change and many can only live in particular environments. The fossils of plants and animals can tell us something about the environment and climate at the time these organisms were living. Furthermore, climate-sensitive sedimentary rocks can be used to interpret past climatic conditions. Desert dunes are typically well sorted and exhibit large-scale cross-bedding. Coals form in freshwater swamps where climatic conditions promote abundant plant growth. Evaporites such as rock salt result when evaporation exceeds precipitation, such as in desert regions or along hot, dry shorelines. Tillites (glacial sediments) result from glacial activity and indicate cold, wet environments. So, by combining all relevant geologic and paleontologic information, geologists can reconstruct what the climate was like in the past and how it has changed over time at a given locality.

In a recently published book titled *Ancient Denvers: Scenes from the Past 300 Million Years of the Colorado Front Range*, Kirk R. Johnson depicts what Denver, Colorado looked like at 13 different time periods in the past. The time slices begin during the Pennsylvanian Period, 300 million years ago, and end with a view of the Front Range amid a spreading wave of houses on the southern edge of metropolitan Denver.

The information for piecing together Denver's geologic past was derived mainly from a 688-m-deep well drilled by the Denver Museum of Nature and Science beneath Kiowa, Colorado, in 1999. Using the information gleaned from the rocks recovered from the well, plus additional geologic evidence from other parts of the area, the museum scientists and artists were able to reconstruct Denver's geologic past.

Beginning 300 million years ago (Pennsylvanian Period), the Denver area had coastlines on its eastern and western borders and a mountain range (not the Rocky Mountains of today). The climate was mostly temperate with lots of seedless vascular plants, like ferns, as well as very tall scale trees related to the modern horsetail rush. Insects such as millipedes, cockroaches, and dragonflies were huge and shared this region with small (approximately 3 m long) fin-backed reptiles and a variety of amphibians.

By 280 million years ago, huge sand seas, much like the Sahara today, covered the area (Figure 1). This change in climate and landscape was the result of the formation of Pangaea. As the continents

Denver Museum of Nature and Science

Figure 1 Denver as it appeared 280 million years ago. As a result of the collision of continents and the formation of Pangaea, the world's climate was generally arid, and Denver was no exception. Great seas of sand probably covered Denver, much as the Sahara is today.

collided, arid and semi-arid conditions prevailed over much of the supercontinent, and the Denver area was no exception.

During the Late Jurassic (150 million years ago), herds of plant-eating dinosaurs such as *Apatosaurus* roamed throughout the Denver area, feasting on the succulent and abundant vegetation. Grasses and flowering plants had not yet evolved, so the dinosaurs ate the ferns and gymnosperms that were abundant at this time.

Beginning around 90 million years ago, tectonic forces started a mountain building episode known as the *Laramide orogeny* that resulted in the present-day Rocky Mountains. Dinosaurs still roamed the land around Denver, and flowering plants began their evolutionary history.

As a result of rising sea level, a warm, shallow sea covered Denver 70 million years ago (Late Cretaceous). Marine reptiles such as plesiosaurs and mosasaurs ruled these seas, while overhead, pterosaurs soared through the skies, looking for food (Figure 2).

By 55 million years ago (Eocene), the world was in the grip of an intense phase of global warming. A subtropical rainforest with many trees that would be recognizable today filled the landscape. Primitive mammals were becoming more abundant, and many warm-climate-loving animals could be found living north of the Arctic Circle.

Although ice caps still covered portions of North America, mammoths and other mammals wandered among the plains of Denver 16,000 years ago (Figure 3). Mastodons, horses, bison, lions, and

giant ground sloths, to name a few, all lived in this region, and their fossils can be found in the sedimentary rocks from this area.

What was once a rainforest, desert, warm, shallow sea, and mountainous region is now home to thousands of people. What the Denver region will be like in the next several million years is anyone's guess. Whereas humans can affect change, what change we will cause is open to debate. Certainly the same forces that have shaped the Denver area in the past will continue to determine its future. With the rise of humans and technology, we, as a species, will also influence what future Denvers will be like. Let us hope that the choices we make are good ones.

Figure 2 Pterosaurs (flying reptiles) soar over Denver 70 million years ago. At this time, Denver was below a warm, shallow sea that covered much of western North America. Marine reptiles, such as plesiosaurs and mosasaurs, swam in these seas in search of schools of fish.

Figure 3 Mammoths and camels wander on the prairie during a summer day 16,000 years ago. Whereas an ice sheet covered much of northern North America, Denver sported pine trees and prairie grass. This area was home to a large variety of mammals, including mammoths, camels, horses, bison, and giant ground sloths. Humans settled in this area approximately 11,000 years ago, hunting the plentiful game that was available.

SUMMARY

- Time is defined by the methods used to measure it. Relative dating places geologic events in sequential order as determined from their position in the geologic record. Absolute dating provides specific dates for geologic rock units or events, expressed in years before the present.

- Early Christian theologians were responsible for formulating the idea that time is linear and that Earth was very young. Archbishop Ussher calculated Earth's age at approximately 6000 years based on his interpretation of the scriptures.

- During the 18th and 19th centuries, attempts were made to determine Earth's age based on scientific evidence rather than revelation. Although some attempts were quite ingenious, they yielded a variety of ages that are now known to be much too young.

- Considering the religious, political, and social climate of the 17th, 18th, and early 19th centuries, it is easy to see why concepts such as neptunism, catastrophism, and a very young Earth were eagerly embraced. As geologic data accumulated, it became apparent that these concepts were not supported by evidence and that Earth must be much older than 6000 years.

- James Hutton, considered by many to be the father of modern geology, thought that present-day processes operating over long periods of time could explain all of the geologic features of Earth. He also viewed Earth history as cyclical and thought Earth to be very old. Hutton's observations were instrumental in establishing the principle of uniformitarianism.

- Uniformitarianism, as articulated by Charles Lyell, soon became the guiding principle of geology. It holds that the laws of nature have been constant through time, and that the same processes operating today have also operated in the past, although not necessarily at the same rates.

- Besides uniformitarianism, the principles of superposition, original horizontality, lateral continuity, and cross-cutting relationships are basic for determining relative geologic ages and for interpreting Earth history.

- Radioactivity was discovered during the late 19th century, and soon thereafter radiometric dating techniques allowed geologists to determine absolute ages for geologic events.

- Absolute ages for rock samples are usually obtained by determining how many half-lives of a radioactive parent element have elapsed since the sample originally crystallized. A half-life is the time it takes for one-half of the radioactive parent element to decay to a new, more stable daughter element.

- The most accurate radiometric dates are obtained from long-lived radioactive isotope pairs in igneous rocks. The five common long-lived radioactive isotope pairs are uranium 238–lead 206, uranium 235–lead 207, thorium 232–lead 208, rubidium 87–strontium 87, and potassium 40–argon 40. The most reliable dates are those obtained by using at least two different radioactive-decay series in the same rock.

- Carbon-14 dating is effective back to about 70,000 years ago and can only be used on organic matter such as wood, bones, and shells. Unlike the long-lived isotopic pairs, the carbon-14 dating technique determines age by the ratio of radioactive carbon 14 to stable carbon 12.

- To reconstruct past climate changes and link them to possible causes, geologists must have a geologic calendar that is as precise and accurate as possible. Thus, they must be able to date geologic events and the onset and duration of climate changes as precisely as possible.

IMPORTANT TERMS

absolute dating, p. 67

carbon-14 dating, p. 77

catastrophism, p. 70

fission-track dating, p. 76

half-life, p. 73

neptunism, p. 69

principle of cross-cutting relationships, p. 69

principle of lateral continuity, p. 69

principle of original horizontality, p. 69

principle of superposition, p. 68

principle of uniformitarianism, p. 70

radioactive decay, p. 72

relative dating, p. 66

tree-ring dating, p. 77

REVIEW QUESTIONS

1. Who is generally considered the founder of modern geology?
   a. _____ Werner; b. _____ Lyell; c. _____ Steno; d. _____ Cuvier; e. _____ Hutton.

2. In which type of radioactive decay is a fast-moving electron emitted from a neutron in the nucleus?
   a. _____ Alpha decay; b. _____ Beta decay; c. _____ Electron capture; d. _____ Fission track; e. _____ None of the previous answers.

3. If a radioactive element has a half-life of 16 million years, what fraction of the original amount of parent material will remain after 96 million years?
   a. _____ 1/2; b. _____ 1/16; c. _____ 1/32; d. _____ 1/4; e. _____ 1/64.

4. Because of the heat and pressure exerted during metamorphism, daughter atoms were driven out of a mineral being analyzed for a radiometric date. The date obtained

from this mineral will, therefore, be _____ its actual age of formation.

a. _____ younger than; b. _____ older than; c. _____ the same as; d. _____ it can't be determined; e. _____ none of the previous answers.

5. How many half-lives are required to yield a mineral with 625,000,000 atoms of thorium 232 and 19,375,000,000 atoms of lead 235?

a. _____ 1; b. _____ 2; c. _____ 3; d. _____ 4; e. _____ 5.

6. Placing geologic events in sequential or chronologic order as determined by their position in the geologic record is

a. _____ absolute dating; b. _____ correlation; c. _____ historical dating; d. _____ relative dating; e. _____ uniformitarianism.

7. If a flake of biotite within a sedimentary rock (such as a sandstone) is radiometrically dated, the date obtained indicates when

a. _____ the biotite crystal formed; b. _____ the sedimentary rock formed; c. _____ the parent radioactive isotope formed; d. _____ the daughter radioactive isotope(s) formed; e. _____ none of the previous answers.

8. The atomic number of an element is determined by the number of _____ in its nucleus.

a. _____ protons; b. _____ neutrons; c. _____ electrons; d. _____ protons and neutrons; e. _____ protons and electrons.

9. What is being measured in radiometric dating is

a. _____ the time when a radioactive isotope formed; b. _____ the time of crystallization of a mineral containing an isotope; c. _____ the amount of the parent isotope only; d. _____ when the dated mineral became part of a sedimentary rock; e. _____ when the stable daughter isotope was formed.

10. As carbon 14 decays back into nitrogen in radiocarbon dating, what isotopic ratio decreases?

a. _____ Nitrogen 14 to carbon 12; b. _____ Carbon 14 to carbon 12; c. _____ Carbon 13 to carbon 12; d. _____ Nitrogen 14 to carbon 14; e. _____ Nitrogen 14 to carbon 13.

11. What is the difference between relative dating and absolute dating?

12. How does metamorphism affect the potential for accurate radiometric dating using any and all techniques discussed in this chapter? How would such radiometric dates be affected by metamorphism, and why?

13. Describe the principle of uniformitarianism according to Hutton and Lyell. What is the significance of this principle?

14. If you wanted to calculate the absolute age of an intrusive body, what information would you need?

15. A volcanic ash fall was radiometrically dated using the potassium 40 to argon 40 and rubidium 87 to strontium 87 isotope pairs. The isotope pairs yielded distinctly different ages. What possible explanation could be offered as to why these two isotope pairs yielded different ages? What would you do to rectify the discrepancy in ages?

16. In addition to the methods mentioned in this chapter, can you think of some other non-radiometric dating methods that could be used to determine the age of Earth? What would be the drawbacks to the methods you propose?

17. Why were Lord Kelvin's arguments and calculations so compelling, and what was the basic flaw in his assumption? What do you think the course of geology would have been if radioactivity had not been discovered?

18. Can the various principles of relative dating be used to reconstruct the geologic history of Mars? Which principles might not apply to interpreting the geologic history of another planet?

19. Why do igneous rocks yield the most accurate radiometric dates? Why can't sedimentary rocks be dated radiometrically? What problems are encountered in dating metamorphic rocks?

20. Why are the fundamental principles used in relative dating so important in geology?

APPLY YOUR KNOWLEDGE

1. How many half-lives are required in which there are presently 1,250,000,000 atoms of rubidium 87 and 38,750,000,000 strontium 87 in its crystal structure? What is the percentage of rubidium 87 remaining after this many half-lives?

2. Given the current debate concerning global warming and the many possible short-term consequences for humans, can you visualize how the world might look in 10,000 years or even 1 million years from now? Use what you have learned about plate tectonics and the direction and rate of movement of plates, as well as how plate movement and global warming will affect ocean currents, weather patterns, weathering rates, and other factors, to make your prediction. Do you think such short-term changes can be extrapolated to long-term trends in trying to predict what Earth will be like in the future?

Sue Monroe

CHAPTER 5

ROCKS, FOSSILS, AND TIME—MAKING SENSE OF THE GEOLOGIC RECORD

These rocks exposed in Monument Valley Navajo Tribal Park on the Utah-Arizona border make up part of the geologic record for the Permian Period. The uppermost rock layer, the one forming the mesas and buttes, is an ancient wind-blown dune deposit, whereas the underlying rocks formed in stream channels and on flood-plains. This scenic area has served as a backdrop for many television shows and movies, especially westerns.

[OUTLINE]

Introduction

Stratigraphy
   Vertical Stratigraphic Relationships
   Lateral Relationships—Facies
   Marine Transgressions and Regressions
   Extent, Rate, and Causes of Marine
   Transgressions and Regressions

Fossils and Fossilization
   How Do Fossils Form?
   Fossils and Telling Time

The Relative Geologic Time Scale

Stratigraphic Terminology
   Lithostratigraphic Units and
   Biostratigraphic Units
   Time Stratigraphic Units and Time Units

Correlation
   Perspective *Monument Valley Navajo
   Tribal Park*

Absolute Dates and the Relative Geologic
Time Scale

Summary

At the end of this chapter, you will have learned that

- To analyze the geologic record, you must first determine the correct vertical sequence of rocks—that is, from oldest to youngest—even if they have been deformed.

- Although rocks provide our only evidence of prehistoric events, the record is incomplete at any one locality because discontinuities are common.

- Stratigraphy is a discipline in geology that is concerned with sedimentary rocks most of which are layered or stratified, but many principles of stratigraphy also apply to igneous and metamorphic rocks.

- Several marine transgressions and regressions occurred during Earth history, at times covering much of the continents and at other times leaving the land above sea level.

- Fossils, the remains or traces of prehistoric organisms, are preserved in several ways, and some types of fossils are much more common than most people realize.

- Distinctive groups of fossils found in sedimentary rocks are useful for determining the relative ages of rocks in widely separated areas.

- Superposition and the principle of fossil succession were used to piece together a composite geologic column, which is the basis for the relative geologic time scale.

- Geologists have developed terminology to refer to rocks and to time.

- Several criteria are used to match up (correlate) similar rocks over large regions or to demonstrate that rocks in different areas are the same age.

- Absolute ages of sedimentary rocks are most often determined by radiometric dating of associated igneous or metamorphic rocks.

Introduction

Among the terrestrial planets—Mercury, Venus, Earth, and Mars—all except Earth and possibly Venus are essentially dead worlds, as is Earth's moon. Earth's internal heat accounts for ongoing volcanism, seismicity, and plate movements, and its hydrosphere, atmosphere, and biosphere continually modify its surface. Earth's size, distance from the Sun, its surface waters, and oxygen-rich atmosphere make it unique among the terrestrial planets. We will have more to say on Earth's distinctive features in Chapter 8. Indeed, Earth is a dynamic planet. The **geologic record,** that is, the evidence for past physical and biological events preserved in rocks, clearly indicates that Earth has evolved throughout its existence, and it continues to do so.

In their efforts to decipher Earth history geologists study all of the families of rocks, but they pay particular attention to sedimentary rocks. But how is it possible to interpret events that no one witnessed? Perhaps an analogy will help. Suppose you are walking in a forest and observe a shattered, charred tree. Considering your knowledge of trees and how they grow you immediately know that it did not always exist in its present form. So, perhaps your observations can be explained by an exploding bomb, but you quickly reject this hypothesis because you see no evidence of a bomb crater or shrapnel (metallic fragments) from a bomb. Finally, you conclude that the tree was hit by lightning, a hypothesis that you can check by comparing with trees known to have been hit by lightning.

In our hypothetical example, you did not witness the event but you determined what happened because evidence of the event having occurred was present. It does not matter whether the lightning struck the tree a few days before or many years ago; the only requirement is that the evidence still exists. Likewise, the geologic record provides evidence of past physical and biological events having occurred (see the chapter opening photo). For instance, when mud dries it shrinks and cracks and waves in shallow water deform sand into ripples (see Chapter 6), which if found in ancient rocks must have formed by the same processes responsible for them now.

An interesting historical note on deciphering Earth history comes from the work of Philip Henry Gosse, a British naturalist, who in 1857 proposed that Earth was created a few thousands of years ago with the appearance of great age. According to this idea, rocks that looked like they cooled from lava flows or that appeared to have been sand that later became sandstone were created with that appearance. And likewise, fossil clams, corals, bones, and so on were also created to resemble these objects.

Most people reasoned that if Gosse were correct, a deceitful creator fabricated the geologic record, a thesis they could not accept. Furthermore, one cannot check Gosse's idea because the geologic record would look exactly the same if he were correct or if it formed over millions of years by natural processes. Scientists opted for the latter—the uniformitarian approach—because they had every reason to think that the features they observed in rocks formed by the same processes operating now. As for Earth's age, even Gosse had to admit that nothing in the geologic record supported his determination. His critics were quick to point out that if his idea was correct, that Earth was created with the appearance of great age, it could have been created at any time during the past, even moments ago along with all records and memories.

Philip Henry Gosse's ideas on Earth history now serve only as a failed attempt to reconcile Earth history with a particular interpretation of scripture. The uniformitarian method offers a much more fruitful approach, but keep in mind that the geologic record is complex, incomplete at any one location, and it requires interpretation. In this chapter, we endeavor to bring some organization to this record.

Stratigraphy

The branch of geology called **stratigraphy** is concerned with the composition, origin, age relationships, and geographic extent of sedimentary rocks, but the principles of stratigraphy apply to any sequence of stratified rocks. Sedimentary rocks are, with few exceptions, stratified (• Figure 5.1), but volcanic rocks, including lava flows and ash beds, as well as many metamorphic rocks are also stratified.

Where sedimentary rocks are well exposed, as in the walls of deep canyons, you can easily determine the vertical relationships among individual layers or strata (singular *stratum*). Lateral relationships are equally important in analyzing the geologic record, but they usually must be determined from a number of separate rock exposures, or what geologists call *outcrops*.

Vertical Stratigraphic Relationships
In vertical successions of sedimentary rocks, surfaces known as *bedding planes* separate individual strata from one another (Figure 5.1), or the strata grade vertically from one rock type into another. The rocks below and above a bedding plane differ in composition, texture, color, or a combination of these features, indicating a rapid change in sedimentation or perhaps a period of nondeposition and/or erosion followed by renewed deposition. In contrast, gradually changing conditions of sedimentation account for those rocks that show a vertical gradation. Regardless of the nature of the vertical relationships among strata, the correct order in which they were deposited, that is, their relative ages, must be determined.

Superposition In Chapter 4, we discussed the *principle of superposition* that resulted from the work of Nicolas Steno during the 17th century. According to this principle, you can determine the correct relative ages of underformed strata by their position in a sequence; the oldest layer is at the bottom of the sequence with successively younger layers upward in the sequence (see Figure 4.2). If strata

• **Figure 5.1 Stratified Sedimentary Rocks**

a These rocks in South Dakota have been deeply eroded, but the stratification (layering) is still clearly visible.

b Stratified rocks in California that have been deformed so that they no longer lie in their original position. It is not obvious from the image but the youngest layer is on the right.

are deformed by faulting, folding, or both, the task is more difficult (Figure 5.1b), but several sedimentary structures and some fossils allow geologists to resolve these kinds of problems (see Chapter 6).

The **principle of inclusions** is yet another way to figure out relative ages, because inclusions, or fragments, in a body of rock must be older than the rock itself. Obviously the sand grains making up sandstone are older than the sandstone, but what about the relative ages of the granite and sandstone in • Figure 5.2? Two interpretations are possible: The granite was intruded into the sandstone and thus is the youngest; the granite was the source of the sand and the sandstone is oldest.

Suppose you encounter a sequence of mostly sedimentary rocks but one layer is made up of basalt (• Figure 5.3). If the basalt is a buried lava flow, you can determine its

• Figure 5.2 **The Principle of Inclusions**

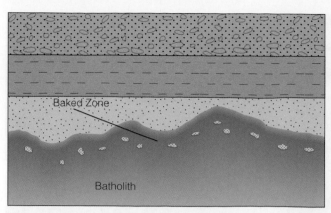

a The batholith is younger than sandstone because the sandstone has been baked at its contact with the granite and the granite has sandstone inclusions.

c Basalt inclusions (black) in granite in Wisconsin show that the basalt is oldest.

James S. Monroe

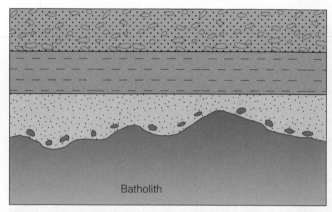

b Granite inclusions in the sandstone indicate that the batholith was the source of the sand and is therefore older.

• Figure 5.3 **How to Determine the Relative Ages of Lava Flows, Sills, and Associated Sedimentary Rocks**

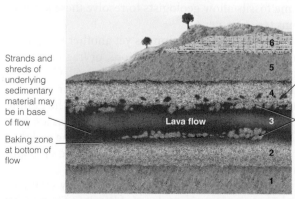

Strands and shreds of underlying sedimentary material may be in base of flow

Baking zone at bottom of flow

Clasts of lava may be present in overlying, younger layer

Rubble zones may be present at top and bottom of flow

a A buried lava flow has baked underlying bed 2, and clasts of the lava were deposited along with other sediment in bed 4. The lava flow is older than beds 4, 5, and 6.

Baking zones on both sides of sill

Inclusions of rock from both layers above and below may exist in sill

b The layers above and below the sill have been baked, so the sill is younger than layers 2 and 4. Can you determine its age relative to beds 5, 6, and 7?

relative age by the principle of superposition, but if it is a sill—a sheet-like intrusive body—it is younger than the layers below it and younger than the layer immediately above it. Study Figure 5.3 closely and note that the principle of inclusions as well as contact metamorphic effects help resolve this problem.

Unconformities So far, we have discussed vertical relationships among **conformable** strata—that is, sequences of rocks in which deposition was more or less continuous. A bedding plane between strata may represent a depositional break of anywhere from minutes to tens or hundreds of years but is inconsequential in the context of geologic time. However, in many sequences of strata, surfaces known as **unconformities** are present that represent times of nondepositation and/or erosion. These unconformities encompass long periods of geologic time, perhaps millions or tens of millions of years. Accordingly, the geologic record is incomplete at that particular location, just as a book with missing pages is incomplete, and the interval of geologic time not represented by strata is a *hiatus* (• Figure 5.4).

Unconformity is a general term that encompasses three distinct types of surfaces called *disconformity*, *nonconformity*, and *angular unconformity*. Furthermore, unconformities of regional extent may change from one type to another, and they do not necessarily encompass equivalent amounts of geologic time everywhere (• Figure 5.5a). Unconformities are common, so the geologic record is incomplete at those locations where unconformities are present. Nevertheless, the geologic time not recorded by rocks in one area is represented by rocks elsewhere.

A **disconformity** is an erosion surface in sedimentary rocks that separates younger rocks from older rocks, both of which are parallel to each other (Figure 5.5b). However, an erosion surface cut into plutonic rocks or metamorphic rocks that is overlain by sedimentary rocks is a **nonconformity** (Figure 5.5c). And finally, an **angular unconformity** is present if the strata below an erosion surface are inclined at some angle to the strata above (Figure 5.5d). In this case, we can infer that the sequence of events included deposition, lithification, deformation, erosion, and, finally, renewed deposition.

• Figure 5.4 **The Origin of an Unconformity and a Hiatus**

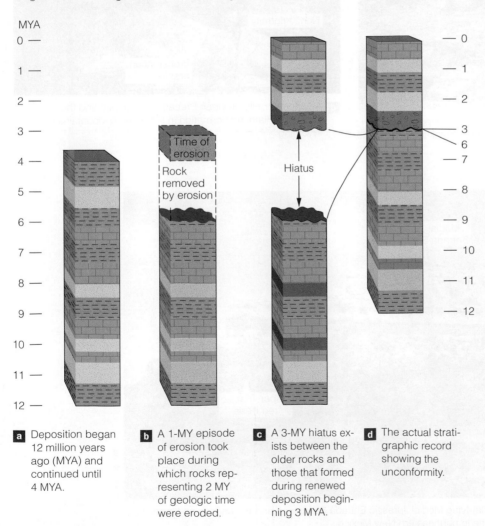

a Deposition began 12 million years ago (MYA) and continued until 4 MYA.

b A 1-MY episode of erosion took place during which rocks representing 2 MY of geologic time were eroded.

c A 3-MY hiatus exists between the older rocks and those that formed during renewed deposition beginning 3 MYA.

d The actual stratigraphic record showing the unconformity.

• Figure 5.5 **Types of Unconformities**

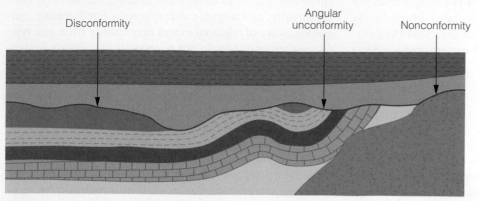

Disconformity

Angular unconformity

Nonconformity

a Cross section showing the regional variation in an unconformity.

Jurassic rocks

Mississippian rocks

James S. Monroe

b A disconformity. The surface between the Mississippian and Jurassic rocks represents a 165-million-year break in the geologic record at this location in Montana.

Cambrian Flathead Sandstone

Nonconformity

Precambrian granite

James S. Monroe

c A nonconformity between Precambrian granite and the Cambrian Flathead Formation in the Bighorn Mountains of Wyoming.

Medial Jurassic Entrada Sandstone

Angular unconformity

Upper Triassic red beds

James S. Monroe

d An angular unconformity between the flat-lying Medial Jurassic Entrada Sandstone and the underlying Upper Triassic red beds at Wedding Cake Butte in northeastern New Mexico.

Lateral Relationships—Facies

In 1669, Nicolas Steno proposed his principle of lateral continuity, meaning that layers of sediment extend outward in all directions until they terminate (see Chapter 4). Rock layers may terminate abruptly where they abut the edge of a depositional basin, where they are eroded, or where they are cut by faults. A rock unit may also become progressively thinner until it *pinches out,* or it splits laterally into thinner units each of which pinches out—a phenomenon known as *intertonguing.* And finally, a rock unit might change by *lateral gradation* as its composition and/or texture become increasingly different (• Figure 5.6).

Both intertonguing and lateral gradation indicate the simultaneous operation of different depositional processes in adjacent environments. For example, on the continental shelf sand may accumulate in the high-energy nearshore environment, while at the same time mud and carbonate deposition takes place in offshore low-energy environments. Deposition in each of these laterally adjacent environments yields a **sedimentary facies,** a body of sediment with distinctive physical, chemical, and biological attributes.

Armanz Gressly, in 1838, was the first to use the term *facies* when he carefully traced sedimentary rocks in the Jura Mountains of Switzerland and noticed lateral changes such as sandstone grading into shale. He reasoned that these changes indicated deposition in different environments that lie next to one another. Any attribute of sedimentary rocks that makes them recognizably different from laterally adjacent rocks of about the same age is sufficient to establish a sedimentary facies.

Marine Transgressions and Regressions

Three rock units exposed in the walls of the Grand Canyon of Arizona consist of sandstone followed upward by shale and finally limestone (• Figure 5.7). These three facies, all with fossils of marine-dwelling trilobites and brachiopods, were deposited on one another. Superposition tells us their relative ages, but what accounts for their presence in Arizona far from the sea, and how were these facies deposited in the order observed? These deposits formed during a time when sea level rose with respect to the land, giving rise to a **marine transgression.** During a marine

• Figure 5.6 **Lateral Termination of Rock Layers**

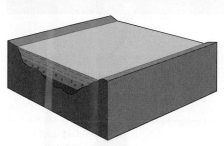

a Lateral termination at the edge of a depositional basin.

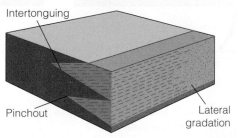

b Lateral termination by intertonguing, pinchout, and lateral gradation.

• Figure 5.7 **Sedimentary Rocks in the Grand Canyon** View of three formations in the Grand Canyon of Arizona. These rocks, representing three facies, were deposited during a marine transgression. Compare with the vertical sequence of rocks in Figure 5.8.

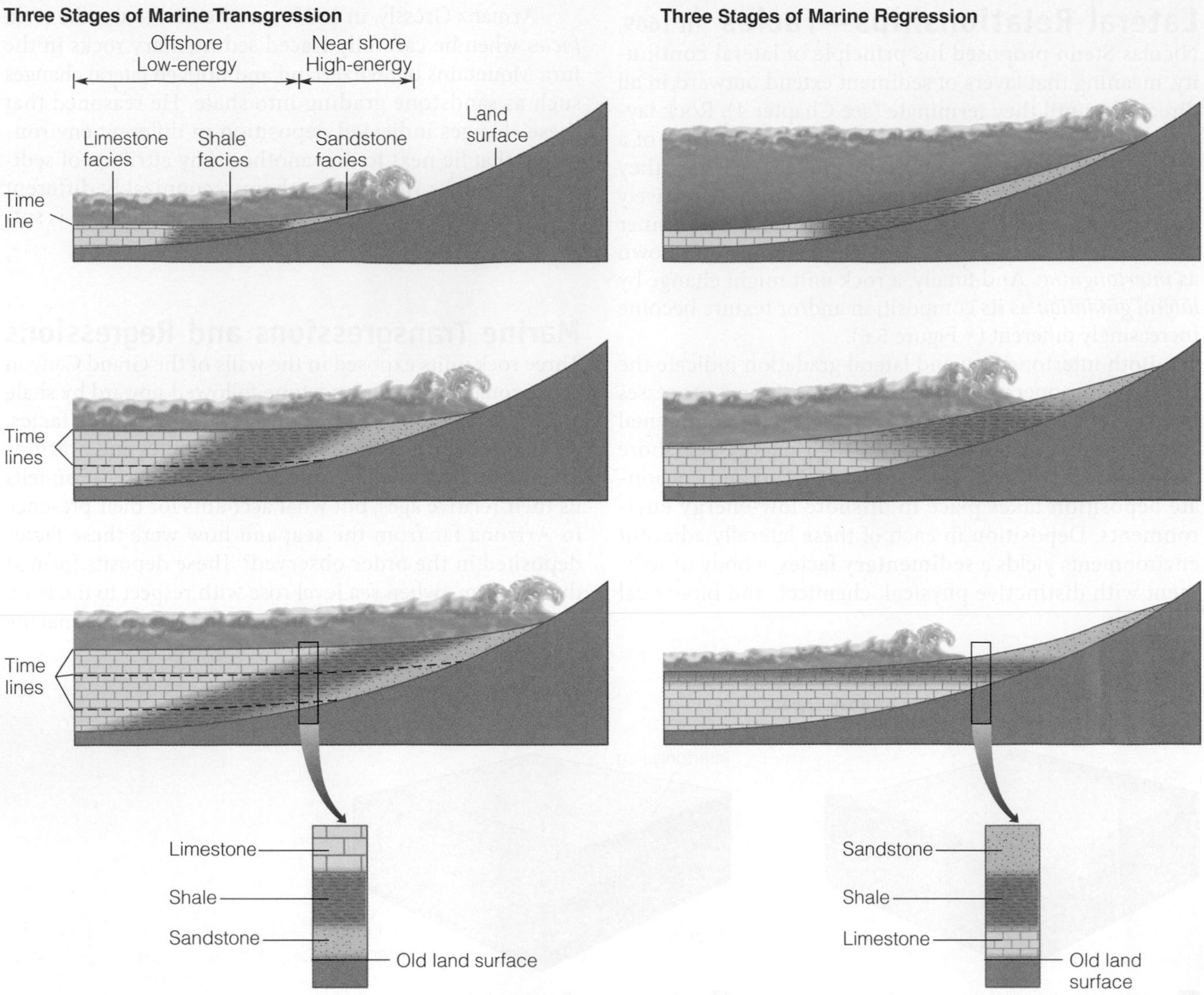

Three Stages of Marine Transgression

Offshore Low-energy — Near shore High-energy

Limestone facies   Shale facies   Sandstone facies   Land surface

Time line

Time lines

Time lines

Limestone
Shale
Sandstone
Old land surface

Three Stages of Marine Regression

Sandstone
Shale
Limestone
Old land surface

• Figure 5.8 **Marine Transgressions and Regressions**

transgression the shoreline migrates landward, as do the environments that parallel the shoreline as the sea progressively covers more and more of a continent(• Figure 5.8). Remember that each laterally adjacent depositional environment is an area where a sedimentary facies develops. So during a marine transgression, the facies that formed in offshore environments become superposed on facies deposited in nearshore environments.

Another aspect of a marine transgression is that the rocks making up each facies become younger in a landward direction. The sandstone, shale, and limestone facies in Figure 5.8 were deposited continuously as the shoreline moved landward, but none of the facies was deposited simultaneously over its entire geographic extent. In other words, the facies are *time transgressive,* meaning that their ages vary from place to place.

Obviously the sea is no longer present in the Grand Canyon area, so the transgression responsible for the rocks in Figure 5.7 must have ended. In fact, it did—during a

marine regression when sea level fell with respect to the continent and the environments that paralleled the shoreline migrated seaward. In other words, a marine regression is the opposite of a transgression and it yields a vertical sequence with nearshore facies overlying offshore facies (Figure 5.8). In this case individual rock units become younger in a seaward direction.

In our discussion of marine transgressions and regressions, we considered both vertical and lateral facies relationships, the significance of which was first recognized by Johannes Walther (1860–1937). When Walther traced rock units laterally, he reasoned, as Gressly had, that each sedimentary facies he encountered was deposited in laterally adjacent environments. In addition, Walther noticed that the same facies he found laterally were also present in a vertical sequence. His observations have since been formulated into **Walther's law,** which holds that the facies seen in a conformable vertical sequence will also replace one another laterally.

The application of Walther's law is well illustrated by the marine transgression and regression shown in Figure 5.8. In practice, it is usually difficult to follow rock units far enough laterally to demonstrate facies changes. It is much easier to observe vertical facies relationships and use Walther's law to work out the lateral relationships. Remember, though, that Walther's law applies only to conformable sequences of rocks; rocks above and below an unconformity are unrelated and Walther's law does not apply.

Extent, Rates, and Causes of Marine Transgressions and Regressions

Geologists carefully analyze sedimentary rocks to determine the maximum extent of marine transgressions and the greatest withdrawal of the sea during regressions. Six major marine transgressions followed by regressions have taken place in North America since the Neoproterozoic, yielding unconformity-bounded rock sequences that provide the stratigraphic framework for our discussions of Paleozoic and Mesozoic geologic history. The paleogeographic maps in Chapters 10, 11, 14, and 16 show the extent to which North America was covered by the sea or exposed during these events.

Shoreline movements during transgressions and regressions probably amount to no more than a few centimeters per year. Suppose that a shoreline moves landward 1000 km in 20 million years, giving 5 cm/yr as the average rate of transgression. Our average is reasonable, but we must point out that large-scale transgressions are not simply events during which the shoreline steadily moves landward. In fact, they are characterized by a number of reversals in the overall transgressive trend, thus accounting for the intertonguing we see among some sedimentary rock units (Figure 5.6b).

Geologists agree that uplift and subsidence (downward movement) of the continents, the amount of water frozen in glaciers, and rates of seafloor spreading are responsible for marine transgressions and regressions. During uplift of a continent, the shoreline moves seaward, and just the opposite takes place during subsidence. Widespread glaciers expanded and contracted during the Pennsylvanian Period (see Chapter 11), which caused several sea level changes and resulted in transgressions and regressions. Indeed, if all of Earth's present-day glacial ice were to melt, sea level would rise by about 70 m.

Geologic evidence indicates that sea level may have been as much as 250 m higher during the Cretaceous Period, and as a result, widespread marine transgressions occurred during which large parts of the continents were invaded by the sea (see Figure 14.6). The probable cause of this event was comparatively rapid seafloor spreading during which heat beneath the mid-oceanic ridges caused them to expand and displace water onto the continents. When seafloor spreading slows, the mid-oceanic ridges subside, increasing the volume of the ocean basins, and the seas retreat from the continents.

Fossils and Fossilization

You now know that superposition, cross-cutting relationships, and original horizontality are essential for interpreting geologic history, but consider the situation in • Figure 5.9. For the rocks in either of the two columns, or what geologists call *columnar sections*, the relative ages are apparent from superposition, but what are the ages of the rocks in one column compared to those in the other column? With the data provided, you cannot resolve this problem. The solution to this problem involves using fossils, if present, and some physical events of short duration such as volcanic ash falls.

Fossils, the remains or traces of prehistoric organisms preserved in rocks, are most common in sedimentary rocks, but they may also be found in volcanic ash and volcanic mudflows. Geologists use fossils extensively to

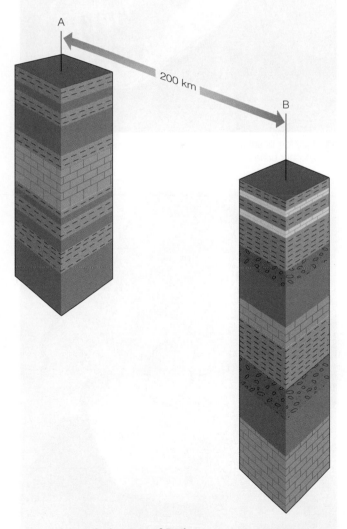

• Figure 5.9 **Relative Ages of Rocks** You can determine the relative ages of rocks in either column A or B, but you cannot tell the relative ages of rocks in A compared to B. The rocks in A may be younger than those in B, the same age as some in B, or older than all of the rocks in B.

determine the relative ages of strata, but fossils also provide useful information for determining environments of deposition (see Chapter 6), and they constitute some of the evidence for the theory of evolution (see Chapter 7). In short, fossils are essential to fully decipher Earth and life history.

Today, it is apparent that bones, teeth, and shells in rocks are the remains of once-living creatures, yet this view is rather recent. Indeed, during most of historic time in the Western world, people variously believed fossils were inorganic objects formed within rocks by some kind of molding force, or even objects placed in rocks by the Creator to test our faith or by Satan to sow seeds of doubt. Some perceptive observers—such as Leonardo da Vinci in 1508, Robert Hook in 1665, and Nicolas Steno in 1667—recognized the true nature of fossils, but their views were largely ignored. By the 18th and 19th centuries, though, it was apparent that fossils were truly the remains of organisms, and it was also clear that many fossils were of organisms now extinct.

How Do Fossils Form? Our definition of *fossil* includes the term "remains," or what are called **body fossils** (• Figure 5.10a and b), consisting mostly of skeletal parts

• Figure 5.10 **Body Fossils and Trace Fossils**

a Body fossils are actual remains of organisms, such as these dinosaur bones.

c These bird tracks are one kind of trace fossil. This slab of rock formed over the actual tracks, so it is a cast of the tracks.

b Shells of invertebrate marine animals such as this ammonite known as *Parapuzosia seppenradensis* are also body fossils. This specimen measures about 1.0 m in diameter.

d Fossilized feces (coprolite) of a carnivorous mammal. The specimen is about 5 cm long and it contains small fragments of bones and teeth.

such as shells, bones, and teeth. However, under some exceptional circumstances soft parts may be preserved by freezing or mummification. In contrast, **trace fossils** are not actual remains but an indication of organic activity such as tracks, trails, burrows, and nests (Figure 5.10c and d). The trace fossil known as a coprolite is fossilized feces that may provide information about the diet and size of the animal that produced it.

The most favorable conditions for preserving body fossils are that an organism has a durable skeleton and lives where rapid burial is likely. Consequently, corals, clams, and brachiopods have good fossil records because they meet these criteria, whereas jellyfish with no skeletons have a poor record. Bats have skeletons but they are delicate and they live where burial is not so likely, so they too have a poor fossil record. Even given these qualifications and the fact that bacterial decay, scavenging, and metamorphism destroy organic remains, fossils of some organisms are common.

For any organism, death is certain but preservation as a fossil is rare. Nevertheless, fossils are more common than most people realize because so many billions of creatures have lived during so many millions of years. Indeed, at many localities you can collect hundreds of fossils of corals and brachiopods, and even the fragmented bones and teeth of dinosaurs are common in some areas. It is true that for fossilization to occur, shells or bones must be buried rapidly in some protective medium such as sand or mud, but rapidly means only before remains are destroyed by physical and chemical processes, which might take many years in some environments.

Thousands of fossils have been recovered from the La Brea Tar Pits in Los Angeles, California; a bone bed in Canada has the remains of hundreds of horned dinosaurs; and hundreds of rhinoceroses, three-toed horses, saber-toothed deer and other animals are in Miocene-age ash deposits in Nebraska. Such phenomenal concentrations of fossils might seem to be unrelated to any everyday process with which we are familiar, but we know of many instances of the first stages of fossilization taking place now. For example, small animals today are trapped in the sticky reside at oil seeps at the La Brea Tar Pits, and wildebeests by the hundreds drown in river crossings, probably much like the horned dinosaurs in Canada did millions of years ago. Burial in volcanic ash should come as no surprise because that is exactly what has happened during recent eruptions. In short, just as with the record of past events preserved in rocks we use the principle of uniformitarianism to determine how fossils were buried and preserved.

Body fossils may be preserved as *unaltered remains* (• Figure 5.11), meaning they retain their original composition and structure, or as *altered remains* (• Figure 5.12), in which case some change has taken place in their composition and/or structure. (Table 5.1). In addition, **molds** and **casts** are common in the fossil record (• Figure 5.13). A mold forms when buried organic remains dissolve, leaving a cavity shaped like a clam or bone, for example. If minerals or sediment should later fill the cavity, it forms a cast—that is, a replica of the original.

The *fossil record,* the record of ancient life, just as the geologic record of which it is a part, must be analyzed and interpreted. As a repository of prehistoric organisms, this record provides our only knowledge of such extinct animals as trilobites and dinosaurs. Furthermore, the study of fossils has several practical applications that we will discuss in the following sections and in Chapter 6.

• Figure 5.11 **Unaltered Remains** Unaltered means that there has been no change in composition or structure of the organism.

a Insects preserved in amber, which is hardened tree resin.

b Frozen baby mammoth found in Russia in 1989. The 44,000-year-old carcass is about 1.0 m tall.

• Figure 5.12 Altered Remains of Organisms

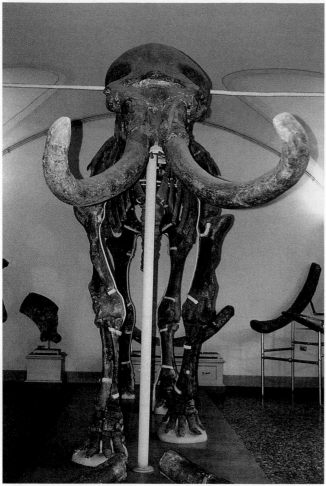

a The bones of this mammoth on display at the Museum of Geology and Paleontology in Florence, Italy, have been permineralized, that is minerals have been added to the pores and cavities in the bones.

b Eocene-age carbonized palm frond on display at the Natural History Museum in Vienna, Austria. c This carbonized insect is from Oligocene-age deposits in Montana.

TABLE 5.1	Types of Fossil Preservation
Body fossils—unaltered remains	Original composition and structure preserved
Freezing	Large Ice Age mammals frozen in sediment
Mummification	Air drying and shriveling of soft tissues
Preservation in amber	Leaves, insects, and small reptiles trapped and preserved in hardened tree resin
Preservation in tar	Bones, insects preserved in asphaltlike substance at oil seeps
Body fossils—altered remains	Change in composition and/or structure of original material
Permineralization	Addition of minerals to pores and cavities in shells and bones
Recrystallization	Change in the crystal structure—for example, aragonite in shells recrystallizes as calcite
Replacement	One chemical compound replaces another—for example, pyrite (FeS_2) replaces calcium carbonate ($CaCO_3$) of shells
Carbonization	Volatile elements lost from organic matter leaving a carbon film; most common for leaves and insects
Trace fossils	Any indication of organic activity such as tracks, trails, burrows, droppings (coprolites), and nests
Molds and casts	Mold—a cavity with the shape of a bone or shell; cast—a mold filled by minerals or sediment

• Figure 5.13 **Origin of a Mold and a Cast**

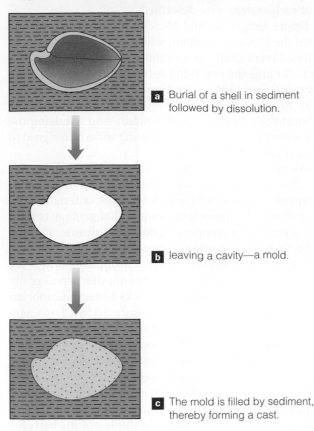

a Burial of a shell in sediment followed by dissolution.

b leaving a cavity—a mold.

c The mold is filled by sediment, thereby forming a cast.

Fossils and Telling Time

We began the section on fossils by referring to the columnar sections in Figure 5.9 and asking how you might determine the relative ages of the strata in these geographically separate areas. Our answer was "fossils"—but exactly how do fossils resolve this problem? To fully understand the usefulness of fossils, we must examine the historical development of an important geologic principle.

The use of fossils in relative dating and geologic mapping was demonstrated during the early 1800s in England and France. William Smith (1769–1839), an English civil engineer who was surveying and building canals in southern England, independently discovered Steno's principle of superposition. He reasoned that in a sequence of strata, the oldest is at the bottom and the youngest is at the top, and he came to the same conclusion regarding any fossils the rocks contained. Smith made numerous observations at outcrops, mines, and quarries and discovered that the sequence of fossils, and especially groups of fossils, is consistent from area to area. In short, he discovered that the relative ages of sedimentary rocks at different locations could be determined by their fossil content (• Figure 5.14).

By recognizing the relationship between strata and fossils, Smith could predict the order in which fossils would appear in rocks at some locality he had not previously visited. In addition, his knowledge of rocks and fossils allowed him to predict the best route for a canal and

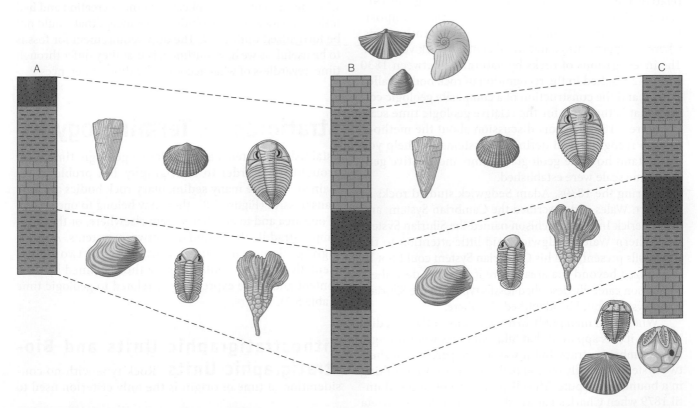

• **Figure 5.14 Applying the Principle of Fossil Succession** This generalized cross section shows how geologists use the principle of fossil succession to determine the relative ages of rocks in widely separated areas. The rocks encompassed by the dashed lines contain similar fossils and are thus the same age. Note that the youngest rocks are in column B, whereas the oldest ones are in column C.

the best areas for bridge foundations. As a result, his services were in great demand.

Smith gets much of the credit for the idea of using fossils to determine relative ages, but other geologists, such as Alexander Brongniart in France, also recognized this relationship. In any case, their observations served as the basis for what we now call the **principle of fossil succession** (also known as the *principle* or *law of faunal succession*). This important principle holds that fossil assemblages (groups of fossils) succeed one another through time in a regular and determinable order.

But why not simply match up similar rock types in Figure 5.9 and conclude they are the same age? Rock type will not work because the same kind of rock—sandstone, for instance—has formed repeatedly through time. Fossils have also formed continuously through time, but because different organisms existed at different times, fossil assemblages are unique. In short, an assemblage of fossils has a distinctive aspect compared with younger or older fossil assemblages. Accordingly, if we match up rocks containing similar fossils, we can assume they are of the same relative age.

The Relative Geologic Time Scale

In Chapter 4 we noted that the 18th-century Neptunists thought they could use composition to determine the relative ages of rocks (see Table 4.1). During the 19th century, however, it became apparent that composition is insufficient to make relative age determinations, whereas superposition and fossil content work very well. The investigations of rocks by naturalists between 1830 and 1842 resulted in the recognition of rock bodies called *systems* and the construction of a composite geologic column that is the basis for the relative geologic time scale (• Figure 5.15a). A short discussion about the methods used to recognize and define the systems will help you understand how the geologic column and relative geologic time scale were established.

During the 1830s, Adam Sedgwick studied rocks in northern Wales and described the Cambrian System, and Sir Roderick Impey Murchison named the Silurian System in southern Wales. Sedgwick paid little attention to the few fossils present, and his Cambrian System could not be recognized beyond the area where it was first described. Murchison carefully described fossils typical of the Silurian, so his system could be identified elsewhere.

When both men published the results of their studies in 1835, it was apparent that Silurian rocks were younger than Cambrian strata, but it was also apparent that their two systems partially overlapped (Figure 5.15b), resulting in a boundary dispute. This dispute was not resolved until 1879 when Charles Lapworth suggested that the strata in the area of overlap be assigned to a new system, the Ordovician. So finally three systems were named, their

stratigraphic positions were known, and distinctive fossils of each system were described.

Before Sedgwick's and Murchison's dispute, they had named the Devonian System, which contained fossils distinctly different from those in the underlying Silurian System rocks and the overlying rocks of the Carboniferous System.* Then in 1841, Murchison visited western Russia, where he identified strata as Silurian, Devonian, and Carboniferous by their fossil content. And overlying the Carboniferous strata were fossil-bearing rocks he assigned to a Permian System (Figure 5.15a).

We need not discuss the specifics of where and when the other systems were established, except to say that superposition and fossil content were the criteria used to define them. The important point is that geologists were piecing together a composite **geologic column,** which is in effect a **relative geologic time scale,** because the systems are arranged in their correct chronologic order (see Figure 1.11). You should be aware of another aspect of the relative geologic time scale: All rocks beneath Cambrian strata are called Precambrian (Figure 5.15a). Long ago, geologists realized that these rocks contain few fossils and that they could not effectively subdivide them into systems. In Chapters 8 and 9, we will discuss the terminology for Precambrian rocks and time.

It is important to realize that although scientists today accept that organic evolution accounts for the differences in organisms through time, evolution is not the basis for the geologic time scale. Indeed, the geologic column and time scale were pieced together by the 1840s by scientists who fully accepted the biblical account of creation and had little or no idea about evolution—a concept that would not be formalized until 1859. The only requirement for fossils to be useful, as we have outlined, is that they differ through time, regardless of what accounts for those differences.

Stratigraphic Terminology

Establishing systems and a relative geologic time scale brought some order to stratigraphy, but problems remained. Because many sedimentary rock bodies are time transgressive (Figure 5.8), they may belong to one system in one area and to another system elsewhere, or they may simply straddle the boundary between systems. Accordingly, geologists now use terminology for two fundamentally different kinds of units: those defined by their content and those expressing or related to geologic time (Table 5.2).

Lithostratigraphic Units and Biostratigraphic Units
Rock type with no consideration of time of origin is the only criterion used to

---

*In North America, two systems, the Mississippian and Pennsylvanian, closely correspond to the Lower and Upper Carboniferous, respectively.

• Figure 5.15 **The Geologic Column and Relative Geologic Time Scale**

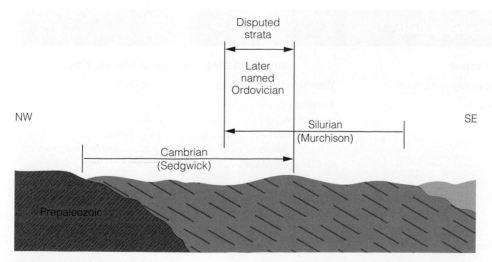

Era	Period	Epoch	
Cenozoic	Quaternary	Recent	
		Pleistocene	1.8
	Tertiary	Pliocene	
		Miocene	
		Oligocene	
		Eocene	
		Paleocene	66
Mesozoic	Cretaceous		146
	Jurassic		200
	Triassic		251
Paleozoic	Permian		299
	Carboniferous	Pennsylvanian	318
		Mississippian	359
	Devonian		416
	Silurian		444
	Ordovician		488
	Cambrian		542
	Precambrian		4600

Map labels:

0 400 km
0 400 mi

Cambrian (Sedgwick, 1835)

Carboniferous (Coneybeare and Phillips, 1822)

Ordovician (Lapworth, 1879)

Silurian (Murchison, 1835)

Permian Basin 1,609 km

Permian (Murchison, 1841)

Devonian (Murchison and Sedgwick, 1840)

Triassic (Von Alberti, 1834)

Jurassic (Von Humboldt, 1795)

Quaternary (Desnoyers, 1829)

Cretaceous (d'Halloy, 1822)

Tertiary (Arduino, 1760)

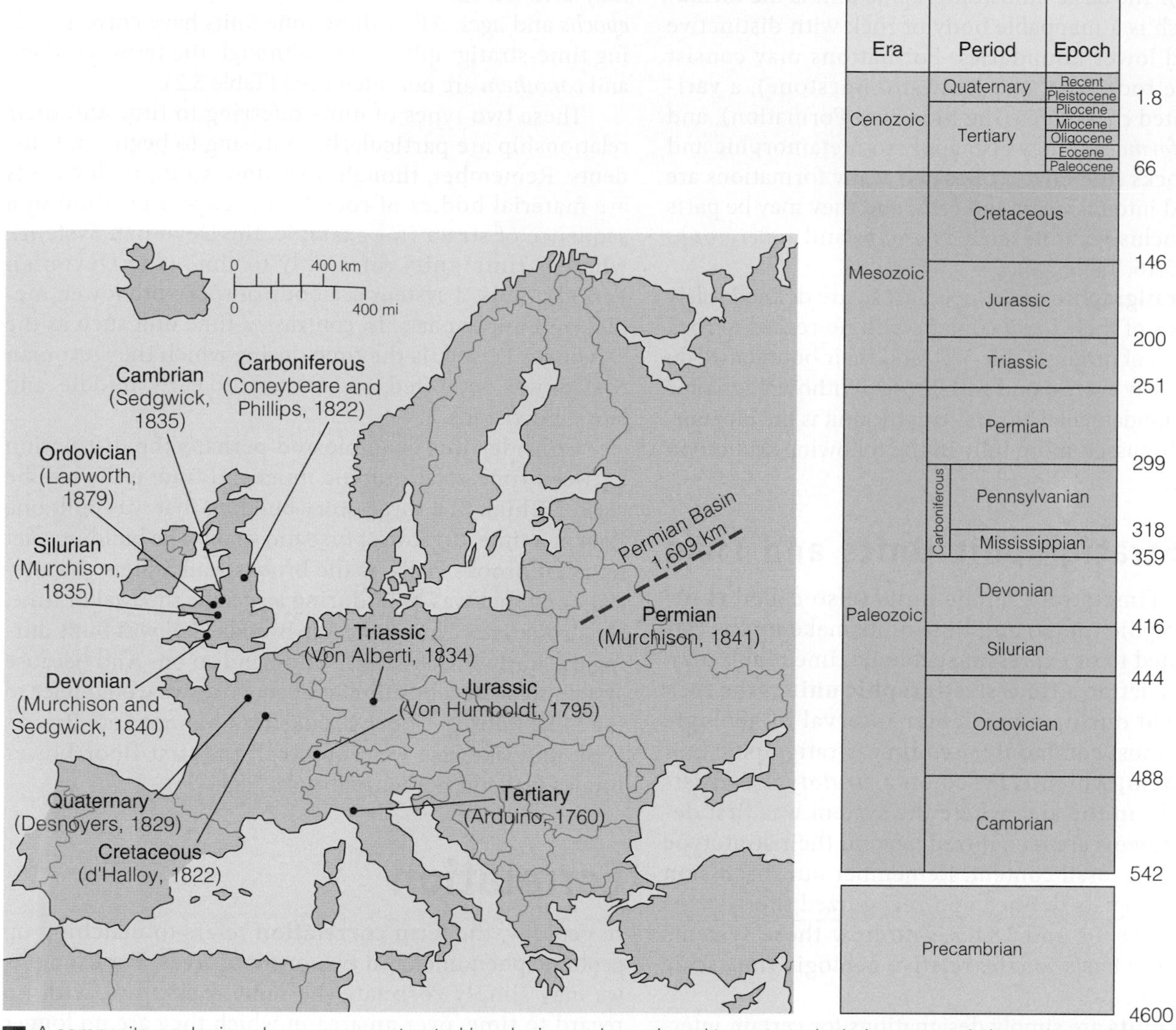

a A composite geologic column was constructed by placing rocks assigned to systems in their correct relative order, so the column is a relative geologic time scale. The absolute ages were added much later. Notice that the Cenozoic consists of the Tertiary and Quarternary periods, which was the terminology used then.

Diagram **b** labels:

Disputed strata

Later named Ordovician

NW SE

Silurian (Murchison)

Cambrian (Sedgwick)

Prepaleozoic

b Adam Sedgwick and Roderick Murchison established the Cambrian and Silurian systems, but as shown here the two systems overlapped. In 1879 Charles Lapworth porposed assigning the area of overlap to the Ordovician System.

define a **lithostratigraphic**[1] **unit**, which has unifying features that set it apart from other lithostratigraphic units (Table 5.2). The basic lithostratigraphic unit is the **formation,** which is a mappable body of rock with distinctive upper and lower boundaries. Formations may consist of a single rock type (the Redwall Limestone), a variety of related rock types (the Morrison Formation), and the term *formation* may also apply to metamorphic and igneous rocks (the Lovejoy Basalt). Many formations are subdivided into *members* and *beds*, and they may be parts of more inclusive units such as *groups* and *supergroups* (• Figure 5.16, Table 5.2).

Biostratigraphic units, in contrast, are defined solely on the basis of their fossil content with no regard to rock type or time of origin (Table 5.2), and their boundaries do not necessarily correspond with those of lithostratigraphic units. The fundamental biostratigraphic unit is the **biozone,** which is discussed more fully in the following section on correlation.

Time-Stratigraphic Units and Time Units

Time-stratigraphic units (also called chronostratigraphic[2] units) and time units make up the category related to or expressing geologic time (Table 5.2). Geologists define a **time-stratigraphic unit** as the rock that formed during a particular interval of geologic time. The most commonly used time-stratigraphic unit is the **system,** which is based on a *stratotype* consisting of rocks in the area where the system was first described. Systems are recognized beyond their stratotype areas by their fossil content. Remember our discussion on how geologists defined and recognized the systems during the 1830s and 1840s, and how these systems served as the basis for the relative geologic time scale (Figure 5.15).

Time units are simply designations for certain intervals of geologic time. The **period** is the most commonly used time unit, but two or more periods may be designated as an *era*, and two or more eras make up an *eon*. Periods may also be subdivided into shorter intervals such as *epochs* and *ages*. All of these time units have corresponding time-stratigraphic units, although the terms *erathem* and *eonothem* are not often used (Table 5.2).

These two types of units referring to time and their relationship are particularly confusing to beginning students. Remember, though, that time-stratigraphic units are material bodies of rock that occupy a position in a sequence of strata (for example, the Devonian System), whereas time units refer only to time (the Devonian Period). Thus, a system is a body of rock with lower, medial, and upper parts. In contrast, a time unit such as the Devonian Period, is the time during which the Devonian System was deposited, and we refer to early, middle, and late subdivisions.

Consider this example and perhaps the distinction between time-stratigraphic units and time units will be clearer. Think of a three-story building that was built one floor at a time, the lowest first and so on. The building—let us call it Brooks Hall—is the Brooks Hall System, a material body that was built during a specific interval of time, the Brook Hall Period. Lower Brooks Hall was built during the Early Brooks Hall Period and so on. And because system refers to a position in a sequence, we would refer to the third floor as Upper Brooks Hall but not Late Brooks Hall, and likewise we would call the first floor Lower Brooks Hall but not Early Brooks Hall.

Correlation

In geology, the term **correlation** refers to matching up geologic phenomena in two or more areas. For example, we may simply correlate the same rock units, with no regard to time, over an area in which they are no longer continuous, in which case we refer to *lithostratigraphic*

TABLE **5.2**	Classification of Stratigraphic Units		
Units Defined by Content		**Units Expressing or Related to Geologic Time**	
Lithostratigraphic Units	**Biostratigraphic Units**	**Time-Stratigraphic Units**	**Time Units**
Supergroup	Biozone	Eonothem . . . . Eon	
Group		Erathem . . . . Era	
Formation		System . . . . Period	
Member		Series . . . . Epoch	
Bed		Stage . . . . Age	

---

[1]*Lith* and *litho* mean "stone" or "stonelike."
[2]*Chron* and *chrono* are added to words to indicate time.

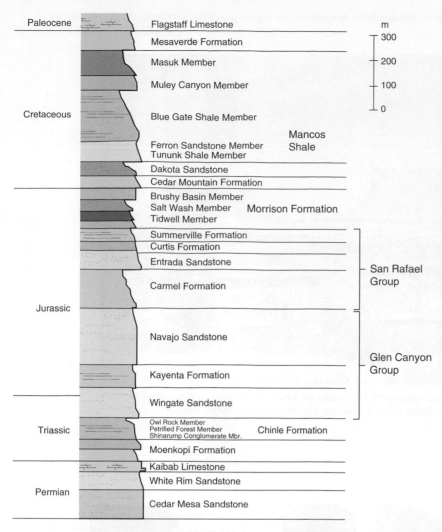

• **Figure 5.16 Graphic Representation of the Lithostratigraphic Units in Capital Reef National Park in Utah** Notice that some of the formations are further divided into members, and some are parts of more inclusive groups.

correlation. In a previous section, we noted that systems were identified beyond their stratotype areas by applying the principle of fossil succession. In this case, we match up rocks of the same relative age, which is a *time-stratigraphic correlation.*

Correlation of lithostratigraphic units involves demonstrating that a formation or group was once continuous over a particular area. If outcrops are adequate, lithostratigraphic units may be traced laterally even if occasional gaps are present (• Figure 5.17a). Other criteria for lithostratigraphic correlation include the presence of a distinctive key bed, position in a sequence, and composition (Figure 5.17b and c); keep in mind that composition indicates only the geographic extent of a rock unit (see Perspective). Rock cores and well cuttings from drilling operations as well as geophysical data are useful for correlating rock units below the surface.

Many sedimentary rock units, such as widespread formations, are time transgressive, so we cannot rely on lithostratigraphic correlation to demonstrate time equivalence. For example, geologists have accurately correlated sandstone in Arizona with similar rocks in Colorado and South Dakota, but the age of the rocks varies from Early Cambrian in the west to Middle Cambrian further east. The most effective way to demonstrate the time-stratigraphic equivalence of sedimentary rocks in different areas involves using biozones, but several other methods are useful too.

The existence of all organisms now extinct marks two points in time—time of origin and time of extinction. One type of biozone, known as the **range zone,** is defined as the geologic range (total time of existence) of a particular fossil group, such as a species or a group of related species called a genus. The most useful fossils are those that are easily identified, geographically widespread, and that had a rather short geologic range. *Lingula* is a type of brachiopod that meets the first two of these criteria, but its geologic range makes it of little use, whereas the brachiopod *Atrypa* and the trilobite *Paradoxides* are **guide fossils** because they are well suited for time-stratigraphic correlation (• Figure 5.18).

Geologists identify several types of *interval zones* which are defined by the first and last occurrence of a species or genus, but one of the most useful is the **concurrent range zone** that is established by plotting the overlapping ranges of two or more fossils with different ranges (• Figure 5.19). Correlating concurrent range zones is probably the most accurate method to determine time equivalence between sedimentary rocks in widely separated areas.

Some physical events are also used to demonstrate time equivalence. Remember that rock composition in most cases is useless in this endeavor, but a few rocks, particularly lava flows and ash falls, form rapidly over their entire extent. Accordingly, correlating an ash fall, for example, is time significant (• Figure 5.20). Furthermore, ash falls are not restricted to a specific environment, so they may extend from continental to marine environments.

The methods of time-stratigraphic correlation work well for Phanerozoic-age rocks, but Precambrian rocks have few fossils and those present are not very useful for correlation. For these rocks absolute ages based on radioactive dating techniques are most useful.

• Figure 5.17 **Lithostratigraphic Correlation** This kind of correlation demonstrates the original geographic continuity of rock unit, such as a formation or group.

a In areas of adequate exposures, geologists trace rocks on the skyline laterally and correlate them even if occasional gaps exist. These rocks in Utah once covered a much larger area.

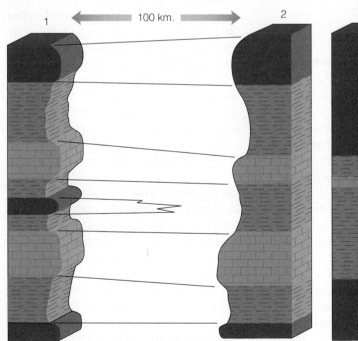

b Lithostratigraphic correlation by similarities in rock type and position in a sequence. We assume the middle sandstone in column 1 intertongues or grades laterally into the shale in column 2.

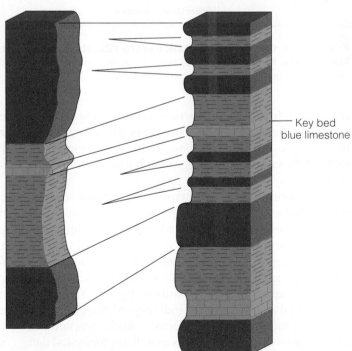

c Lithostratigraphic correlation using a key bed, a distinctive blue limestone in this case.

Key bed
blue limestone

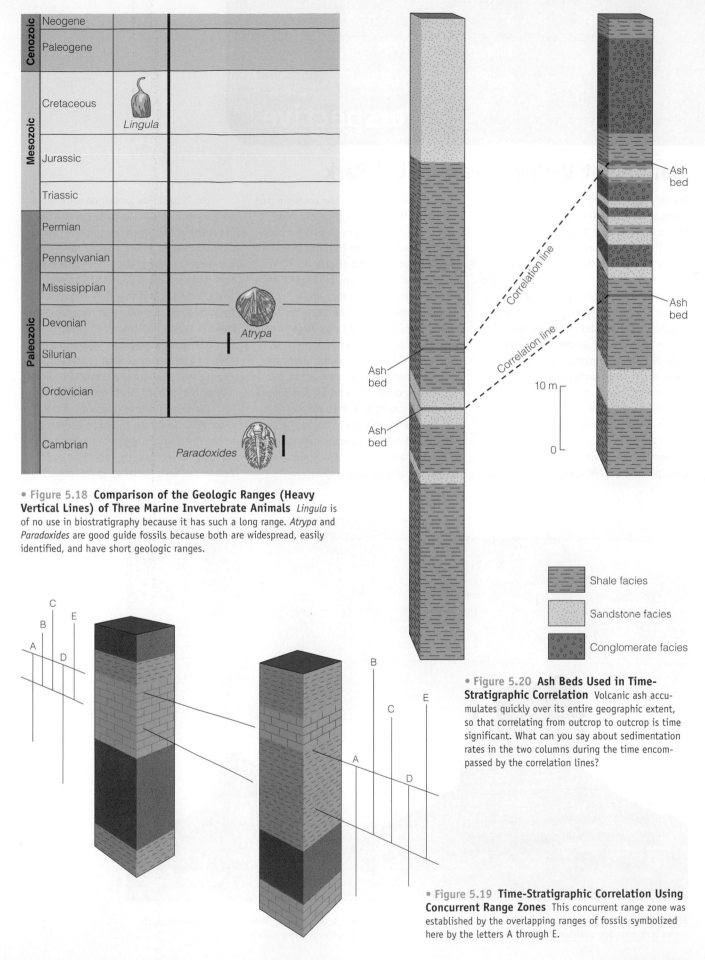

• Figure 5.18 **Comparison of the Geologic Ranges (Heavy Vertical Lines) of Three Marine Invertebrate Animals** *Lingula* is of no use in biostratigraphy because it has such a long range. *Atrypa* and *Paradoxides* are good guide fossils because both are widespread, easily identified, and have short geologic ranges.

Ash bed
Ash bed
Correlation line
Correlation line

10 m
0

Ash bed
Ash bed

Shale facies
Sandstone facies
Conglomerate facies

• Figure 5.20 **Ash Beds Used in Time-Stratigraphic Correlation** Volcanic ash accumulates quickly over its entire geographic extent, so that correlating from outcrop to outcrop is time significant. What can you say about sedimentation rates in the two columns during the time encompassed by the correlation lines?

• Figure 5.19 **Time-Stratigraphic Correlation Using Concurrent Range Zones** This concurrent range zone was established by the overlapping ranges of fossils symbolized here by the letters A through E.

Perspective

Monument Valley Navajo Tribal Park

Tsé Bii' Ndzisgaii, meaning Valley of the Rocks according to the Navajo, is better known as Monument Valley Navajo Tribal Park. It is not a valley at all but rather a broad flat part of the Colorado Plateau where deep erosion has yielded some of the most magnificent scenery in the Southwest (Figure 1, chapter opening photo). The Navajo Nation administers the park, which lies on the Arizona-Utah border and is accessible by a single unpaved road. You can see many of the park's geologic features from the main highway in the area, but a visit to the park itself is well worth the time. Not only is it a scenic marvel that has served as the backdrop for many television shows and movies, especially westerns, it is also a good place to experience geology up close.

A notable feature in the southern part of the park is El Capitan (Agathla Peak), a 457-m-high volcanic neck that once led to a much larger volcanic edifice that has long since been eroded (Figure 2). However, the main attractions in the park are the numerous mesas, buttes, spires, and hoodoos formed by erosion along vertical joints (fractures) in the Cedar Mesa Sandstone that once covered the entire area (Figure 1). The terms *mesa* and *butte* are used interchangeably in many areas, but we use mesa to designate a broad, flat-topped erosional remnant of a resistant rock layer, whereas a butte is more pinnacle-like. However, very slender erosional remnants go by several names such as spire, pinnacle, monument, and splinter (Figure 3).

The Cedar Mesa Sandstone is the most obvious rock unit in the park, but also present is the underlying Halgaito Shale, which forms the slopes at the bases of the mesas and buttes. In many places the Cedar Mesa Sandstone is capped by the Organ Rock Shale, although it is very thin in most areas. Detailed studies of the Cedar Mesa Sandstone indicate that it was deposited mostly as a vast blanket of wind-blown sand dunes that covered a large area of what is now northern Arizona and southern Utah. Small amounts of iron oxides account for its red color.

The original lateral continuity of the Cedar Mesa Sandstone should be obvious even to beginners in geology, because it is easy to match up or correlate the rocks from one outcrop to the next. Matching up rocks based on

Figure 1 These buttes in Monument Valley Navajo Tribal Park are West Mitten (on the left) and East Mitten. The rock making up the buttes is Permian-age sandstone, whereas the rock forming the slopes beneath the buttes is softer shale and sandstone.

Figure 2 The 25-million-year-old volcanic neck of Agatha Peak rises 457 m above the desert floor. It is composed of tuff breccia that solidified in the conduit leading to a large volcano that has since been eroded.

Figure 3 Just as the buttes in Figure 1, the Totem Pole (foreground) and Yie Bi Chei are erosional remnants also made up of sandstone and they formed in the same way. Slender features such as these go by a variety of names, but here we will simply refer to them as pinnacles.

their composition and position in a sequence is lithostratigraphic correlation. In fact, about 50 km to the north, the same sequence of rocks is exposed in another scenic area known as Valley of the Gods. Keep in mind, though, that this kind of correlation demonstrates only the original geographic extent of the rocks in question. The age of the rocks, about 270 million years old (Permian), was established using completely different criteria.

Monument Valley Navajo Tribal Park is scenic and in an area where you can see other features of geologic interest. Furthermore, its well exposed, flat-lying rocks are ideal for beginners to understand how erosion yields mesas, buttes, and so on, and to appreciate some of the concepts in stratigraphy.

Absolute Dates and the Relative Geologic Time Scale

For about 150 years, geologists have known that Ordovician rocks are younger than those of the Cambrian and older than Silurian rocks. But how old are they? That is, when did the Ordovician begin and end? Beginners might think it simple to obtain radiometric ages for fossil-bearing Ordovician rocks and thus resolve the problem. Unfortunately, absolute ages determined for minerals in sedimentary rocks give only the age of the source that supplied the minerals, not the age of the rock itself. For instance, sandstone that formed 10 million years ago might contain sand grains 3 billion years old.

One mineral, however does form in some sedimentary rocks shortly after deposition, and it can be dated by the potassium–argon method (see Chapter 4). This greenish mineral known as *glauconite* easily loses argon (the daughter product), so absolute ages determined are generally too young, giving only a minimum age for the host sedimentary rock. In most cases, absolute ages for sedimentary rocks must be determined indirectly by dating associated igneous and metamorphic rocks.

According to the principle of cross-cutting relationships, a dike must be younger than the rock it cuts, so an absolute age for a dike gives a minimum age for the host rock and a maximum age for any overlying rocks (Figure 5.21a). Absolute dates obtained from regionally metamorphosed rocks also give a maximum age for overlying sedimentary rocks.

Lava flows and ash falls interbedded with sedimentary rocks are the most useful for determining absolute ages (Figure 5.21b). Both are time-equivalent, providing a maximum age for any rocks above and a minimum age for rocks below. Multiple lava flows or ash fall deposits in a stratigraphic sequence are particularly useful in this effort.

Accurate radiometric dates are now available for many lava flows, ash falls, plutons, and metamorphic rocks with associated fossil-bearing sedimentary rocks. Geologists have added these absolute ages to the geologic time scale, so we now can answer questions such as when the Ordovician Period began and ended (Figure 5.15a). In addition, we now know when a particular organism lived. For example, the *Baculites reesidei* biostratigraphic zone in the Bearpaw Formation in Saskatchewan, Canada, is about 72 to 73 million years old because absolute ages have been determined for associated volcanic ash layers (• Figure 5.22).

• Figure 5.21 **Determining the Absolute Ages of Sedimentary Rocks** Absolute ages for sedimentary rocks are most often found by radiometric dating of associated igneous and metamorphic rocks.

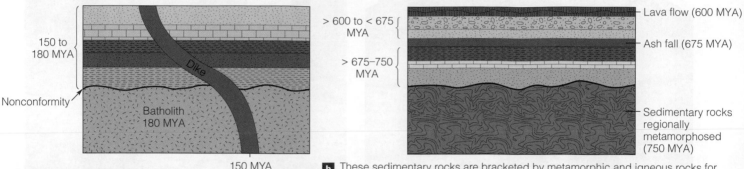

a We know that these sedimentary rocks are younger than the batholith (180 MY) and older than the dike (150 MY).

b These sedimentary rocks are bracketed by metamorphic and igneous rocks for which absolute ages are known.

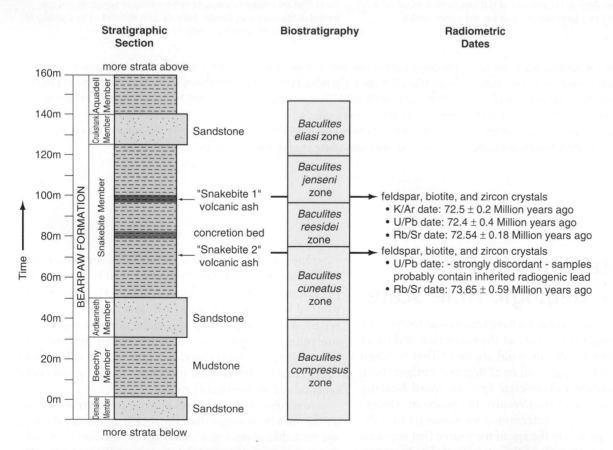

• Figure 5.22 **Rocks and Fossils of the Bearpaw Formation in Saskatchewan, Canada** The column on the left shows formations and members (lithostratigraphic units). Notice that the biozone boundaries do not coincide with lithostratigraphic boundaries. The absolute ages for the two volcanic ash layers indicate that the *Baculites reesidei* zone is about 72 to 73 million years old.

SUMMARY

- The first step in deciphering the geologic history of a region involves determining the correct relative ages of the rocks present. Thus, the vertical relationships among rock layers must be ascertained even if they have been complexly deformed.

- The geologic record is an accurate chronicle of ancient events, but at any single location it has discontinuities or unconformities representing times of nondeposition, erosion, or both.

- Simultaneous deposition in adjacent but different environments yields sedimentary facies, which are bodies of sediment, or sedimentary rock, with distinctive lithologic and biologic attributes.

- According to Walther's law, the facies in a conformable vertical sequence replace one another laterally.

- During a marine transgression, a vertical sequence of facies results with offshore facies superposed over nearshore facies. Just the opposite facies sequence results from a marine regression.

- Uplift and subsidence of continents, the amount of water frozen in glaciers, and the rate of seafloor spreading are responsible for marine transgressions and regressions.

- Most fossils are found in sedimentary rocks, although they might also be in volcanic ash and volcanic mudflows, but rarely in other rocks.

- Fossils of some organisms are common, but the fossil record is strongly biased toward those organisms that have durable skeletons and that lived where burial was likely.

- The work of William Smith, among others, is the basis for the principle of fossil succession that holds that fossil assemblages succeed one another through time in a predictable order.

- Superposition and fossil succession were used to piece together a composite geologic column, which was the basis for the relative geologic time scale.

- To bring order to stratigraphic terminology, geologists recognize units based on content (lithostratigraphic and biostratigraphic units) and those related to time (time-stratigraphic and time units).

- Lithostratigraphic correlation involves demonstrating the original continuity of a rock unit over a given area even though it may not now be continuous over this area.

- Correlation of biostratigraphic zones, especially concurrent range zones, demonstrates that rocks in different areas, even though they may differ in composition, are of the same relative age.

- The best way to determine absolute ages of sedimentary rocks and their contained fossils is to obtain absolute ages for associated igneous rocks and metamorphic rocks.

IMPORTANT TERMS

angular unconformity, p. 89
biostratigraphic unit, p. 100
biozone, p. 100
body fossil, p. 94
cast, p. 95
concurrent range zone, p. 101
conformable, p. 89
correlation, p. 100
disconformity, p. 89
formation, p. 100
fossil, p. 93

geologic column, p. 98
geologic record, p. 86
guide fossil, p. 101
lithostratigraphic unit, p. 100
marine regression, p. 92
marine transgression, p. 91
mold, p. 95
nonconformity, p. 89
period, p. 100
principle of fossil succession, p. 98
principle of inclusions, p. 87

range zone, p. 101
relative geologic time scale, p. 98
sedimentary facies, p. 91
stratigraphy, p. 87
system, p. 100
time-stratigraphic unit, p. 100
time unit, p. 100
trace fossil, p. 95
unconformity, p. 89
Walther's law, p. 92

REVIEW QUESTIONS

1. The most commonly used time-stratigraphic unit is the
   a. _____ system; b. _____ period; c. _____ epoch; d. _____ member; e. _____ formation.

2. The idea that the facies seen in a vertical sequence replace one another laterally is
   a. _____ Smith's dictum; b. _____ Walter's law; c. _____ Steno's principle; d. _____ Gressly's model; e. _____ Darwin's theory.

3. A group is a lithostratigraphic unit made up of
   a. _____ at least three periods; b. _____ four systems; c. _____ a disconformity and a nonconformity; d. _____ biozones and members; e. _____ two or more formations.

4. According to the principle of fossil succession
   a. _____ a dike is older than the sedimentary rock it cuts through; b. _____ time-stratigraphic units are defined by rock type; c. _____ fossil assemblages succeed one another in a regular and predictable order; d. _____ a marine regression takes place when the sea rises and invades a continent; e. _____ the geologic column and time scale are based on the theory of evolution.

5. Which one of the following terms is used to describe an erosion surface cut into metamorphic rocks that is covered by sedimentary rocks?
   a. _____ Stratigraphic break; b. _____ Guide fossil; c. _____ Geologic anomaly; d. _____ Nonconformity; e. _____ Sedimentary facies.

6. Which one of the following statements is not correct?
   a. _____ Among other things, a guide fossil must be geographically widespread; b. _____ An era consists of two or more periods; c. _____ Biozone boundaries do not necessarily coincide with lithostratigraphic boundaries;

d. _____ Most fossils are found in igneous and metamorphic rocks; e. _____ Offshore facies are superposed on the nearshore facies during a marine transgression.

7. The principle of inclusions holds that
a. _____ all aspects of the fossil record are important in deciphering Earth history; b. _____ Walter's law applies only to conformable sequences of strata; c. _____ fragments in a layer of rock are older than the layer itself; d. _____ an eon is made up of two or more eras; e. _____ concurrent range zones are useful in time-stratigraphic correlation.

8. The geologic column and relative geologic time scale were established by the 1840s based on
a. _____ the theory of evolution; b. _____ the principle of unconformities; c. _____ superposition and faunal succession; d. _____ lithostratigraphic biozones; e. _____ the rate of radioactive decay.

9. Geologists use the principle of superposition to determine
a. _____ how long ago a fossil organism lived; b. _____ the duration of a marine regression; c. _____ absolute ages for geologic events; d. _____ whether fossil remains have been altered; e. _____ the relative ages of rocks in a vertical sequence.

10. Which one of the following is a trace fossil?
a. _____ clam shell; b. _____ dinosaur bone; c. _____ worm burrow; d. _____ disconformity; e. _____ biozone.

11. How were the principles of superposition and faunal succession used to piece together the relative geologic time scale?

12. You encounter an outcrop where you see a layer of sandstone overlying granite. What kinds of evidence would allow you to determine which of the two rocks is youngest?

13. What kinds of features would you see in the geologic record to indicate that a marine regression had taken place?

14. How do time units and time-stratigraphic units compare?

15. How does the lithostratigraphic unit differ from a time-stratigraphic unit?

16. Refer to Figure 5.9. If you correlate the limestone (blue) in the left column with the lower limestone in the right column, can you assume they are of the same relative age? Explain.

17. What criteria are necessary for a fossil to be a good guide fossil?

18. How can you use the principle of uniformitarianism to interpret events that no one witnessed, such as the origin of an ancient wind-blown dune deposit?

19. Unconformities are common in the geologic record, and some are found over large areas, but none of them are worldwide. Why?

20. Explain how the principle of uniformitarianism helps us understand how fossils were buried and preserved.

APPLY YOUR KNOWLEDGE

1. You are among the first astronauts to visit an Earth-like planet that has widespread fossil-bearing sedimentary rocks including conglomerate, sandstone, mudstone, and limestone. How would you go about establishing a geologic column and relative geologic time scale? Also, how could you determine the absolute ages of the fossil-bearing rocks?

2. While visiting one of our national parks, you observe the following sequence of rocks; marine fossil-bearing limestone, shale, and sandstone, all inclined at 50 degrees. Lying over these rocks is a horizontal layer of mudstone baked by basalt overlying it. Finally, at the top of the sequence is a sandstone with inclusions of basalt. Decipher the geologic history of this area.

3. Suppose that the upper ash layer in Figure 5.20 formed 1.7 million years ago and the lower one is 1.9 million years old. What was the average sedimentation rate in cm/yr for the interval between the ash beds for the column on the right? Do you think this average is accurate for the actual rate of sedimentation? Explain.

4. In the image below showing light-colored granite and dark-colored basalt, which type of rock is oldest and how do you know?

Sue Monroe

Copyright and photograph by Dr. Parvinder S. Sethi

CHAPTER 6

SEDIMENTARY ROCKS—THE ARCHIVES OF EARTH HISTORY

View of Delicate Arch in the Jurassic-age Entrada Sandstone in Arches National Park near Moab, Utah. The park is famous for its arches, pillars, and spires, all of which formed by weathering and erosion along large fractures called joints. At other locations in the park there are collapsed arches as well as arches in the earliest stages of formation.

[OUTLINE]

Introduction

Sedimentary Rock Properties
 Composition and Texture
 Sedimentary Structures
 Geometry of Sedimentary Rocks
 Fossils—The Biologic Content of
 Sedimentary Rocks

Depositional Environments
 Continental Environments
 Transitional Environments
 Marine Environments

Perspective Evaporites—What We Know
and Don't Know

Interpreting Depositional Environments

Paleogeography

Summary

At the end of this chapter, you will have learned that

- Sedimentary rocks have a special place in deciphering Earth history because they preserve evidence of surface processes responsible for deposition, and many contain fossils.

- Geologists recognize three broad areas of deposition—continental, transitional, and marine—each with several specific depositional environments.

- The distinctive attributes of sedimentary rocks result from processes operating in specific depositional environments.

- Textures such as sorting and rounding provide evidence about depositional processes.

- Three-dimensional geometry of sedimentary rock bodies and fossils are also useful for determining depositional environments.

- Sedimentary structures such as ripple marks, cross-bedding, and mud cracks taken with other rock features allow geologists to make environmental interpretations with a high degree of confidence.

- The interpretations of sedimentary rocks in the chapters on geologic history are based on the considerations reviewed in this chapter.

- Interpretation of how and where rocks formed, especially sedimentary rocks, is the basis for determining Earth's ancient geographic features.

Introduction

The Jurassic-age Navajo Sandstone of the southwestern United States was deposited as a vast blanket of sand dunes (• Figure 6.1), whereas the Cambrian-age Muav Limestone in Arizona was deposited in a warm, shallow sea during a marine transgression (see Figure 5.7). As we noted previously, no one witnessed these events, so how do geologists make such determinations? In the last chapter, we noted that geologists use the principle of uniformitarianism to decipher events that no one witnessed, which is exactly how they have determined the circumstances under which these rock formations were deposited.

We also mentioned previously that sedimentary rocks have a special place in deciphering Earth history, but we should also emphasize that all rocks are important in this endeavor. Certainly igneous rocks help geologists identify ancient convergent plate boundaries and areas of ancient volcanism, and most of our absolute dates for sedimentary

rocks come from associated igneous rocks (see Chapter 5). Metamorphic rocks tell us something of the temperatures and pressures that prevailed in Earth's crust, especially where plates collided and mountains formed.

Nevertheless, our main emphasis in this chapter is on sedimentary rocks because (1) they preserve evidence of surface processes responsible for deposition of sediment by running water, wind, waves, glaciers, and so on, and (2) many contain fossils, our only record of prehistoric life. So, we focus on areas or environments where sediment accumulates—that is, *depositional environments*—as well as the criteria used to recognize deposits of specific depositional environments. Accordingly, geologists examine the distinctive properties of sedimentary rocks, keeping in mind that the processes responsible for these features are the same as those operating now (Figure 6.1).

Sedimentary rocks are also important in any discussion of geology because many are resources—sand and gravel used in construction, for example—or they contain resources such as petroleum and natural gas. Coal is certainly an important resource, most of which is used at electrical-power generating plants, and phosphorus-rich sedimentary rocks are used in fertilizers, animal feed supplements, metallurgy, matches, ceramics, and preserved foods. Furthermore, a type of sedimentary rock known as banded iron formation is the main source of the world's iron ores (see Chapter 9).

Sue Monroe

• **Figure 6.1 Jurassic-age Navajo Sandstone in Zion National Park, Utah** Geologists conclude that the Navajo Sandstone was deposited as a vast blanket of sand dunes based on several features of these rocks as well as comparisons with present-day deposits. Much of this chapter covers the criteria geologists use to make such determinations. This rock exposure is called Checkerboard Mesa, so named because of the pattern formed where vertical fractures intersect the layering in the rocks.

Sedimentary Rock Properties

The first step in investigating sedimentary rocks—or any other rock type, for that matter—is observation and data gathering by visiting rock exposures (outcrops) and carefully examining textures, composition, fossils (if present), thickness, and relationships to other rocks. During these initial field studies, geologists may make some preliminary interpretations. For example, color is a useful feature of some sedimentary rocks: Red rocks likely were deposited on land (see the chapter opening photo), whereas greenish ones more likely were deposited in a marine environment. However, exceptions are numerous, so color must be used with caution.

After completing fieldwork, geologists study the rocks more carefully by microscopic examination, chemical analyses, fossil identification, and construction of diagrams that show vertical and lateral facies relationships. A particularly important aspect of these studies is to compare the features of sedimentary rocks with those in present-day sediment accumulations that formed in known depositional environments. When all the data have been analyzed, geologists make an environmental interpretation.

Determining how sedimentary rocks were deposited and examining their fossils satisfies our curiosity about Earth and life history. If you live in Kansas you might be interested to know that a widespread shallow sea populated by shelled animals known as ammonites as well as sharks and huge marine reptiles were present there during the Cretaceous Period. In addition, there are economic reasons to determine environments of deposition in exploration for mineral deposits, petroleum, and natural gas.

Composition and Texture
Detrital sedimentary rocks contain more than 100 minerals, but only quartz, feldspars, and clay minerals are very common. Rock composition depends mostly on the composition of the rocks in the area from which the detrital sediment was derived (the source area), but it tells little about how deposition took place. Quartz, for example, may have been deposited in a stream channel, in desert dunes, or on a beach, so composition taken alone is not very useful for environmental determinations.

Among the chemical sedimentary rocks, limestone (composed of calcite [$CaCO_3$]) and dolostone (composed of dolomite [$CaMg(CO_3)_2$]) are the most common, and both are usually deposited mostly in warm, shallow seas, but a small amount forms in lakes. Evaporites such as rock salt (composed of halite [$NaCl$]) and rock gypsum (made up of gypsum [$CaSO_4 \cdot 2H_2O$]*) invariably indicate arid environments where evaporation rates are high.

Grain size in detrital sedimentary rocks tells us something of the conditions of transport and deposition. Conglomerate is made up of gravel, so we can be sure that it was transported by energetic processes such as swiftly flowing streams, by waves, or by glaciers. Sand also requires high-energy transport and tends to be deposited in stream channels, desert dunes, and on beaches, but silt and clay are transported by weak currents and are deposited in low-energy environments such as lagoons, lakes, and river floodplains.

Texture refers to the size, size distribution, shape, and arrangement of clasts in detrital sedimentary rocks. Recall from Chapter 2 that gravel, sand, silt, and clay are simply size designations for detrital particles (see Figure 2.13). The degree to which detrital particles have had their sharp edges and corners smoothed off by abrasion is called **rounding** (• Figure 6.2a). Gravel tends to become rounded very quickly

---
*Gypsum ($CaSO_4 \cdot 2H_2O$) is the common sulfate mineral precipitated from seawater, but when deeply buried, gypsum loses its water and is converted to anhydrite ($CaSO_4$).

• Figure 6.2 **Rounding and Sorting in Sediments**

a Beginners often mistake rounding to mean spherical or ball-shaped. All of these stones are rounded—that is, their sharp corners and edges have been worn smooth—but only the one in the upper left is spherical.

b Sorting in this deposit is only moderate, but all of the stones are well rounded.

c Angular, poorly sorted gravel.

as particles collide with one another, and with considerable transport, sand is also rounded, but smaller particles are carried suspended in water and usually are not so well rounded.

Another textural feature is **sorting**, which refers to the size variation in a sedimentary deposit or rock. If most of the particles are of about the same size, the sediment or rock is well sorted, but if a wide range of sizes is present, the material is poorly sorted (Figure 6.2c). Wind has a limited capacity to transport and deposit sediment so its deposits tend to be well sorted, whereas glaciers transport anything supplied to them and their deposits are poorly sorted.

Sedimentary Structures
Most sedimentary rocks have features known as **sedimentary structures** that formed during deposition or shortly thereafter. All are manifestations of the physical and biological processes that take place in depositional environments, so analyses of these structures is the single most important aspect of sedimentary rocks for determining how deposition took place.* The origin of sedimentary structures is well known because geologists see them forming today and many have been produced experimentally (Table 6.1).

With very few exceptions, sedimentary rocks have a layered aspect called **stratification** or **bedding** (see the chapter opening photo and ● Figure 6.3); layers less than 1 cm thick

TABLE 6.1	Summary of Sedimentary Structures	
Physical Sedimentary Structures (produced by processes such as currents)	**Characteristics**	**Origin**
Laminations (or laminae)	Layers less than 1 cm thick	Form mostly as particles settle from suspension
Beds	Layers more than 1 cm thick	Form as particles settle from suspension and from moving sediment as sand in a stream channel
Graded-bedding	Individual layers with an upward decrease in grain size	Deposition by turbidity currents or during the waning stages of floods
Cross-bedding	Layers deposited at an angle to the surface on which they accumulated	Deposition on a sloping surface at the downwind side of a sand dune
Ripple marks	Small (<3 cm high) ridges and troughs on bedding planes	
Current ripple marks	Asymmetric ripple marks	Result from deposition by water or air currents flowing in one direction
Wave-formed ripple marks	Symmetric ripple marks; generally with a sharp crest and broad though	Formed by oscillating currents (waves)
Mud cracks	Intersecting cracks in clay-rich sediments	Drying and shrinkage of mud along a lakeshore, a floodplain, or on tidal flats
Biogenic Sedimentary Structures (produced by organisms)		
Trace fossils	Tracks, trails, tubes, and burrows	Indications of organic activity. Intense activity results in *bioturbation* involving disruption of sediment

● **Figure 6.3 Stratification in Sedimentary Rock**

a These rocks in Valley of the Gods in Utah are layered or stratified. They were originally continuous over this entire area but have been deeply eroded.

b Laminations, layers less than 1 cm thick, in the chemical sedimentary rock called banded iron formation in Michigan. The pencil measures about 14 cm long.

*There are structures that form long after deposition, mostly by chemical processes that are of no use for determining depositional environments.

are la*minations*, whereas *beds* are thicker. Laminations are most common in the mudrocks and form as silt and clay settle from suspension, but they are also found in sandstones and limestones. However, coarser grained rocks, sandstone and conglomerate, as well as many limestones are frequently bedded. The surfaces separating one bed from another are *bedding planes*, above and below which the rocks differ in composition, texture, color, or a combination of these features.

Some of the deposition that takes place in the seas marginal to continents results from *turbidity currents*, which are sediment–water mixtures that move along the seafloor because they are denser than seawater. When the flow velocity in these currents diminishes, they deposit large particles followed by progressively smaller ones thus forming **graded bedding** in which the grain size decreases upward (• Figure 6.4). Graded bedding is also found in some stream channels where flow velocity decreases rapidly as during flash floods.

Deposition that takes place on the downwind, sloping side of a sand dune or a similar feature in a stream channel produces **cross-bedding,** which accumulates at an angle to the surface on which deposition occurs (Table 6.1, • Figure 6.5). Transport and deposition by either wind or water produces cross-bedding, and individual cross beds

• Figure 6.4 **Turbidity Currents and Graded Bedding**

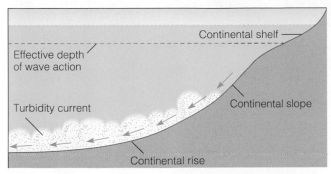

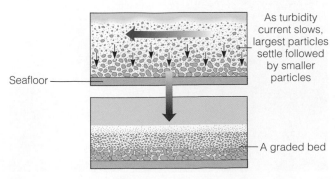

a Sediment–water mixtures called turbidity currents flow along the seafloor (or lake bottom) because they are more dense than sediment-free water.

b When a turbidity current slows, graded bedding forms as deposition of smaller and smaller particles takes place.

• Figure 6.5 **Cross-Bedding** Cross-bedding forms when individual beds or strata are deposited at an angle to the surface upon which they accumulate.

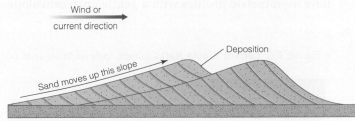

a Origin of cross-bedding by deposition on the sloping surface of a desert dune. Cross-bedding also forms in dune-like structures in stream and river channels and on the continental shelf.

b Horizontal bedding and cross-bedding in the Upper Cambrian St. Peter Sandstone at Wisconsin Dells, Wisconsin. About 10 m of strata are visible in this image.

are inclined downward or dip in the direction of flow. Accordingly, cross-bedding indicates the direction of flow of ancient currents.

Small-scale alternating ridges and troughs common on bedding planes, especially in sand, are **ripple marks** (Table 6.1). Ripple marks, or simply ripples, form either by unidirectional flow of wind or water, and they form by the to and fro motion of waves. *Current* ripples, those formed by the flow of wind or water, have asymmetric profiles with a gentle upstream slope and a seep downstream slope (• Figure 6.6). Because of their asymmetry they indicate the original flow direction. *Wave-formed ripple marks* tend to have symmetrical profiles (• Figure 6.7).

• **Figure: 6.6 Current Ripple Marks** Current ripple marks are small (<3 cm high) sedimentary structures with an asymmetric profile.

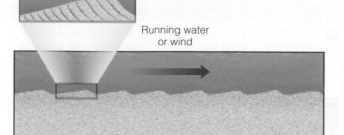

Running water or wind

a Current ripple marks form where water or wind flows in one direction over sand. The enlargement shows the internal cross-bedding in one ripple mark.

b Current ripple marks that formed in a stream channel. Which way did the current flow?

c Current ripple marks formed by wind.

• **Figure 6.7 Wave-Formed Ripple Marks** Just as with current ripple marks, these are small-scale sedimentary structures, but they tend to have symmetric profiles.

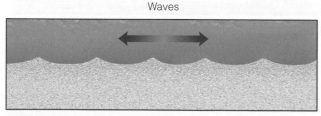

Waves

a Wave-formed ripples form where waves move to and fro.

b Wave-formed ripples in shallow seawater.

Clay-rich sediments tend to shrink as they dry and crack into polygonal forms called **mud cracks** (Table 6.1, • Figure 6.8). If you see mud cracks in ancient rocks you can be sure that deposition took place in an environment where alternating wetting and drying took place, such as along a lakeshore, on a river floodplain, or where mud is exposed at low tide along a seashore.

The sedimentary structures discussed so far all form by physical processes, but biological processes yield **biogenic sedimentary structures** and include tracks, trails, and burrows (Table 6.1, • Figure 6.9). Remember from Chapter 5 that these biologically produced features are also called *trace fossils*. Extensive burrowing by organisms, or what is known as **bioturbation,** may so thoroughly disrupt sediments that other structures are destroyed.

Sedimentary structures are important for environmental analyses, but be aware that no single structure is unique to a specific environment—current ripples can form in desert dunes, in stream channels, and in tidal channels in the sea, for example. However, associations of sedimentary structures taken along with other sedimentary rock properties allow geologists to make environmental interpretations with a high degree of confidence.

We noted in Chapter 5 that to decipher the geologic history of any area geologists use the principle of superposition to determine the relative ages of sedimentary rocks, but if the rocks have been deformed the task is more difficult. Several sedimentary structures as well as some fossils allow geologists to resolved these kinds of problems (• Figure 6.10).

• **Figure 6.8 Mud Cracks Form in Clay-Rich Sediments When They Dry and Contract**

James S. Monroe

a Mud cracks in a present-day environment.

Alan L. Mayo

b Ancient mud cracks in Glacier National Park, Montana. Notice that the cracks have been filled in by sediment.

• **Figure 6.9 Bioturbation Results from Organisms Burrowing Through Sediments**

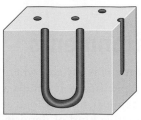

a U-shaped burrows

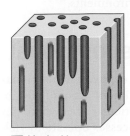

b Vertical burrows

Sue Monroe

c The vertical dark-colored areas in this rock are sediment-filled burrows

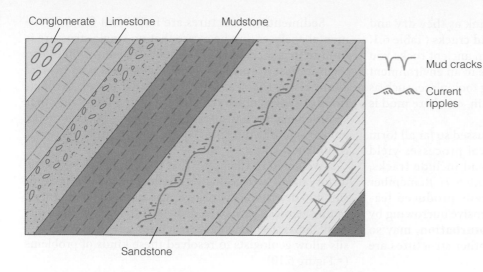

Conglomerate Limestone   Mudstone

∨∨ Mud cracks

∼∼ Current ripples

Sandstone

• Figure 6.10 **Using Sedimentary Structures to Determine the Relative Ages of Deformed Sedimentary Rocks** According to the principle of original horizontality, these strata were not deposited in their present position but rather deposited more or less horizontally and then deformed. To determine which layer is oldest and youngest notice the sedimentary structures. The mud cracks form a "V" that opens up toward the younger strata, and the shape of the current ripple marks indicate that the youngest layer is the one in the lower right.

Geometry of Sedimentary Rocks The three-dimensional shape, or geometry, of a sedimentary rock body may be helpful in environmental analyses, but it must be used with caution because the same geometry can be found in more than one environment. Moreover, geometry can be modified by sediment compaction during lithification, and by erosion and deformation. Nevertheless, it is useful when considered in conjunction with other features.

Some of the most extensive sedimentary rocks in the geologic record resulted from marine transgressions and regressions (see Figure 5.8). These rocks cover hundreds or thousands of square kilometers but are perhaps only a few tens to hundreds of meters thick. That is, they are not very thick compared to their dimensions of length and width and thus, have a *blanket* or *sheet geometry.*

Some sand deposits have an *elongate* or *shoestring geometry,* especially those deposited in stream channels or barrier islands. Delta deposits tend to be lens shaped when viewed in cross profile or long profile, but lobate when observed from above. Buried reefs are irregular, but many are long and narrow, although rather circular ones are known, too.

Fossils—The Biologic Content of Sedimentary Rocks We defined *fossils* as the remains or traces of prehistoric organisms, and we discussed how geologists use fossils in some aspects of stratigraphy—to establish biostratigraphic units, for example (see Chapter 5). Fossils are also important constituents of some rocks, especially limestones that may be composed largely of shells of marine-dwelling animals such as brachiopods, clams, and corals (see Figure 2.15a), or even the droppings (pellets) of these organisms. Fossils are not present in all sedimentary rocks, but if they are, they are important for determining depositional environments.

We must consider two factors when using fossils in environmental analyses. First, did the organisms in question live where they were buried, or were their remains transported there? Fossil dinosaurs, for example, usually indicate deposition in some land environment such as a river floodplain, but if their bones are found in rocks with clams, corals, and sea lilies, we must assume a carcass was washed out to sea. Second, what kind of habitat did the organisms originally occupy? Studies of a fossil's structure and its living relatives, if any, are helpful. Clams with heavy, thick shells typically live in shallow, turbulent water, whereas those with thin shells are found in low-energy environments. Most corals live in warm, clear, shallow marine environments where symbiotic bacteria can carry out photosynthesis.

Microfossils are particularly useful because geologists can recover many individuals from small rock samples. In oil-drilling operations, small rock chips called *well cuttings* are brought to the surface. These cuttings rarely contain complete fossils of large organisms, but they may have thousands of microfossils that aid in relative age determinations and environmental analyses. Even some trace fossils, which of course are not transported from their place of origin, are also characteristic of particular environments.

Depositional Environments

We defined a **depositional environment** as any area where sediment accumulates, but more specifically it entails a particular area where physical, chemical, and biological processes operate to yield a distinctive kind of deposit. Geologists recognize three broad areas of deposition—*continental, transitional,* and *marine*—each of which has several specific environments (• Figure 6.11).

Continental Environments Deposition on the continents—that is, on land—takes place in fluvial

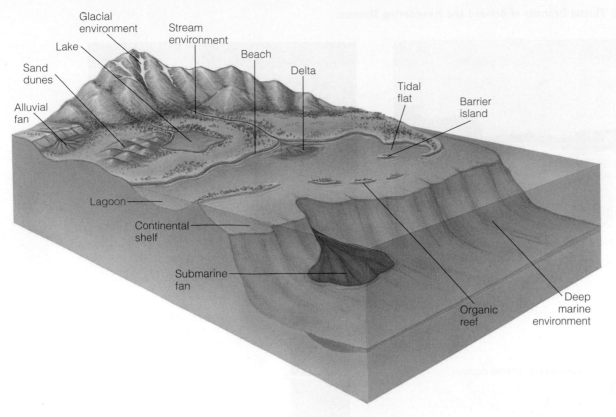

• **Figure 6.11 Depositional Environments** Continental environments are shown in red type. The environments along the seashore, shown in blue type, are transitional from continental to marine. The others, shown in black type, are marine environments.

systems (rivers and streams), lakes, deserts, and areas covered by or adjacent to glaciers (Table 6.2). The sedimentary deposits in each of these environments show combinations of features that allow us to determine that they were in fact deposited on land and to distinguish one from the other.

The term **fluvial** refers to river and stream activity and to their deposits. These fluvial deposits accumulate in either braided stream systems or meandering stream systems. **Braided streams** have multiple broad, shallow channels in which mostly sheets of gravel and cross-bedded sand are deposited, and mud is conspicuous by its near absence (• Figure 6.12a and b). In contrast, **meandering streams** have a single sinuous channel. They deposit mostly fine-grained sediments on floodplains but do have cross-bedded sand bodies, each with a shoestring geometry (Figure 6.12c and d). One of the most distinctive features of meandering streams is *point bar* deposits consisting of a sand body overlying an erosion surface that developed on the convex side of a meander loop (Figure 6.12c).

Desert environments are commonly inferred from an association of features found in **sand dune, alluvial fan,** and **playa lake** deposits (• Figure 6.13). Alluvial fans form best along the margins of desert basins where streams and debris flows discharge from mountains onto a valley floor and form a triangular deposit of sand

and gravel (Figure 6.13a). Wind-blown dunes are typically composed of well-sorted, well-rounded sand with cross-beds meters to tens of meters high (Figure 6.13b). And, of course, any fossils are those of land-dwelling plants and animals. The more central part of a desert basin might be the site of a temporary lake (playa lake), in which laminated mud and evaporites accumulate (Figure 6.13c).

All sediment deposited in glacial environments is collectively called **drift,** but we must distinguish between two kinds of drift. **Till** is poorly sorted, nonstratified drift deposited directly by glacial ice, mostly in ridgelike deposits called *moraines* (• Figure 6.14). A second type of drift, called **outwash,** is mostly sand and gravel deposited by braided streams issuing from melting glaciers. The association of these deposits along with scratched (striated) and polished bedrock indicates that glaciers were responsible for deposition. Another feature that may be helpful is glacial lake deposits showing alternating dark and light laminations; each dark–light couplet is a **varve** (Figure 6.14d). Varves represent yearly accumulations of sediment—the light layers form in spring and summer and consist of silt and clay, whereas the dark layers consist of clay and organic matter that settled when the lake froze over. Dropstones liberated from icebergs might also be present in glacial lake deposits (Figure 6.14d).

• Figure 6.12 **Fluvial Deposits of Braided and Meandering Streams**

a The White River in South Dakota has sand bars that divide its channel into several parts, so it is a braided stream.

c This stream in Yellowstone National Park in Wyoming has a single, sinuous channel, so it is a meandering stream.

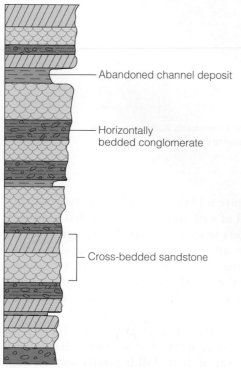

- Abandoned channel deposit
- Horizontally bedded conglomerate
- Cross-bedded sandstone

b Braided stream deposits are mostly gravel and cross-bedded sand with subordinate mud.

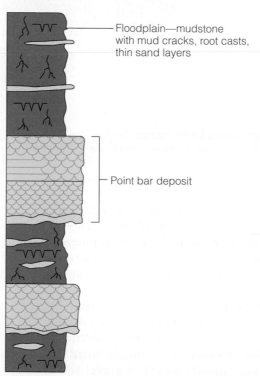

Floodplain—mudstone with mud cracks, root casts, thin sand layers

Point bar deposit

d The deposits of meandering streams are mostly fine-grained floodplain deposits with subordinate sand bodies.

• Figure 6.13 **Deposits in Deserts**

a Huge alluvial fan along the base of the Panamint Range in Death Valley, California.

b Sand dunes also in Death Valley

c A playa lake near Fallon, Nevada. Notice the deposits of rock salt that were forming when this photo was taken in June 2007.

Environment	Dominant Type of Rocks
Continental Environments	
Fluvial	
Braided stream	Mostly horizontally bedded conglomerate and cross-bedded sandstone; mudrocks not common
Meandering stream	Mostly mudrocks deposited on floodplains; subordinate but distinctive lenticular sandstones deposited in point bars
Desert	
Alluvial fan	Poorly sorted conglomerate from debris flows and sandstone- conglomerate-filled channels
Sand dune	Well sorted, rounded sandstone with large-scale cross-beds
Playa lake	Laminated mudstone/siltstone; evaporites, rock salt, rock gypsum, and others
Glacial	
Outwash	Much like braided stream deposits
Moraines	Unsorted, nonstratified deposits of sand and gravel
Transitional Environments	
Delta (marine)	Mudrocks and sandstone in coarsening upward sequences; associated rocks of marine origin; fossils of marine and land-dwelling organisms
Beach	Rounded sandstone with variable soring, commonly with shells or shell fragments, wave-formed ripple marks, and small-scale cross-bedding
Barrier Island	
Beach	As above
Sand dunes	Much like desert dunes but with sand-sized shell fragments
Tidal flat	Mudstone and sandstone in fining-upward sequences; distinctive herringbone cross-bedding in sandstone
Marine Environments	
Continental shelf	
Inner Shelf	Mostly cross-bedded sandstone with wave-formed ripples, marine fossils, and bioturbation
Outer Shelf	Mostly mudrocks with subordinate sandstone; marine fossils and bioturbation
Continental slope and rise	Turbidite sequences in submarine fans with graded bedding in sandstone and mudrocks
Carbonate shelf	Limestone (dolostone). Limestone varies from coquina (made of shell fragments) to oolitic limestone to micrite (carbonate mud). Cross-beds, mud cracks, ripple marks common; marine fossils.
Deep-ocean basin	Pelagic clay and calcareous and siliceous oozes.
Evaporite environments	Rock salt and rock gypsum the most common, but others including potassium and magnesium salts may be present.

Transitional Environments Transitional environments include those in which both marine processes and processes typical of continental environments operate (Figure 6.11, Table 6.2). For instance, deposition where a river or stream (fluvial system) enters the sea yields a body of sediment called a **delta,** but the deposit is modified by marine processes, especially waves and tides. Small deltas in lakes are also common, but the ones along seashores are much larger, far more complex, and many deltas are important economically because they form the reservoir rocks for petroleum and natural gas.

The simplest deltas, those in lakes, have a threefold subdivision of bottomset, foreset, and topset beds much like the one shown in • Figure 6.15. As delta sedimentation takes place, bottomset beds are deposited some distance offshore, over which foreset beds come to rest. And finally, deposition also takes place in stream channels forming the topset beds. As a result, the delta builds seaward (or lakeward), a phenomenon called **progradation** and forms a vertical sequence of rocks that become coarser-grained from bottom to top. In addition, the bottomset beds may contain marine fossils (or fossils of lake-dwelling animals), whereas the topset beds usually have fossils of land plants or animals.

Marine deltas rarely conform precisely to this simple threefold division, because they are strongly influenced by fluvial processes, as well as by waves and tides. Thus, a

• Figure 6.14 **Glaciers and Their Deposits**

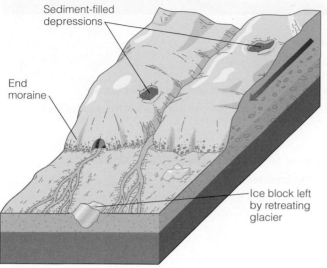

a Glaciers transport huge quantities of sediment that is deposited as moraines and outwash.

c This end moraine is made up of till that shows no sorting or stratification.

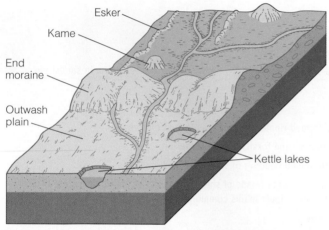

b The same area as shown in **a** but with the ice gone.

d Dropstone in glacial lake deposits showing varves in Canada.

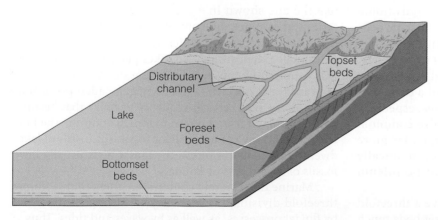

• Figure 6.15 **Origin of a Delta** Surface view and internal structure of the simplest type of prograding delta.

• Figure 6.16 **Marine Deltas**

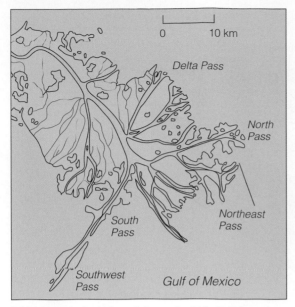

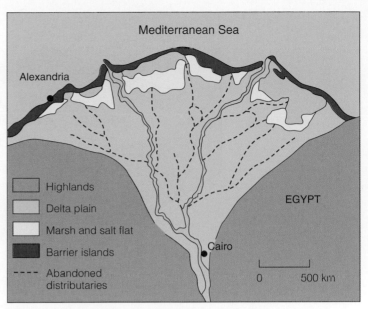

a The Mississippi River delta on the U.S. Gulf Coast is stream-dominated.

b The Nile delta of Egypt is wave-dominated.

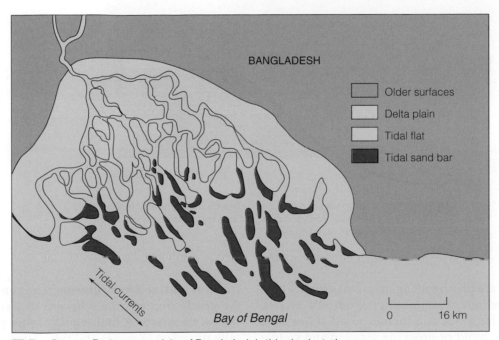

c The Ganges-Brahmaputra delta of Bangladesh is tide-dominated.

stream/river-dominated delta has long distributary channels extending far seaward, as in the Mississippi River delta (• Figure 6.16a). Wave-dominated deltas also have distributary channels, but as their entire seaward margin progrades, it is simultaneously modified by wave action (Figure 6.16b). Tidal sand bodies that parallel the direction of tidal flow are characteristic of tide-dominated deltas (Figure 6.16c).

On broad continental margins with abundant sand, long **barrier islands** lie offshore separated from the mainland by a lagoon. These islands are common along the Gulf Coast, and much of the Atlantic Coast of the United States and many ancient deposits in those areas formed in this environment. Notice from • Figure 6.17a that a barrier island complex has subenvironments in which beach sands grade offshore into finer deposits of the shoreface, dune sands that are differentiated from desert dune sands by the presence of shell fragments, and fine-grained lagoon deposits with marine fossils and bioturbation.

• **Figure 6.17** **Barrier Islands and Tidal Flats**

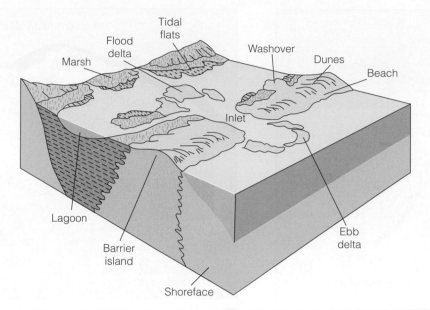

a A barrier island is made up mostly of sand, but it has subenvironments where mud or sand–mud mixtures are deposited.

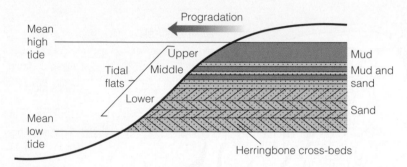

b These tidal flat deposits formed where a shoreline prograded into the sea. Notice the distinctive herringbone cross-beds that dip in opposite directions because of reversing tidal currents.

Tidal flats are present along many coastlines where part of the shoreline environment is periodically covered by seawater at high tide and then exposed at low tide. Many tidal flats build or prograde seaward and yield a sequence of rocks grading upward from sand to mud. One of their most distinctive features is herringbone cross-bedding—sets of cross-beds that dip in opposite directions (Figure 6.17b).

Marine Environments
Marine environments include the continental shelf, slope, and rise, and the deep seafloor (• Figure 6.18, Table 6.2). Much of the detritus eroded from continents is eventually deposited in marine environments, but other types of sediments are found here, too. Much of the limestone in the geologic record, as well as many evaporites, were deposited in shallow marine environments.

Detrital Marine Environments The **continental shelf,** the gently sloping area adjacent to a continent, consists of two parts. A high-energy inner part is periodically stirred up by waves and tidal currents. Its sediment is mostly sand shaped into large cross-bedded dunes, and bedding planes are marked by wave-formed ripple marks. Of course, marine fossils are typical, and so is bioturbation. The low-energy outer part of the shelf has mostly mud with marine fossils, but at the transition between the inner and outer shelf, layers of sand and mud intertongue.

Much of the sediment derived from continents crosses the continental shelf and is funneled into deeper water through submarine canyons. It eventually comes to rest on the **continental slope** and **continental rise** as a series of overlapping submarine fans (Figure 6.18). Once sediment passes the outer margin of the shelf, turbidity currents transport it, so sands with graded bedding are common—but so is mud that settled from seawater (Figure 6.4).

Beyond the continental shelf, the seafloor is nearly covered by fine-grained deposits known as *pelagic clay* and *ooze* (• Figure 6.19). The notable exceptions are the mid-oceanic ridges, which are where new oceanic crust forms. Sand and gravel are notable by their absence on the deep seafloor, because few mechanisms transport coarse-grained sediment far from land. However, ice rafting accounts for a band of sand and gravel adjacent to Greenland and Antarctica.

The main sources of deep-sea sediments are dust blown from continents or oceanic islands, volcanic ash, and shells of microorganisms that dwelled in the ocean's surface waters. One type of sediment called pelagic clay consists of clay covering most of the deeper parts of the seafloor. Calcareous ($CaCO_3$) and siliceous (SiO_2) oozes, in contrast, are made up of microscopic shells (Figure 6.19).

Carbonate Depositional Environments

Limestone and dolostone are the only widespread carbonate rocks, and we already know that most dolostone is altered limestone. Thus, our discussion focuses on the deposition of sediment that, when lithified, becomes limestone.

In some respects limestone is similar to detrital sedimentary rocks—many limestones are made up of gravel- and sand-sized grains, and microcrystalline carbonate mud called *micrite*. However, these materials form in the environment of deposition rather than being transported there. Nevertheless, local currents and waves sort them and account for sedimentary structures such as cross-bedding and ripple marks, and micrite along shorelines commonly shows mud cracks.

Some limestone forms in lakes, but by far most of it was deposited in warm, shallow seas on carbonate shelves (• Figure 6.20). In any case, deposition occurs where little detrital sediment is present, especially mud. Carbonate barriers form in high-energy areas and may be reefs, banks of skeletal particles, or accumulations of spherical carbonate grains known as *ooids*—limestone composed mostly of ooids is called *oolitic limestone* (see Figure 2.15b). Reef rock tends to be structureless, composed of skeletons of corals, mollusks, sponges, and many other organisms, whereas carbonate banks are made up of layers with horizontal beds, cross-beds, and wave-formed ripple marks. The lagoons, however, tend to have micrite with marine fossils and bioturbation.

Evaporite Environments

Evaporites consisting mostly of rock salt and rock gypsum* are found in playa lakes and saline lakes, but most of the extensive deposits formed in the seas. Although locally abundant, evaporites are not nearly as common as sandstone, mudrocks, and limestone. Deposits more than 2 km thick lie beneath the floor of the Mediterranean Sea (see Perspective), and large deposits are also known in Michigan, Ohio, and New York as well as in several Gulf Coast states and western Canada.

Exactly how some of these deposits originated is controversial, but all agree that they formed where evaporation rates were high enough for seawater to become sufficiently concentrated for minerals to precipitate from solution. These conditions are best met in coastal environments in arid regions such as the present-day Persian Gulf. If the area shown in Figure 6.20 were an arid region, inflow of normal seawater into the lagoon would be restricted, and evaporation might lead to increased salinity and evaporite deposition.

---

*Remember that when gypsum ($CaSO_4 \cdot 2H_2O$) is deeply buried, it loses its water and is converted to anhydrite ($CaSO_4$).

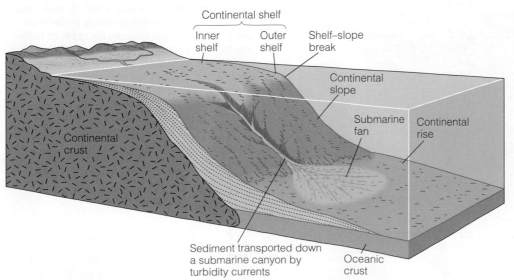

• **Figure 6.18 Depositional Environments on the Continental Shelf, Slope, and Rise** Submarine canyons are the main avenues of sediment transport across the shelf. Turbidity currents carry sediments beyond the shelf, where they accumulate as submarine fans.

Continental shelf
Inner shelf
Outer shelf
Shelf–slope break
Continental slope
Submarine fan
Continental rise
Continental crust
Sediment transported down a submarine canyon by turbidity currents
Oceanic crust

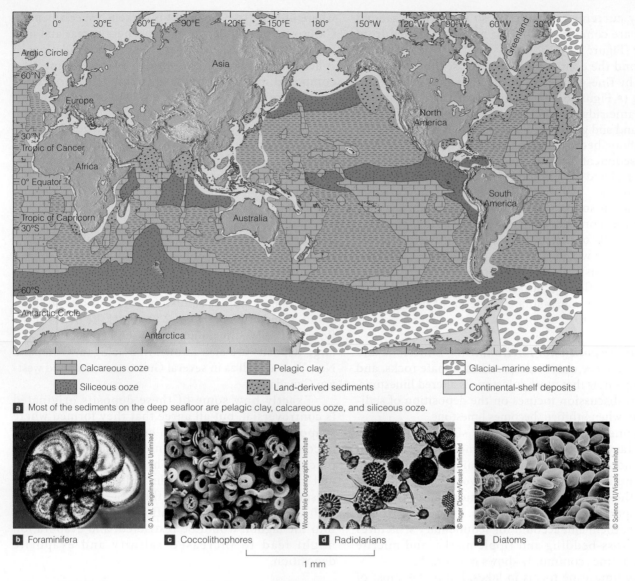

Calcareous ooze   **Pelagic clay**   **Glacial–marine sediments**

Siliceous ooze   **Land-derived sediments**   **Continental-shelf deposits**

a Most of the sediments on the deep seafloor are pelagic clay, calcareous ooze, and siliceous ooze.

b Foraminifera   **c** Coccolithophores   **d** Radiolarians   **e** Diatoms

1 mm

• **Figure 6.19 Sediments on the Deep Seafloor** The particles making up the calcareous ooze are skeletons of **b** foraminifera (floating single-celled animals) and **c** coccolithophores (floating single-celled plants), whereas siliceous ooze is composed of skeletons of **d** radiolarians (single-celled floating animals) and **e** diatoms (single-celled floating plants.)

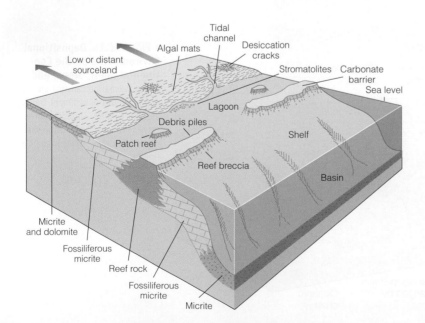

Low or distant sourceland
Algal mats
Tidal channel
Desiccation cracks
Stromatolites
Carbonate barrier
Sea level
Lagoon
Debris piles
Patch reef
Shelf
Reef breccia
Basin
Micrite and dolomite
Fossiliferous micrite
Reef rock
Fossiliferous micrite
Micrite

• **Figure 6.20 Carbonate Depositional Environments** Although some limestone forms in lakes, most of it is deposited on carbonate shelves similar to the one in this illustration. However, the carbonate barrier may be reefs of corals, clams, and algae, or it may be made up of oolitic sand or skeletal-fragment sand. Deposition of limestone in environments much like these is now taking place in southern Florida and the Persian Gulf.

EVAPORITES—WHAT WE KNOW AND DON'T KNOW

Geologists fully understand that for evaporites such as rock gypsum and rock salt to form, water must become concentrated by evaporation until minerals precipitate from solution. They also know of several environments where evaporites are forming now, such as tidal flats in arid regions and playa lakes (Figure 6.13c). Although the origin of these evaporites is well known, they are rather restricted geographically and not found in abundance in the geologic record. There are, however, vast evaporite deposits in Canada, several U.S. states, as well as in a large area in Europe, and beneath the Mediterranean Sea. There is little doubt that these rocks formed in marine environments, but the exact conditions of their deposition have not been fully resolved.

Scientists have known since 1849 that if 1000 m of seawater evaporates, 15 m of evaporites will form. And furthermore, the

first to precipitate from this evaporating solution is limestone ($CaCO_3$), followed by rock gypsum ($CaSO_4 \cdot 2H_2O$), rock salt (NaCl), and finally a few others in very small quantities. Some in this latter category, including magnesium and potassium chlorides, are important resources but not germane to our discussion here. In the chapters on Paleozoic geologic history, we address several evaporite deposits that formed on shallow marine shelves. Here we focus on evaporites that probably formed in deep marine basins, particularly the evaporites beneath the Mediterranean Sea.

If it were not for the connection between the Atlantic Ocean and the Mediterranean Sea at the Strait of Gibraltar, the latter would eventually dry up and become a vast desert lying far below sea level (Figure 1). The reason for this is that the Mediterranean loses more water to evaporation than it gains form rainfall

and river inflow. And yet there are 2.0-km-thick evaporites in the Mediterranean basin that some geologists think were deposited in shallow water rather than in a deep, water-filled basin as the Mediterranean is now. According to one model, the Mediterranean was cut off from the Atlantic and evaporated to near dryness, and in the process evaporites were deposited (Figure 2).

This event, called the Messinian Salinity Crisis, took place during the Late Miocene Epoch, about 6.0 million years ago. If the Mediterranean were to evaporate to dryness now, only about 25 m of evaporites would form, and yet the Messinian evaporites are 2.0 km thick. Accordingly, several periods of isolation of the Mediterranean basin must have alternated with periods during which an oceanic connection was reestablished, perhaps as many as 40 times. When the oceanic connection existed,

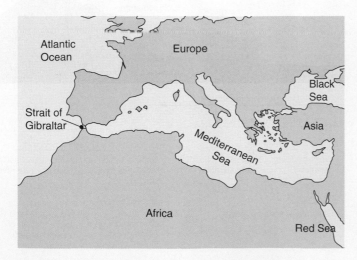

Figure 1 The only connection between the Atlantic Ocean and the Mediterranean Sea is at the Strait of Gibraltar. Because the Mediterranean is in an arid environment where seawater lost to evaporation exceeds additions by rainfall and river runoff, the Mediterranean would evaporate to dryness if this connection were severed.

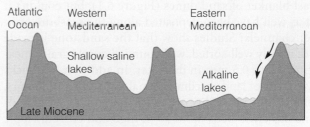

a Isolation of the Mediterranean and evaporation to near dryness.

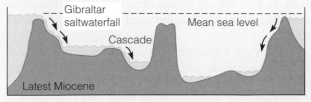

b Refilling stage.

Figure 2 The deep basin, shallow water model for the Mediterranean evaporites is one in which many cycles of isolation and evaporation followed by refilling stages must have occurred.

gravel and sand were deposited around the basin margin, and deep-sea sedimentation took place near the basin center. When the oceanic connection was lost, though, limestone followed by rock gypsum and rock salt was deposited in a body of water that became progressively shallower and more saline.

More evidence for a Mediterranean desert comes from southern European and North African rivers for which the Mediterranean is base level, the lowest level to which they can erode. Beneath these rivers are buried valleys that were deeply eroded into bedrock far below the present level of the sea. For instance, a buried channel lies more than 900 m below the surface of the Rhone delta in Europe, and the Nile River in Egypt has a buried valley that lies 200 m below present sea level 465 km upstream from its terminus. It seems that these rivers eroded deeply when base level was much lower,

and their valleys were later filled when base level rose.

This deep-basin, shallow-water model for the Messinian evaporites is widely but not unanimously accepted. Another idea has the Mediterranean and Atlantic

continuously connected, but because the Mediterranean is in an arid environment it eventually reached the saturation point and precipitation of evaporites took place. So, this alternative is a deep-basin, deep-water model (Figure 3).

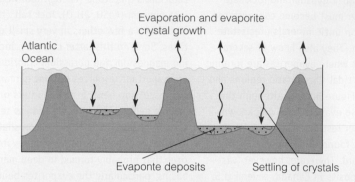

Figure 3 Unlike the model in Figure 2, the deep basin, deep water model for Mediterranean evaporites shown here holds that the Mediterranean was never isolated from the Atlantic Ocean.

Interpreting Depositional Environments

We began this chapter by stating that the Navajo Sandstone of the southwestern United States was deposited as a vast blanket of sand dunes (Figure 6.1), but couldn't it just as well have been deposited along the seashore or in river channels? Studies show that the sandstone is made up of mostly well-sorted, well-rounded quartz grains measuring 0.2 to 0.5 mm in diameter. In addition, the rocks have tracks of land-dwelling animals, including dinosaurs, excluding the possibility of a marine origin, and the sandstone has cross-beds up to 30 m high and current ripple marks like those produced by wind. All the evidence taken together justifies our interpretation of a wind-blown dune deposit. In fact, the cross-beds dip generally southwest, so we can confidently say that the wind blew mostly from the northeast.

Another good example of using the combined features of sedimentary rocks and comparisons with present-day deposits is Lower Silurian strata exposed in New Jersey and Pennsylvania. The features of these rocks are summarized in • Figure 6.21, all of which lead to the conclusion that they were deposited in braided streams that flowed generally from east to west.

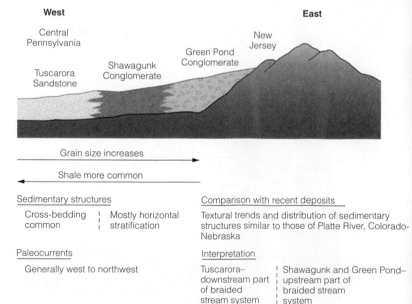

• **Figure 6.21 Lower Silurian Strata in the Eastern United States** Simplified cross section showing the probable lateral relationships for the Green Pond Conglomerate, Shawangunk Conglomerate, and Tuscarora Sandstone, and the criteria used to interpret these rocks as braided stream deposits.

The preceding examples were taken from continental depositional environments, but the same reasoning is used for marine deposits. For example, geologists have evaluated vertical facies relationships, rock types, sedimentary structures, and fossils in Ordovician rocks in Arkansas and conclude that they formed as transgressive shelf carbonate deposits (● Figure 6.22).

In several of the following chapters we refer to fluvial deposits, ancient deltas, carbonate shelf deposits, and others as we examine the geologic record, especially for North America. We cannot include the supporting evidence for all of these interpretations, but we can say that they are based on the kinds of criteria discussed in this chapter.

Paleogeography

As noted, the Navajo Sandstone is a wind-blown desert dune deposit, so we know that during the Jurassic Period a vast desert was present in what is now the southwestern United States (Figure 6.1). We also know that a marine transgression took place from the Proterozoic to Middle Cambrian in North America as the shoreline migrated inland from east and west (see Figure 5.7). In short, Earth's geography—that is, the distribution of its various surface features—has varied through time.

Paleogeography deals with Earth's geography of the past. In the chapters on geologic history, paleogeographic maps show the distribution of continents at various times; they have not remained fixed because of plate movements. The distribution of land and sea constitutes Earth's first-order features, but paleogeography also applies on a more local scale. For example, detailed studies of rocks in several western states allow us to determine with some accuracy how the area appeared during the Late Cretaceous (● Figure 6.23a). According to this paleogeographic map, a broad coastal plain sloped gently eastward from a mountainous region to the sea. The rocks also tell us that many animals lived here, including dinosaurs and

mammals. Later, however, vast lakes, river floodplains, and alluvial fans covered much of this area, and the sea had withdrawn from the continent (Figure 6.23b).

● Figure 6.22 **Middle and Upper Ordovician Strata in Northern Arkansas**

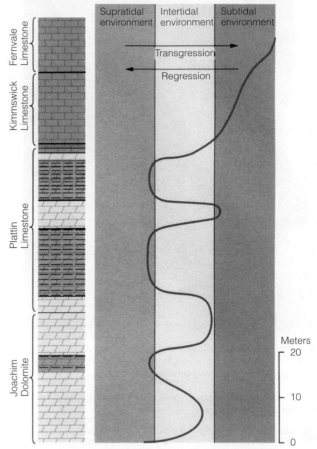

a Vertical stratigraphic relationships and inferred environments of deposition. Note that the trend was transgressive even though several regressions also occurred.

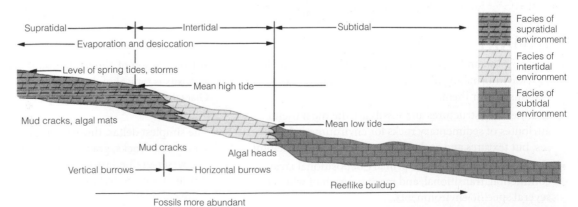

b Simplified cross section showing lateral facies relationships. Inferring these lateral relationships from a vertical sequence of facies is a good example of the application of Walther's law (see Chapter 5).

• Figure 6.23 **Paleogeographic Maps**

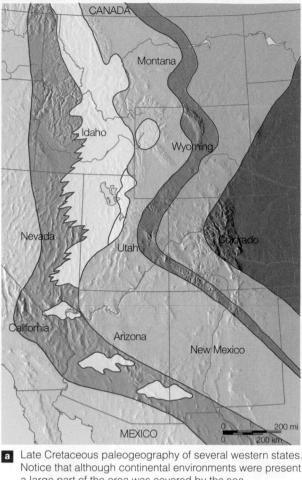

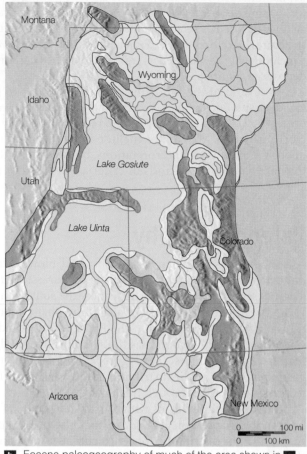

a Late Cretaceous paleogeography of several western states. Notice that although continental environments were present, a large part of the area was covered by the sea.

b Eocene paleogeography of much of the area shown in **a**. All of the deposition taking place at this time was in continental environments.

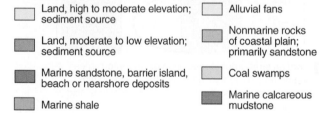

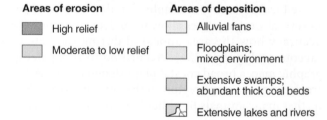

☐ Land, high to moderate elevation; sediment source

☐ Land, moderate to low elevation; sediment source

☐ Marine sandstone, barrier island, beach or nearshore deposits

☐ Marine shale

☐ Alluvial fans

☐ Nonmarine rocks of coastal plain; primarily sandstone

☐ Coal swamps

☐ Marine calcareous mudstone

Areas of erosion

☐ High relief

☐ Moderate to low relief

Areas of deposition

☐ Alluvial fans

☐ Floodplains; mixed environment

☐ Extensive swamps; abundant thick coal beds

☐ Extensive lakes and rivers

SUMMARY

- The physical and biologic features of sedimentary rocks provide information about the depositional processes responsible for them.

- Sedimentary structures and fossils are the most useful attributes of sedimentary rocks for environmental analyses, but textures and rock body geometry are also helpful.

- Geologists recognize three primary depositional areas—continental, transitional, and marine—each of which has several specific environments.

- Braided streams deposit mostly sand and gravel, whereas deposits of meandering streams are mostly mud and subordinate sand bodies with shoestring geometry.

- An association of alluvial fan, sand dune, and playa lake deposits is typical of desert depositional environments. Glacial deposits consist mostly of till in moraines and outwash.

- The simplest deltas, those in lakes, consist of a three-part sequence of rocks, grading from finest at the base upward to coarser-grained rocks. Marine deltas dominated by fluvial processes, waves, or tides are much larger and more complex.

- A barrier island system includes beach, dune, and lagoon subenvironments, each characterized by a unique association of rocks, sedimentary structures, and fossils.

- Inner shelf deposits are mostly sand, whereas those of the outer shelf are mostly mud; both have marine fossils and bioturbation. Much of the sediment from land crosses the shelves and is deposited on the continental slope and rise as submarine fans.
- Either pelagic clay or oozes derived from the shells of microscopic floating organisms cover most of the deep seafloor.
- Most limestone originates in shallow, warm seas where little detrital mud is present. Just like detrital rocks, car-

bonate rocks may possess cross-beds, ripple marks, mud cracks, and fossils that provide information about depositional processes.
- Evaporites form in several environments, but the most extensive ones were deposited in marine environments. In all cases, they formed in arid regions with high evaporation rates.
- With information from sedimentary rocks, as well as other rocks, geologists determine the past distribution of Earth's surface features.

IMPORTANT TERMS

alluvial fan, p. 117
barrier island, p. 121
biogenic sedimentary structure, p. 115
bioturbation, p. 115
braided stream, p. 117
continental rise, p. 122
continental shelf, p. 122
continental slope, p. 122
cross-bedding, p. 113
delta, p. 119

depositional environment, p. 116
drift, p. 117
fluvial, p. 117
graded bedding, p. 113
meandering stream, p. 117
mud crack, p. 115
outwash, p. 117
paleogeography, p. 127
playa lake, p. 117
progradation, p. 119

ripple mark, p. 113
rounding, p. 111
sand dune, p. 117
sedimentary structure, p. 112
sorting, p. 112
stratification (bedding), p. 112
tidal flat, p. 122
till, p. 117
varve, p. 117

REVIEW QUESTIONS

1. Braided streams deposit mostly
   a. _____ sheets of sand and gravel; b. _____ evaporites; c. _____ turbidity current sequences; d. _____ limestone and pelagic ooze; e. _____ submarine fans.
2. The process whereby organisms burrow through and thoroughly mix sediment is
   a. _____ lithification; b. _____ sedimentation; c. _____ bioturbation; d. _____ sorting; e. _____ rounding.
3. Deltas form where
   a. _____ the shells of microscopic organisms settle from suspension; b. _____ rivers and streams spread across their floodplains; c. _____ glaciers deposit till and outwash; d. _____ sediment is transported through submarine canyons; e. _____ a fluvial system flows into a standing body of water.
4. Cross-beds that dip in opposite directions are found in _____ environments.
   a. _____ desert; b. _____ tidal flat; c. _____ playa lake; d. _____ meandering stream; e. _____ alluvial fan.
5. Much of the graded bedding in the geologic record is accounted for by deposition by(on)
   a. _____ tidal currents; b. _____ mud settling from suspension; c. _____ point bars; d. _____ turbidity currents; e. _____ wave-dominated deltas.
6. A well-sorted detrital sedimentary rocks is one in which
   a. _____ 50% of the grains are more than 2 mm across; b. _____ at least four minerals are present; c. _____ micrite and shell fragments are found in equal amounts; d. _____ most of the grains are of about the same size; e. _____ magnesium replaced the calcium in limestone.

7. Which one of the following is not a sedimentary structure?
   a. _____ Outwash; b. _____ Ripple mark; c. _____ Mud crack; d. _____ Cross-bed; e. _____ Lamination.
8. A sand body with a blanket geometry that has large-scale cross beds, wave-formed ripple marks, and bioturbation probably was deposited on(in)
   a. _____ a braided stream system; b. _____ desert dunes; c. _____ inner continental shelf; d. _____ barrier island complex; e. _____ submarine fans.
9. The alternating dark- and light-colored laminations that form in glacial lakes are
   a. _____ graded beds; b. _____ current ripples; c. _____ tidal flat muds; d. _____ varves; e. _____ tills.
10. Which one of the following statements is correct?
    a. _____ Rounding refers to how nearly spherical sedimentary grains are; b. _____ The sand in desert dunes is poorly sorted; c. _____ The deep seafloor is covered by sand and gravel; d. _____ Cross-beds are good indicators of ancient current directions. e. _____ Limestone made up of broken shells is called micrite.
11. Under what conditions are evaporites deposited, and what are the most common evaporite rocks?
12. How and why do the sedimentary deposits on the inner and outer continental shelf differ?
13. What association of deposits would indicate deposition in a desert environment?
14. How do the deposits of braided streams and meandering streams differ?
15. What are the similarities and differences between wave-formed ripples marks and current ripple marks?

16. What kinds of sedimentary structures allow geologists to determine ancient current directions?
17. Describe the vertical sequence of rocks that form as a result of deposition of the simplest type of prograding delta.
18. Explain how graded bedding forms.

19. Why are sets of cross-beds dipping in opposite directions found on tidal flats?
20. Describe the sedimentary properties of rounding and sorting. Are they useful in environmental analyses?

APPLY YOUR KNOWLEDGE

1. No one was present millions of years ago to record data about climate, geography, and geologic processes. So how is it possible to decipher unobserved past events? In other words, what features in rocks, and especially in sedimentary rocks, would you look for to determine what happened in the far distant past?

2. You live in the continental interior where sedimentary rocks are well exposed. A local resident tells you of a nearby area where sandstone and mudstone with dinosaur fossils are overlain first by seashell-bearing sandstone, followed upward by shale, and finally by limestone with fossil corals and sea lilies. How would you summarize the events leading to the deposition of these rocks?

3. The cross section below shows folded sedimentary rocks. Your task is to determine which of the rock layers is the oldest and the youngest. How did you come to your conclusion?

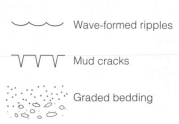

Wave-formed ripples

Mud cracks

Graded bedding

James S. Monroe

4. This image shows an outcrop at Natural Bridges National Monument in Utah. What kind of bedding is present in the layer indicated, how did it form, and can you infer anything about ancient current directions?

M. de Lamarck

CHAPTER 7

EVOLUTION—THE THEORY AND ITS SUPPORTING EVIDENCE

The French botanist-geologist Jean-Baptiste Pierre Antoine de Lamarck (1744–1829) in 1809 proposed the first formal, widely accepted theory of evolution. Although his mechanism to account for evolutionary change has since been discredited, he nevertheless made important contributions to science.

[OUTLINE]

Introduction

Evolution: What Does It Mean?

    Jean-Baptiste de Lamarck and His Ideas on Evolution

    The Contributions of Charles Darwin and Alfred Wallace

    Natural Selection—What Is Its Significance?

    Perspective *The Tragic Lysenko Affair*

Mendel and the Birth of Genetics

    Mendel's Experiments

    Genes and Chromosomes

The Modern View of Evolution

    What Brings About Variation?

    Speciation and the Rate of Evolution

Divergent, Convergent, and Parallel Evolution

Microevolution and Macroevolution

Cladistics and Cladograms

Mosaic Evolution and Evolutionary Trends

Extinctions

What Kinds of Evidence Support Evolutionary Theory?

    Classification—A Nested Pattern of Similarities

    How Does Biologial Evidence Support Evolution?

    Fossils: What Do We Learn from Them?

    Missing Links—Are They Really Missing?

    The Evidence—A Summary

Summary

At the end of this chapter, you will have learned that

- The central claim of the theory of evolution is that today's organisms descended, with modification, from ancestors that lived in the past.

- Jean-Baptiste de Lamarck in 1809 proposed the first widely accepted mechanism—inheritance of acquired characteristics—to account for evolution.

- In 1859, Charles Darwin and Alfred Wallace simultaneously published their views on evolution and proposed natural selection as a mechanism to bring about change.

- Experiments carried out by Gregor Mendel during the 1860s demonstrated that variations in populations are maintained rather than blended during inheritance, as previously thought.

- In the modern view of evolution, variation is accounted for mostly by sexual reproduction and by mutations in sex cells.

- The fossil record provides many examples of macroevolution—that is, changes that result in the origin of new species, genera, and so on—but these changes are simply the cumulative effect of microevolution, which involves changes within a species.

- Fossils showing the transition from ancestor to descendant groups, as in fish to amphibians and reptiles to mammals, are well known.

- A number of evolutionary trends, such as size increase or changing configuration of shells, teeth, or limbs, are well known for organisms for which sufficient fossil material is available.

- The theory of evolution is truly scientific, because we can think of observations or experiments that would support it or render it incorrect.

- Fossils are important as evidence for evolution, but additional evidence comes from classification, biochemistry, molecular biology, genetics, and geographic distribution.

Introduction

A rugged group of islands belonging to Ecuador lies in the Pacific Ocean about 1000 km west of South America. Called the Archipelago de Colon after Christopher Columbus, the group is better known as the Galápagos Islands (• Figure 7.1). During Charles Robert Darwin's five-year voyage (1831–1836) as an unpaid naturalist aboard the research vessel HMS *Beagle*, he visited the Galápagos Islands, where he made important observations that changed his ideas about the then widely held concept called *fixity of species* (• Figure 7.2). According to this idea, all present-day species had been created in their present form and had changed little or not at all.

Darwin began his voyage not long after graduating from Christ's College of Cambridge University with a degree in theology, and although he was rather indifferent to religion, he fully accepted the biblical account of creation. During the voyage, though, his ideas began to change. For example, some of the fossil mammals he collected in South America were similar to present-day llamas, sloths, and armadillos yet also differed from them. He realized that these animals had descended with modification from ancestral species, and so he began to question the idea of fixity of species.

Darwin postulated that the 13 species of finches living on the Galápagos Islands had evolved from a common ancestor species that somehow reached the islands as accidental immigrants from South America. Indeed, their ancestor was very likely a single species resembling the blue-back grassquit finch now living along South America's Pacific Coast. The islands' scarcity of food accounts for the ancestral species evolving different physical characteristics, especially beak shape, to survive (• Figure 7.3).

Darwin became convinced that organisms descended with modification from ancestors that lived during the past, which is the central claim of the **theory of evolution.** So why should you study evolution? For one thing, evolution involving inheritable changes in organisms through time is fundamental to biology and **paleontology,** the study of life history as revealed by fossils. Furthermore, like plate tectonic theory, evolution is a unifying theory that explains an otherwise encyclopedic collection of facts. And finally, it serves as the framework for discussions of life history in the following chapters. Unfortunately, many people have a poor understanding of the theory of evolution and hold a number of misconceptions, some of which we address in this and later chapters.

Evolution: What Does It Mean?

Evolution is the process whereby organisms have changed since life originated—that is, they have descended with modification from ancestors that lived during the past. It is important to note that the theory of evolution does not address how life originated, only how it has changed and diversified through time. There are theories that attempt to explain life's origin (see Chapter 8), but they must be evaluated on their own merits.

The idea that organisms have evolved is commonly attributed solely to Charles Darwin, but it was seriously considered long before he was born, even by some ancient Greeks and by philosophers and theologians during the Middle Ages (A.D. 476–1453). Nevertheless, the

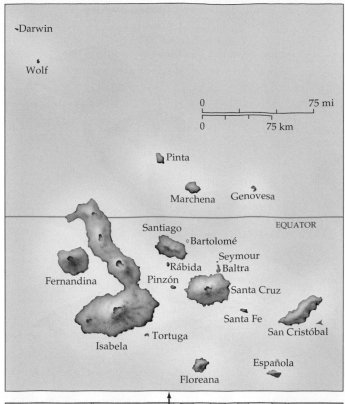

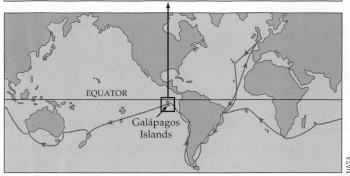

• **Figure 7.1 The Galápagos Islands** The Galápagos Islands are specks of land composed of basalt in the Pacific Ocean west of Ecuador. The red line in the map shows the route followed by Charles Darwin when he was aboard the HMS *Beagle* from 1831 to 1836.

• **Figure 7.2 Charles Robert Darwin in 1840** During his five-year voyage aboard the HMS *Beagle,* Darwin collected fossils and made observations that would later be instrumental in his ideas about evolution.

prevailing belief among Europeans well into the 1700s was that the works of Aristotle (384–322 B.C.) and the first two chapters of the book of Genesis contained all-important knowledge. Literally interpreted, Genesis was taken as the final word on the origin and variety of organisms, as well as much of Earth history. To question any aspect of this interpretation was heresy, which was usually dealt with harshly.

The social and intellectual climate changed in 18th century Europe, and the absolute authority of the Christian church in all matters declined. And ironically, the very naturalists trying to find physical evidence supporting Genesis found more and more evidence that could not be reconciled with a literal reading of scripture. For example, observations of sedimentary rocks previously attributed to

a single worldwide flood led naturalists to conclude that they were deposited during a vast amount of time, in environments like those existing now. They reasoned that this evidence truly indicated how the rocks originated, for to infer otherwise implied a deception on the part of the Creator (see Chapter 5 Introduction).

In this changing intellectual atmosphere, scientists gradually accepted the principle of uniformitarianism and Earth's great age, and the French zoologist Georges Cuvier demonstrated that many types of plants and animals had become extinct. In view of the accumulating evidence, particularly from studies of living organisms, scientists became convinced that change from one species to another actually took place. However, they lacked a theoretical framework to explain evolution.

Jean-Baptiste de Lamarck and His Ideas on Evolution

Jean-Baptiste de Lamarck (1744–1829) was not the first to propose a mechanism to account for evolution, but in 1809 he was the first to be taken seriously. Lamarck contributed greatly to our understanding of the natural world, but unfortunately he is best remembered for his theory of **inheritance of acquired characteristics.** According to this idea, new traits arise in organisms because of their needs, and somehow these characteristics are passed on to their descendants. In an

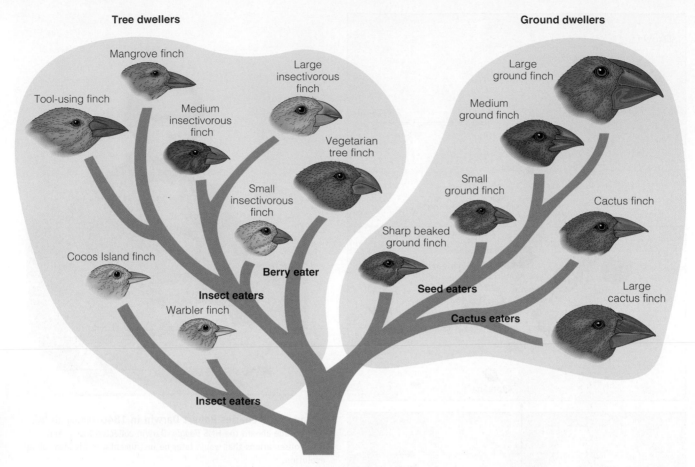

Tree dwellers

Mangrove finch

Tool-using finch

Large insectivorous finch

Medium insectivorous finch

Vegetarian tree finch

Small insectivorous finch

Cocos Island finch

Berry eater

Insect eaters

Warbler finch

Insect eaters

Ground dwellers

Large ground finch

Medium ground finch

Small ground finch

Cactus finch

Sharp beaked ground finch

Seed eaters

Large cactus finch

Cactus eaters

• Figure 7.3 **Darwin's Finches of the Galápagos Islands** The finches, which differ in size, color, beak shape, and diet, are arranged to show evolutionary relationships. The birds evolved from accidental immigrants much like the blue-back grassquit finch that lives along the west coast of South America.

ancestral population of short-necked giraffes, for instance, neck stretching to browse in trees gave them the capacity to have offspring with longer necks. In short, Lamarck thought that characteristics acquired during an individual's lifetime were inheritable.

Considering the information available at the time, Lamarck's theory seemed logical and was accepted by some as a viable mechanism for evolution. Indeed, it was not totally refuted until decades later, when scientists discovered that the units of heredity known as *genes* cannot be altered by any effort by an organism during its lifetime.

In fact, all attempts to demonstrate inheritance of acquired characteristics have been unsuccessful (see Perspective).

The Contributions of Charles Darwin and Alfred Wallace

In 1859, Charles Robert Darwin (1809–1882) (Figure 7.2) published *On the Origin of Species,* in which he detailed his ideas on evolution and proposed a mechanism whereby evolution could take place. Although 1859 marks the beginning of modern evolutionary thought, Darwin actually formulated his ideas more than 20 years earlier but, being aware of the furor they would cause, was reluctant to publish them.

Darwin had concluded during his 1831–1836 voyage aboard the *Beagle* that species are not immutable and fixed, but he had no idea what might bring about change. However, his observations of selection practiced by plant and animal breeders and a chance reading of Thomas Malthus's essay on population gave him the elements necessary to formulate his theory.

Plant and animal breeders practice **artificial selection** by selecting those traits they deem desirable and then breed plants and animals with those traits, thereby bringing about a great amount of change (• Figure 7.4). The fantastic variety of plants and animals so produced made Darwin wonder if a process selecting among variant types in nature could also bring about change. He came to fully appreciate the power of selection when he read in Malthus's essay that far more animals are born than reach maturity, yet the adult populations remain rather constant. Malthus reasoned that competition for resources resulted in a high infant mortality rate, thus limiting population size.

In 1858, Darwin received a letter from Alfred Russel Wallace (1823–1913), a naturalist working in southern Asia, who had also read Malthus's essay and had come to the same conclusion: A natural process was selecting only a few individuals for survival. Darwin's and Wallace's idea,

Wild mustard

broccoli

cauliflower

kale

cabbage

• **Figure 7.4** **Artificial Selection** Humans have practiced artificial selection for thousands of years, thereby giving rise to dozens of varieties of domestic dogs, pigeons, sheep, cereal crops, and vegetables. Wild mustard was selectively bred to yield broccoli, cauliflower, kale, and cabbage, and two other vegetables not shown here.

called *natural selection,* was presented simultaneously in 1859 to the Linnaean Society in London.

Natural Selection—What Is Its Significance?

We can summarize the salient points on **natural selection,** a mechanism that accounts for evolution, as follows:

1. Organisms in all populations possess heritable variations—size, speed, agility, visual acuity, digestive enzymes, color, and so forth.

2. Some variations are more favorable than others; that is, some variant types have a competitive edge in acquiring resources and/or avoiding predators.

3. Those with favorable variations are *more likely* to survive and pass on their favorable variations.

In colloquial usage, natural selection is sometimes expressed as "survival of the fittest," which is misleading because it reduces to "the fittest are those that survive and are thus the fittest." But it actually involves inheritable variations leading to differential reproductive success. Having favorable variations does not guarantee survival for an individual, but in a population of perhaps thousands, those with favorable variations are more likely to survive and reproduce.

Natural selection works on the existing variation in a population, thus giving a competitive edge to some individuals. So, evolution by natural selection and evolution by inheritance of acquired characteristics are both testable, but evidence supports the former, whereas attempts to verify the latter have failed (see Perspective).

Darwin was not unaware of potential problems for his newly proposed theory of natural selection. In fact, in *On the Origin of Species,* he said:

If it could be demonstrated that any complex organ existed which could not possibly have been formed by numerous, successive, slight modifications, my theory would absolutely break down. (p. 171)

Critics have cited eyes, bird's wings, and many other features that they claim are too complex to have evolved by natural selection, because, according to them, anything less than the structure as it exists now would be useless. Eyes of vertebrate animals are a favorite example—after all, of what use is half an eye? There are no half eyes, all eyes are fully developed and functional, even eyes that are no more than light-sensitive spots, crude image makers, or more sophisticated image makers such as those of reptiles, birds, and mammals.

The "too complex to have evolved" argument fails on another score as well, because it assumes that structures have always been used exactly as they are now and anything

The Tragic Lysenko Affair

When Jean-Baptiste de Lamarck (1744–1829) proposed his *theory of inheritance of acquired characteristics* some scientists accepted it as a viable explanation to account for evolution. In fact, it was not fully refuted for several decades, and even Charles Darwin, at least for a time, accepted some kind of Lamarckian inheritance. Nevertheless, the notion that acquired characteristics could be inherited is now no more than an interesting footnote in the history of science. There was, however, one instance in recent times when the idea enjoyed some popularity—but with tragic consequences.

One of the most notable examples of adherence to a disproved scientific theory involved Trofim Denisovich Lysenko (1898–1976), who became president of the Soviet Academy of Agricultural Sciences in 1938. He lost this position in 1953 but in the same year was appointed director of the Institute of Genetics, a post he held until 1965. Lysenko endorsed Lamarck's theory of inheritance of acquired characteristics because he thought plants and animals could

be changed in desirable ways by exposing them to a new environment. Lysenko reasoned that seeds exposed to dry conditions or to cold would acquire a resistance to drought or cold weather, and these traits would be inherited by future generations.

Lysenko accepted inheritance of acquired characteristics because he thought it was compatible with Marxist-Leninist philosophy. Beginning in 1929, Soviet scientists were encouraged to develop concepts consistent with this philosophy.

As president of the Academy of Agricultural Sciences and with the endorsement of the Central Committee of the Soviet Union, Lysenko did not allow any other research concerning inheritance. Those who publicly disagreed with him lost their jobs or were sent to labor camps. In fact, beginning in 1948 it was illegal to teach or do research in genetic theories developed in the West, and even high school textbooks had no information on the role of chromosomes in heredity.

Unfortunately for the Soviet people, inheritance of acquired characteristics

had been discredited more than 50 years earlier. The results of Lysenko's belief in the political correctness of this theory and its implementation were widespread crop failures and famine and the thorough dismantling of Soviet genetic research in agriculture. In fact, it took decades for Soviet genetic research programs to fully recover from this setback, after scientists finally became free to experiment with other theories of inheritance.

Lysenko's ideas on inheritance were not mandated in the Soviet Union because of their scientific merit but rather because they were deemed compatible with a belief system. The fact that Lysenko retained power and only his type of genetic research was permitted for more than 25 years is a testament to the absurdity of basing scientific theories on philosophic or political beliefs. In other words, a government mandate does not validate a scientific theory, nor does a popular vote or a decree by some ecclesiastic body. Legislating or decreeing that acquired traits are inheritable does not make it so.

less would be incomplete and functionless. Mature partridges can fly, although they prefer to stay on the ground. However, the wings of young partridges are small and useless for flying, so are their wings somehow incomplete and of no use? Hardly. Young partridges flap their wings vigorously when running up trees, which they cannot do if their wings are taped down. In short, this *vertical running,* as it is called, is a mechanism to escape predators, so their tiny wings are actually quite useful.

One misconception about natural selection, and especially the phrase "survival of the fittest," is that among animals only the biggest, strongest, and fastest are likely to survive. Indeed, size and strength are important when male bighorn sheep compete for mates, but remember that females pass along their genes, too. Speed is certainly an advantage to some predators, but weasels and skunks are not very fast and survive quite nicely. In fact, natural selection may favor the

smallest where resources are limited, as on small islands, or the most easily concealed, or those that adapt most readily to a new food source, or those having the ability to detoxify some natural or human-made substance.

Actually, the idea of evolution was accepted rather quickly, but the Darwin–Wallace theory of natural selection faced some resistance, mainly because so little was known about inheritance.

Mendel and the Birth of Genetics

Critics of natural selection were quick to point out that Darwin and Wallace could not account for the origin of variations or explain how variations were maintained in populations. These critics reasoned that should a variant trait arise, it would blend with other traits and would be lost. Actually, the answers to these criticisms existed even then, but they remained in obscurity until 1900.

Mendel's Experiments

During the 1860s, Gregor Mendel, an Austrian monk, performed a series of controlled experiments with true-breeding strains of garden peas (strains that when self-fertilized always display the same trait). He concluded from these experiments that traits such as flower color are controlled by a pair of factors, or what we now call **genes,** and that these factors (genes) that control the same trait occur in alternate forms, now called **alleles.** He further realized that one allele may be dominant over another and that offspring receive one allele of each pair from each parent. For example, in • Figure 7.5,

R represents the allele for red flower color, and r is the allele for white flowers, so any offspring may receive the RR, Rr, or rr combinations of alleles. And because R is dominant over r, only those offspring with the rr combination of alleles will have white flowers.

We can summarize the most important aspects of Mendel's work as follows: The factors (genes) that control traits do not blend during inheritance; and even though traits may not be expressed in each generation, they are not lost (Figure 7.5). Therefore, some variation in populations is accounted for by alternate expressions of genes (alleles), because traits do not blend, as previously thought. Mendelian genetics explains much about heredity, but we now know the situation is much more complex than he realized. Our discussion focused on a single gene controlling a trait, but most traits are controlled by many genes, and some genes show incomplete

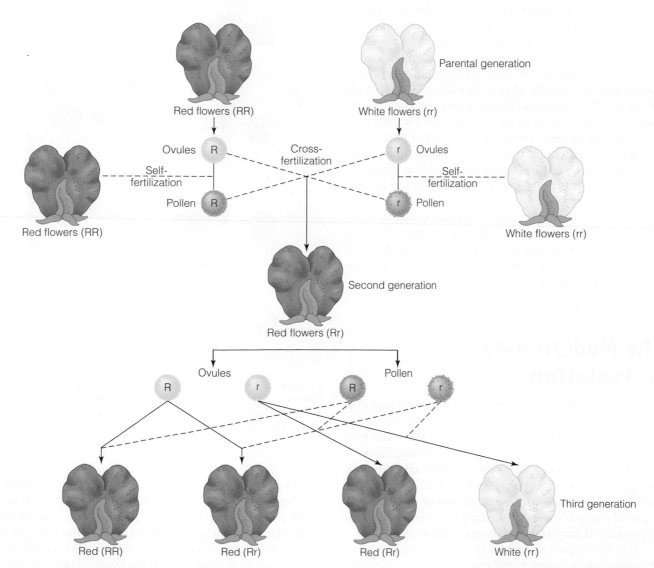

• **Figure 7.5 Mendel's Experiments with Flower Color** In his experiments with flower color in garden peas, Mendel used true-breeding strains. These plants, shown as the parental generation, when self-fertilized always yield offspring with the same trait as the parent. If the parental generation is cross-fertilized, however, all plants in the second generation receive the Rr combination of alleles; these plants will have red flowers because R is dominant over r. The second generation of plants produces ovules and pollen with the alleles shown and, when left to self-fertilize, produces a third generation with a ratio of three red-flowered plants to one white-flowered plant.

dominance. Mendel was unaware of mutations (changes in the genetic material), chromosomes, and the fact that some genes control the expression of other genes. Nor had he heard of Hox genes, which were discovered during the 1980s, that regulate the development of major body segments. Nevertheless, Mendel's work provided the answers Darwin and Wallace needed, but his research was published in an obscure journal and went unnoticed until 1900 when three independent researchers rediscovered them.

Genes and Chromosomes

Complex, double-stranded, helical molecules of **deoxyribonucleic acid (DNA),** called **chromosomes,** are found in the cells of all organisms (• Figure 7.6). Specific segments or regions of the DNA molecule are the hereditary units: the genes. The number of chromosomes is specific for a single species but varies among species. For instance, fruit flies have 8 chromosomes (4 pairs), humans have 46, and domestic horses have 64. Remember, chromosomes are always found in pairs carrying genes controlling the same traits.

In sexually reproducing organisms, the production of *sex cells* (pollen and ovules in plants and sperm and eggs in animals) results when cells undergo a type of cell division called **meiosis.** This process yields cells with only one chromosome of each pair, so all sex cells have only half the chromosome number of the parent cell (• Figure 7.7a). During reproduction, a sperm fertilizes an egg (or pollen fertilizes an ovule), yielding an egg (or ovule) with the full set of chromosomes typical for that species (Figure 7.7b).

As Mendel deduced from his experiments, half of the genetic makeup of a fertilized egg comes from each parent. The fertilized egg, however, develops and grows by a cell division process called **mitosis** during which cells are simply duplicated—that is, there is no reduction in the chromosome number as in meiosis (Figure 7.7c).

The Modern View of Evolution

During the 1930s and 1940s, the ideas developed by paleontologists, geneticists, population biologists, and others were merged to form a **modern synthesis** or neo-Darwinian view of evolution. The chromosome theory of inheritance was incorporated into evolutionary thinking, changes in genes (mutations) were seen as one source of variation in populations, Lamarck's idea of inheritance of acquired characteristics was completely rejected, and the importance of natural selection was reaffirmed.

It is also important to realize that according to modern evolutionary theory, populations rather than individuals evolve. Individuals develop according to the genetic blueprint they inherit from their parents, but they may have favorable inherited variations not present in most members of their population. In short, individuals in

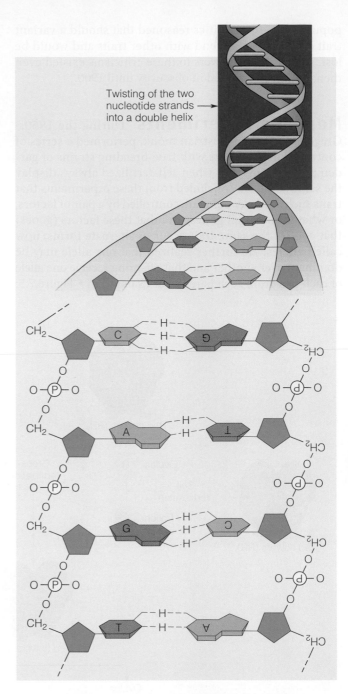

Twisting of the two nucleotide strands into a double helix →

• **Figure 7.6 The Double-Stranded, Helical DNA Molecule** The molecule's two strands are joined by hydrogen bonds (H). Notice the bases A, G, C, and T, combinations of which code for amino acids during protein synthesis in cells.

genetically varying populations are more likely to survive and reproduce if their variations are favorable, and as a result descendant populations possess these variations in greater frequency. A good example is the changes in Galápagos Islands finches, about 80% of which died during a drought during the 1970s. Most of the survivors were larger birds with heavy beaks that allowed them to crack the tough seeds available. Researchers found that, on average, individuals in the next generation were larger and had heavier beaks.

• Figure 7.7 **Meiosis and Mitosis**

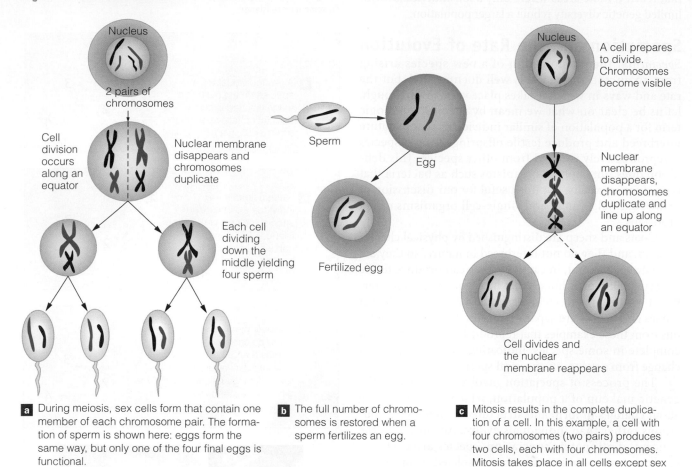

Nucleus

2 pairs of chromosomes

Cell division occurs along an equator

Nuclear membrane disappears and chromosomes duplicate

Each cell dividing down the middle yielding four sperm

Sperm

Egg

Fertilized egg

Nucleus

A cell prepares to divide. Chromosomes become visible

Nuclear membrane disappears, chromosomes duplicate and line up along an equator

Cell divides and the nuclear membrane reappears

a During meiosis, sex cells form that contain one member of each chromosome pair. The formation of sperm is shown here: eggs form the same way, but only one of the four final eggs is functional.

b The full number of chromosomes is restored when a sperm fertilizes an egg.

c Mitosis results in the complete duplication of a cell. In this example, a cell with four chromosomes (two pairs) produces two cells, each with four chromosomes. Mitosis takes place in all cells except sex cells. Once an egg has been fertilized, the developing embryo grows by mitosis.

What Brings About Variation?

Natural selection works on variations in populations, most of which results from reshuffling of genes from generation to generation during sexual reproduction. Given that each of thousands of genes might have several alleles, and that offspring receive half of their genes from each parent, the potential for variation is enormous. However, this variation was already present, so new variations arise by **mutations**—that is, changes in the chromosomes or genes. In other words, a mutation is a change in the hereditary information. Whether a *chromosomal mutation* (affecting a large segment of a chromosome) or a *point mutation* (a change in a particular gene), as long as it takes place in a sex cell, it is inheritable.

To fully understand mutations, we must explore them further. For one thing, they are random with respect to fitness, meaning they may be beneficial, neutral, or harmful. If a species is well adapted to its environment, most mutations would not be particularly useful and perhaps be harmful. Some mutations are absolutely harmful, but some that were harmful can become useful if the environment changes. For instance, some plants have developed a resistance for contaminated soils around mines. Plants of the same species from the normal environment do poorly or

die in contaminated soils, whereas contaminant-resistant plants do very poorly in the normal environment. Mutations for contaminant resistance probably occurred repeatedly in the population, but they were not beneficial until contaminated soils were present.

How can a mutation be neutral? In cells, information carried on chromosomes directs the formation of proteins by selecting the appropriate amino acids and arranging them into a specific sequence. However, some mutations have no effect on the type of protein synthesized. In other words, the same protein is synthesized before and after the mutation, and thus the mutation is neutral.

What causes mutations? Some are induced by *mutagens,* agents that bring about higher mutation rates. Exposure to some chemicals, ultraviolet radiation, X-rays, and extreme temperature changes might cause mutations. But some mutations are spontaneous, taking place in the absence of any known mutagen.

Sexual reproduction and mutations account for much of the variation in populations, but another factor is *genetic drift,* a random change in the genetic makeup of a population due to chance. Genetic drift is significant in small populations but probably of little or no importance in large ones. Good examples of genetic drift include evolution in small populations that reach remote areas where only a few individuals with limited genetic diversity rebuilt a larger population.

Speciation and the Rate of Evolution

Speciation, the phenomenon of a new species arising from an ancestral species, is well documented, but the rate and ways in which it takes place vary. First, though, let us be clear on what we mean by **species,** a biologic term for a population of similar individuals that in nature interbreed and produce fertile offspring. Thus, a species is reproductively isolated from other species. This definition does not apply to organisms such as bacteria that reproduce asexually, but it is useful for our discussion of plants, animals, fungi, and single-cell organisms called protistans.

Goats and sheep are distinguished by physical characteristics, and they do not interbreed in nature, so they are separate species. Yet, in captivity, they can produce fertile offspring. Domestic horses that have gone wild can interbreed with zebras to yield a *zebroid,* which is sterile; yet horses and zebras are separate species. It should be obvious from these examples that reproductive barriers are not complete in some species, indicating varying degrees of change from a common ancestral species.

The process of speciation involves a change in the genetic makeup of a population, which also may bring about changes in form and structure—remember, populations rather than individuals evolve. According to the concept of **allopatric speciation,** species arise when a small part of a population becomes isolated from its parent population (• Figure 7.8). Isolation may result from a rising mountain barrier that effectively separates a once-interbreeding species, or a few individuals may somehow get to a remote area and no longer exchange genes with the parent population (Figure 7.3). Another example involves at least 15 iguanas that in 1995 were rafted 280 km on floating vegetation from Guadeloupe to Anguilla in the Caribbean Sea when the trees they were on were blown out to sea by hurricane winds. Given these conditions and the fact that different selective pressures are likely, they may eventually give rise to a reproductively isolated species.

Widespread agreement exists on allopatric speciation, but scientists disagree on how rapidly a new species may evolve. According to the modern synthesis, the gradual accumulation of minor changes eventually brings about the origin of a new species, a phenomenon called **phyletic gradualism.** Another view, known as **punctuated equilibrium,** holds that little or no change takes place in a species during most of its existence, and then evolution occurs rapidly, giving rise to a new species in perhaps as little as a few thousands of years.

Proponents of punctuated equilibrium argue that few examples of gradual transitions from one species to another are found in the fossil record. Critics, however, point out that neither Darwin nor those who formulated the modern synthesis insisted that all evolutionary change was gradual and continuous, a view shared by many

• Figure 7.8 **Allopatric Speciation** An example of allopatric speciation on some remote islands.

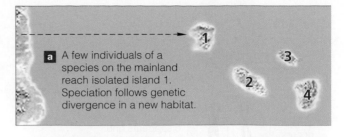

a A few individuals of a species on the mainland reach isolated island 1. Speciation follows genetic divergence in a new habitat.

b Later in time, a few individuals of the new species colonize nearby island 2. In this new habitat, speciation follows genetic divergence.

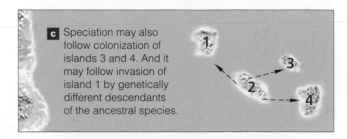

c Speciation may also follow colonization of islands 3 and 4. And it may follow invasion of island 1 by genetically different descendants of the ancestral species.

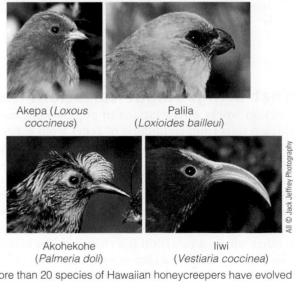

Akepa (*Loxous coccineus*)

Palila (*Loxioides bailleui*)

Akohekohe (*Palmeria doli*)

Iiwi (*Vestiaria coccinea*)

All © Jack Jeffrey Photography

d More than 20 species of Hawaiian honeycreepers have evolved from a common ancestor, four of which are shown here, as they adapted to diverse resources on the islands.

present-day biologists and paleontologists. Furthermore, deposition of sediments in most environments is not continuous; thus the lack of gradual transitions in many cases is simply an artifact of the fossil record. And finally, despite the incomplete nature of the fossil record, a number of examples of gradual transitions from ancestral to descendant species are well known.

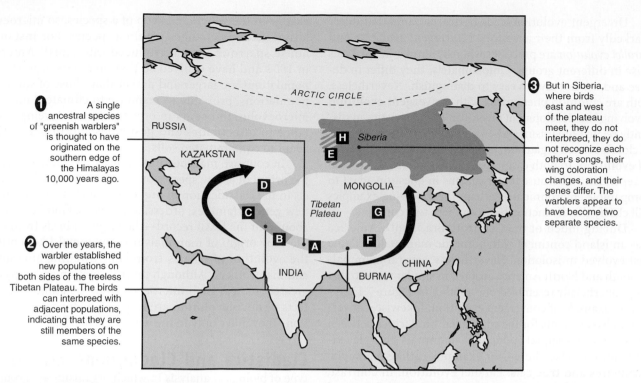

① A single ancestral species of "greenish warblers" is thought to have originated on the southern edge of the Himalayas 10,000 years ago.

② Over the years, the warbler established new populations on both sides of the treeless Tibetan Plateau. The birds can interbreed with adjacent populations, indicating that they are still members of the same species.

③ But in Siberia, where birds east and west of the plateau meet, they do not interbreed, they do not recognize each other's songs, their wing coloration changes, and their genes differ. The warblers appear to have become two separate species.

• Figure 7.9 **Speciation in Songbirds** Two species of Eurasian songbirds appear to have evolved from an ancestral species. Adjacent populations of these birds. A and B, or F and G, for instance, can interbreed even though they differ slightly. However, where populations E and H overlap in Siberia, they cannot interbreed.

If speciation proceeds as we have described, evidence of it taking place should be available from present-day organisms, and indeed it is. This is especially true in some new plant species that have arisen by *polyploidy* when their chromosome count doubled. Speciation, or at least incipient speciation, has also taken place in populations of mosquitoes, bees, mice, salamanders, fish, and birds that are isolated or partially isolated from one another (• Figure 7.9).

Divergent, Convergent, and Parallel Evolution

Paleontologists refer to the phenomenon of an ancestral species giving rise to diverse descendants adapted to various aspects of the environment as **divergent evolution.** An impressive example involves the mammals whose diversification from a common ancestor during the Late Mesozoic gave rise to such varied animals as platypuses, armadillos, rodents, bats, primates, whales, and rhinoceroses (• Figure 7.10).

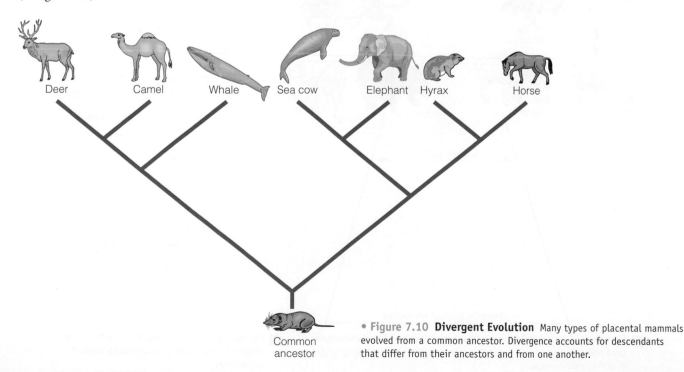

• Figure 7.10 **Divergent Evolution** Many types of placental mammals evolved from a common ancestor. Divergence accounts for descendants that differ from their ancestors and from one another.

Divergent evolution leads to descendants that differ markedly from their ancestors. *Convergent evolution* and *parallel evolution* are processes whereby similar adaptations arise in different groups. Unfortunately, they differ in degree and are not always easy to distinguish. Nevertheless, both are common phenomena with **convergent evolution** involving the development of similar characteristics in distantly related organisms. When similar characteristics arise in closely related organisms, however, it is considered **parallel evolution.** In both cases, similar characteristics develop independently, because the organisms in question adapt to comparable environments. Perhaps the following examples will clarify the distinction between these two concepts.

During much of the Cenozoic Era, South America was an island continent with a unique mammalian fauna that evolved in isolation. Nevertheless, several mammals in South and North America adapted in similar ways so that they superficially resembled one another (• Figure 7.11a)—a good example of convergent evolution. Likewise, convergence also accounts for the superficial similarities between Australian marsupial (pouched) mammals and placental mammals elsewhere—for example, catlike marsupial carnivores and true cats. Parallel evolution, in contrast, involves closely related organisms, such as jerboas and kangaroo rats, that independently evolved comparable features (Figure 7.11b).

Microevolution and Macroevolution

Micro and *macro* are prefixes that mean "small" and "large," respectively. Actually, **microevolution** is any change in the genetic makeup of a species, so microevolution involves changes within a species. For instance, house sparrows were introduced into North America in 1852 and have evolved so that members of northern populations are larger and darker than those of southern populations, probably in response to climate. Likewise, microevolution occurs in organisms that develop resistance to insecticides and pesticides and in plants that adapt to contaminated soils.

In contrast, **macroevolution** involves changes such as the origin of a new species or changes at even higher levels in the classification of organisms, such as the origin of new genera, families, orders, and classes. Good examples abound in the fossil record—the origin of birds from reptiles, the origin of mammals from mammal-like reptiles, the evolution of whales from land-dwelling ancestors, and many others. Although macroevolution encompasses greater changes than microevolution, the cumulative effects of microevolution are responsible for macroevolution. They differ only in the degree of change.

Cladistics and Cladograms
Cladistics is a type of biological analysis in which organisms are grouped together based on derived as opposed to primitive characteristics. For instance, all land-dwelling vertebrates have bone and paired limbs, so these characteristics are primitive and of little use in establishing relationships among them. However, hair and three middle ear bones are derived characteristics, sometimes called *evolutionary novelties,* because only one subclade, the mammals, has

• Figure 7.11 **Convergent and Parallel Evolution**

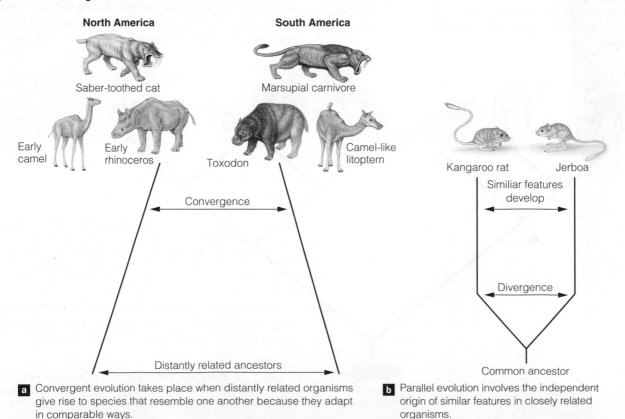

a Convergent evolution takes place when distantly related organisms give rise to species that resemble one another because they adapt in comparable ways.

b Parallel evolution involves the independent origin of similar features in closely related organisms.

Figure 7.12 Phylogenetic Tree and Cladogram

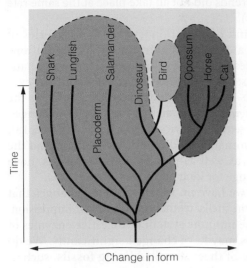

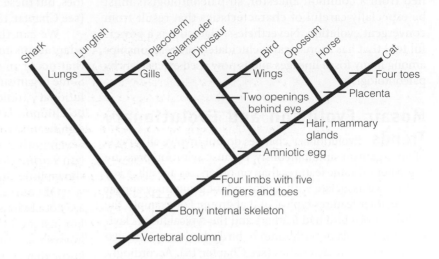

a A phylogenetic tree showing the relationships among various vertebrate animals.

b A cladogram showing inferred relationships. Some of the characteristics used to construct this cladogram are indicated.

them. If you consider only mammals, hair and middle ear bones are primitive characteristics, but live birth is a derived characteristic that serves to distinguish most mammals from the egg-laying mammals.

Traditionally, scientists have depicted evolutionary relationships with *phylogenetic trees,* in which the horizontal axis represents anatomic differences and the vertical axis denotes time (• Figure 7.12a). The patterns of ancestor–descendant relationships shown are based on a variety of characteristics, although the ones used are rarely specified. In contrast, a **cladogram** shows the relationships among members of a *clade,* a group of organisms including its most recent common ancestor (Figure 7.12b).

Any number of organisms can be depicted in a cladogram, but the more shown, the more complex and difficult it is to construct. • Figure 7.13 shows three cladograms, each with a different interpretation of the relationships among bats, birds, and dogs. Bats and birds fly, so we might conclude that they are more closely related to one another than to dogs (Figure 7.13a). On the other hand, perhaps birds and dogs are more closely related than either is to bats (Figure 7.13b). However, if we concentrate on evolutionary novelties, such as hair and giving birth to live young, we conclude that the cladogram in Figure 7.13c shows the most probable relationships—that is, bats and dogs are more closely related than they are to birds.

Scientists use cladograms to solve a variety of problems. For instance, if whales are more closely related to even-toed hoofed mammals (deer, hippos, antelope, etc.), as cladistic analyses indicate, they should possess some anatomical characteristics verifying this interpretation. And in fact some ancient whales possess the unique "double pulley" ankle typical of this group of mammals, even though the rear appendages of today's whales are only vestiges. Scientists using cladistic analysis of Pacific

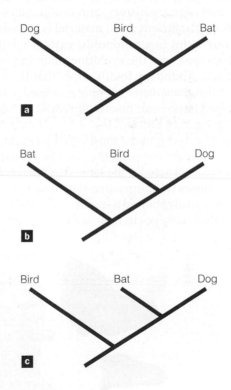

• Figure 7.13 **Cladograms for Dogs, Birds, and Bats** Cladograms showing three hypotheses for the relationships among dogs, birds, and bats. Derived characteristics such as hair and giving birth to live young indicate that dogs and bats are most closely related, as shown in **c**.

yews discovered a hard-to-recover compound used to treat cancer, and assumed that the same compound may be in closely related species. Indeed it was, and now the compound is widely available.

When applied to fossils, care must be taken in determining what are primitive versus derived characteristics, especially in groups with poor fossil records. Furthermore, cladistic analysis depends solely on characteristics inherited from a common ancestor, so paleontologists must be especially careful of characteristics that result from convergent evolution. Nevertheless, cladistics is a powerful tool that has more clearly elucidated the relationships among many fossil lineages and is now used extensively by paleontologists.

Mosaic Evolution and Evolutionary Trends

Evolutionary changes do not involve all aspects of an organism simultaneously, because selection pressure is greater on some features than on others. As a result, a key feature we associate with a descendant group may appear before other features typical of that group. For example, the oldest known bird had feathers and the typical fused clavicles (the furcula or wishbone) of birds, but it also retained many reptile characteristics (see Chapter 15). Accordingly, it represents the concept of **mosaic evolution,** meaning that organisms have recently evolved characteristics as well as some features of their ancestral group.

Paleontologists determine in some detail the *phylogeny,* or evolutionary history, and *evolutionary trends* for organisms if sufficient fossil material is available. For instance, one trend in ammonoids, extinct relatives of squid and octopus, was the evolution of an increasingly complex shell. Abundant fossils show that the Eocene mammals called *titanotheres* not only increased in size but also developed large nasal horns; moreover, the shape of their skull changed (• Figure 7.14).

Size increase is one of the most common evolutionary trends, but trends are complex, they may be reversed, and trends may not all proceed at the same rate. One evolutionary trend in horses was larger size, but some now-extinct horse species actually showed a size decrease. Other trends in horses include changes in their teeth and skull, as well as lengthening of their legs and reduction in the number of toes, but these trends did not all take place at the same rate (see Chapter 18).

We can think of evolutionary trends as a series of adaptations to a changing environment or adaptations that occur in response to exploitation of new habitats. Some organisms, however, show little evidence of any evolutionary trends for long periods. A good example is the brachiopod *Lingula,* whose shell at least has not changed significantly since the Ordovician Period (see Figure 5.18). Several other animals including horseshoe crabs and the fish *Latimeria,* as well as plants such as ginkgos, have shown little change for long periods (• Figure 7.15).

Of course, we have no way to evaluate changes that are not obvious in shells, teeth, bones, and leaf impression. For instance, the immune system or digestive enzymes of horseshoe crabs may have changed significantly. We do know that some of these so-called **living fossils,** such as opossums, are generalized animals, meaning that they can live under a wide variety of environmental conditions and, therefore, are not so sensitive to environmental changes. Nevertheless, the remarkable stasis of living fossils is one aspect of the organic world that paleontologists and evolutionary biologists do not yet fully understand.

But isn't evolution by natural selection a random process? If so, how is it possible for a trend to continue long enough to account, just by chance, for such complex structures as eyes, wings, and hands? Actually, evolution by natural selection is a two-step process, and only the first step involves chance. First, variation must be present or arise in some population. Whether or not variations arising by mutations are favorable is indeed a matter of chance, but the second step, natural selection, is not, because only individuals with favorable variations are most likely to survive and reproduce. In one sense, then, natural selection is a process of elimination, weeding out those individuals not as well adapted to a particular set of environmental circumstances.

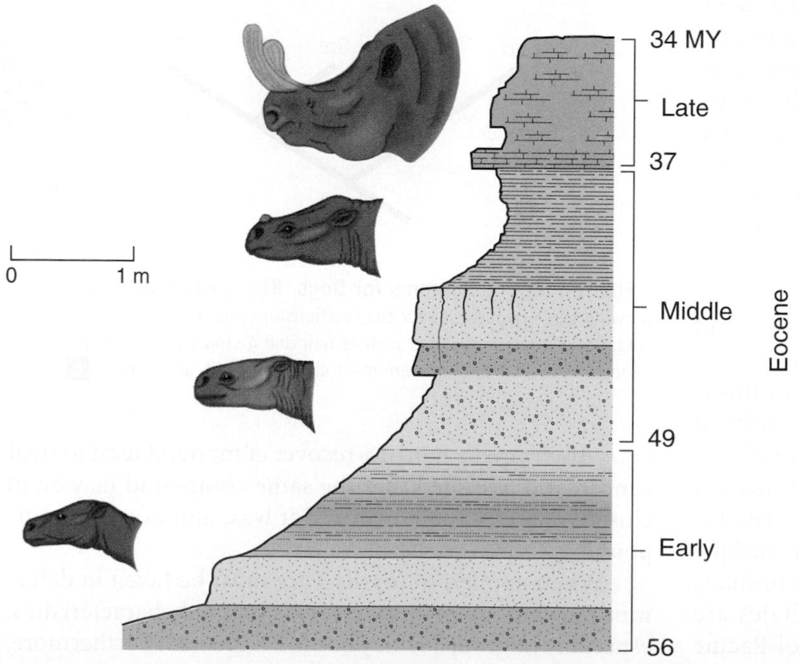

Extinctions

Judging from the fossil record, most organisms that ever existed are now extinct—perhaps as many as 99% of all species. Now, if species actually evolve as natural selection favors certain traits, shouldn't organisms be evolving toward some kind of higher order of perfection or greater complexity? Vertebrate animals are more complex, at least in overall organization, than are bacteria, but complexity does not necessarily mean they are superior in some survival sense—after all, bacteria have persisted for at least 3.0 billion years. Actually,

• Figure 7.14 **Evolutionary Trends in Titanotheres** These relatives of horses and rhinoceroses existed for about 20 million years during the Eocene Epoch. During that time they evolved from small ancestors to giants standing 2.4 m at the shoulder. In addition, they developed large horns, and the shape of the skull changed. Only 4 of the 16 known genera of titanotheres are shown here.

• Figure 7.15 **Two Examples of So-Called Living Fossils**

a *Latimeria* belongs to a group of fish known as coelacanths thought to have gone extinct at the end of the Mesozoic Era. One was caught off the east coast of Africa in 1938, and since then many more have been captured.

b These fan-shaped leaves from a present-day gingko tree look much like those of their ancient ancestors.

natural selection does not yield some kind of perfect organism, but rather those adapted to a specific set of circumstances at a particular time. Thus a clam or lizard existing now is not somehow superior to those that lived millions of years ago.

The continual extinction of species is referred to as *background extinction,* to clearly differentiate it from a **mass extinction,** during which accelerated extinction rates sharply reduce Earth's biotic diversity. Extinction is a continuous occurrence, but so is the evolution of new species that usually, but not always, quickly exploit the opportunities created by another species' extinction. When the dinosaurs and their relatives died out, the mammals began a remarkable diversification as they began occupying the niches left temporarily vacant.

What Kinds of Evidence Support Evolutionary Theory?

When Charles Darwin proposed his theory of evolution, he cited supporting evidence such as classification, embryology, comparative anatomy, geographic distribution, and,

to a limited extent, the fossil record. He had little knowledge of the mechanism of inheritance, and both biochemistry and molecular biology were unknown during his time. Studies in these areas, coupled with a more complete and much better understood fossil record, have convinced scientists that the theory is as well supported by evidence as any other major theory.

But is the theory of evolution truly scientific? That is, can testable predictive statements be made from it? First we must be clear on what a theory is and what we mean by "predictive." Scientists propose hypotheses for natural phenomena, test them, and in some cases raise them to the status of a theory—an explanation of some natural phenomenon well supported by evidence from experiments and/or observations.

Almost everything in the sciences has some kind of theoretical underpinning—optics, heredity, the nature of matter, the present distribution of continents, diversity in the organic world, and so on. Of course, no theory is ever proven in some final sense, although it might be supported by substantial evidence; all are always open to question, revision, and occasionally to replacement by a more comprehensive theory. In his book, *Why Darwin Matters,* (pp. 1–2) Michael Shermer noted that

> A 'theory' is not just someone's opinion or a wild guess made by a scientist. A theory is a well-supported and well-tested generalization that explains a set of observations. Science without theory is useless.

Prediction is commonly taken to mean to foresee some future event such as the next solar eclipse, but not all predictions are about the future. For instance, according to plate tectonic theory oceanic crust is continually produced at spreading ridges as plates diverge. If this is true, it is logical to predict that oceanic crust should become increasingly older with distance from a spreading ridge, and that sediment thickness on the seafloor should also increase with distance from a ridge. Both of these predictive statements have been verified (see Figure 3.12).

Evolutionary theory cannot make predictions about the far distant future, but it does make many predictions about the present-day biological world as well as the fossil record that should be consistent with the theory if it is correct. It is just as important that we be able to think of observations or experiments that would be negative evidence for the theory, or any other theory. In other words, if we could not think of any conceivable observation or experiment that would falsify a theory it would be useless as an explanation for natural phenomena. A good example is Philip Henry Gosse's idea that Earth was created a few thousand years ago with the appearance of great age (see Chapter 5).

If the theory of evolution is correct, closely related species such as wolves and coyotes should be similar not only in their anatomy but also in terms of biochemistry, genetics, and embryonic development (number 3 in Table 7.1). Suppose that they differed in their biochemical mechanisms as well as their embryology. Obviously our prediction would fail, and we would at least have to modify

1. If evolution has taken place, the oldest fossil-bearing rocks should have remains of organisms very different from those existing now, and more recent rocks should have fossils more similar to today's organisms.

2. If today's organisms descended with modification from ones in the past, there should be fossils showing characteristics connecting orders, classes, and so on.

3. If evolution is true, closely related species should be similar not only in details of their anatomy, but also in their biochemistry, genetics, and embryonic development, whereas distantly related species should show fewer similarities.

4. If the theory of evolution is correct—that is, living organisms descended from a common ancestor—classification of organisms should show a nested pattern or similarities.

5. If evolution actually took place, we would expect cave-dwelling plants and animals to most closely resemble those immediately outside their respective caves rather than being most similar to those in caves elsewhere.

6. If evolution actually took place, we would expect land-dwelling organisms on oceanic islands to most closely resemble those of nearby continents rather than those on other distant islands.

7. If evolution occurred, we would expect mammals to appear in the fossil record long after the appearance of the first fish. Likewise, we would expect reptiles to appear before the first mammals or birds.

8. If we examine the fossil record of related organisms such as horses and rhinoceroses, we should find that they were quite similar when they diverged from a common ancestor but became increasingly different as their divergence continued.

our theory. If other predictions also failed—say, mammals appear in the fossil record before fishes—we would have to abandon our theory and find a better explanation for our observations.

Classification—A Nested Pattern of Similarities

Carolus Linnaeus (1707–1778) proposed a classification system in which organisms are given a two-part genus and species name; the coyote, for instance, is *Canis latrans*. Table 7.2 shows Linnaeus's classification scheme, which is a hierarchy of categories that becomes more inclusive as one proceeds up the list. The coyote (*Canis latrans*) and the wolf (*Canis lupus*) share numerous characteristics, so they are members of the same genus, whereas both share some but fewer characteristics with the red fox (*Vulpes fulva*), and all three are members of the family Canidae (• Figure 7.16). All canids share some characteristics with cats, bears, and weasels and are grouped together in the order Carnivora, which is one of 18 living orders of the class Mammalia, all of whom are warm-blooded, possess fur or hair, and have mammary glands.

Linnaeus recognized shared characteristics among organisms, but his intent was simply to categorize species he thought were specially created and immutable. Following the publication of Darwin's *On the Origin of Species* in 1859, however, scientists quickly realized that shared characteristics constituted a strong argument for evolution. After all, if present-day organisms actually descended from ancient species, we should expect a pattern of similarities between closely related species and fewer between more distantly related ones.

How Does Biological Evidence Support Evolution?

Life is incredibly diverse—scientists have described at least 1.5 million species and estimate that there may be as many as several tens of millions not yet

TABLE **7.2**	Expanded Linnaen Classification Scheme (the animal classified in this example is the coyote, *Canis latrans*)
Kingdom	**Animalia**—Multicelled organisms; cells with nucleus (see Chapter 9); reproduce sexually; ingest preformed organic molecules for nutrients
Phylum	**Chordata**—Possess notochord, gill slits, dorsal hollow nerve cord at some time during life cycle (see Chapter 13)
Subphylum	**Vertebrata**—Those chordates with a segmented vertebral column (see Chapter 13)
Class	**Mammalia**—Warm-blooded vertebrates with hair or fur and mammary glands (see Chapter 15)
Order	**Carnivora**—Mammals with teeth specialized for a diet of meat (see Chapter 18)
Family	**Canidae**—The doglike carnivores (coyotes, wolves, and foxes)
Genus	***Canis***—Made up only of closely related species—coyotes and wolves (also includes domestic dogs)
Species	***latrans***—Consists of similar individuals that in nature can interbreed and produce fertile offspring

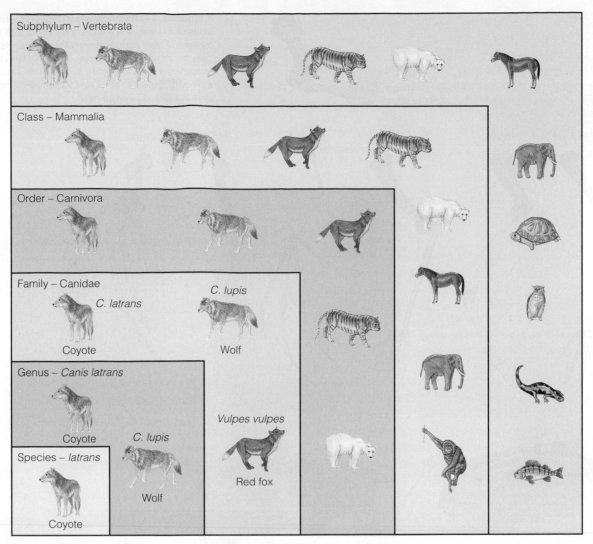

• **Figure 7.16** **Classification of Organisms Based on Shared Characteristics** All members of the subphylum Vertebrata—fishes, amphibians, reptiles, birds, and mammals—have a segmented vertebral column. Subphyla belong to more inclusive phyla (singular, phylum), and these in turn belong to kingdoms (see Table 7.2). Among vertebrates, only warm-blooded animals with hair or fur and mammary glands are mammals. Eighteen orders of mammals exist, including the order Carnivora shown here. The family Canidae includes only doglike carnivores, and the genus *Canis* includes only closely related species. The coyote, *Canis latrans,* stands alone as a species.

recognized. And yet, if existing organisms evolved from common ancestors, there should be similarities among all life-forms. As a matter of fact, all living things, be they bacteria, redwood trees, or whales, are composed mostly of carbon, nitrogen, hydrogen, and oxygen. Furthermore, their chromosomes consist of DNA, and all cells synthesize proteins in essentially the same way.

Studies in biochemistry also provide evidence for evolutionary relationships. Blood proteins are similar among all mammals but also indicate that humans are most closely related to great apes, followed in order by Old World monkeys, New World monkeys, and lower primates such as lemurs. Biochemical tests support the idea that birds descended from reptiles, a conclusion supported by evidence in the fossil record.

The forelimbs of humans, whales, dogs, and birds are superficially dissimilar (• Figure 7.17). Yet all are made up of the same bones; have basically the same arrangement of

muscles, nerves, and blood vessels; are similarly arranged with respect to other structures; and have a similar pattern of embryonic development. These **homologous structures,** as they are called, are simply basic vertebrate forelimbs modified for different functions; that is, they indicate derivation from a common ancestor. However, some similarities are unrelated to evolutionary relationships. For instance, wings of insects and birds serve the same function but are quite dissimilar in both structure and development and are thus termed **analagous structures** (• Figure 7.18).

Why do dogs have tiny remnants of toes, called dewclaws, on their forefeet or hind feet or, in some cases, on all four feet (• Figure 7.19a)? These dewclaws are **vestigial structures**—that is, remnants of structures that were fully functional in their ancestors. Ancestral dogs had five toes on each foot, each of which contacted the ground. As they evolved, though, dogs became toe walkers with only four digits on the ground, and the thumbs and big toes were

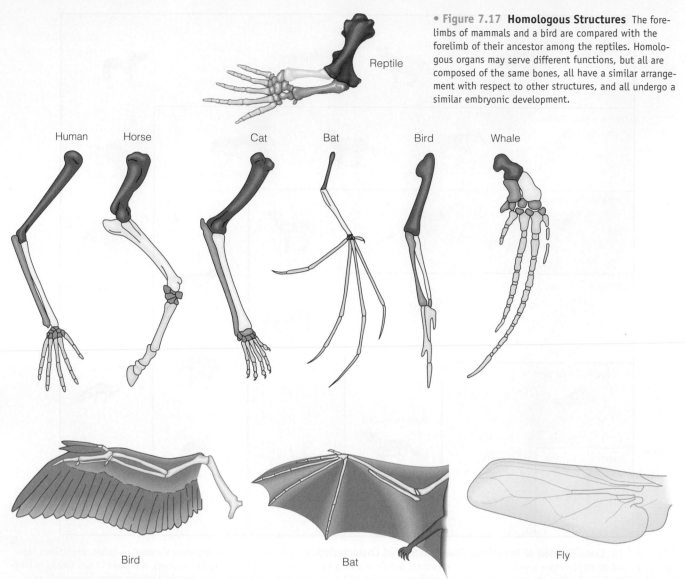

• Figure 7.17 **Homologous Structures** The forelimbs of mammals and a bird are compared with the forelimb of their ancestor among the reptiles. Homologous organs may serve different functions, but all are composed of the same bones, all have a similar arrangement with respect to other structures, and all undergo a similar embryonic development.

Reptile

Human Horse Cat Bat Bird Whale

Bird Bat Fly

• Figure 7.18 **Analogous Structures** The fly's wings serve the same function as wings of birds and bats. Thus, they are analogous, because they have a different structure and different embryologic development.

• Figure 7.19 **Vestigial Structures**

Dewclaw

a Notice the dewclaw, a vestige of the thumb, on the forefoot of this dog.

Sue Monroe

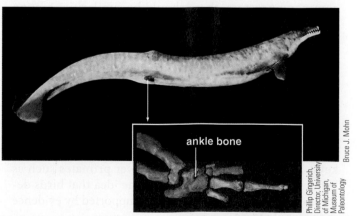

ankle bone

Phillip Gingerich, Director, University of Michigan, Museum of Paleontology

Bruce J. Mohn

b The Eocene-age whale *Basilosaurus* had tiny vestigial back limbs, but it did not use limbs to support its body weight. Even today, whales have a vestige of a pelvis, and on a few occasions, whales with rear limbs have been caught.

either lost or reduced to their present state. Although vestigial structures need not be totally functionless, many are, or at best have a reduced function. Remnants of toes are found in pigs, deer, and horses, and horses are occasionally born with extra toes. Whales and some snakes have a pelvis but no rear limbs (Figure 7.19b), except for a rare whale with hind limbs and two Cretaceous-age fossil snakes with back legs.

No one would seriously argue that a dog's dewclaws are truly necessary, and it would be difficult to convince most people that wisdom teeth in humans are essential. After all, some people never have wisdom teeth, and in those that do, they commonly never erupt from the gums, or they come in impacted and cause significant pain. The reason? The human jaw is too short to accommodate the ancestral number of teeth. In some cases, vestigial structures are fully functional, but perform a totally different function from what they did in ancestral condition. For instance, the incus and malleus of the mammalian middle ear were derived from the articular and quadrate bones that formed the joint between the jaw and skull in mammal-like reptiles (see Chapter 15).

Another type of evidence for evolution is observations of microevolution in living organisms. We have already mentioned the adaptations of some plants to contaminated soils. As a matter of fact, small-scale changes take place rapidly enough that new insecticides and pesticides must be developed continually because insects and rodents develop resistance to existing ones. Development of antibiotic-resistant strains of bacteria is a continuing problem in medicine. Whether the variation in these populations previously existed or was established by mutations is irrelevant. In either case, some variant types were more likely to survive and reproduce, bringing about a genetic change.

Fossils: What Do We Learn From Them?

Fossil marine invertebrates found far from the sea, and even high in mountains, led early naturalists to conclude that the fossil-bearing rocks were deposited during a worldwide flood. In 1508, Leonardo da Vinci realized that the fossil distribution was not what one would expect from a rising flood, but the flood explanation persisted, and John Woodward (1665–1728) proposed a testable hypothesis. According to Woodward, fossils were sorted out as they settled from floodwaters based on their shape, size, and density, but this hypothesis was quickly rejected because observations did not support it. Fossils of various sizes, shapes, and densities are found throughout the fossil record.

The fossil record does show a sequence, but not one based on density, size, shape, agility, or habitat. Rather, the sequence consists of first appearances of various organisms through time. One-celled organisms appear before multicelled ones, plants before animals, and invertebrates before vertebrates. Among vertebrates, fish appear first, followed in succession by amphibians, reptiles, mammals, and birds (• Figure 7.20).

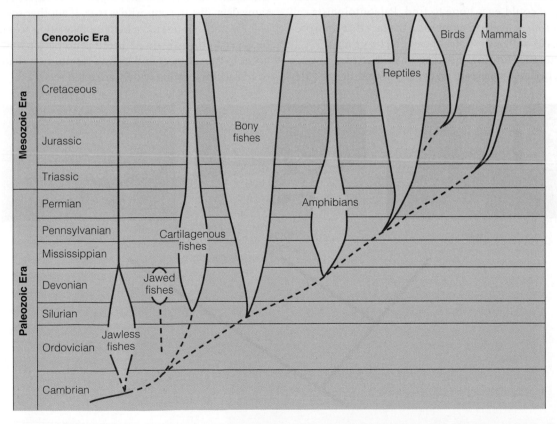

• **Figure 7.20 The Fossil Record and Evolution** The spindles indicate the time of appearance and the relative abundance of the members of each group of vertebrates. So, the jawless fishes appeared first and there are far more species of fish than there are of amphibians, reptiles, birds, and mammals. A mass extinction at the end of the Mesozoic Era is shown by the marked constriction of the reptile spindle.

Fossils are much more common than many people realize, so we might ask, "Where are the fossils showing the diversification of horses, rhinoceroses, and tapirs from a common ancestor; or the origin of birds from reptiles; or the evolution of whales from a land-dwelling ancestor?" It is true that the origin and initial diversification of a group is the most poorly represented in the fossil record, but in these cases, as well as many others, fossils of the kind we would expect are known.

Horses, rhinoceroses, and tapirs may seem an odd assortment of animals, but fossils and studies of living animals indicate that they, along with the extinct titanotheres and chalicotheres, share a common ancestor (• Figure 7.21). If this statement is correct, then we can predict that as we trace these animals back in the fossil record, differentiating one from the other should become increasingly difficult (Table 7.1). And, in fact, the earliest members of each are remarkably similar, differing mostly in size and details of their teeth. As their diversification proceeded, though, differences became more apparent.

Missing Links—Are They Really Missing?

The most profound misconception about the fossil record is that paleontologists supposedly have no fossils bridging the gaps between presumed ancestor and descendant groups of organisms. In short, these missing links, as they are popularly called, are claimed by some outside the sciences to invalidate the theory of evolution. Indeed, there are missing links, but there are also many fossils about as intermediate between groups as we could ever hope to find. Paleontologists call these *intermediate fossils* to emphasize the fact that they possess features of both ancestral and descendant groups.

It is true that finding "missing links" between species is difficult, because species in an evolutionary lineage differ so little from one another. However, when you consider genera, families, orders, and classes, transitional forms become much more common. For instance, the transition from fish to amphibians is exceptionally well documented by fossils (see Chapter 13), and some of the most advanced mammal-like reptiles and earliest mammals are so similar that distinguishing one from the other is difficult. Dugongs and manatees, collectively called sea cows, are thoroughly aquatic mammals with limbs modified into paddles. Fossils from Jamaica show that the ancestors of sea cows had four well-developed limbs and a pelvis that supported the animals on land—that is, they were sea cows with legs.

The Evidence—A Summary

As already noted, scientists agree that the theory of evolution is as well supported by evidence as any other major theory. For example, the transitional fossils we discussed in the preceding section provide compelling evidence for evolution, but fossils, although important, are not the only evidence. Indeed, much of the evidence comes from comparative anatomy, biogeography, molecular biology, genetics, embryology, and biochemistry, that is from converging lines of evidence.

Certainly scientists disagree on specific issues such as the rate of evolution, the significance of some fossils, and precise relationships among organisms, but they are overwhelmingly united in their support of the theory. And despite stories holding that scientists are unyieldingly committed to this theory and unwilling to investigate other explanations for the same data, just the opposite is true. If another scientific (testable) hypothesis offered a better explanation scientists would be eager to investigate it. There is no better way in the sciences to gain respect and lasting recognition than to modify or replace an existing widely accepted theory.

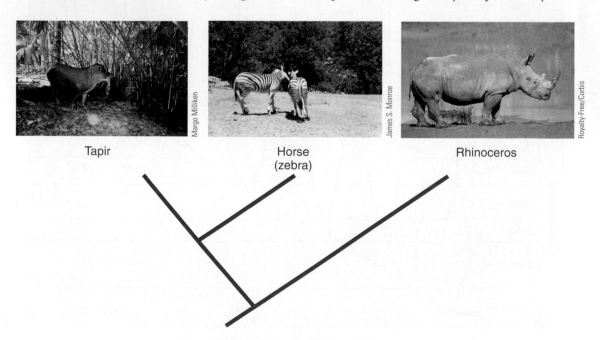

Tapir

Horse (zebra)

Rhinoceros

• Figure 7.21 **Cladogram Showing the Relationships Among Horses, Rhinoceroses, and Tapirs (Order Perissodactyla)** These relationships are well established by fossils (see Chapter 18) and by studies of living animals. When these animals first diverged from a common ancestor, all were remarkably similar, but as divergence continued, they became increasingly different.

SUMMARY

- The central claim of the theory of evolution is that all organisms have descended with modification from ancestors that lived during the past.

- The first widely accepted mechanism to account for evolution—inheritance of acquired characteristics—was proposed in 1809 by Jean-Baptiste de Lamarck.

- Charles Darwin's observations of variation in populations and artificial selection, as well as his reading of Thomas Malthus's essay on population, helped him formulate his idea of natural selection as a mechanism for evolution.

- In 1859, Charles Darwin and Alfred Wallace published their ideas of natural selection, which hold that in populations of organisms, some have favorable traits that make it more likely that they will survive and reproduce and pass on these traits.

- Gregor Mendel's breeding experiments with garden peas provided some of the answers regarding variation and how it is maintained in populations. Mendel's work is the basis for present-day genetics.

- Genes, specific segments of chromosomes, are the hereditary units in all organisms. Only the genes sex cells are inheritable.

- Sexual reproduction and mutations (changes in chromosomes or genes) account for most variation in populations.

- Evolution by natural selection is a two-step process. First, variation must exist or arise and be maintained in interbreeding populations, and second, organisms with favorable variants must be selected for survival so that they can reproduce.

- Many species evolve by allopatric speciation, which involves isolation of a small population from its parent population that is then subjected to different selection pressures.

- When diverse species arise from a common ancestor, it is called divergent evolution. The development of similar adaptive features in different groups of organisms results from convergent evolution and parallel evolution.

- Microevolution involves changes within a species, whereas macroevolution encompasses all changes above the species level. Macroevolution is simply the outcome of microevolution over time.

- Scientists have traditionally used phylogenetic trees to depict evolutionary relationships, but now they more commonly use cladistic analyses and cladograms to show these relationships.

- Background extinction occurs continually, but several mass extinctions have also taken place, during which Earth's biotic diversity has decreased markedly.

- The theory of evolution is truly scientific because we can think of observations that would falsify it, that is, prove it wrong.

- Much of the evidence supporting evolutionary theory comes from classification, comparative anatomy, embryology, genetics, biochemistry, molecular biology, and present-day examples of microevolution.

- The fossil record also provides evidence for evolution in that it shows a sequence of different groups appearing through time, and some fossils show the features we would expect in the ancestors of birds, mammals, horses, whales, and so on.

IMPORTANT TERMS

allele, p. 137
allopatric speciation, p. 140
analogous structure, p. 147
artificial selection, p. 134
chromosome, p. 138
cladistics, p. 142
cladogram, p. 143
convergent evolution, p. 142
deoxyribonucleic acid (DNA), p. 138
divergent evolution, p. 141
gene, p. 137

homologous structure, p. 147
inheritance of acquired characteristics, p. 133
living fossil, p. 144
macroevolution, p. 142
mass extinction, p. 145
meiosis, p. 138
microevolution, p. 142
mitosis, p. 138
modern synthesis, p. 138
mosaic evolution, p. 144

mutation, p. 139
natural selection, p. 135
paleontology, p. 132
parallel evolution, p. 142
phyletic gradualism, p. 140
punctuated equilibrium, p. 140
species, p. 140
theory of evolution, p. 132
vestigial structure, p. 147

REVIEW QUESTIONS

1. Genes for the same trait come in alternative forms called
   a. _____ meiosis; b. _____ alelles; c. _____ analogous structures; d. _____ mutations; e. _____ symbiosis.

2. Meiosis is the type of cell division that yields
   a. _____ deoxyribonucleic acids; b. _____ vestigial structures; c. _____ eggs and sperm; d. _____ cladograms; e. _____ inheritance of acquired characteristics.

3. Large-scale evolution resulting in the origin of new species and genera is an example of
   a. _____ phyletic gradualism; b. _____ Wallace's law; c. _____ binomial nomenclature; d. _____ cladistics; e. _____ macroevolution.

4. Inheritance of acquired characteristics was proposed in 1809 by
   a. _____ Alfred Russel Wallace; b. _____ Charles Lyell; c. _____ Jean-Baptiste de Lamarck; d. _____ Charles Robert Darwin; e. _____ Gregor Mendel.

5. The study of the history of life as revealed by the fossil record is
   a. _____ paleontology; b. _____ archaeology; c. _____ mineralogy; d. _____ physiology; e. _____ herpetology.

6. The forelimbs of horses, whales, and birds are examples of
   a. _____ inheritance of acquired characteristics; b. _____ convergent speciation; c. _____ artificial selection; d. _____ homologous structures; e. _____ parallel evolution.

7. A structure possessed by an animal that no longer serves any function, has a reduced function, or does something completely different than in the ancestral condition is a(n)
   a. _____ superfluous organ; b. _____ extra appendage; c. _____ parallel chromosome; d. _____ living fossil; e. _____ vestigial structure.

8. According to the concept of punctuated equilibrium,
   a. _____ species remain unchanged during most of their existence then evolve "rapidly"; b. _____ organisms from different classes interbreed but their offspring are infertile; c. _____ natural selection favors the largest and strongest for survival; d. _____ organisms evolve slowly but continuously; e. _____ animals and plants can pass on characteristics acquired during their lifetimes.

9. A time during which Earth's biotic diversity is reduced significantly is called
   a. _____ synthetic evolution; b. _____ allopatric speciation; c. _____ mass extinction; d. _____ phyletic explosion; e. _____ artificial selection.

10. Complex, helical molecules of deoxyribonucleic acid (DNA) are known as
    a. _____ paleofossils; b. _____ chromosomes; c. _____ allopatric cells; d. _____ embryos; e. _____ inherited features.

11. Does natural selection really mean that only the biggest, fastest, and strongest are likely to survive?

12. Discuss the concept of allopatric speciation, and give two examples of how it might take place.

13. What criteria are important in defining a vestigial structure? Give some examples.

14. How do divergent and convergent evolution compare? Give examples of both.

15. What are the concepts of phyletic gradualism and punctuated equilibrium? Is there any evidence supporting either?

16. Give three examples of predictions from evolutionary theory.

17. Explain what is meant by macroevolution, and give two examples from the fossil record.

18. Construct three cladograms showing the possible relationships among sharks, whales, and bears. Which one best depicts the relationships among these animals, and what criteria did you use to make your decision?

19. What kinds of evidence should we find in the fossil record if the theory of evolution is correct?

20. How did the experiments carried out by Gregor Mendel help answer the inheritance questions that plagued Darwin and Wallace?

APPLY YOUR KNOWLEDGE

1. Suppose that a powerful group in Congress were to mandate that all future genetic research had to conform to strict guidelines—specifically, that plants and animals should be exposed to particular environments so that they would acquire characteristics that would allow them to live in otherwise inhospitable areas. Furthermore, this group enacted legislation that prohibited any other kind of genetic research. Why would it be unwise to implement this research program and place restrictions on the type of research?

2. According to scientists, present-day tapirs, horses, and rhinoceroses are more closely related to one another than they are to any other living hoofed mammal. What kinds of evidence from biology and the fossil record would support this conclusion?

3. What if someone were to tell you that "evolution is only a theory that has never been proven," and that "the fossil record shows a sequence of organisms in older to younger rocks that was determined by their density, shape, and habitat." Why is the first statement irrelevant to theories in general, and what kinds of evidence would you cite to refute the second statement?

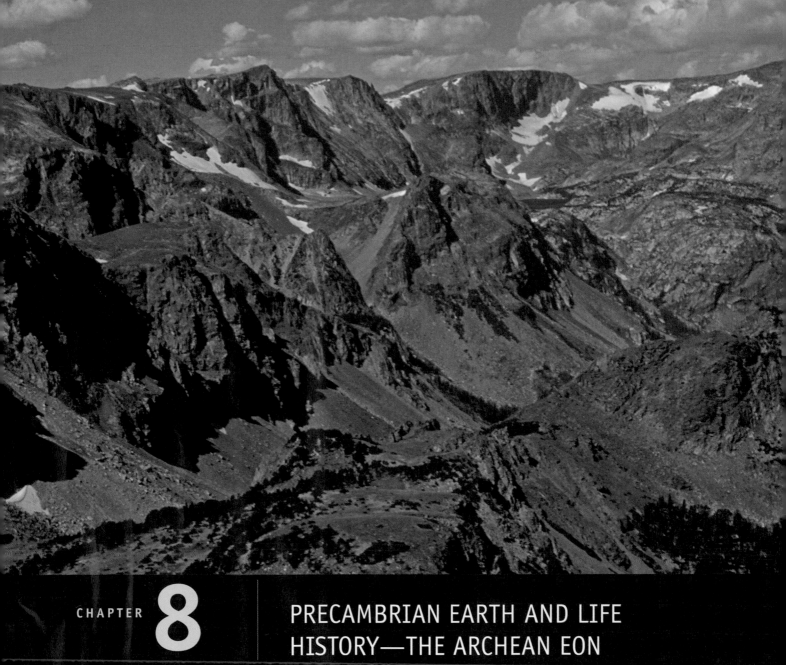

© Ralph Maughan

CHAPTER **8**

PRECAMBRIAN EARTH AND LIFE HISTORY—THE ARCHEAN EON

View of some of the rugged topography in the Beartooth Mountains on the Wyoming–Montana Border. Much of the area is more than 3000 m above sea level. The mountain range takes its name from a particularly jagged peak that looks much like a bear's tooth. The Beartooth Scenic Byway is advertised as one of the most beautiful in the United States. Most of the rocks of the Beartooth Mountains are Archean-age gneisses, and are among the oldest in North America.

[OUTLINE]

Introduction

What Happened During the Eoarchean?

Continental Foundations—Shields, Platforms, and Cratons

  Archean Rocks

  Greenstone Belts

  Evolution of Greenstone Belts

  Perspective *Earth's Oldest Rocks*

Archean Plate Tectonics and the Origin of Cratons

The Atmosphere and Hydrosphere

  How Did the Atmosphere Form and Evolve?

  Earth's Surface Waters—The Hydrosphere

The Origin of Life

  Experimental Evidence and the Origin of Life

  Submarine Hydrothermal Vents and the Origin of Life

  The Oldest Known Organisms

Archean Mineral Resources

Summary

At the end of this chapter, you will have learned that

- Precambrian time, which accounts for most of geologic time, is divided into two eons: the Archean and the younger Proterozoic.

- The Archean geologic record is difficult to interpret because many of the rocks are metamorphic, deformed, deeply buried, and they contain few fossils.

- Each continent has at least one area of exposed Precambrian rocks called a shield and a buried extension of the shield known as a platform. A shield and its platform make up a craton.

- All cratons show evidence of deformation, metamorphism, and emplacement of plutons, but they have been remarkably unaffected by these activities since Precambrian time.

- The two main associations of Archean rocks are granite-gneiss complexes, which are by far the most common, and greenstone belts that consist mostly of igneous rocks, but sedimentary rocks are present primarily in their upper parts.

- Greenstone belts likely formed in several tectonic settings, but many appear to have evolved in back-arc marginal basins and in rifts within continents.

- Plate tectonics was taking place during the Archean, but plates probably moved faster, and igneous activity was more common then because Earth possessed more heat from radioactive decay.

- Gases released during volcanism were responsible for the origin of the hydrosphere and atmosphere, but this early atmosphere had little or no free oxygen.

- The oldest known fossil organisms are single-celled bacteria and chemical traces of bacteria-like organisms known as archaea. Bacteria known as blue-green algae produced irregular mats and moundlike structures called stromatolites.

- Resources found in Archean rocks include gold, platinum, copper, zinc, and iron.

Introduction

We think of time from the human perspective of years or a few decades, so we have no frame of reference for time measured in millions or billions of years. Accordingly, the Precambrian, which lasted for more than 4.0 billion years, is longer than we can even imagine, and Earth has existed for 4.6 billion years. Consider this. Suppose that 1 second equals 1 year. If that were the case, and you were to count to 4.6 billion, the task would take you and your descendants nearly 146 years! Or suppose that a 24-hour clock represented 4.6 billion years, in which case the Precambrian alone would be more than 21 hours long and constitute about 88% of all geologic time (● Figure 8.1).

Precambrian is a widely used term that refers to both time and rocks. As a time term, it includes all geologic time from Earth's origin 4.6 billion years ago to the beginning of the Phanerozoic Eon 542 million years ago. The term also refers to all rocks lying below rocks of the Cambrian System. Unfortunately, no rocks are known for the first 600 million years of geologic time, so our geologic record actually begins about 4.0 billion years ago with the oldest known rocks on Earth except meteorites. The geologic record we do have for the Precambrian, particularly the older part of the Precambrian, is difficult to decipher because (1) many of the rocks have been metamorphosed and complexly deformed; (2) in many areas, they are deeply buried beneath younger rocks; and (3) they contain few fossils of any use in biostratigraphy.

Because of the complexities of the rocks and the scarcity of fossils, establishing formal subdivisions of the Precambrian is difficult. In 1982, in an effort to standardize terminology, the North American Commission on Stratigraphic Terminology proposed two eons for the Precambrian, the *Archean*, and the *Proterozoic*, both of which are now widely used. More recent work in 2004 by the International Commission on Stratigraphy (ICS) has refined the Precambrian subdivision as shown in ● Figure 8.2. It is important to note that these subdivisions are based on absolute ages rather than time-stratigraphic units, which is a departure from standard practice. Remember from Chapter 5 that the systems, such as the Cambrian System, are based on stratotypes, whereas the Precambrian designations are not.

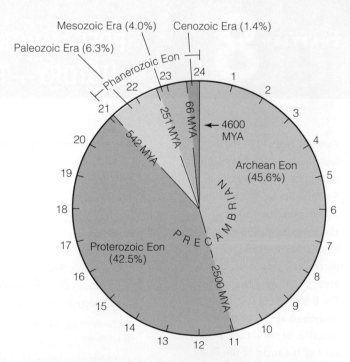

● Figure 8.1 **Geologic Time Represented by a 24-hour Clock**
If 24 hours represented all geologic time, the Precambrian would be more than 21 hours long—this is more than 88% of the total.

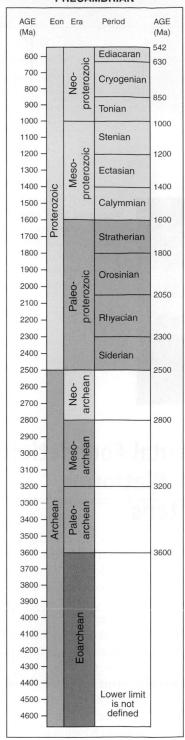

PRECAMBRIAN

• **Figure 8.2 The Precambrian Geologic Time Scale** This most recent version of the geologic time scale was published by the International Commission on Stratigraphy (ICS) in 2004. Notice the use of the prefixes *eo* (early or dawn), *paleo* (old or ancient), *meso* (middle), and *neo* (new or recent). The age columns on the left and right sides of the time scale are in hundreds and thousands of millions of years (1800 million years = 1.8 billion years, for example). See Figure 1.11 for the complete geologic time scale.

One important reason for studying Precambrian Earth and life history is that it constitutes most of geologic time, and many events that took place then set the stage for further evolution of the planet and life. It was during the Precambrian

that the Earth systems we introduced in Chapter 1 (see Figure 1.1) became operative, although not all at the same time or necessarily in their present form. Earth did not differentiate into a core and mantle until millions of years after it formed. Once it did, though, internal heat was responsible for moving plates (plate tectonics) and the origin and continuing evolution of the continents. Earth's early atmosphere evolved from a carbon dioxide-rich one to one with free oxygen and an ozone layer, organisms appeared as much as 3.5 billion years ago, and surface waters began to accumulate.

In short, Earth was very different when it first formed (• Figure 8.3), but during the Precambrian, it began to evolve and became increasingly more like it is now. Our task in this and the following chapters is to investigate the geologic record for the intervals of geologic time designated Precambrian (Archean and Proterozoic), Paleozoic, Mesozoic, and Cenozoic.

What Happened During the Eoarchean?

Eoarchean encompasses all geologic time from Earth's origin until the onset of the Paleoarchean 3.6 billion years ago (Figure 8.2). Unfortunately, Earth's oldest known body of rocks, the Acasta Gneiss in Canada, is about 4.0 billion years old (see Perspective), so we have no geologic record for much of the Archean. Nevertheless, it was during this time that Earth accreted from planetesimals and differentiated into a core and mantle (see Figures 1.5 and 1.6), and at least some continental crust formed perhaps as much as 4.4 billion years ago. Like the other terrestrial planets and the Moon, Earth was bombarded by meteorites, the rocky debris from the origin of the solar system, until about 3.8 billion years ago. Unlike the Moon, however, the evidence of this period of impacts has been obliterated on Earth by weathering and erosion, as well as by plate movements and mountain building. In addition to the accretion of planetesimals, the Eoarchean Earth was probably hit by a Mars-sized planetesimal 4.4 to 4.6 billion years ago, causing an ejection of a huge amount of hot material that eventually coalesced and formed the Moon.

When Earth first formed, it retained much of the residual heat from its origin and it had more heat generated by radioactive decay, and as result, volcanism was ubiquitous. An early atmosphere formed, but it was one very unlike the oxygen-rich one present now, and when the planet cooled sufficiently, surface waters began to accumulate. If we could somehow go back and visit early Earth, we would see a rapidly rotating, hot, barren, waterless planet bombarded by meteorites and comets. We would also see no continents, intense cosmic radiation, no organisms, and widespread volcanism (Figure 8.3). The age of the oldest continental crust is uncertain, but we can be sure that at least some was present by 4.0 billion years ago, and detrital sedimentary rocks in Australia with zircons ($ZrSiO_4$) 4.4 billion years old indicate that source rocks that old must have existed.

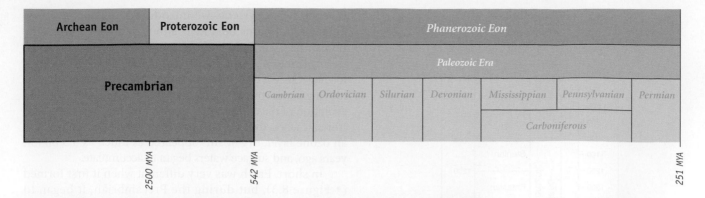

Archean Eon	Proterozoic Eon	Phanerozoic Eon						
		Paleozoic Era						
Precambrian		Cambrian	Ordovician	Silurian	Devonian	Mississippian	Pennsylvanian	Permian
						Carboniferous		

2500 MYA   542 MYA   251 MYA

• **Figure 8.3 Earth As It May Have Appeared Soon After It Formed** No rocks are known from this early time in Earth history, but geologists can make reasonable inferences about the nature of the newly-formed planet.

We know from studies in geology and astronomy that Earth's rate of rotation is slowing because of tidal effects. That is, friction caused by the Moon on Earth's oceanic waters, as well as its landmasses, causes its rate of rotation to slow very slightly every year. When Earth first formed, it probably rotated in as little as 10 hours. Another effect of the Earth—Moon tidal interaction is the recession of the Moon from Earth at a few centimeters per year. Accordingly, during the Eoarchean, the view of the Moon would have been spectacular.

Geologists agree that shortly after Earth formed, it was exceedingly hot and volcanism was widespread. However, rather than being a fiery orb for half a billion years as was formerly accepted, some geologists now think that Earth cooled sufficiently by 4.4 billion years ago for surface waters to accumulate. They base this conclusion on oxygen-18 to oxygen-16 ratios in tiny inclusions of oxygen trapped in zircon crystals that indicate reactions with surface water.

The first crust formed as upwelling mantle currents of mafic magma disrupted the surface, and numerous subduction zones developed to form the first island arcs (• Figure 8.4a). Weathering of the mafic rocks of island arcs yielded sediments richer in silica, and some of the magma in the arcs also became more enriched in silica. Collisions between island arcs eventually formed a few continental cores as silica-rich materials were metamorphosed (Figure 8.4b). Larger groups of merged island arcs, or protocontinents, grew faster by accretion along their margins than smaller ones did, and eventually the first continental nuclei or cratons formed (Figure 8.4c).

Continental Foundations— Shields, Platforms, and Cratons

Most of the present-day continents are above sea level, but parts of them lie beneath the seas. So, continents are not simply areas above sea level, but rather they consist of rocks with an overall composition similar to that of granite, whereas Earth's oceanic crust is made up of basalt and gabbro. In addition, continental crust is much thicker and less dense than oceanic crust. Furthermore, a **Precambrian shield** made up of an area or areas of exposed ancient rocks is found on all continents. Continuing outward from shields are broad **platforms** of buried Precambrian rocks that underlie large parts of all continents. A shield and its platform make up a **craton**, which we can think of as a continent's ancient nucleus (• Figure 8.5).

The cratons are the foundations of the continents, and along their margins more continental crust was added by accretion as they evolved to their present sizes and shapes (• Figure 8.6). In North America, for example, the Superior, Hearne, Rae, and Slave cratons, all within the **Canadian shield,** amalgamated along deformation belts to form a larger cratonic unit during the Proterozoic Eon (see Chapter 9). Both Archean- and Proterozic-age rocks are found in cratons, many of which indicate several episodes of deformation accompanied by igneous activity, metamorphism, and mountain building. However, most of

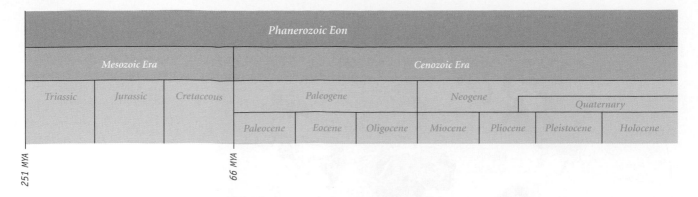

Phanerozoic Eon								
Mesozoic Era			Cenozoic Era					
Triassic	Jurassic	Cretaceous	Paleogene			Neogene		
							Quaternary	
			Paleocene	Eocene	Oligocene	Miocene	Pliocene	Pleistocene	Holocene

251 MYA

66 MYA

• Figure 8.4 **Origin of Granitic Continental Crust**

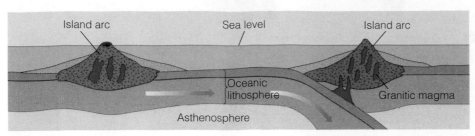

a An andesitic island arc forms by subduction of oceanic lithosphere and partial melting of basaltic oceanic crust. Partial melting of andesite yields granitic magma.

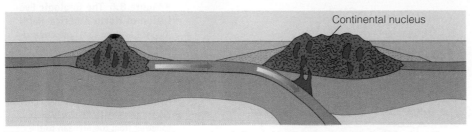

b The island arc in **a** collides with a previously formed island arc, thereby forming a continental core.

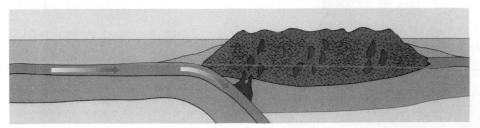

c The process occurs again when the island arc in **b** collides with the evolving continent, thereby forming a craton, the nucleus of a continent.

the areas within the cratons have experienced remarkably little deformation since the Precambrian.

In North America, the exposed part of the craton is the Canadian shield, which occupies most of northeastern Canada, a large part of Greenland, the Adirondack Mountains of New York, and parts of the Lake Superior region of Minnesota, Wisconsin, and Michigan (Figure 8.5). Much of the Canadian shield is an area of subdued topography, numerous lakes, exposed Archean and Proterozoic rocks thinly covered in places by Pleistocene glacial deposits. The rocks are volcanic, plutonic, and sedimentary, many of which have been altered to varying degrees by metamorphism.

Drilling and geophysical evidence indicate that Precambrian rocks, that is, rocks of the platform, underlie much of North America as well as other continents, but beyond the shields they are seen only in areas of erosion and uplift. For instance, Archean and Proterozoic rocks are present in the deeper parts of the Grand Canyon and in many ranges of the Rocky Mountains and the Appalachian Mountains (See chapter opening photo and • Figure 8.7).

Archean Rocks Only 22% of Earth's exposed Precambrian crust is Archean, with the largest exposures in Africa and North America. Archean crust is made up of a variety of rocks, but we characterize most of them as greenstone belts and granite-gneiss complexes, the latter being by far the most common. **Granite-gneiss complexes** are actually composed of a variety of rocks, with granitic gneiss and granitic plutonic rocks being the most common, that were probably derived from plutons emplaced in volcanic island arcs (Figure 8.4). Nevertheless, there are other rocks ranging from peridotite to sedimentary rocks, all of which have been metamorphosed. Greenstone belts are subordinate, accounting for only 10% of Archean rocks, and yet they are important in unraveling some of the complexities of Archean tectonic events.

Greenstone Belts A **greenstone belt** has three main rock association; its lower and middle parts are mostly volcanic, whereas the upper rocks are mostly sedimentary (• Figure 8.8a). Greenstone belts typically have a synclinal structure, measure anywhere from 40 to 250 km wide and 120 to 800 km long, and have been intruded by granitic

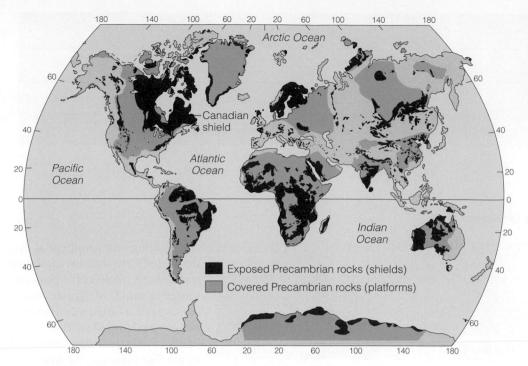

• Figure 8.5 **The Distribution of Precambrian Rocks** Areas of exposed Precambrian rocks constitute the shields, whereas the platforms consist of buried Precambrian rocks. A shield and its adjoining platform make up a craton.

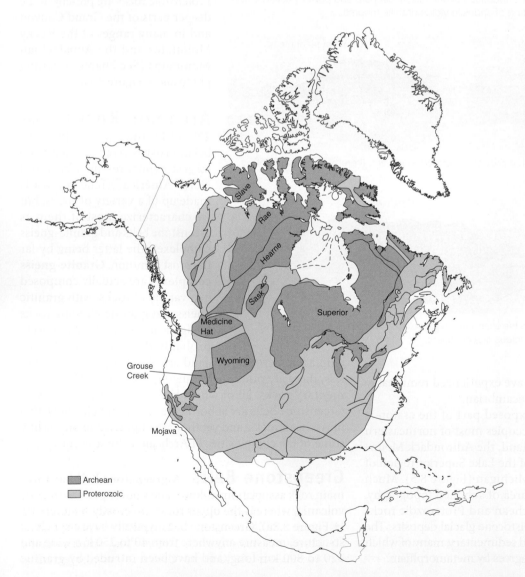

• Figure 8.6 **The Geologic Evolution of North America** North America evolved to its present size and shape by the amalgamation of Archean cratons that served as nuclei around which younger continental crust was added. We can think of North America, and the other continents, as a mosaic that was pieced together from the Archean to the present. The sequence of Precambrian events in North America's history are discussed in more detail in Chapter 9.

a Outcrop of gneiss cut by a granite dike from a granite-gneiss complex in Ontario, Canada.

b Shell Creek in the Bighorn Mountains of Wyoming has cut a gorge into this 2.9-billion-year-old (Mesoarchean) granite.

• **Figure 8.8** **Greenstone Belts and Granite-Gneiss Complexes**

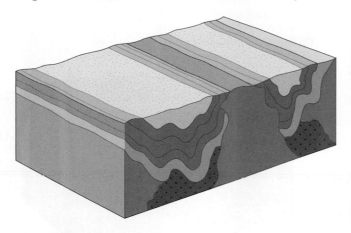

Greenstone belt succession →

▪ Granitic intrusives

▫ Upper sedimentary unit: sandstones and shales most common

▪ Middle volcanic unit: mainly basalt

▪ Lower volcanic unit: mainly peridotite and basalt

▪ Granite–gneiss complex

a Two adjacent greenstone belts. Older belts—those more than 2.8 billion years old—have an ultramafic unit overlain by a basaltic unit. In younger belts, the succession is from a basaltic unit. In younger belts, the succession is from a basaltic lower unit to an andesite-rhyolite unit. In both cases, sedimentary rocks are found mostly in the uppermost unit.

b Pillow lava of the Ishpeming greenstone belt in Michigan.

c Gneiss from a granite-gneiss complex in Ontario, Canada.

magma and cut by thrust faults. Many of the igneous rocks are greenish, because they contain green minerals such as chlorite, actinolite, and epidote that formed during low-grade metamorphism.

Thick accumulations of pillow lava are common in greenstone belts, indicating that much of the volcanism was subaqueous (Figure 8.8b). Pyroclastic materials, in contrast, almost certainly formed by subaerial eruptions where large volcanic centers built above sea level. The most interesting igneous rocks in greenstone belts are *komatiites* that cooled from ultramafic lava flows, which are rare in rocks younger than Archean, and none occur now.

To erupt, ultramafic magma (one with less than 45% silica) requires near-surface magma temperatures of more than 1600°C; the highest recorded surface temperature for recent lava flows is 1350°C. During its early history, however, Earth possessed more radiogenic heat, and the mantle was as much as 300°C hotter than it is now. Given these conditions, ultramafic magma could reach the surface, but as Earth's radiogenic heat production decreased, the mantle cooled and ultramafic flows no longer occurred.

Sedimentary rocks are found throughout greenstone belts, but they predominate in the upper unit (Figure 8.8a). Many of these rocks are successions of *graywacke* (sandstone with abundant clay and rock fragments) and *argillite* (slightly metamorphosed mudrocks). Small-scale cross-bedding and graded bedding indicate these rocks represent turbidity current deposition (see Figure 6.4).

Other sedimentary rocks are also present, including sandstone, conglomerate, chert, and carbonates, although none are very abundant. Iron-rich rocks known as *banded iron formations* are also found, but they are more typical of Proterozoic deposits, so we discuss them in Chapter 9.

Before leaving this topic entirely, we should mention that the oldest large, well-preserved greenstone belts are in South Africa, dating from about 3.6 billion years ago. In North America, most greenstone belts are found in the Superior and Slave cratons of the Canadian shield (• Figure 8.9), but they are also found in Michigan, Wyoming, and several other areas. Most formed between 2.7 and 2.5 billion years ago.

Evolution of Greenstone Belts
Geologists know that greenstone belts are found mostly in Archean and Proterozoic terrains, and that they occur in multiple, parallel belts, each separated from the next by granite-gneiss complexes (Figure 8.8a). As to their origin, however, most geologists would probably agree that greenstone belts formed in several tectonic settings. One appealing model for the origin of some Archean greenstone belts relies on plate tectonics and the development of greenstone belts in **back-arc marginal basins** that subsequently close (• Figure 8.10). The present-day Sea of

• **Figure 8.10** **Origin of a Greenstone Belt in a Back-Arc Marginal Basin**

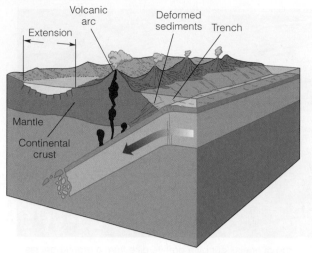

a Rifting on the continent side of a volcanic arc forms a back-arc marginal basin. Partial melting of subducted oceanic lithosphere supplies andesite and diorite magmas to the island arc.

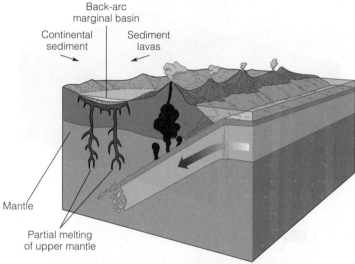

b Basalt lavas and sediment derived from the island arc and continent fill the back-arc marginal basin.

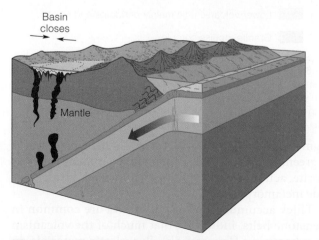

c Closure of the back-arc marginal basin, compression, and deformation. A syncline-like structure forms which is intruded by granitic magma.

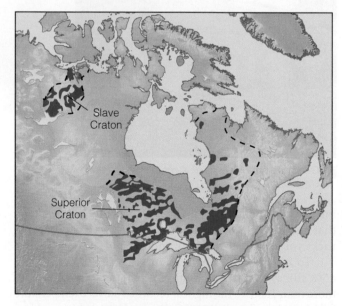

• **Figure 8.9** **Greenstone Belts in North America** Archean greenstone belts (shown in dark green) of the Canadian shield are mostly in the Superior and Slave cratons.

Japan lying between a volcanic island arc and mainland Asia is a good example of a back-arc marginal basin.

According to this model, there is an early stage of extension as the back-arc marginal basin forms, accompanied by volcanism, emplacement of plutons, and sedimentation, followed by an episode of compression when the basin closes. During this latter stage, the evolving greenstone belt rocks are deformed, metamorphosed, and intruded by granitic magma. Proponents of this model suggest that multiple episodes of the opening and closing of back-arc marginal basins account for the parallel arrangement of greenstone belts.

The back-arc marginal basin model for the origin of greenstone belts is well accepted by many geologists, but others think that some of these belts form in intracontinental rifts above rising mantle plumes (• Figure 8.11). As the plume rises beneath sialic (silica- and aluminum-rich) crust, it spreads and generates tensional forces that cause rifting. The mantle plume is the source of the lower and middle volcanic units of the greenstone belt, and the upper sedimentary unit results from erosion of the volcanic rocks along the flanks of the rift. And finally, there is an episode of closure of the rift, deformation, low-grade metamorphism, and emplacement of plutons (Figure 8.11).

• Figure 8.11 **Origin of a Greenstone Belt in an Intracontinental Rift**

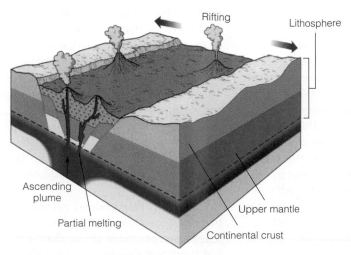

a An ascending mantle plume causes rifling and volcanism.

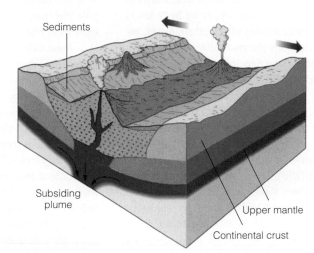

b As the plume subsides, erosion of the rift flanks yields sediments.

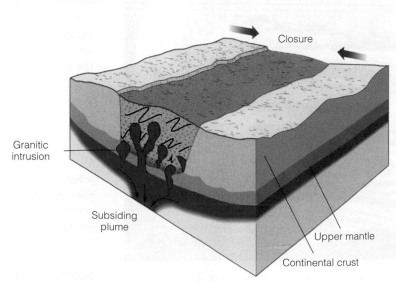

c Closure of the rift causes compression and deformation. Granitic magma intrudes the greenstone belt.

Perspective

Earth's Oldest Rocks

You already know that every continent has an ancient nucleus or craton around which new continental material has been added by accretion, a process that began during the Precambrian and continues even now. For instance, the Yakutat terrane, a block of crust measuring about 220 km by 75 km, is currently docking with southern Alaska. We discuss this phenomenon of continental accretion more fully in Chapter 9. In any event, each continent's most ancient rocks, except for meteorites, are found within their cratons.

Meteorites represent the oldest material in the solar system, and their ages determined by radioactive dating techniques cluster around 4.5 to 4.6 billion years old (BYO). The Allende meteorite that hit northern Mexico in 1869 is 4.563 BYO plus or minus 3.0 million years, and the oldest Moon rock is 4.45 BYO. Presumably, the entire solar system formed at the same time,

about 4.6 billion years ago, but, unfortunately, there are no rocks on Earth this old. Nevertheless, scientists have discovered that the Acasta Gneiss in Canada, Earth's oldest known rock outcrop, is 4.03 to 4.055 BYO, and because the rocks are metamorphic their parent rocks must have been even older (Figure 1).

These rocks from Canada are the most ancient ones discovered so far, but based on evidence from the mineral zircon even older rocks existed. Many rocks contain the mineral zircon ($ZrSiO_4$) as an accessory, especially granitic rocks and silica-rich (felsic) volcanic rocks, as well as some metamorphic and sedimentary rocks. Zircon is an extremely durable mineral that commonly shows concentric zones around a central core, each zone indicating an episode in the mineral's history; obviously the central part is the oldest. Zircon is the birthstone for December, and it is

mined for abrasives, the metal zirconium, and crucibles used to fuse platinum.

Geologists collect zircon-bearing rocks and grind them to separate individual mineral grains. Then, they separate the zircons from other minerals by settling in a fluid (zircon is denser than most other minerals in these rocks), or by its property of diamagnetism (it is repelled by a magnetic field). As noted, zircons are commonly zoned, that is, made up of zones of slightly different composition and different ages. Scientists analyze these zones to find isotopic composition, and because zircons contain small amounts of uranium that decays to lead scientists determine the absolute ages of the zones.

Scientists have analyzed zircons from the Yilgarn craton of Australia that were deposited in stream channels and found that they are 4.3 to 4.4 BYO, indicating that some continental crust was present by then. Finding evidence of older rocks is possible but unlikely, because much of Earth's ancient crust has been metamorphosed, recycled, deformed, or otherwise disrupted. For example, from the time of its origin until about 3.8 billion years ago, Earth, as well as the other terrestrial planets and the Moon, was heavily impacted by meteorites and comets.

Of course, any dating method does not work under all circumstances, but the remarkable thing about radiometric dating is its consistency. The age of the Acasta Gneiss noted above was determined by five independent dating methods. And furthermore, detailed analyses indicate that the gneiss experienced at least four episodes of igneous activity or metamorphism between 3.4 and 4.0 billion years ago. Perhaps older rocks will be found one day, but for the present, the Acasta Gneiss is the record holder for Earth's oldest rock.

<small>Copyright and photograph by Dr. Paravinder S. Sethi</small>

Figure 1 At about 4.0 billion years old, the Acasta Gneiss from the Northwest Territories of Canada is the oldest known rock on Earth, except for meteorites.

Archean Plate Tectonics and the Origin of Cratons

Certainly, the present plate tectonic regime of opening and closing ocean basins has been a primary agent in Earth evolution since the Paleoproterozoic. Most geologists are convinced that some kind of plate tectonic activity took place during the Archean as well, but it differed in detail from what is going on now. With more residual heat from Earth's origin and more radiogenic heat, plates must have moved faster and magma was generated more rapidly. As a result, continents no doubt grew more rapidly along their margins, a process called **continental accretion,** as plates collided with island arcs and other plates. Also, ultramafic lava flows (komatiites) were more common.

There were, however, marked differences between the Archean world and the one that followed. We have little evidence of Archean rocks deposited on broad, passive continental margins, but associations of passive continental margin sediments are widespread in Proterozoic terrains. Deformation belts between colliding cratons indicate that Archean plate tectonics was active, but the *ophiolites* so typical of younger convergent plate boundaries are rare.

Nevertheless, geologists have reported probable Neoarchean ophiolites from several areas. One in Russia is about 2.8 billion years old, whereas the Dongwanzi complex of China is 2.5 billion years old. However, not all agree that this one in China is actually an ophiolite sequence; the rocks have been complexly deformed and intruded by magma possibly as young as Mesozoic. Another possible ophiolite in the Wind River Mountains of Wyoming is also highly deformed and difficult to interpret. Whether these rocks truly represent ophiolites may be an open question, but there is no doubt that ophiolite sequences were present by the Paleoproterozoic (see Chapter 9).

In any case, several small cratons were present during the Archean and grew by accretion along their margins (Figure 8.6). By the end of the Archean, perhaps 30% to 40% of the present volume of continental crust had formed. Remember, though, that these cratons amalgamated into a larger unit during the Proterozoic (see Chapter 9). A plate tectonic model for the Archean crustal evolution of the southern Superior craton of Canada relies on the evolution of greenstone belts, plutonism, and deformation (• Figure 8.12). We can take this as a model for Archean crustal evolution in general.

The events responsible for the origin of the southern Superior craton (Figure 8.12b) are part of a much more extensive orogenic episode that took place during the Mesoarchean and Neoarchean. Deformation was responsible for the origin of some of the Archean rocks in several parts of the Canadian shield as well as in Wyoming, Montana, and the Mississippi River Valley. In the northwestern part of the Canadian shield, deformation along the Snowbird tectonic zone yielded metamorphic rocks 3.2 and 2.6 billion years old that belong to the granulite metamorphic facies, which form at very high temperatures, at least 700°C, and some form at more than 1000°C. By the time this Archean event had ended, several cratons had formed that are now found in the older parts of the Canadian shield (see Chapter 9).

The Atmosphere and Hydrosphere

In Chapter 1, we emphasized the interactions among systems, two of which, the atmosphere and hydrosphere, have had a profound impact on Earth's surface (see Figure 1.1). Shortly after Earth formed, its atmosphere and hydrosphere,

• Figure 8.12 **Origin of the Southern Superior Craton**

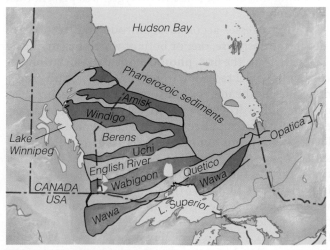

a Geologic map showing greenstone belts (dark green) and areas of granite-gneiss complexes (light green).

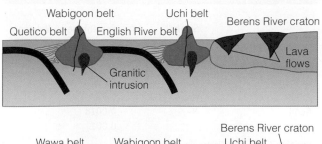

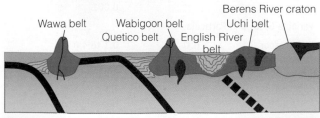

b Plate tectonic model for the evolution of the southern Superior craton. The figure is a north-south cross section, and the upper diagram shows an earlier stage of development than the lower one.

although present, were quite different from the way they are now. They did, however, play an important role in the development of the biosphere.

How Did the Atmosphere Form and Evolve?

Today, Earth's atmosphere is quite unlike the noxious one we described earlier. Now it is composed mostly of nitrogen and abundant free oxygen (O_2), meaning oxygen not combined with other elements as in carbon dioxide (CO_2) and water vapor (H_2O). It also has small but important amounts of other gases such as ozone (O_3) (Table 8.1), which, fortunately for us, is common enough in the upper atmosphere to block most of the Sun's ultraviolet radiation.

Earth's very earliest atmosphere was probably composed of hydrogen and helium, the most abundant gases in the universe. If so, it would have quickly been lost into space, for two reasons. First, Earth's gravitational attraction is insufficient to retain gases with such low molecular weights. And second, before Earth differentiated, it had no core or magnetic field. Accordingly, it lacked a *magnetosphere*, the area around the planet within which the magnetic field is confined, so a strong solar wind, an outflow of ions from the Sun, would have swept away any atmospheric gases. Once Earth had differentiated and a magnetosphere was present, though, an atmosphere began accumulating as a result of **outgassing** involving the release of gases from Earth's interior during volcanism (• Figure 8.13).

Water vapor is the most common gas emitted by volcanoes today, but they also emit carbon dioxide, sulfur dioxide, carbon monoxide, sulfur, hydrogen, chlorine, and nitrogen. No doubt Archean volcanoes emitted the same

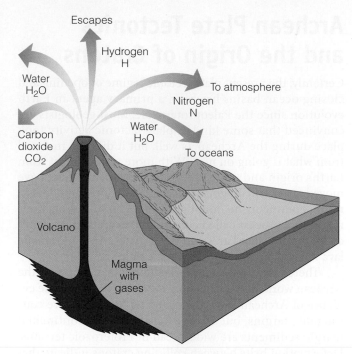

• Figure 8.13 **Outgassing and Earth's Early Atmosphere** Erupting volcanoes emit mostly water vapor, carbon dioxide, and several other gases but no free oxygen, that is, oxygen not combined with other elements. In addition to the gases shown here, chemical reactions in the early atmosphere probably yielded methane (CH_4) and ammonia (NH_3).

gases, and thus an atmosphere developed, but one lacking free oxygen and an ozone layer. It was, however, rich in carbon dioxide, and gases reacting in this early atmosphere probably formed ammonia (NH_3) and methane (CH_4).

This early oxygen-deficient but carbon dioxide–rich atmosphere persisted throughout the Archean. Some of the evidence for this conclusion comes from detrital deposits containing minerals such as pyrite (FeS_2) and uraninite (UO_2), both of which oxidize rapidly in the presence of free oxygen. So, the atmosphere was a chemically reducing one rather than an oxidizing one. However, oxidized iron becomes increasingly common in Proterozoic rocks, indicating that at least some free oxygen was present then (see Chapter 9).

Two processes account for introducing free oxygen into the atmosphere, one or both of which began during the Eoarchean. The first, **photochemical dissociation,** involves ultraviolet radiation from the Sun disrupting water molecules in the upper atmosphere, thus releasing their oxygen and hydrogen (• Figure 8.14). This process may eventually have supplied 2% of the present-day oxygen level, but with this amount of free oxygen in the atmosphere ozone forms, creating a barrier against ultraviolet radiation and the formation of more ozone. Even more important was **photosynthesis,** a metabolic process in which organisms use carbon dioxide and water to make organic molecules, and release oxygen as a waste product (Figure 8.14). Even so, probably no more than 1% of the free oxygen level of today was present by the end of the Archean, 2.5 billion years ago.

TABLE **8.1**	Composition of Earth's Present-Day Atmosphere	
	Symbol	Percentage by Volume*
Nonvariable Gases		
Nitrogen	N_2	78.08
Oxygen	O_2	20.95
Argon	Ar	0.93
Neon	Ne	0.0018
Others		0.001
Variable Gases and Particulates		
Water vapor	H_2O	0.1 to 4.0
Carbon dioxide	CO_2	0.038
Ozone	O_3	0.000004
Other gases		Trace
Particulates		Normally trace

*Percentages, except for water vapor, are for dry air.

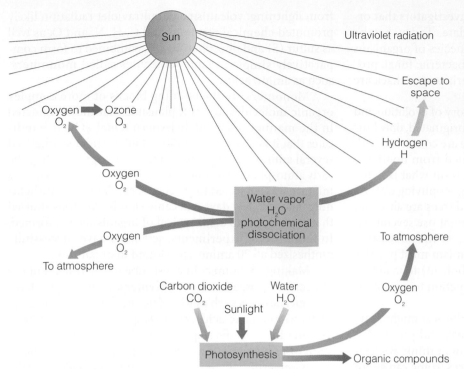

• Figure 8.14 **Evolution of the Atmosphere** Photochemical dissociation occurs when ultraviolet radiation disrupts water molecules that release hydrogen (H_2) and free oxygen (O_2), some of which is converted to ozone (O_3) that blocks most of the ultraviolet radiation. In the presence of sunlight, photosynthesizing organisms use carbon dioxide (CO_2) and water (H_2O) to make organic molecules and in the processes release free oxygen as a waste product.

Earth's Surface Waters—The Hydrosphere

All water on Earth is part of the hydrosphere, but most of it—more then 97%—is in the oceans. Where did it come from, and how has it changed? Certainly, outgassing released water vapor from Earth's interior, and once the plant cooled sufficiently, water vapor condensed and surface waters began to accumulate. Another source of water vapor, and eventually liquid water, was meteorites and especially icy comets (Figure 8.3). It is not known whether outgassing or meteorites and comets was most important, but we do know that oceans were present by the Eoarchean, although their volumes and geographic extent cannot be determined. Nevertheless, we can envision an Archean Earth with numerous erupting volcanoes and an early episode of intense meteorite and comet bombardment accounting for a rapid rate of surface water accumulation.

Following Earth's early episode of meteorite and comet bombardment, which ended about 3.8 billion years ago, these extraterrestrial bodies have added little to the accumulating surface waters. However, volcanoes continue to erupt and expel water vapor (much of it recycled surface water), so is the volume of the oceans increasing? Probably it is, but at a considerably reduced rate, because much of Earth's residual heat from its origin has dissipated and the amount of radioactive decay to generate internal heat has diminished, so volcanism is not nearly as commonplace (• Figure 8.15). Accordingly, the amount of water added to the oceans now is trivial compared to their volumes.

Recall from Chapter 4 that one early attempt to determine Earth's age was to calculate how long it took for the oceans to reach their current salinity level—assuming, of course, that the oceans formed soon after Earth did, that

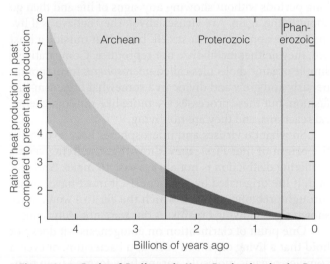

• Figure 8.15 **Ratio of Radiogenic Heat Production in the Past Compared to the Present** The colored band encloses the ratios according to different models, but all show an exponential decay of radioactive elements through time. Heat production 4 billion years ago was three to six times as great as it is now, whereas radiogenic heat production at the beginning of the Phanerozoic Eon was only slightly higher than at present.

they were freshwater to begin with, and that their salinity increased at a uniform rate. None of these assumptions is correct, so the ages determined were vastly different. We now know that the very early oceans were salty, probably about as salty as they are now. That is, very early in their history, the oceans reached chemical equilibrium and have remained in near-equilibrium conditions ever since.

The Origin Of Life

The oldest known fossils are in rocks perhaps 3.5 billion years old, and chemical evidence in 3.8-billion-year-old rocks in Greenland convince some investigators that organisms were present at that early date. Today, Earth's biosphere is made up of millions of species of organisms assigned to six kingdoms—archaea*, bacteria, fungi, protists, plants, and animals. Only bacteria and archaea are known from Archean rocks.

In Chapter 7, we discussed the theory of evolution and noted that it does not address how life originated, only how life has changed through time. Here we are concerned with **abiogenesis,** that is, how life originated from nonliving matter. However, first we must be clear on what is living and nonliving. In most cases, the living–nonliving distinction is easy enough—dogs, worms, and trees are alive, but rocks and water are not. A biologist might use several criteria to make this distinction, such as growth and reaction to stimuli, but minimally a living organism must practice some kind of chemical activity (metabolism) to maintain itself, and it must be capable of reproduction to ensure the long-term survival of its group.

This metabolism/reproduction criterion might seem sufficient to decide if something is alive, and yet the distinction is not always as clear as you might think. Bacteria are living, but under some circumstances, some can go for long periods without showing any signs of life and then go about living again. Are viruses alive? They behave as if living in the appropriate host cell, but when outside a host cell, they neither metabolize nor reproduce. Comparatively simple organic molecules called *microspheres* form spontaneously and grow and divide in a somewhat organism-like fashion, but these processes are more like random chemical reactions, and they are not living.

So what do viruses and microspheres have to do with the origin of life? First, they show that the living versus nonliving distinction is not always easy to make. And second, if life originated by abiogenesis, it must have passed through prebiotic stages in which the entities would have shown signs of living organisms but were not truly living.

One point of clarification on abiogenesis—it does not hold that a living organism such as a bacterium, or even a complex organic molecule, sprang fully formed from nonliving matter. Rather than one huge step from nonliving to living, the origin of life involved several small steps, each leading to an increase in organization and complexity.

Experimental Evidence and the Origin of Life

Investigators disagree on many aspects of the origin of life, but they do agree that two requirements were necessary for abiogenesis: (1) a source of the appropriate elements for organic molecules and (2) energy to promote chemical reactions. All organisms are composed mostly of carbon, hydrogen, nitrogen, and oxygen, all of which were present in Earth's early atmosphere as carbon dioxide (CO_2), water vapor (H_2O), and nitrogen (N_2), and possibly methane (CH_4) and ammonia (NH_3). Energy

from lightning, volcanism, and ultraviolet radiation likely promoted chemical reactions and C, H, N, and O, as well as sulfur (S) and phosphorous (P), combined to form comparatively simple organic molecules called **monomers,** such as amino acids.

Monomers are the building blocks of more complex organic molecules, but is it plausible that they originated in the manner postulated? Experimental evidence indicates that it is. During the 1950s, Stanley Miller synthesized several amino acids by circulating gases approximating the early atmosphere in a closed glass vessel (● Figure 8.16). This mixture was subjected to an electric spark to simulate lightning, and in a few days it became cloudy. Analysis showed that several amino acids typical of organisms had formed. In more recent experiments, scientists have successfully synthesized all 20 amino acids found in organisms.

Making monomers in a test tube is one thing, but the molecules of organisms are **polymers,** including proteins and nucleic acids such as RNA (ribonucleic acid) and DNA (deoxyribonucleic acid), consisting of monomers linked together in a specific sequence. So, how did this phenomenon of polymerization take place? This is more difficult to answer, especially if it occurred in water, which usually causes depolymerization. However, researchers have synthesized small molecules known as *proteinoids* or *thermal proteins,* some of which consist of more than 200 linked amino acids (● Figure 8.17a), when dehydrated concentrated amino acids were heated. Under these conditions, the concentrated amino acids spontaneously polymerized to form thermal proteins. Perhaps similar conditions for polymerization existed on early Earth.

At this stage we refer to these molecules as *protobionts,* which are intermediate between inorganic chemical compounds and living organisms. These protobionts would have ceased to exist, though, if some kind of outer

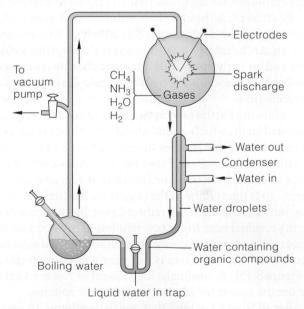

● **Figure 8.16 Stanley Miller's Experimental Apparatus** Miller synthesized several amino acids used by organisms. The spark discharge was to simulate lightning, and the gases approximated Earth's early atmosphere.

---

*Archaea includes microscopic organisms that resemble bacteria, but differ from them genetically and biochemically.

• Figure 8.17 **Experimental Production of Thermal Proteins and Microspheres**

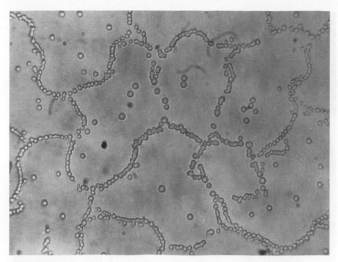

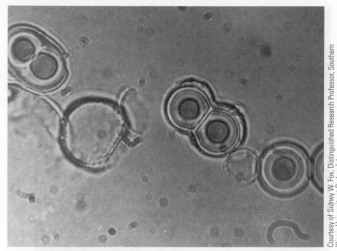

a Bacteria-like thermal proteins

b Thermal proteins aggregate to form microspheres.

membrane had not developed. In other words, they had to be self-contained, as cells are now. Experiments show that thermal proteins spontaneously aggregate into microspheres (Figure 8.17b) that are bounded by a cell-like membrane and grow and divide much as bacteria do.

The origin-of-life experiments are interesting, but how can they be related to what may have occurred on Earth more than 3.5 billion years ago? Monomers likely formed continuously and by the billions, and accumulated in the early oceans, which was, as the British biochemist J. B. S. Haldane called it, a hot, dilute soup. The amino acids in this "soup" may have washed up onto a beach or perhaps cinder cones, where they were concentrated by evaporation and polymerized by heat. The polymers were then washed back into the ocean, where they reacted further.

Not much is known about the next critical step in the origin of life: the development of a reproductive mechanism. The microspheres shown in Figure 8.17b divide, and some experts think they represent a protoliving system. However, in today's cells, nucleic acids, either RNA or DNA, are necessary for reproduction. The problem is that nucleic acids cannot replicate without protein enzymes, and the appropriate enzymes cannot be made without nucleic acids. Or so it seemed until fairly recently.

Now we know that small RNA molecules can replicate without the aid of protein enzymes. In view of this evidence, the first replicating systems may have been RNA molecules. In fact, some researchers propose an early "RNA world" in which these molecules were intermediate between inorganic chemical compounds and the DNA-based molecules of organisms. However, RNA cannot be easily synthesized under conditions that probably prevailed on early Earth, so how they were naturally synthesized remains an unsolved problem.

Intractable problems such as RNA replication have a way of eventually being solved. For instance, for many years scientists thought that the ribose of RNA was "impossible" to make by prebiotic synthesis from any conceivable molecular mixture that was assumed to exist on early

Earth. Finally, however, they found that it could indeed form naturally under conditions that were likely to have existed during the Eoarchean Eon.

Researchers agree on some of the basic requirements for the origin of life, but the exact steps involved and the significance of some experimental results are debated. Indeed, our discussion has focused on the role of atmospheric gases in the origin of monomers and polymers, but some scientists think that life originated near hydrothermal vents on the seafloor.

Submarine Hydrothermal Vents and the Origin of Life

Seawater seeps down into the oceanic crust at or near spreading ridges, becomes heated by magma, and rises and discharges onto the seafloor as **submarine hydrothermal vents.** Some of these vents, which are now known from the Atlantic, Pacific, and Indian Oceans, are called **black smokers** because they discharge water saturated with dissolved minerals, giving them the appearance of black smoke (• Figure 8.18a). Because Earth had more radiogenic heat during its early history it is reasonable to infer that submarine hydrothermal vents were more common then.

Submarine hydrothermal vents are interesting because several minerals containing zinc, copper, and iron precipitate around them, and they support communities of organisms previously unknown to science (Figure 8.18b). In recent years, some investigators have proposed that the first self-replicating molecules came into existence near these vents on the seafloor. The necessary elements (C, H, O, and N), as well as sulfur (S) and phosphorous (P), are present in seawater, and heat could have provided the energy necessary for monomers to form. Those endorsing this hypothesis hold that polymerization took place on the surfaces of clay minerals, and, finally, protocells were deposited on the seafloor. In fact, amino acids have been detected in some hydrothermal vent emissions.

• Figure 8.18 **Submarine Hydrothermal Vents**

a This black smoker on the East Pacific Rise is at a depth of 2800 m. The plume of "black smoke" is heated seawater with dissolved minerals.

b Several types of organisms including these tube worms, which may be up to 3 m long, live near black smokers. Bacteria that oxidize sulfur compounds lie at the base of the food chain.

Not all scientists are convinced this is a viable environment for the origin of life. They point out that polymerization would be inhibited in this hot, aqueous environment, and further claim that life adapted to this hostile environment after it originated elsewhere. So, this alternate location for the origin of life remains controversial, at least for now.

The Oldest Known Organisms

The first organisms were bacteria and archaea, both of which consist of **prokaryotic cells**—that is, cells that lack an internal, membrane-bounded nucleus and other structures typical of *eukaryotic cells.* (We discuss eukaryotic cells in Chapter 9.) Members of the archaea are single-celled, as are bacteria, and in fact they look much like bacteria but differ from them in the details of their genetics and chemical makeup. In addition, archaea can live in extremely hot, acidic, and saline environments where even bacteria cannot survive. In any case, prior to the 1950s, scientists assumed that the fossils common in Cambrian-age rocks had a long, earlier history, but the fossil record offered little to support this idea. In fact, it appeared that Precambrian rocks were so devoid of fossils that this interval of time was formerly called the *Azoic,* meaning "without life."

During the early 1900s, Charles Walcott described layered moundlike structures from the Paleoproterozoic-age Gunflint Chert of Ontario, Canada. He proposed that these structures, now called **stromatolites,** represented reefs constructed by algae (• Figure 8.19), but not until 1954 did paleontologists demonstrate that stromatolites are the product of organic activity. Present-day stromatolites form and grow as sediment grains are trapped on sticky mats of photosynthesizing cyanobacteria (blue-green algae), although now they are restricted to environments where snails cannot live.

Currently, the oldest known undisputed stromatolites are found in 3.0-billion-year-old rocks in South Africa, but probable ones are also known from the 3.3- to 3.5-billion-year-old Warrawoona Group in Australia. And, as noted previously, chemical evidence in rocks 3.8 billion years old in Greenland indicates life was perhaps present then. These oldest known cyanobacteria were photosynthesizing organisms, but photosynthesis is a complex metabolic process that must have been preceded by a simpler type of metabolism.

No fossils are known of these earliest organisms, but they must have resembled tiny bacteria that needed no free oxygen, so they were **anaerobic.** In addition, they must have been totally dependent on an external source of nutrients—that is, they were **heterotrophic,** as opposed to **autotrophic** organisms, which make their own nutrients, as in photosynthesis. And, finally, they were all prokaryotic cells. The nutrient source for these earliest organisms was most likely adenosine triphosphate (ATP), which was used to drive the energy-requiring reactions in cells. ATP can easily be synthesized from simple gases and phosphate, so it was no

• Figure 8.19 **Stromatolites**

Courtesy of Phillip E. Playford, Geological Survey of Western Australia

a Present-day stromatolites exposed at low tide at Shark Bay, Australia.

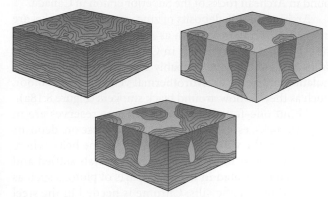

b Different types of stromatolites include irregular mats, isolated columns, and columns linked by mats.

doubt available in the early Earth environment. These organisms must have acquired their ATP from their surroundings, but this situation could not have persisted for long as more and more cells competed for the same resources. The first organisms to develop a more sophisticated metabolism probably used *fermentation* to meet their energy needs. Fermentation is an anaerobic process in which molecules such as sugars are split, releasing carbon dioxide, alcohol, and energy. In fact, most living prokaryotic cells ferment.

Other than the origin of life itself, the most significant biologic event of the Archean was the development of the autotrophic process of photosynthesis as much as 3.5 billion years ago. The cells were still prokaryotic and anaerobic, but as autotrophs, they were no longer dependent on preformed organic

molecules as a source of nutrients. The Archean microfossils in • Figure 8.20 are anaerobic, autotrophic prokaryotes.

Archean Mineral Resources

Of the several types of mineral deposits found in Archean rocks, gold is the one most commonly associated with them, although it is also found in Proterozoic and Phanerozoic rocks. This soft yellow metal has been the cause of feuds and wars and was one incentive for European exploration of the Americas. It is prized for jewelry, but it is or has also been used as a monetary standard, and it is still used in glass making, electrical circuitry, and the chemical industry. Since 1886, Archean and Proterozoic rocks in South Africa have

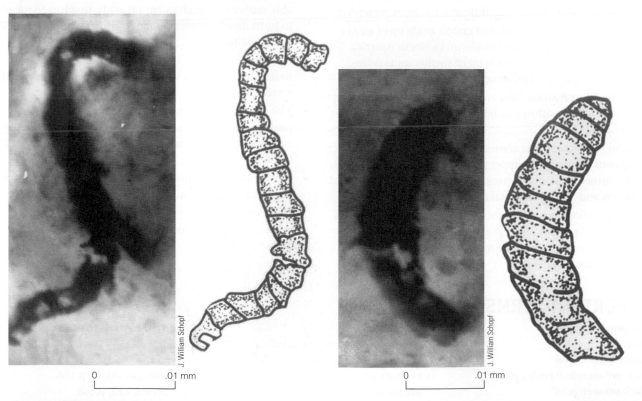

J. William Schopf

• Figure 8.20 **Probable Archean Microfossils** These fossils and their restorations are from the Warrawoona Group of Western Australia resemble bacteria. They were found in layers of chert between lava flows that have been dated; the chert layer is 3.465 billion years old plus or minus 5 million years.

yielded about half the world's gold. Gold mines are also found in Archean rocks of the Superior craton in Canada.

Archean sulfide deposits of zinc, copper, and nickel are known in several areas, such as Australia, Zimbabwe, and the Abitibi greenstone belt in Ontario, Canada. At least some of these deposits probably formed as mineral accumulations adjacent to hydrothermal vents on the seafloor, much as they do now around black smokers (Figure 8.18a).

About one-fourth of Earth's chrome reserves are in Archean rocks, especially in Zimbabwe. These ore deposits are found in the volcanic units of greenstone belts, where they probably formed when mineral crystals settled and became concentrated in the lower parts of plutons such as mafic and ultramafic sills. Chrome is needed in the steel industry, but the United States has very few commercial chrome deposits so must import most of what it uses.

One chrome deposit in the United States is in the Stillwater Complex in Montana. Low-grade ores were mined there during both World Wars as well as during the Korean War, but they were simply stockpiled and never refined for their chrome content. These same rocks are also a source of platinum, one of the precious metals, which in the automotive industry is used in catalytic converters. It is also used in the chemical industry and for cancer chemotherapy. Most of it mined today comes from the Bushveld Complex in South Africa.

Most of Earth's banded iron formations, the primary source of iron ore, were deposited during the Proterozoic, but about 6% of them are of Archean age. Some Archean-age banded iron formations are mined but they are neither as thick nor as extensive as those of the Proterozoic (see Chapter 9).

The term *pegmatite* refers to very coarsely crystalline plutonic rock, most of which approximates the composition of granite and has little commercial value. But some Archean pegmatites, such as those in the Herb Lake District in Manitoba, Canada, and the Rhodesian Province of Africa, contain valuable minerals. In addition to minerals of gem quality, Archean pegmatites also contain minerals mined for their lithium, beryllium, rubidium, and cesium.

SUMMARY

Table 8.2 provides a summary of the Archean geologic and biologic events discussed in the text.

- All geologic time from Earth's origin to the beginning of the Phanerozoic Eon is included in the Precambrian. Precambrian also refers to rocks lying stratigraphically below Cambrian-age rocks.
- Geologists divide the Precambrian into two eons, the Archean and the Proterozoic, each of which has further subdivisions.
- Rocks from the latter part of the Eoarchean indicate that crust existed then, but very little of it has been preserved.
- All continents have an ancient craton made up of an exposed shield and a buried platform. In North America, the Canadian shield is made up of smaller units delineated by their ages and structural trends.
- Archean greenstone belts are linear, syncline-like bodies of rock found within much more extensive granite-gneiss complexes.
- Ideal greenstone belts consist of two lower units of mostly igneous rocks and an upper sedimentary unit. They probably formed in several settings, including back-arc marginal basins and intracontinental rifts.
- Many geologists are convinced that Archean plate tectonics took place, but plates probably moved faster, and igneous activity was more common then because Earth had more radiogenic heat.

- Outgassing was responsible for the early atmosphere and hydrosphere, but comets also contributed to the later. However, the atmosphere that formed lacked free oxygen but contained abundant carbon dioxide and water vapor.
- Models for the origin of life require an oxygen-deficient atmosphere, the necessary elements for organic molecules, and energy to promote the synthesis of organic molecules.
- The first naturally formed organic molecules were probably monomers, such as amino acids, that linked together to form more complex polymers, including nucleic acids and proteins.
- RNA molecules may have been the first self-replicating molecules. However, the method whereby a reproductive system formed is not yet known.
- The only known Archean fossils are of single-celled, prokaryotic bacteria such as blue-green algae, but chemical compounds in some Archean rocks may indicate the presence of archaea.
- Stromatolites that formed by the activities of photosynthesizing bacteria are found in rocks as much as 3.5 billion years old.
- Archean mineral resources include gold, chrome, zinc, copper, and nickel.

IMPORTANT TERMS

abiogenesis, p.166
anaerobic, p.168
autotrophic, p.168
back-arc marginal basin, p.160
black smoker, p.167
Canadian shield, p.156
continental accretion, p.163

craton, p.156
granite-gneiss complex, p.157
greenstone belt, p.157
heterotrophic, p.168
monomer, p.166
outgassing, p.164
photochemical dissociation, p.164

photosynthesis, p.164
platform, p.156
polymer, p.166
Precambrian shield, p.156
prokaryotic cell, p.168
stromatolite, p.168
submarine hydrothermal vent, p.167

REVIEW QUESTIONS

1. The origin of life from nonliving matter is known as
   a. _____ outgassing; b. _____ abiogenesis; c. _____ cratonization; d. _____ biotic accretion; e. _____ polymerization.

2. The ancient, stable part of a continent made up of a shield and platform is called a
   a. _____ stromatolite; b. _____ greenstone belt; c. _____ craton; d. _____ black smoker; e. _____ komatiite.

3. Photochemical dissociation is a process whereby
   a. _____ plants synthesize organic molecules; b. _____ carbon dioxide forms as a metabolic waste product of animal respiration; c. _____ continents grow along their margins by accretion; d. _____ gases emitted from Earth's interior release methane and ammonia into the atmosphere; e. _____ water molecules are disrupted to yield hydrogen and oxygen.

4. Stromatolites are produced by cyanobacteria which is also known as
   a. _____ blue-green algae; b. _____ eukaryotic cells; c. _____ black smokers; d. _____ heterotrophs; e. _____ polymers.

5. Granite-gneiss complexes are
   a. _____ the most widespread Archean-age rocks; b. _____ found at oceanic spreading ridges; c. _____ most likely turbidite deposits; d. _____ noted for their fossil plants and animals; e. _____ green because they contain the minerals chlorite and epidote.

6. Many scientists think that the first self-replicating system was a(n)
   a. _____ ATP cell; b. _____ autotroph; c. _____ proteinoid; d. _____ RNA molecule; e. _____ stromatolite.

7. Ultramafic lava flows (komatiites) are found mostly in
   a. _____ sandstone deposits of passive continental margins; b. _____ banded iron formations; c. _____ the lower volcanic units in greenstone belts; d. _____ pyroclastic-gneiss rift deposits; e. _____ Cenozoic volcanic deposits.

8. The exposed part of the craton in North America is called the
   a. _____ Canadian shield; b. _____ Wyoming province; c. _____ Adirondack terrane; d. _____ Michigan basin; e. _____ Midcontinent platform.

9. Which one of the following sequences of geologic time designations is in the correct order from oldest to youngest?
   a. _____ Archean-Phanerozoic-Proterozoic; b. _____ Proterozoic-Phanerozoic-Archean; c. _____ Phanerozoic-Archean-Proterozoic; d. _____ Archean-Proterozoic-Phanerozoic; e. _____ Proterozoic-Archean-Phanerozoic.

10. The origin of greenstone belts is not fully resolved, but many geologists think that some of them formed in
    a. _____ continental shelf environments; b. _____ back-arc marginal basins; c. _____ carbonate-evaporite depositional areas; d. _____ transform boundary shear zones; e. _____ river floodplain environments.

11. What are black smokers, and what is their economic and biologic significance?

12. Describe the process of continental accretion and explain how it occurs.

13. Explain the two processes that accounted for adding free oxygen to Earth's early atmosphere? Which one was the most important?

14. What are polymers and monomers and how might they have formed naturally?

15. Describe the succession of rock types in a typical greenstone belt, and explain how a greenstone belt might originate in a back-arc marginal basin.

16. Some of the rocks in greenstone belts are described as greywacke and argillite. What are these rocks and how were they likely deposited?

17. How do stromatolites form and what is their significance?

18. Why is it so difficult to apply the principle of superposition and to determine time-stratigraphic relationships among Archean rocks?

19. Fully explain what it means to say that the earliest organisms were heterotrophic, anaerobic, prokaryotic cells.

APPLY YOUR KNOWLEDGE

1. Summarize the evidence that indicates some Eoarchean curst was present. Why do you think that so few rocks that formed during this time have been preserved?

2. Why are ultramafic lava flows (komatiites) found in Archean rocks but only rarely thereafter?

3. As a high school teacher you must explain to your students that an initially hot Earth has cooled and at some time in the far distant future its internally driven processes will cease. In other words, why are internal heat-driven processes such as volcanism, seismicity, plate movements, and mountains building, slowing down?

4. Precambrian rocks have few fossils, and those present are not very useful in stratigraphy. So how would you demonstrate that Archean rocks in different areas are the same age? Also, could you use the principles of superposition, original horizontality, and inclusions to decipher Precambrian Earth history?

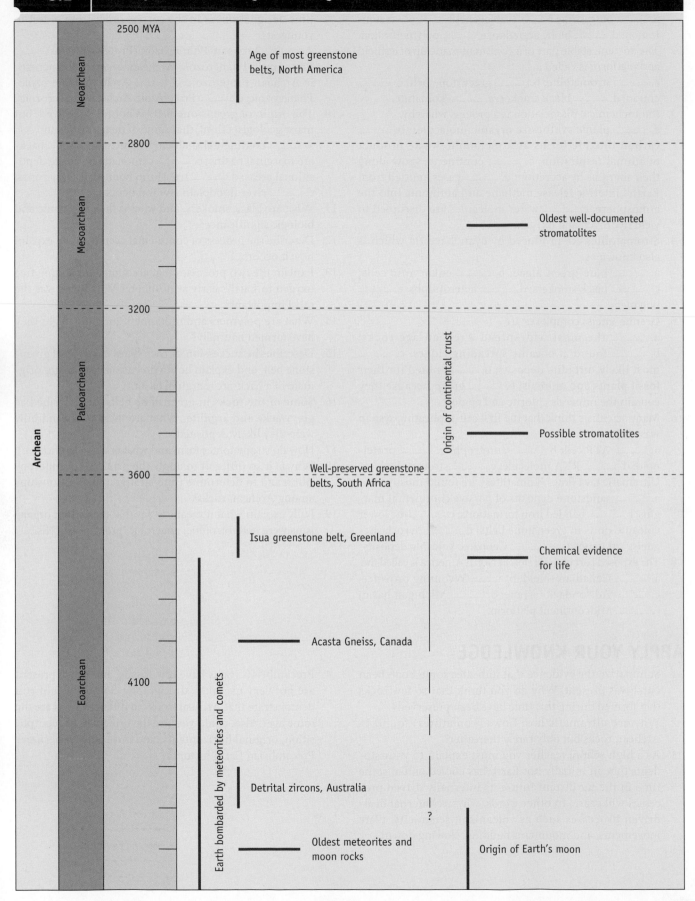

Polly Tripp

CHAPTER **9**

PRECAMBRIAN EARTH AND LIFE HISTORY—THE PROTEROZOIC EON

Proterozoic rocks in Glacier National Park in Montana. These rocks belong to the Belt Supergroup which is widespread in the northern Rocky Mountians. In fact, the same rocks are found in Alberta, Canada, but there they belong to the Purcell Supergroup. These ancient rocks were deposited as sediments in marine and terrestrial environments then they were deformed during the Cretaceoous to Early Cenozoic Laramide orogeny. Their present-day aspect resulted from erosion, especially by valley glaciers.

[OUTLINE]

Introduction

Evolution of Proterozoic Continents

Paleoproterozoic History of Laurentia

Perspective *The Sudbury Meteorite Impact and Its Aftermath*

Mesoproterozoic Accretion and Igneous Activity

Mesoproterozoic Orogeny and Rifting

Meso- and Neoproterozoic Sedimentation

Proterozoic Supercontinents

Ancient Glaciers and Their Deposits

Paleoproterozoic Glaciers

Glaciers of the Neoproterozoic

The Evolving Atmosphere

Banded Iron Formations (BIFs)

Continental Red Beds

Life of the Proterozoic

Eukaryotic Cells Evolve

Endosymbiosis and the Origin of Eukaryotic Cells

The Dawn of Multicelled Organisms

Neoproterozoic Animals

Proterozoic Mineral Resources

Summary

At the end of this chapter, you will have learned that

- A large landmass called Laurentia, made up mostly of Greenland and North America, formed by the amalgamation of Archean cratons along deformation belts during the Paleoproterozoic.

- Following its initial stage of amalgamation, Laurentia grew by accretion along its southern and eastern margins.

- The Mesoproterozoic of Laurentia was a time of widespread igneous activity, orogenies, and rifting.

- Widespread Proterozoic assemblages of sandstone, shale, and carbonate rocks look much like the rocks deposited now on passive continental margins.

- Plate tectonics, essentially like that occurring now, was operating during the Proterozoic and one or possibly two supercontinents formed.

- The presence of banded iron formations and continental red beds indicate that at least some free oxygen was present in the atmosphere.

- Extensive glaciation took place during the Paleoproterozoic and the Neoproterozoic.

- The first eukaryotic cells, that is, cells with a nucleus and other internal structures, evolved from prokaryotic cells by 1.2 billion years ago.

- Impressions of multicelled animals are found on all continents except Antarctica.

- Banded iron formations, as sources of iron ore, are important Proterozoic resources, as are deposits of gold, copper, platinum, and nickel.

Introduction

The Proterozoic Eon lasted 1.958 billion years, accounting for 42.5% of all geologic time (see Figure 8.1), and yet we review this unimaginably long interval of Earth and life history in a single chapter. We devote 10 chapters to the more familiar Phanerozoic Eon made up of the Paleozoic, Mesozoic, and Cenozoic eras, but this seemingly disproportionate treatment is justified when you consider that we know so much more about the events recorded in this more recent part of the geologic record. Remember that Precambrian rocks are deeply buried in many areas, and many, particularly those of Archean age, have been metamorphosed and complexly deformed. Although some Proterozoic rocks are metamorphic, there are vast areas of exposed rocks that show little or no metamorphism, but the few fossils they contain, although important, are not very useful for stratigraphic analyses (• Figure 9.1).

Recent work by the International Commission on Stratigraphy (ICS) shows the Proterozoic Eon consisting of three eras with the prefixes *Paleo* (old or ancient), *Meso* (middle), and *Neo* (new or recent), and each era is further subdivided into periods (see Figure 8.2). Just as with Archean terminology, the Proterozoic terms are not based on time-stratigraphic units with stratotypes. The Archean–Proterozoic boundary at 2.5 billion years ago marks the approximate time of changes in the style of crustal evolution, but we must emphasize "approximate," because Archean-style crustal evolution was not completed at the same time in all areas.

What do we mean by a change in style of crustal evolution? First, the Archean was characterized by the origin of granite-gneiss terrains and greenstone belts that were shaped into cratons (see Figure 8.8). Although these same

• **Figure 9.1 Proterozoic Rocks** Some Proterozoic-age rocks are metamorphic, but many show little or no alteration from their original condition.

Gerald and Buff Corsi/Visual Unlimited

a The Vishnu Schist (black) in the Grand Canyon of Arizona was intruded by the Zoraster Granite about 1.7 billion years ago. The Vishnu Schist was originally lava flows and sedimentary rocks.

James S. Monroe

b This 1.0-billion-year-old outcrop of sandstone and mudstone in Glacier National Park in Montana has been only slightly altered by metamorphism.

rock associations continued to form during the Proterozoic, they did so at a considerably reduced rate. Furthermore, many Archean rocks are metamorphic, whereas vast exposures of Proterozoic rocks are unaltered or nearly so (see the Chapter opening photo), and in many areas Archean and Proterozoic rocks are separated by an unconformity. And finally, widespread associations of sedimentary rocks of passive continental margins were deposited during the Proterozoic, and the plate tectonic style was essentially the same as it is now.

In addition to a change in style of crustal evolution, the Proterozoic was also an important time in the evolution of the atmosphere and biosphere, as well as the origin of some important natural resources. Oxygen-dependant organisms and the types of cells that make up most of today's organisms evolved during this time. Near the end of this long interval of time the first multicelled organisms and animals made their appearances. The fossil record is still poor compared to that of the Phanerozoic, but it is much better than the one for the Archean Eon.

Evolution of Proterozoic Continents

In Chapter 8, we noted that Archean cratons assembled during collisions of island arcs and minicontinents (see Figure 8.4). These cratons provided the nuclei around which Proterozoic crust accreted, thereby forming much larger landmasses. Accretion around these nuclei took place much as it does today where plates converge, but it no doubt occurred more rapidly because Earth possessed more radiogenic heat and plates moved faster. Our focus here is on the evolution of **Laurentia,** a landmass made up of what are now North America, Greenland, parts of northwestern Scotland, and perhaps some of the Baltic shield of Scandinavia.

Most greenstone belts formed during the Archean, but they continued to form during the Proterozoic, and at least one is known from Cambrian-age rocks in Australia. They were not as common, though, and they did differ in one important detail: the near absence of extrusive ultramafic rocks (komatiite). As we discussed in Chapter 8, this detail is no doubt a result of Earth's decreasing rate of radiogenic heat production (see Figure 8.15).

Paleoproterozoic History of Laurentia

As you might suspect, the evolution of Laurentia was long, complex, and is still not fully understood. Nevertheless, geologists have made remarkable progress in deciphering this episode in Earth history, which involves the amalgamation of cratons, the accretion of volcanic arcs and oceanic terranes, and extensive plutonism, metamorphism, and volcanism. The time between 2.0 and 1.8 billion years ago (BYA) was especially important, because this is when collisions among Archean cratons formed several **orogens,** which are linear to arcuate deformation belts in which many of the rocks have been metamorphosed and intruded by magma that cools to form plutons, especially batholiths. As a result, these ancient cratons were sutured along these orogens, thereby forming a much larger landmass (• Figure 9.2a), which now makes up much of Greenland, central Canada, and the north-central United States.

Excellent examples of these craton-forming processes are recorded in rocks of the Thelon orogen (1.92–1.96 BYA) in northwestern Canada where the Slave and Rae cratons collided, and the Trans-Hudson orogen (1.82–1.84 BYA), in the United States and Canada where the Superior, Hearne, and Wyoming cratons were sutured (Figure 9.2a). In addition, the Penokean orogen formed along the southern margin of Laurentia over tens of millions of years, although the most intense episode of this event was about 1.85 BYA.

Sedimentary rocks in the Wopmay orogen of northwestern Canada record the opening and closing of an ocean basin, or what geologists call a Wilson cycle. A complete **Wilson cycle,** named after the Canadian geologist J. Tuzo Wilson, includes rifting and the opening of an ocean basin with passive continental margins on both sides (• Figure 9.3). As a result of rifting, an expansive ocean basin forms, but eventually it begins to close, and subduction zones and volcanic island arcs form on both sides of the ocean basin. Finally, a continent-continent collision takes place resulting in deformation of the passive margin deposits as well as rocks of the oceanic crust (Figure 9.3f).

Some of the rocks in the Wopmay orogen consist of a suite of sedimentary rocks called a **sandstone-carbonate-shale assemblage** that forms on passive continental margins. Remember that passive margin deposits are rare or absent in Archean rocks, but they become common during the Proterozoic and thereafter. This assemblage of rocks is also well represented in the Penokean orogen of the Great Lakes region of the United States and Canada (• Figure 9.4).

Following the initial episodes of amalgamation of cratons, additions to Laurentia occurred along its southeastern margin as it collided with volcanic island arcs and oceanic terranes. As a consequence, the Yavapai and Mazatzal orogens were added to the evolving continent between about 1.65 and 1.76 BYA (Figure 9.2b). Of course these were major orogenies, so the rocks have been deformed, altered by metamorphism, intruded by granitic batholiths, and incorporated into Laurentia.

The Paleoproterozoic events and rocks just reviewed are important in the evolution of Laurentia. However, this was also the time when most of Earth's *banded iron formations* (BIF), and the first *continental red beds*—sandstone and shale with oxidized iron—were deposited. We will have more to say about BIFs and red beds in the section headed "The Evolving Atmosphere."

Archean Eon	Proterozoic Eon	Phanerozoic Eon							
		Paleozoic Era							
Precambrian		*Cambrian*	*Ordovician*	*Silurian*	*Devonian*	*Mississippian*	*Pennsylvanian*	*Permian*	
						Carboniferous			

2500 MYA · 542 MYA · 251 MYA

• **Figure 9.2 Proterozoic Evolution of Laurentia** These three illustrations show the overall trends in the Proterozoic evolution of Laurentia, but they do not show many of the details of this long, complex episode in Earth history.

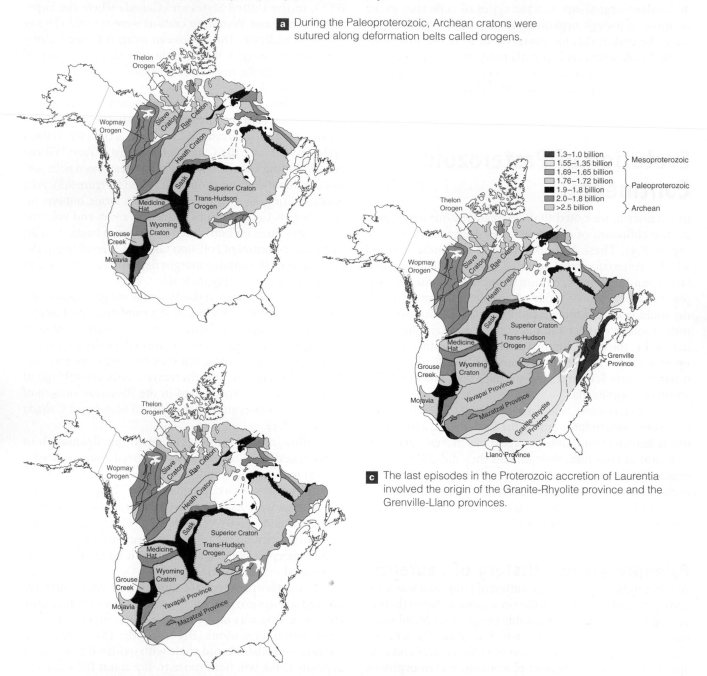

a During the Paleoproterozoic, Archean cratons were sutured along deformation belts called orogens.

1.3–1.0 billion ⎫
1.55–1.35 billion ⎬ Mesoproterozoic
1.69–1.65 billion ⎫
1.76–1.72 billion ⎬ Paleoproterozoic
1.9–1.8 billion
2.0–1.8 billion ⎫
>2.5 billion ⎬ Archean

c The last episodes in the Proterozoic accretion of Laurentia involved the origin of the Granite-Rhyolite province and the Grenville-Llano provinces.

b Laurentia grew along its southeastern margin by accretion of the Yavapai and Mazatzal provinces.

251 MYA

99 MYA

• *Active* Figure 9.3 **The Wopmay Orogen and the Wilson Cycle** Some geologists think that the Wopmay orogen in Canada respresents a complete Wilson cycle, which is shown here diagrammatically. Visit the Geology Resource Center to view this and other active figures at www.cengage.com/sso.

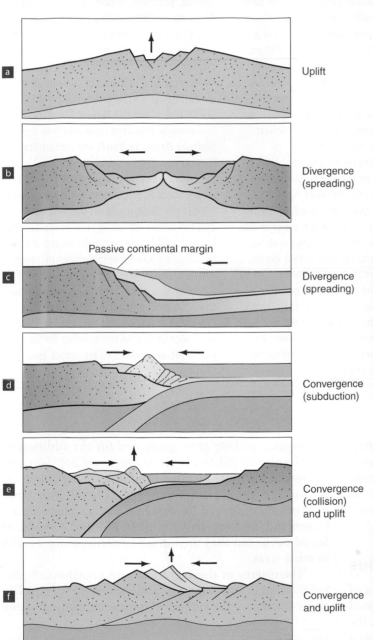

a   Uplift

b   Divergence (spreading)

Passive continental margin

c   Divergence (spreading)

d   Convergence (subduction)

e   Convergence (collision) and uplift

f   Convergence and uplift

a-f In fact, some of the sedimentary rocks in the Wopmay orogen were probably deposited on a passive continental margin as shown in **c**.

• Figure 9.4 **Paleoproterozoic Sedimentary Rocks**

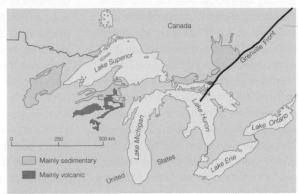

Canada

Lake Superior

Grenville Front

Lake Huron

Lake Michigan

Lake Ontario

Lake Erie

United States

0   250   500 km

☐ Mainly sedimentary
■ Mainly volcanic

a   Distribution of Paleoproterozoic-age rocks in the Great Lakes region.

James S. Monroe

b   Outcrop of the Mesnard Quartzite. The crests of the ripple marks point toward the observer.

James S. Monroe

c   The Kona Dolomite. The bulbous structures are stromatolites that resulted from the activities of cyanobacteria (blue-green algae).

The Sudbury Meteorite Impact and its Aftermath

Earth is a dynamic planet, so erosion, volcanism, moving plates, and mountain building have obliterated the evidence of the intense meteorite bombardment that ended about 3.8 billion years ago. Nevertheless, collisions continued although at a considerably reduced rate. A collision 65 million years ago may have caused or hastened the extinction of dinosaurs and their relatives (see Chapter 15); an impact 35 million years ago occurred where Chesapeake Bay is now located; and about 50,000 years ago a 100-meter diameter meteorite formed Meteor Crater in Arizona. An especially notable impact crater, or astrobleme, is one measuring about 60 km long and 30 km wide in Ontario, Canada, although it may have originally measured 250 km across. The impact took place 1.85 billion years ago, during the Paleoproterozoic, and formed what we now call the Sudbury Basin.

An impact of the Sudbury meteorite was no doubt spectacular, but the only organisms to witness the event were bacteria and archaea. The blast and heat generated by the impact most likely killed everything within hundreds of kilometers, and

a seismic event would have been recorded around the world if there had been any instruments to record it. Debris was scattered as far as 800 km from the impact site, thousands of tons of rock must have been vaporized, releasing copious quantities of gases into the atmosphere, and many hundreds of cubic kilometers of rock melted. Certainly the impact was a catastrophic event of great magnitude, but it was also a fortunate incident for us because the impact probably accounts for the fact that the Sudbury Basin is one of the richest mining areas in the world.

One hypothesis for the concentration of metals in the Sudbury Basin is that ores were concentrated from molten rock following the high-speed meteorite impact. Indeed, many of the rocks in the basin are mafic intrusive rocks such as gabbro and a similar rock called norite, as well as granophyre and diorite, which are plutonic rocks similar to granite. Overlying the igneous rocks, collectively called the Sudbury Igneous Complex, are hundreds of meters of impact breccias and sedimentary rocks that partly filled the basin after the impact.

The first mineral discoveries occurred in 1883, and since then the 90 or so mines around the margins of the basin have produced hundreds of tons of copper, palladium, platinum, nickel, gold, and silver. All of these commodities are essential to industrialized societies. The United States imports some copper but it also produces copper at mines in several western states. However, much of the platinum-group metals (platinum and palladium) come from South Africa and the United Kingdom, but some is imported from mines in the Sudbury Basin. In 2007, the United States had no domestic production of nickel so it depends on imports; more than 40% of the nickel the U.S. imports comes from Canada.

If the Sudbury Basin was originally 250 km across, it would rank as the second largest astrobleme on Earth; at 300 km in diameter, the Vredefort crater in South Africa is the largest. This crater in South Africa is also older—it formed slightly more than 2.0 billion years ago. However, the Chicxulub crater on the Yucatán Peninsula of Mexico is also in the running for second place—it measures 170 km across (see Chapter 15).

Another significant Paleoproterozoic event was a huge meteorite impact that took place in northern Ontario, Canada (see Perspective). And finally, we should mention that some Proterozoic-age rocks and associated features provide evidence for widespread glaciation (also covered later in this chapter).

Mesoproterozoic Accretion and Igneous Activity
Following a lull of several millions of years, tectonism and continental accretion resumed along Laurentia's southeastern margin. In addition, the interval from about 1.35 to 1.55 billion years ago was a time of extensive igneous activity, some of it unrelated

to orogenic activity that accounted for the addition of the Granite-Rhyolite province (Figure 9.2c). Some of the igneous activity resulted in plutons being emplaced in already existing continental crust. The resulting igneous rocks are exposed in eastern Canada, extend across Greenland, and are also found in the Baltic shield of Scandinavia, but they are buried beneath younger rocks in most areas.

The origins of these granitic and anorthosite* plutons, calderas and their fill, and vast sheets of rhyolite and ashflows are the subject of debate. According to one

---

*Anorthosite is a plutonic rock composed almost entirely of plagioclase feldspars.

hypothesis, large-scale upwelling of magma beneath a Proterozoic supercontinent was responsible for these rocks. According to this hypothesis, the mantle temperature beneath a Proterozoic supercontinent would have been considerably higher than beneath later supercontinents because radiogenic heat production within Earth has decreased. Accordingly, nonorogenic igneous activity would have occurred following the amalgamation of the first supercontinent.

Mesoproterozoic Orogeny and Rifting

A Mesoproterozoic orogenic event in Laurentia, the 1.3- to 1.0-billion-year-old **Grenville orogeny,** took place in the eastern part of the evolving continent (Figure 9.2c). Grenville rocks are well exposed in the present-day northern Appalachian Mountains (• Figure 9.5), as well as in eastern Canada, Greenland, and Scandinavia. The Llano province in Texas is probably a westward extension of the Grenville orogenic belt. Many geologists think the Grenville orogen resulted from closure of an ocean basin, the final stage in a Wilson cycle. Indeed, the Grenville orogeny may have been the final episode in the assembly of the supercontinent Rodina which persisted into the Neoproterozoic.

Whatever the cause of the Grenville orogeny, it was the final stage in the Proterozoic continental accretion of Laurentia (Figure 9.2c). By then, about 75% of present-day North America existed. The remaining 25% accreted along its margins, particularly its eastern and western margins, during the Phanerozoic Eon.

Beginning about 1.1 billion years ago, tensional forces opened the **Midcontinent rift,** a long, narrow trough bounded by faults that outline two branches; one branch extended southeast as far as Kansas, and one extended southeasterly through Michigan and into Ohio (Figures 9.2c and • Figure 9.6). The rift cuts through Archean and Proterozoic rocks and terminates against the Grenville orogen in the east. Although not all geologists agree, many think that the Midcontinent rift is a failed rift where Laurentia began splitting apart. Had this rifting continued, Laurentia would have split into two separate landmasses, but the rifting ceased after about 20 million years.

Most of the Midcontinent rift is buried beneath younger rocks except in the Lake Superior region, where igneous and sedimentary rocks are well exposed (Figure 9.6c and d). The central part of the rift contains numerous overlapping basalt lava flows, forming a volcanic pile several kilometers thick. Along the rift's margins, conglomerate was deposited in large alluvial fans that grade into sandstone and shale with increasing distance from the sediment source.

Meso- and Neoproterozoic Sedimentation

Remember the Grenville orogeny (Figure 9.2c), was the final episode of large-scale deformation in Laurentia until the Ordovician Period (see Chapter 10). Nevertheless, important geologic events were taking place, such as sediment deposition in what is now the eastern United States and Canada, in the Death Valley region of California and Nevada, and in three huge basins in the west (• Figure 9.7a).

Meso- and Neoproterozoic-age sedimentary rocks are exceptionally well exposed in the northern Rocky Mountains of Montana, Idaho, and Alberta, Canada. Indeed, their colors, deformation features, and erosion by Pleistocene and recent glaciers have yielded some fantastic scenery. Like the Paleoproterozoic rocks in the Great Lakes region, they are mostly sandstones, shales, and stromatolite-bearing carbonates (Figure 9.7b). Sedimentary rocks of Proterozoic age are also found in Utah (Figure 9.7c).

Proterozoic rocks of the Grand Canyon Supergroup lie unconformably on Archean rocks and in turn are overlain unconformably by Phanerozoic-age rocks (Figure 9.7d). The rocks, consisting mostly of sandstone, shale, and dolostone, were deposited in shallow-water marine and fluvial environments. The presence of stromatolites and carbonaceous impressions of algae in some of these rocks indicate probable marine deposition.

To summarize this entire section on The Evolution of Proterozoic Continents, in which we emphasized the geologic history of Laurentia, the events described did take place in the order discussed. But, as you might expect for such a long episode in Earth evolution, the actual history is much more complex than indicated here. We did not mention the multiple episodes of deformation, plutonism, metamorphism, and the suturing of small crustal blocks to Laurentia, nor did we talk about the vast amounts of erosion that occurred when mountains formed and were subsequently worn away. In short, we compressed nearly 2.0 billion years of Earth history into a few pages of text.

R. V. Dietrich

• Figure 9.5 **Rocks of the Grenville Orogen** At this outcrop near Alexandria, New York, the metamorphic rocks of the Grenville orogen in the lower part of the image are unconformably overlain by the Upper Cambrian Potsdam Formation.

• Figure 9.6 **The Midcontinent Rift**

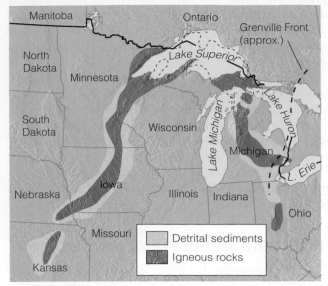

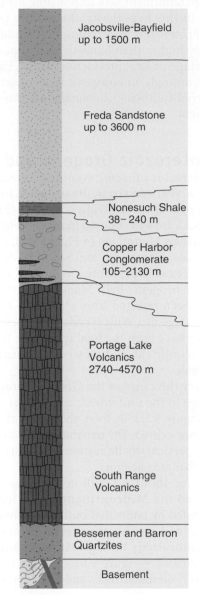

a The rocks filling the rift are exposed around Lake Superior in the United States and Canada, but they are deeply buried elsewhere.

Detrital sediments

Igneous rocks

Jacobsville-Bayfield
up to 1500 m

Freda Sandstone
up to 3600 m

Nonesuch Shale
38–240 m

Copper Harbor
Conglomerate
105–2130 m

Portage Lake
Volcanics
2740–4570 m

South Range
Volcanics

Bessemer and Barron
Quartzites

Basement

c The Copper Harbor Conglomerate in Michigan.

b Columnar section showing vertical relationships among the rocks in the Midcontinent rift.

d The Portage Lake Volcanics in Michigan.

Proterozoic Supercontinents

A *continent* is a landmass made up of granitic crust, with much of its surface above sea level, but the geologic margin of a continent—that is, where granitic continental crust changes to basaltic oceanic crust—is beneath sea level. A **supercontinent,** in contrast, consists of at least two continents merged into one, and in the context of Earth history, we usually refer to a supercontinent as one composed of all or most of Earth's landmasses, other than oceanic islands. You are already aware of the supercontinent Pangaea that existed at the end of the Paleozoic Era, but you probably have not heard of previous supercontinents.

• Figure 9.7 **Proterozoic Rocks in the Western Untied States and Canada**

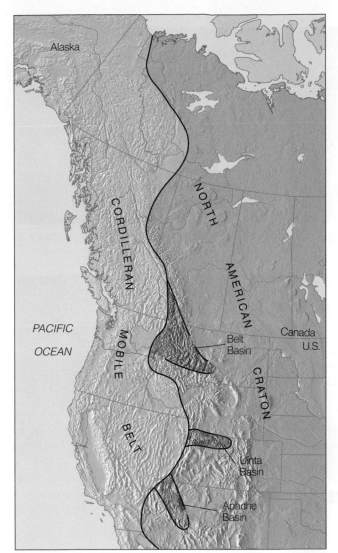

c Neoproterozoic rocks of the Uinta Mountain Group in Utah.

a Meso- to Neoproterozoic basins in the western United States and Canada.

d Neoproterozoic sandstone in the Grand Canyon.

b Meso- and Neoproterozoic rocks in the Belt basin of Glacier National Park, Montana.

Before we specifically address supercontinents, it is important to note that the present style of plate tectonics involving opening and then closing ocean basins had certainly been established by the Paleoproterozoic. In fact, *ophiolites* that provide evidence for convergent plate boundaries are known from Neoarchean and Paleoproterozoic rocks of Russia and probably China, respectively. Furthermore, these ancient ophiolites compare closely with younger, well-documented ophiolites, such as the highly deformed but complete Jormua mafic-ultramafic complex in Finland (• Figure 9.8).

Supercontinents may have existed as early as the Neoarchean, but if so we have little evidence of them. The first that geologists recognize with some certainty, known as **Rodinia** (• Figure 9.9), assembled between 1.3 and 1.0 billion years ago and then began fragmenting 750 million years ago. Judging by the large-scale deformation called the *Pan-African orogeny* that took place in what are now the Southern Hemisphere continents, Rodinia's separate pieces reassembled about 650 million years ago and formed another supercontinent,

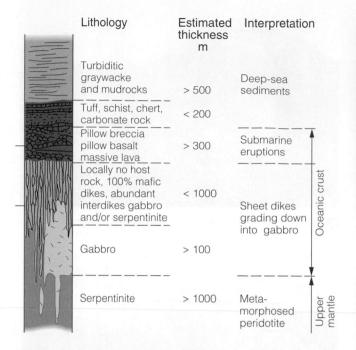

Lithology	Estimated thickness m	Interpretation
Turbiditic graywacke and mudrocks	> 500	Deep-sea sediments
Tuff, schist, chert, carbonate rock	< 200	
Pillow breccia pillow basalt massive lava	> 300	Submarine eruptions
Locally no host rock, 100% mafic dikes, abundant interdikes gabbro and/or serpentinite	< 1000	Sheet dikes grading down into gabbro
Gabbro	> 100	
Serpentinite	> 1000	Metamorphosed peridotite

Oceanic crust / Upper mantle

a Stratigraphic column showing the sequence of units in the ophiolite.

Courtesy of Asko Kontinen, Geological Survey of Finland

b A metamorphosed basaltic pillow lava. The code plate is about 12 cm long.

Courtesy of Asko Kontinen, Geological Survey of Finland

c Metamorphosed gabbro between mafic dikes. The hammer shaft is 65 cm long.

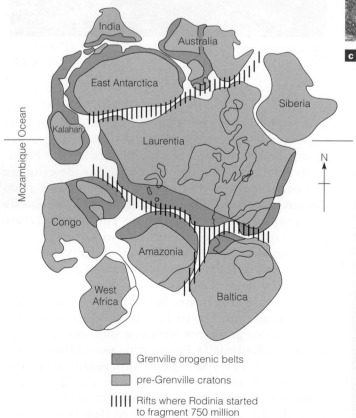

Grenville orogenic belts

pre-Grenville cratons

|||| Rifts where Rodinia started to fragment 750 million years ago

• Figure 9.9 **Rodinia** Possible configuration of the Neoproterozoic supercontinent Rodinia before it began fragmenting about 750 million years ago.

this one known as **Pannotia.** And, finally, by the latest Neoproterozoic, about 550 million years ago, fragmentation was under way, giving rise to the continental configuration that existed at the onset of the Phanerozoic Eon (see Chapter 10).

Ancient Glaciers and Their Deposits

Very few times of widespread glaciation have occurred during Earth history. The most recent one, during the Pleistocene Epoch (1.8 million to 10,000 years ago), is the best known, but we also have evidence for Pennsylvanian glaciers (see Chapter 11) and two major episodes of Proterozoic glaciation. But how can we be sure there were Proterozoic glaciers? After all, their most common deposit, called *tillite,* is simply a type of conglomerate that may look much like conglomerate that originated by other processes.

• Figure 9.10 **Paleo- and Neoproterozoic Glaciers**

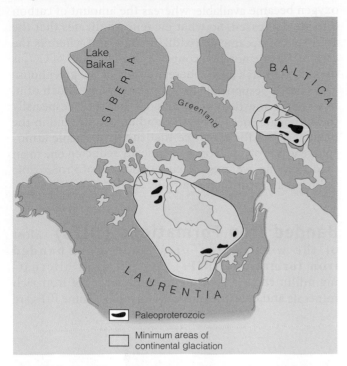

a Paleoproterozoic glacial deposits in the United States and Canada indicate that Laurentia had an extensive ice sheet.

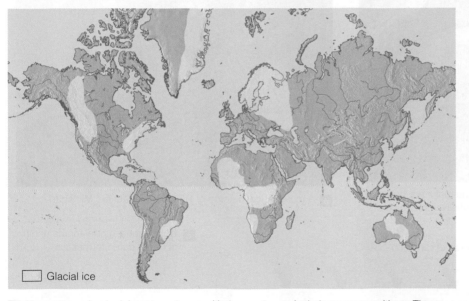

b Neoproterozoic glacial centers shown with the continents in their present positions. The extent of the glaciers is approximate.

Paleoproterozoic Glaciers Tillite or tillitelike deposits are known from at least 300 Precambrian localities, and some of these are undoubtedly not glacial deposits. But the extensive geographic distribution of others and their associated features, such as striated and polished bedrock, are distinctive. Based on this kind of evidence, geologists are now convinced that widespread glaciation took place during the Paleoproterozoic.

Tillites of about the same age in Michigan, Wyoming, and Quebec indicate that North America may have had an ice sheet centered southwest of Hudson Bay (• Figure 9.10a). Tillites of about this age are also found in Australia and South Africa, but dating is not precise enough to determine if there was a single widespread glacial episode or a number of glacial events at different times in different areas. One tillite in the Bruce Formation in Ontario, Canada, is about 2.7 billion years ago, thus making it Neoarchean.

Glaciers of the Neoproterozoic

Tillites and other glacial features dating from between 900 and 600 million years ago are found on all continents except Antarctica. Glaciation was not continuous during this entire time but was episodic, with four major glacial episodes so far recognized. Figure 9.10b shows the approximate distribution of these glaciers, but we must emphasize "approximate," because the actual geographic extent of these glaciers is unknown. In addition, the glaciers covering the area in Figure 9.10b were not all present at the same time. Despite these uncertainties, this Neoproterozoic glaciation was the most extensive in Earth history. In fact, glaciers seem to have been present even in near-equatorial areas.

Because Neoproterozoic glacial deposits are so widespread, some geologists think that glaciers covered all land and the seas were frozen—a *snowball Earth*, as it has come to be known. The snowball Earth hypothesis is controversial, but proponents claim that the onset of this glacial episode may have been triggered by the near-equatorial location of all continents, and as a result, accelerated weathering would absorb huge quantities of carbon dioxide from the atmosphere. With little CO_2 in the atmosphere, glaciers would form and reflect solar radiation back into space and more glacial ice would form.

So if there actually was a snowball Earth, why wouldn't it stay frozen? Of course, volcanoes would continue to erupt spewing volcanic gases, which includes the greenhouse gases carbon dioxide and methane, which would warm the atmosphere and end the glacial episode. In fact, proponents of this hypothesis note that several such snowball Earths may have occurred until the continents moved into higher latitudes. One criticism of the hypothesis is that if all land was ice covered and the sea froze, how would life survive? Several suggestions have been made to account for this—life persisted at hydrothermal vents on the seafloor; even photosynthesis can take place beneath thin glacial ice; perhaps life persisted in subglacial lakes as it does now in Antarctica; and there may have been pools of liquid water near active volcanoes.

The Evolving Atmosphere

Geologists agree that the Archean atmosphere contained little or no free oxygen (see Chapter 8), so the atmosphere was not strongly oxidizing as it is now. Photochemical dissociation and photosynthesis were adding free oxygen to the atmosphere, but the amount present at the beginning of the Proterozoic was probably no more than 1% of that present now. In fact, it might not have exceeded 10% of present levels even at the end of the Proterozoic. Remember from our previous discussions that cyanobacteria (blue-green algae) were present during the Archean, but the structures they formed, called *stromatolites,* did not become common until about 2.3 billion years ago— that is, during the Paleoproterozoic. These photosynthesizing organisms and, to a lesser degree, photochemical dissociation both added free oxygen to the evolving atmosphere (see Figure 8.14).

In Chapter 8 we cited some of the evidence indicating that Earth's early atmosphere had little or no free oxygen but abundant carbon dioxide. Here we contend that more oxygen became available, whereas the amount of carbon dioxide decreased. So what evidence indicates that the atmosphere became an oxidizing one, and where is the carbon dioxide now? Of course, a small amount of CO_2 is present in today's atmosphere; it is one of the greenhouse gases partly responsible for global warming. Much of it, however, is now tied up in minerals and rocks, especially the carbonate rocks limestone and dolostone, and in the biosphere. As for evidence that the Proterozoic atmosphere was evolving from a chemically reducing one to an oxidizing one, we must discuss two types of Proterozoic sedimentary rocks that we already alluded to briefly.

Banded Iron Formations (BIFs)

Most of the world's iron ores come from **banded iron formations (BIFs)** consisting of alternating millimeter- to centimeter-thick layers of iron-rich minerals and chert (● Figure 9.11a and b). Some BIFs are

● Figure 9.11 **Paleoproterozoic Banded Iron Formation (BIF)**

a At this outcrop in Ishpeming, Michigan, the rocks are brilliantly colored alternating layers of red chert and silver iron minerals.

b A more typical outcrop of BIF near Negaunee, Michigan.

c Depositional model for the origin of banded iron formations.

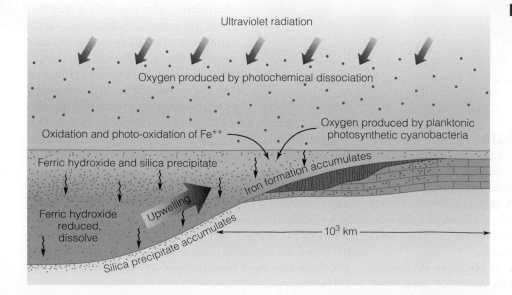

Ultraviolet radiation

Oxygen produced by photochemical dissociation

Oxidation and photo-oxidation of Fe++

Oxygen produced by planktonic photosynthetic cyanobacteria

Ferric hydroxide and silica precipitate

Iron formation accumulates

Upwelling

Ferric hydroxide reduced, dissolve

Silica precipitate accumulates

10^3 km

found in Archean-age rocks, but these are small deposits mostly in greenstone belts and appear to have formed near submarine volcanoes. However, most BIFs, fully 92%, were deposited in shallow-water shelf environments during the interval from 2.5 to 2.0 billion years ago—that is, during the earlier part of the Paleoproterozoic. These deposits are much more extensive than those of the Archean and they have important implications for the evolving atmosphere.

The iron in the Proterozoic BIFs is iron oxide in the minerals hematite (Fe_2O_3) and magnetite (Fe_3O_4), and of course the chert layers are mostly silicon dioxide (SiO_2). Iron is a highly reactive element, and in an oxidizing atmosphere it combines with oxygen to form rustlike oxides that do not readily dissolve in water. If oxygen is absent, though, iron is easily taken into solution and can accumulate in large quantities in the oceans, which it probably did during the Archean. The Archean atmosphere was deficient in free oxygen, so it is doubtful that seawater had very much dissolved oxygen. However, as free oxygen accumulated in the oceans as waste from photosynthesizing organisms, it triggered the precipitation of iron oxides and silica and thus the origin of BIFs.

One popular model that accounts for the details of BIF precipitation involves a Paleoproterozoic ocean with an upper oxygenated layer above a large volume of oxygen-deficient (anoxic) water that contained dissolved iron and silica. Some of the dissolved iron probably came from the weathering of rocks on land, but a likely source for much of it was submarine hydrothermal vents similar to those on the seafloors today (see Figure 8.18). Transfer of water from the anoxic zone to the surface, or upwelling, brought iron- and silica-rich waters onto the developing shallow marine shelves where iron and silica combined with oxygen and widespread precipitation of BIFs took place (Figure 9.11c). Precipitation of these rocks continued until the iron in seawater was largely depleted, and the atmosphere now contained some free oxygen so iron was no longer taken into solution in large quantities.

Continental Red Beds Obviously, the term **continental red beds** refers to red rocks on the continents, but more specifically it means red sandstone or shale colored by iron oxides, especially hematite (Fe_2O_3) (Figure 9.7b and d). These deposits first appear in the geologic record about 1.8 billion years ago, increase in abundance throughout the rest of the Proterozoic, and are quite common in rocks of Phanerozoic age.

The onset of red bed deposition coincides with the introduction of free oxygen into the Proterozoic atmosphere. But the atmosphere at that time may have had only 1% or perhaps 2% of present levels. Is this sufficient to account for oxidized iron in sediment? Probably not, but we must also consider other attributes of this atmosphere.

No ozone (O_3) layer existed in the upper atmosphere before free oxygen (O_2) was present. But as photosynthesizing organisms released free oxygen into the atmosphere, ultraviolet radiation converted some of it to elemental oxygen (O) and ozone (O_3), both of which oxidize minerals more effectively than does O_2. Once an ozone layer became established, most ultraviolet radiation failed to penetrate to the surface, and O_2 became the primary agent for oxidizing minerals.

Life of the Proterozoic

The fossil record of the Archean is sparse, consisting of bacteria and archaea, although there were undoubtedly many types of these organisms. Likewise, the Paleoproterozoic fossil record is characterized by these same organisms (• Figure 9.12a), although stromatolites, structures produced by cyanobacteria, became common (Figures 9.4c and 9.12b). Actually, the lack of biotic diversity is not too surprising, because prokaryotic cells

• Figure 9.12 **Proterozoic Fossil Bacteria and Stromatolites**

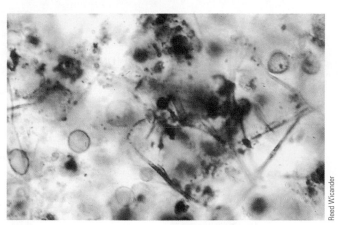

Reed Wicander

a These spherical and filamentous microfossils from the Gunflint Chert of Ontario, Canada, resemble bacteria living today. The filaments measure about 1/1000th of a millimeter across.

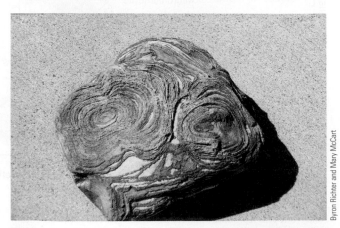

Byron Richter and Mary McCart

b This rock has been eroded but it clearly shows two stromatolites that have grown together (see Figure 8.19b).

reproduce asexually. As a result, they do not share their genetic material[1] as sexually reproducing organisms do, so evolution was a comparatively slow process. Organisms that reproduced sexually probably evolved by the Mesoproterozoic, and the tempo of evolution increased markedly, although from our perspective it was still exceedingly slow.

Eukaryotic Cells Evolve
Eukaryotic cells are much larger than prokaryotic cells, they have an internal membrane-bounded nucleus that contains the chromosomes, and they have other internal structures not found in prokaryotes (• Figure 9.13). Furthermore, many eukaryotes are multicelled and aerobic, so they could not have existed until some free oxygen was present in the atmosphere. And lastly, most of them reproduce sexually.

The distinction of prokaryotic cells from eukaryotic cells is one of the most fundamental in the entire biotic realm. Biologists recognized six kingdoms of organisms—archaea, bacteria, protists, fungi, plants, and animals (Table 9.1), but a system of classification that has become increasingly popular recognizes three broad groups of *domains* or all living things (• Figure 9.14). Two of the domains consist of organisms with prokaryotic cells and the other group has members with eukaryotic cells.

Mesoproterozoic rocks 1.2 billion years old in Canada contain fossils of the oldest known eukaryotes. These tiny organisms called *Bangiomorpha* were single celled, probably reproduced sexually, and look remarkably similar to living red algae (• Figure 9.15a). Other Proterozoic rocks have yielded even older fossils, but their affinities are unknown. For instance, the 2.1-billion-year-old Negaunee Iron Formation in Michigan has fossils known as *Grypania*, the oldest known megafossil, but it was very likely a single-celled bacterium or some kind of algae (Figure 9.15b).

Although *Bangiomorpha* is the oldest accepted eukaryote, cells larger than 60 microns appear in abundance at least 1.4 billion years ago. Many of them show an increase in organizational complexity, and an internal membrane-bounded nucleus is present in some. In addition, hollow fossils known as *acritarchs* that were probably cysts of planktonic algae become common during the Meso- and Neoproterozoic (• Figure 9.16a and b).

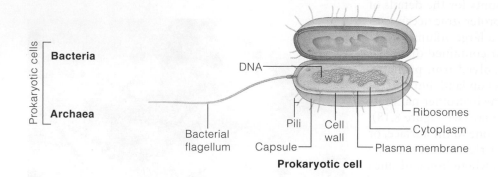

• **Figure 9.13 Prokaryotic and Eukaryotic Cells** Eukaryotic cells have a cell nucleus containing the genetic material and organelles such as mitochondria and plastids, as well as chloroplasts in plant cells. In contrast, prokaryotic cells are smaller and not nearly as complex as eukaryotic cells.

Prokaryotic cell

Bacteria

Archaea

Prokaryotic cells

DNA

Bacterial flagellum

Pili

Capsule

Cell wall

Ribosomes

Cytoplasm

Plasma membrane

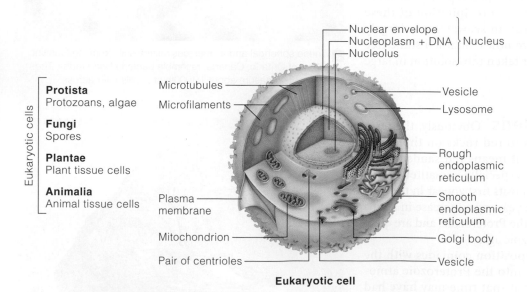

Eukaryotic cell

Eukaryotic cells

Protista
Protozoans, algae

Fungi
Spores

Plantae
Plant tissue cells

Animalia
Animal tissue cells

Microtubules

Microfilaments

Plasma membrane

Mitochondrion

Pair of centrioles

Nuclear envelope

Nucleoplasm + DNA

Nucleolus

Nucleus

Vesicle

Lysosome

Rough endoplasmic reticulum

Smooth endoplasmic reticulum

Golgi body

Vesicle

[1]Prokaryotic cells reproduce by binary fission and give rise to two genetically identical cells. Under some conditions, they engage in conjugation during which some genetic material is transferred from cell to cell.

Kingdom	Characteristics	Examples
Archaea	Prokaryotic cells; single celled, differ from bacteria in genetics and chemistry	Methanogens, halophiles, thermophiles
Bacteria	Prokaryotic cells; single celled, cell walls different from archaea and eukaryotic cells	Cyanobacteria (also called blue-gree algae), mycoplasmas
Protista	Eukaryotic cells; single celled; greater internal complexity than bacteria	Various types of algae, diatoms, protozoans
Fungi	Eukaryotic cells; multicelled; major decomposers and nutrient recyclers	Fungi, yeasts, molds, mushrooms
Plantae	Eukaryotic cells; multicelled; obtain nutrients by photosynthesis	Trees, grasses, roses, rushes, palms, broccoli, poison ivy
Animalia	Eukaryotic cells; multicelled; obtain nutrients by ingestion of preformed organic molecules	Worms, clams, corals, sponges, jellyfish, fishes, amphibians, reptiles, birds, mammals

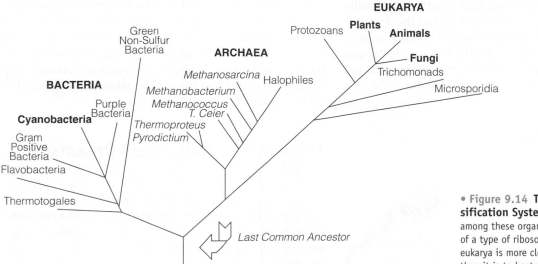

• Figure 9.14 **The Three Domain Classification System** The inferred re1ationships among these organisms are based on analyses of a type of ribosomal RNA. Notice that the eukarya is more closely related to the archaea than it is to bacteria.

• Figure 9.15 **The Oldest Known Eukaryote and Megafossil**

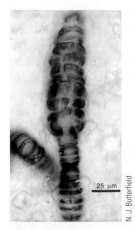

a At 1.2 billion years old, *Bangiomorpha* is the oldest known eukaryotic organism. It is only a few millimeters long.

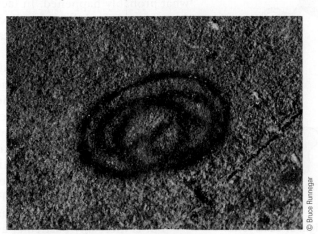

b *Grypania*, at 2.1 billion years old, is the oldest known megafossil. It was probably a bacterium or some kind of algae. It is about 1 cm across.

Other intriguing evidence of Neoproterozoic life comes from the Grand Canyon of Arizona where vase-shaped microfossils tentatively identified as cysts of algae have been found (Figure 9.16c).

Endosymbiosis and the Origin of Eukaryotic Cells

According to a widely accepted theory, eukaryotic cells formed from prokaryotic cells that entered into a symbiotic relationship. Symbiosis involving a prolonged association of two or more dissimilar organisms, is common today. In many cases both symbionts benefit from the association, as in

• Figure 9.16 **Proterozoic Fossils** These are probably from eukaryotic organisms and measure only 40 to 50 microns across.

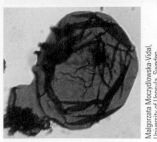
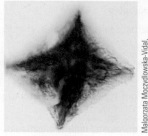

a The acritarch *Tappania piana* is from Mesoproterozoic rocks in China.

b This acritarch, known as *Octoedryxium truncatum*, was found in Neoproterozoic rocks in Sweden.

c This vase-shaped microfossil from Neoproterozoic rocks in the Grand Canyon in Arizona is a cyst from some kind of algae.

lichens, which were once thought to be plants but actually are symbiotic associations between fungi and algae.

In a symbiotic relationship, each symbiont is usually capable of metabolism and reproduction, but the degree of dependence in some relationships is such that one or both symbionts cannot live independently. This may have been the case with Proterozoic symbiotic prokaryotes that became increasingly interdependent until the unit could exist only as a whole. In this relationship, though, one symbiont lived within the other, which is a special type of symbiosis called **endosymbiosis** (• Figure 9.17). Endosymbiosis

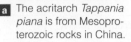

• **Figure 9.17 Endosymbiosis and the Origin of Eukaryotic Cells** An aerobic bacterium and a larger host bacterium united to form a mitochondria-containing amoeboid. An amoeboflagellate was formed by a union of the amoeboid and a bacterium of the spirochete group; this amoeboflagellate gave rise to the animal fungi, and protistan kingdoms. The plant kingdom originated when blue-green algae became plastids within an amoeboflagellate.

was proposed as early as 1905, but it was the work of Lynn Margulis and her 1970 publication that convinced scientists that it was a viable theory accounting for the origin of eukaryotic cells.

Supporting evidence for the endosymbiosis theory comes from studies of living eukaryotic cells that contain internal structures, or organelles, such as mitochondria and plastids. These organelles have their own genetic material and they synthesize proteins just as prokaryotic cells do, so they probably represent free-living bacteria that entered into a symbiotic relationship, eventually giving rise to eukaryotic cells. That such a relationship among cells is viable was demonstrated in 1966, when a microbiologist studying amoeba found that his samples had been infected by a bacterium that invaded and killed most of the amoeba. After several months, though, the amoeba and bacterium coexisted in an endosymbiotic relationship; the amoeba could no longer live without the bacterium that produces an essential protein.

The Dawn of Multicelled Organisms
We certainly live in a world dominated by microscopic organisms, but we are most familiar with multicelled plants and animals. However, **multicelled organisms** are not simply composed of many cells, but also have cells specialized to perform specific functions such as reproduction and respiration. We know that multicelled organisms were present by the Neoproterozoic, but the fossil record does not tell us how this transition took place. Nevertheless, studies of present-day organisms give some clues about what probably happened. In fact, there are some living organisms that while multicelled, possess as few as four identical cells, all of which are capable of living on their own (• Figure 9.18).

Suppose that a single-celled organism divided and formed a group of cells that did not disperse, but remained together as a colony. The cells in some colonies may have become somewhat specialized, as are the cells of some colonial organism today. Further specialization might have led to simple multicelled organisms such as sponges, consisting of cells that carry out functions such as reproduction, respiration, and food gathering. Carbonaceous impressions of Proterozoic multicelled algae are known from many areas (Figure 9.18b).

But is there any particular advantage to being multicelled? After all, for about 1.5 billion years all organisms were single-celled and life seems to have thrived. In fact, single-celled organisms are quite good at what they do, but what they do is limited. For example, they can't become very large, because as size increases,

Figure labels for 9.17: Higher plants, Animals Fungi Protista, Amoeboflagellate, Blue-green algae, Mitochondria-containing amoeboid, Spirochetes, Aerobic bacteria, Prokaryote host

• Figure 9.18 **Single Celled and Multicelled Organisms**

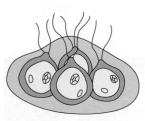
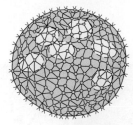

Gonium   *Volvox*

a *Gonium* consists of as few as four cells, all cells are alike and can reproduce a new colony. *Volvox* has some cells specialized for specific functions, so it has crossed the threshold that separates single celled and multicelled organisms.

Courtesy of Robert Hordyski

b This carbonaceous impression in the Little Belt Mountains of Montana may be from multicelled algae.

proportionately less of a cell is exposed to the external environment in relation to its volume. So as volume increases, the proportion of surface area decreases and transferring materials from the exterior to the interior becomes less efficient. Also, multicelled organisms live longer, because cells can be replaced and more offspring can be produced. And finally, cells have increased functional efficiency when they are specialized into organs with specific functions.

Neoproterozoic Animals

Criteria such as method of reproduction and type of metabolism set forth by biologists allow us to easily distinguish between animals and plants. Or, so it would seem—but some present-day organisms blur this distinction, and the same is true for some Proterozoic fossils. Nevertheless, the first fairly controversy-free fossils of animals come from the Ediacaran fauna of Australia and similar faunas of about the same age elsewhere.

The Ediacaran Fauna In 1947, an Australian geologist, R. C. Sprigg, discovered impressions of soft-bodied animals in the Pound Quartzite in the Ediacara Hills of South

Australia. Additional discoveries by others turned up what appeared to be impressions of algae and several animals, many bearing no resemblance to any existing now (• Figure 9.19). Before these discoveries, geologists were perplexed by the apparent absence of fossil-bearing rocks predating the Phanerozoic. They had assumed the fossils so common in Cambrian rocks must have had a long previous history but had little evidence to support this conclusion. This discovery and subsequent ones have not answered all questions about pre-Phanerozoic animals, but they have certainly increased our knowledge about this chapter in the history of life.

Some investigators think that three present-day phyla are represented in the Ediacaran fauna: jellyfish and sea pens (phylum Cnidaria), segmented worms (phylum Annelida), and primitive members of the phylum Arthropoda (the phylum with insects, spiders, crabs, and others). One Ediacaran fossil, *Spriggina,* has been cited as a possible ancestor of trilobites (Figure 9.19b), and another may be a primitive member of the phylum Echinodermata. However, some scientists think these Ediacara animals represent an early evolutionary group quite distinct from the ancestry of today's invertebrate animals.

The **Ediacaran fauna,** the collective name for fossil associations similar to those in the Ediacara Hills, is now known from all continents except Antarctica. For example, excellent fossils from this 545- to 600-million-year-old fauna have been found in Namibia, Africa, and in Newfoundland, Canada (• Figure 9.20). Thus, Ediacaran animals were widespread, but because they lacked durable skeletons, their fossils are not very common.

Other Proterozoic Animal Fossils Although scarce, a few animal fossils older than those of the Ediacaran fauna are known. A jellyfish-like impression is present in rocks 2000 m below the Pound Quartzite, and, in many areas, burrows, presumably made by worms, are found in rocks at least 700 million years old. Some possible fossil worms from China may be 700 to 900 million years old (• Figure 9.21a).

All known Proterozoic animals were soft-bodied, but there is some evidence that the earliest stages in the evolution of skeletons were under way. Even some Ediacaran animals may have had a chitinous carapace, and others appear to have spots or small areas of calcium carbonate. Small branching tubes preserved in 590- to 600-million-year-old rocks in China may be early relatives of corals (Figure 9.21b). The odd creature known as *Kimberella* from the Neoproterozoic of Russia had a tough outer covering similar to that of some present-day marine invertebrates (Figure 9.21c). Exactly what *Kimberella* was is uncertain; although some think it was a mollusk.

Minute scraps of shell-like material and small toothlike denticles and spicules, presumably from sponges, indicate that several animals with skeletons, or at least partial skeletons, existed by latest Neoproterozoic time. Nevertheless, more durable skeletons of

• Figure 9.19 **The Edlacaran Fauna of Australia**

a The affinities of *Tribachidium* remain uncertain. It may be either a primitive echinoderm or a cnidarian.

b *Spriggina* was originally thought be a segmented worm (annelid), but now it appears more closely related to arthopods, possibly even an ancestor of trilobites.

c Shield-shaped *Parvanconrina* is perhaps related to the arthropods.

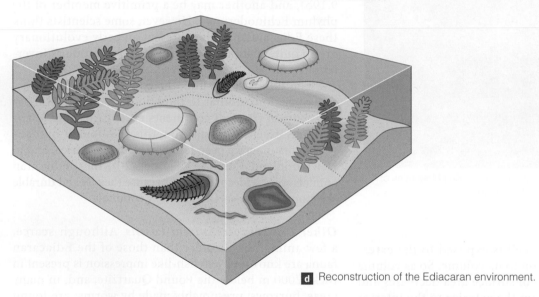

d Reconstruction of the Ediacaran environment.

• Figure 9.20 **Ediacaran-type Fossils from the Mistaken Point Formation of Newfoundland**

a These fossils have not been given scientific names. The most obvious ones are informally called *spindles,* whereas the one toward the lower center is a *feather duster.*

b This fossil, known as *Bradgatia,* has also been found in England. These fossils are about 575 million years old.

• Figure 9.21 Neoproterozoic Fossils

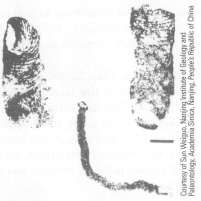

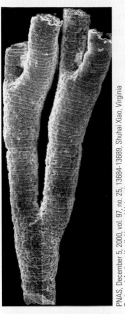

Courtesy of Sun Weiguo, Nanjing Institute of Geology and Paleontology, Academia Sinica, Nanjing, People's Republic of China

a Possible worm fossils from 700- to 900-million-year-old rocks in China.

Ben Waggoner and Mikhail A. Fedonkin

c Many think that *Kimberella,* of Neoproterozoic age from Russia, was some kind of mollusk.

PNAS, December 5, 2000, vol. 97, no. 25, 13684-13689, Shuhai Xiao, Virginia Tech and Andrew Knoll, Harvard University

b These small branching tubes measure only 0.1 to 0.3 mm across. The animals that made the tubes may have been early relatives of corals.

silica, calcium carbonate, and chitin (a complex organic substance) did not appear in abundance until the beginning of the Phanerozoic Eon 542 million years ago (see Chapter 12).

Proterozoic Mineral Resources

In an earlier section, we mentioned that most of the world's iron ore comes from Paleoproterozoic banded iron formations (Figure 9.11) and that Canada and the United States have large deposits of these rocks in the Lake Superior region and in eastern Canada (• Figure 9.22). Both rank among the ten leading nations in iron ore production.

In the Sudbury mining district in Ontario, Canada, nickel and platinum are extracted from Proterozoic rocks (see Perspective). Nickel is essential for the production of nickel alloys such as stainless steel and Monel metal (nickel plus copper), which are valued for their strength and resistance to corrosion and heat. The United States must import more than 50% of all nickel used, mostly from the Sudbury mining district.

Some platinum for jewelry, surgical instruments, and chemical and electrical equipment is also exported to the United States from Canada, but the major exporter is South Africa. The Bushveld Complex of South Africa is a layered

• Figure 9.22 Iron Mining in the Lake Superior Region

James S. Monroe

a The Empire Mine at Palmer Michigan. The iron ore comes from the Paleoproterozoic Negaunee Iron Formation.

Sue Monroe

b The Iron ore is refined and shaped into pellets containing about 65% iron. The pellets are about 1 cm across.

complex of igneous rocks from which both platinum and chromite, the only ore of chromium, are mined. Much chromium used in the United States is imported from South Africa; it is used mostly in manufacturing stainless steel.

Economically recoverable oil and gas have been discovered in Proterozoic rocks in China and Siberia, arousing some interest in the Midcontinent rift as a potential source of hydrocarbons (Figure 9.6). So far, land has been leased for exploration, and numerous geophysical studies have been done. However, even though some rocks within the rift are known to contain petroleum, no producing oil or gas wells are operating.

A number of Proterozoic pegmatites are important economically. The Dunton pegmatite in Maine, whose age is generally considered Neoproterozoic, has yielded magnificent gem-quality specimens of tourmaline and other minerals. Other pegmatites are mined for gemstones as well as for tin; industrial minerals, such as feldspars, micas, and quartz; and minerals containing such elements as cesium, rubidium, lithium, and beryllium.

Geologists have identified more than 20,000 pegmatites in the country rocks adjacent to the Harney Peak Granite in the Black Hills of South Dakota. These pegmatites formed about 1.7 billion years ago when the granite was emplaced as a complex of dikes and sills. A few have been mined for gemstones, tin, lithium, and micas, and some of the world's largest known mineral crystals were discovered in these pegmatites.

SUMMARY

Table 9.2 provides a summary of the Proterozoic geologic and biologic events discussed in the text.

- The crust-forming processes that yielded Archean granite-gneiss complexes and greenstone belts continued into the Proterozoic but at a considerably reduced rate.

- Paleoproterozoic collisions between Archean cratons formed larger cratons that served as nuclei, around which crust accreted. One large landmass so formed was Laurentia, consisting mostly of North America and Greenland.

- Paleoproterozoic amalgamation of cratons, followed by Mesoproterozoic igneous activity, the Grenville orogeny, and the Midcontinent rift, were important events in the evolution of Laurentia.

- Ophiolite sequences marking convergent plate boundaries are first well documented from the Neoarchean and Paleoproterozoic, indicating that a plate tectonic style similar to that operating now had become established.

- Sandstone-carbonate-shale assemblages deposited on passive continental margins were very common by Proterozoic time.

- The supercontinent Rodinia assembled between 1.3 and 1.0 billion years ago, fragmented, and then reassembled to form Pannotia about 650 million years ago, which began fragmenting about 550 million years ago.

- Glaciers were widespread during both the Paleoproterozoic and the Neoproterozoic.

- Photosynthesis continued to release free oxygen into the atmosphere, which became increasingly rich in oxygen through the Proterozoic.

- Fully 92% of Earth's iron ore deposits in the form of banded iron formations were deposited between 2.5 and 2.0 billion years ago.

- Widespread continental red beds dating from 1.8 billion years ago indicate that Earth's atmosphere had enough free oxygen for oxidation of iron compounds.

- Most of the known Proterozoic organisms are single-celled prokaryotes (bacteria).When eukaryotic cells first appeared is uncertain, but they were probably present by 1.2 billion years ago. Endosymbiosis is a widely accepted theory for their origin.

- The oldest known multicelled organisms are probably algae, some of which might date back to the Paleoproterozoic.

- Well-documented multicelled animals are found in several Neoproterozoic localities. Animals were widespread at this time, but because all lacked durable skeletons their fossils are not common.

- Most of the world's iron ore production is from Proterozoic banded iron formations. Other important resources include nickel and platinum.

IMPORTANT TERMS

banded iron formation (BIF), p. 184
continental red beds, p. 185
Ediacaran fauna, p. 189
endosymbiosis, p. 188
eukaryotic cell, p. 186
Grenville orogeny, p. 179

Laurentia, p. 175
Midcontinent rift, p. 179
multicelled organism, p. 188
orogen, p. 175
Pannotia, p. 182
Rodinia, p. 181

sandstone-carbonate-shale assemblage, p. 175
supercontinent, p. 180
Wilson cycle, p. 175

REVIEW QUESTIONS

1. One type of Proterozoic rock that indicates some free oxygen was present in the atmosphere is
   a. _____ continental red beds; b. _____ carbon-conglomerate assemblages; c. _____ ultramafic lava flows; d. _____ Wilson cycle deposits; e. _____ prokaryotic accumulates.

2. A large landmass composed mostly of Greenland and North America that evolved during the Proterozoic is called
   a. _____ Grenvillia; b. _____ Ediacara; c. _____ Laurentia; d. _____ Pannotia; e. _____ Romania.

3. Cells with a membrane-bounded nucleus and internal structures called organelles are called _____ cells
   a. _____ komatiitic; b. _____ endosymbiotic; c. _____ porphyritic; d. _____ aphanitic; e. _____ eukaryotic.

4. Which one of the following was a supercontinent during the Proterozoic?
   a. _____ Pangaea; b. _____ Rodinia; c. _____ Ediacara; d. _____ Laurasia; e. _____ Mesoamerica.

5. A plate tectonic cycle involving the opening of an ocean basin and it subsequent closure is known as a(n)
   a. _____ ophiolite sequence; b. _____ intracontinental rift; c. _____ Wilson cycle; d. _____ separation orogeny; e. _____ collision orogen.

6. A sequence of rocks on land made up of mantle rocks overlain by oceanic crust and deep sea sediments is a(n)
   a. _____ granite-gneiss complex; b. _____ turbidite deposit; c. _____ ophiolite; d. _____ continental red bed; e. _____ supercycle.

7. The oldest known animal fossils are found in the _____ fauna of Australia
   a. _____ Ediacaran; b. _____ Pannotian; c. _____ Wilsonian; d. _____ Stromatolinian; e. _____ Grenvillian.

8. Columnar masses of rock resulting from the activities of cyanobacteria (blue-green algae) are
   a. _____ heterotrophs; b. _____ endosymbionts; c. _____ orogens; d. _____ stromatolites; e. _____ trilobites.

9. The Mesoproterozoic of Laurentia was a time of
   a. _____ widespread glaciation; b. _____ origin of animals with skeletons; c. _____ igneous activity unrelated to orogenic activity; d. _____ origin of the oldest known greenstone belts; e. _____ formation of Pangaea.

10. The widely accepted theory explaining the origin of eukaryotic cells holds that these cells formed by
    a. _____ endosymbiosis; b. _____ parthenogenesis; c. _____ binary fission; d. _____ pangenesis; e. _____ autotrophism.

11. When did the first animals appear in the fossil record, and why are their fossils not very common?

12. Where is the Midcontinent rift, how did it form, and what kinds of rocks are found in it?

13. What evidence indicates that eukaryotic cells evolved from prokaryotic cells?

14. How do banded iron formations and continental red beds provide evidence about changes in Earth's Proterozoic atmosphere?

15. Outline the events that led to the development of Laurentia during the Proterozoic.

16. What is a Wilson cycle? Is there any evidence for these cycles in Precambrian rocks?

17. You encounter an outcrop of what appears to be tillite. What kinds of evidence would be useful for concluding that this material was deposited by glaciers?

18. What association of rocks is typically found on passive continental margins, and when did they become common in the geologic record?

19. How did the style of crustal evolution for the Archean and Proterozoic differ?

APPLY YOUR KNOWLEDGE

1. Proterozoic sedimentary rocks in the northern Rocky Mountains are 4000 m thick and were deposited between 1.45 billion and 850 million years ago. What was the average rate of sediment accumulation in centimeters per year? Why is this figure unlikely to represent the actual rate of sedimentation?

2. Suppose you were to visit a planet that, like Earth, has continents and vast oceans. What kinds of evidence would indicate that this hypothetical planet's continents formed like those on Earth?

3. The illustration below shows the 40- to 90-m-thick, 1.45-billion-year-old Purcell Sill in Glacier National Park, Montana. What evidence convinces you that this is a sill rather than a buried lava flow? What can you say about the absolute ages of the rocks above and below the sill? The minerals in the central part of the sill are larger than those near its upper and lower boundaries. Why?

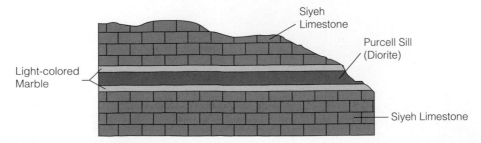

Siyeh Limestone

Purcell Sill (Diorite)

Light-colored Marble

Siyeh Limestone

4. The stratigraphic column shows some of the rocks in the Grand Canyon. Answer these questions:
   a. Which is oldest, the Vishnu Schist or the Zoraster Granite? How do you know?
   b. What kind of unconformity lies between the Vishnu-Zoraster and the overlying Grand Canyon Supergroup?
   c. What kind of unconformity lies between the Grand Canyon Supergroup and the Tapeats Sandstone?
   d. How can you account for the vertical sequence of facies in the Tonto Group, that is, sandstone overlain by shale, which is overlain by limestone?

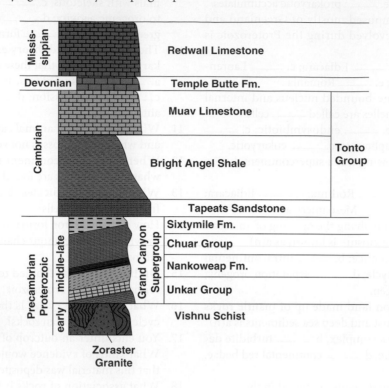

Table 9.2
Summary of the Proterozoic Geologic and Biologic Events Discussed in the Text

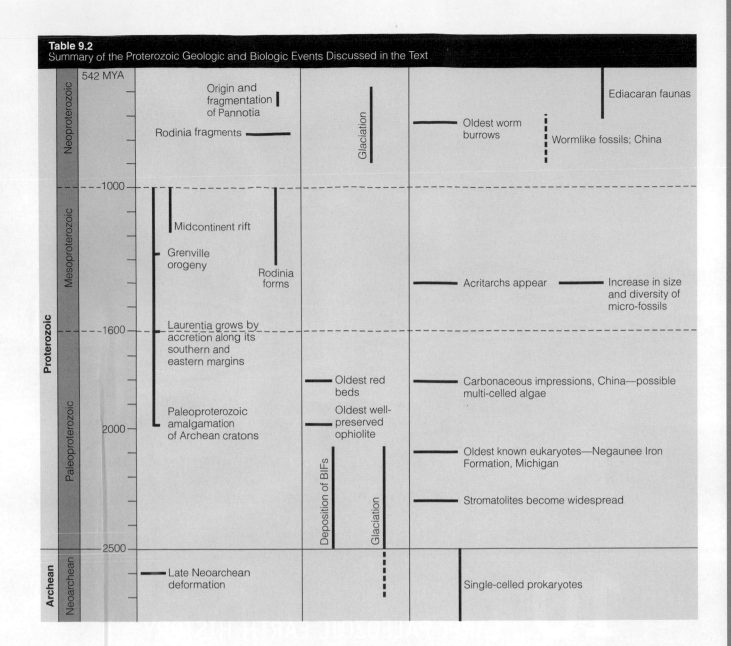

CHAPTER 10

EARLY PALEOZOIC EARTH HISTORY

The need to cheaply transport coal from where it was mined to where it was needed resulted in widespread canal building in England during the late 1700s and early 1800s. William Smith, who started his career mapping various coal mines, and later produced the first geologic map of England, was instrumental in helping to find the most efficient canal routes to bring coal to market. Canals like the Grand Junction Canal, shown here in this 1819 woodcut, were critical not only for transporting coal from the mines to market, but also for the movement of people and goods.

[OUTLINE]

Introduction

Continental Architecture: Cratons and Mobile Belts

Paleozoic Paleogeography
    Early Paleozoic Global History

Early Paleozoic Evolution of North America

The Sauk Sequence
    Perspective *The Grand Canyon— A Geologist's Paradise*
    The Cambrian of the Grand Canyon Region: A Transgressive Facies Model

The Tippecanoe Sequence
    Tippecanoe Reefs and Evaporites
    The End of the Tippecanoe Sequence

The Appalachian Mobile Belt and the Taconic Orogeny

Early Paleozoic Mineral Resources

Summary

At the end of this chapter, you will have learned that

- Six major continents were present at the beginning of the Paleozoic Era, and plate movement during the Early Paleozoic resulted in the first of several continental collisions leading to the formation of the supercontinent Pangaea at the end of the Paleozoic.

- The Paleozoic history of North America can be subdivided into six cratonic sequences, which represent major transgressive–regressive cycles.

- During the Sauk Sequence, warm, shallow seas covered most of North America, leaving only a portion of the Canadian shield and a few large islands above sea level.

- Like the Sauk Sequence, the Tippecanoe Sequence began with a major transgression resulting in widespread sandstones, followed by extensive carbonate and evaporite deposition.

- During Tippecanoe time, an oceanic–continental convergent plate boundary formed along the eastern margin of North America (known as the Appalachian mobile belt) resulting in the Taconic orogeny, the first of several orogenies to affect this area.

- Lower Paleozoic rocks contain a variety of important mineral resources.

Introduction

August 1, 1815, is an important date in the history of geology. On that date William Smith, a canal builder, published the world's first true geologic map. Measuring more than eight feet high and six feet wide, Smith's hand painted geologic map of England represented more than 20 years of detailed study of England's rocks and fossils.

England is a country rich in geologic history. Five of the six Paleozoic geologic systems (Cambrian, Ordovician, Silurian, Devonian, and Carboniferous) were described and named for rocks exposed in England (see Chapter 5). The Carboniferous coal beds of England helped fuel the Industrial Revolution, and the need to transport coal cheaply from where it was mined to where it was used set off a flurry of canal building during the late 1700s and early 1800s. During this time, William Smith, who was mapping various coal mines, first began to notice how rocks and fossils repeated themselves in a predictable manner. During the ensuing years, Smith surveyed the English countryside for the most efficient canal routes to bring the coal to market. Much of his success was based on his ability to predict what rocks the canal diggers would encounter. Realizing that his observations allowed him to unravel the geologic history of an area and correlate rocks from one region to another, William Smith set out to make the first geologic map of an entire country!

The story of how William Smith came to publish the world's first geologic map is a fascinating tale of determination and perseverance. However, instead of finding fame and success, Smith found himself, slightly less than four years later, in debtors' prison, and—upon his release after spending more than two months in prison—homeless. If such a story can have a happy ending, however, William Smith at least lived long enough to finally be recognized and honored for the seminal contribution he made to the then fledgling science of geology.

Just as William Smith applied basic geologic principles in deciphering the geology of England, we use these same principles in the next two chapters to interpret the geology of the Paleozoic Era. In these chapters, we use the geologic principles and concepts discussed in earlier chapters to help explain how Earth's systems and its associated geologic processes interacted during the Paleozoic to lay the groundwork for the distribution of continental landmasses, ocean basins, and the topography we have today.

The Paleozoic history of most continents involves major mountain-building activity along their margins and numerous shallow-water marine transgressions and regressions over their interiors. These transgressions and regressions were caused by global changes in sea level that most probably were related to rates of seafloor spreading and glaciation.

In the next two chapters, we provide an overview of the geologic history of the world during the Paleozoic Era in order to place into context the geologic events taking place in North America during this time. We then focus our attention on the geologic history of North America—not in a period-by-period chronology but in terms of the major transgressions and regressions taking place on the continent as well as the mountain building activity occurring during this time. Such an approach allows us to place the North American geologic events within a global context.

Continental Architecture: Cratons and Mobile Belts

During the Precambrian, continental accretion and orogenic activity led to the formation of sizable continents. Movement of these continents during the Neoproterozoic resulted in the formation of a single Pangaea-like supercontinent geologists refer to as Pannotia (see Chapter 9). This supercontinent began breaking apart sometime during the latest Neoproterozoic, and by the beginning of the Paleozoic Era, six major continents were present. Each continent can be divided into two major components: a craton and one or more mobile belts.

Archean Eon	Proterozoic Eon	Phanerozoic Eon							
		Paleozoic Era							
Precambrian		Cambrian	Ordovician	Silurian	Devonian	Mississippian	Pennsylvanian	Permian	
						Carboniferous			

2500 MYA · 542 MYA · 251 MYA

Cratons are the relatively stable and immobile parts of continents and form the foundation on which Phanerozoic sediments were deposited (• Figure 10.1). Cratons typically consist of two parts: a shield and a platform.

Shields are the exposed portion of the crystalline basement rocks of a continent and are composed of Precambrian metamorphic and igneous rocks that reveal a history of extensive orogenic activity during the Precambrian (see Chapters 8 and 9). During the Phanerozoic, however, shields were extremely stable and formed the foundation of the continents.

Extending outward from the shields are buried Precambrian rocks that constitute a *platform,* another part of the craton. Overlying the platform are flat-lying or gently dipping Phanerozoic detrital and chemical sedimentary rocks that were deposited in widespread shallow seas that transgressed and regressed over the craton. These seas, called **epeiric seas,** were a common feature of most Paleozoic cratonic histories. Changes in sea level caused primarily by continental glaciation as well as by rates of seafloor spreading were responsible for the advance and retreat of these epeiric seas.

Whereas most Paleozoic platform rocks are still essentially flat lying, in some places they were gently folded into regional arches, domes, and basins (Figure 10.1). In many cases, some of these structures stood out as low islands during the Paleozoic Era and supplied sediments to the surrounding epeiric seas.

Mobile belts are elongated areas of mountain-building activity. They are located along the margins of continents where sediments are deposited in the relatively shallow waters of the continental shelf and the deeper waters at the base of the continental slope. During plate convergence along these margins, the sediments are deformed and intruded by magma, creating mountain ranges.

Four mobile belts formed around the margin of the North American craton during the Paleozoic: the Franklin, **Cordilleran, Ouachita,** and **Appalachian**

• **Figure 10.1 Major Cratonic Structures and Mobile Belts** The major cratonic structures and mobile belts of North America that formed during the Paleozoic Era.

Phanerozoic Eon										
Mesozoic Era			Cenozoic Era							
Triassic	Jurassic	Cretaceous	Paleogene			Neogene		Quaternary		
			Paleocene	Eocene	Oligocene	Miocene	Pliocene	Pleistocene	Holocene	

251 MYA   66 MYA

mobile belts (Figure 10.1). Each was the site of mountain building in response to compressional forces along a convergent plate boundary and formed mountain ranges such as the Appalachians and Ouachitas.

Paleozoic Paleogeography

One result of plate tectonics is that Earth's geography is constantly changing. The present-day configuration of the continents and ocean basins is merely a snapshot in time. As the plates move about, the location of continents and ocean basins constantly changes. One of the goals of historical geology is to provide paleogeographic reconstructions of the world during the geologic past. By synthesizing all of the pertinent paleoclimatic, paleomagnetic, paleontologic, sedimentologic, stratigraphic, and tectonic data available, geologists construct paleogeographic maps. Such maps are simply interpretations of the geography of an area for a particular time in the geologic past. The majority of paleogeographic maps show the distribution of land and sea, possible climatic regimes, and such geographic features as mountain ranges, swamps, and glaciers.

The paleogeographic history of the Paleozoic Era, for example, is not as precisely known as for the Mesozoic and Cenozoic eras, in part because the magnetic anomaly patterns preserved in the oceanic crust were destroyed when much of the Paleozoic oceanic crust was subducted during the formation of Pangaea. Paleozoic paleogeographic reconstructions are, therefore, based primarily on structural relationships, climate-sensitive sediments such as red beds, evaporites, and coals, as well as the distribution of plants and animals.

At the beginning of the Paleozoic, six major continents were present. Besides these large landmasses, geologists have also identified numerous microcontinents such as *Avalonia* (composed of parts of present-day Belgium, northern France, England, Wales, Ireland, the Maritime Provinces and Newfoundland of Canada, as well as parts of the New England area of the United States) and various island arcs associated with microplates. We are primarily concerned, however, with the history of the six major continents and their relationships to each other. The six

major Paleozoic continents are **Baltica** (Russia west of the Ural Mountains and the major part of northern Europe), **China** (a complex area consisting of at least three Paleozoic continents that were not widely separated and are here considered to include China, Indochina, and the Malay Peninsula), **Gondwana** (Africa, Antarctica, Australia, Florida, India, Madagascar, and parts of the Middle East and southern Europe), **Kazakhstania** (a triangular continent centered on Kazakhstan but considered by some to be an extension of the Paleozoic Siberian continent), **Laurentia** (most of present North America, Greenland, northwestern Ireland, and Scotland), and **Siberia** (Russia east of the Ural Mountains and Asia north of Kazakhstan and south of Mongolia). The paleogeographic reconstructions that follow (• Figure 10.2) are based on the methods used to determine and interpret the location, geographic features, and environmental conditions on the paleocontinents.

Early Paleozoic Global History

In contrast to today's global geography, the Cambrian world consisted of six major continents dispersed around the globe at low tropical latitudes (Figure 10.2a). Water circulated freely among ocean basins, and the polar regions were mostly ice free. By the Late Cambrian, epeiric seas had covered areas of Laurentia, Baltica, Siberia, Kazakhstania, and China, whereas highlands were present in northeastern Gondwana, eastern Siberia, and central Kazakhstania.

During the Ordovician and Silurian periods, plate movement played a major role in the changing global geography (Figure 10.2b and c). Gondwana moved southward during the Ordovician and began to cross the South Pole as indicated by Upper Ordovician tillites found today in the Sahara Desert. During the Early Ordovician, the microcontinent Avalonia separated from Gondwana and began moving northeastward, where it would finally collide with Baltica during the Late Ordovician–Early Silurian. In contrast to the passive continental margin Laurentia exhibited during the Cambrian, an active convergent plate boundary formed along its eastern margin during the Ordovician, as

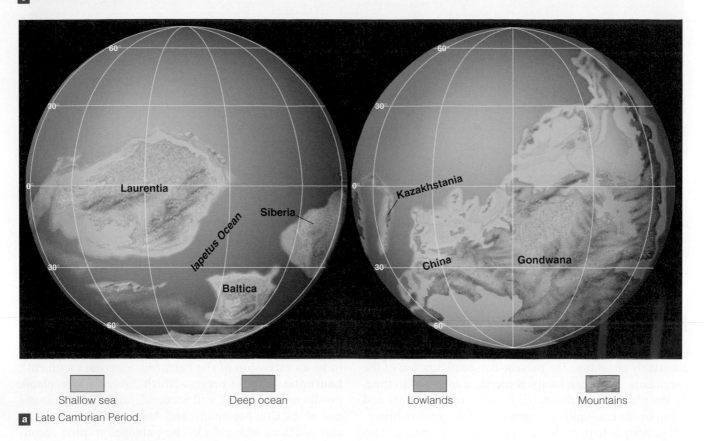

Shallow sea   Deep ocean   Lowlands   Mountains

a Late Cambrian Period.

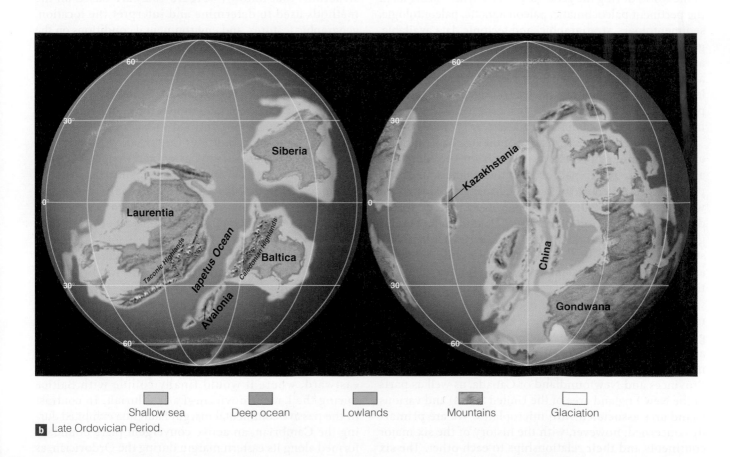

Shallow sea   Deep ocean   Lowlands   Mountains   Glaciation

b Late Ordovician Period.

• Figure 10.2 **(cont.)**

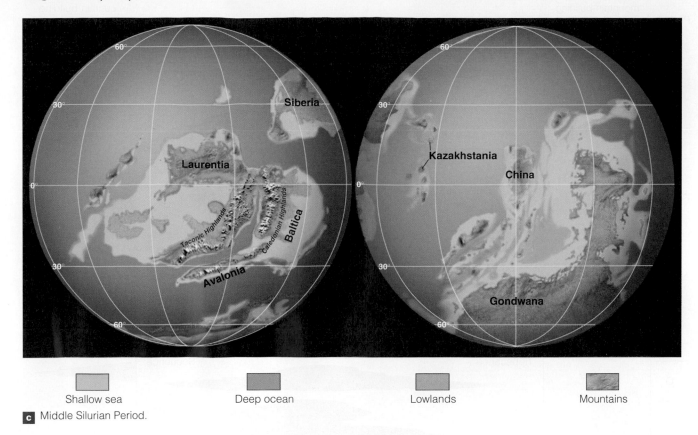

Shallow sea   Deep ocean   Lowlands   Mountains

c Middle Silurian Period.

indicated by the Late Ordovician Taconic orogeny that occurred in New England.

During the Silurian, Baltica, along with the newly attached Avalonia, moved northwestward relative to Laurentia and collided with it to form the larger continent of Laurasia. This collision, which closed the northern Iapetus Ocean, is marked by the Caledonian orogeny. After this orogeny, the southern part of the Iapetus Ocean still remained open between Laurentia and Avalonia-Baltica (Figure 10.2c). Siberia and Kazakhstania moved from a southern equatorial position during the Cambrian to north temperate latitudes by the end of the Silurian Period.

With this plate tectonics overview in mind, we now focus our attention on North America (Laurentia) and its role in the Early Paleozoic geologic history of the world.

Early Paleozoic Evolution of North America

It is convenient to divide the geologic history of the North American craton into two parts: the first dealing with the relatively stable continental interior over which epeiric seas transgressed and regressed, and the other with the mobile belts where mountain building occurred.

In 1963, the American geologist Laurence Sloss subdivided the sedimentary rock record of North America into six cratonic sequences. A **cratonic sequence** is a large-scale (greater than supergroup) lithostratigraphic unit representing a major transgressive–regressive cycle bounded by craton-wide unconformities (• Figure 10.3). The transgressive phase, which is usually covered by younger sediments, commonly is well preserved, whereas the regressive phase of each sequence is marked by an unconformity. Where rocks of the appropriate age are preserved, each of the six unconformities can be shown to extend across the various sedimentary basins of the North American craton and into the mobile belts along the cratonic margin.

Geologists have also recognized major unconformity-bounded sequences in cratonic areas outside North America. Such global transgressive and regressive cycles of sea level changes are thought to result from major tectonic and glacial events.

The realization that rock units can be divided into cratonic sequences, and that these sequences can be further subdivided and correlated, provides the foundation for an important concept in geology that allows high-resolution analysis of time and facies relationships within sedimentary rocks. **Sequence stratigraphy** is the study of rock relationships within a time–stratigraphic framework of related

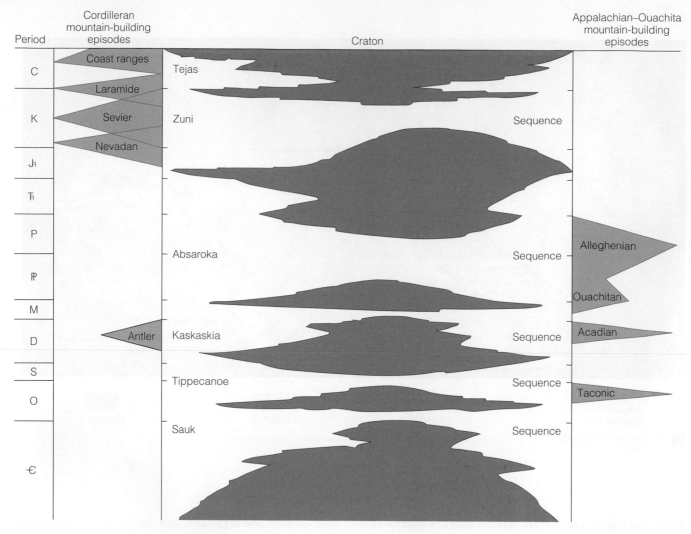

Figure 10.3 Cratonic Sequences of North America A cratonic sequence is a large-scale lithostratigraphic unit representing a major transgressive–regressive cycle and bounded by craton-wide unconformities. The white areas represent sequences of rocks separated by large-scale unconformities (brown areas). The major Cordilleran orogenies are shown on the left side, and the major Appalachian orogenies are on the right.

facies bounded by erosional or nondepositional surfaces. The basic unit of sequence stratigraphy is the *sequence*, which is a succession of rocks bounded by unconformities and their equivalent conformable strata. Sequence boundaries result from a relative drop in sea level. Sequence stratigraphy is an important tool in geology because it allows geologists to subdivide sedimentary rocks into related units that are bounded by time–stratigraphic significant boundaries. Geologists use sequence stratigraphy for high-resolution correlation and mapping, as well as interpreting and predicting depositional environments.

The Sauk Sequence

Rocks of the **Sauk Sequence** (Neoproterozoic–Early Ordovician) record the first major transgression onto

the North American craton (Figure 10.3). During the Neoproterozoic and Early Cambrian, deposition of marine sediments was limited to the passive shelf areas of the Appalachian and Cordilleran borders of the craton. The craton itself was above sea level and experiencing extensive weathering and erosion. Because North America was located in a tropical climate at this time and there is no evidence of any terrestrial vegetation, weathering and erosion of the exposed Precambrian basement rocks must have proceeded rapidly.

During the Middle Cambrian, the transgressive phase of the Sauk began with epeiric seas encroaching over the craton (see Perspective). By the Late Cambrian, these epeiric seas had covered most of North America, leaving only a portion of the Canadian shield and a few large islands above sea level (• Figure 10.4). These islands, collectively named the **Transcontinental Arch,** extended

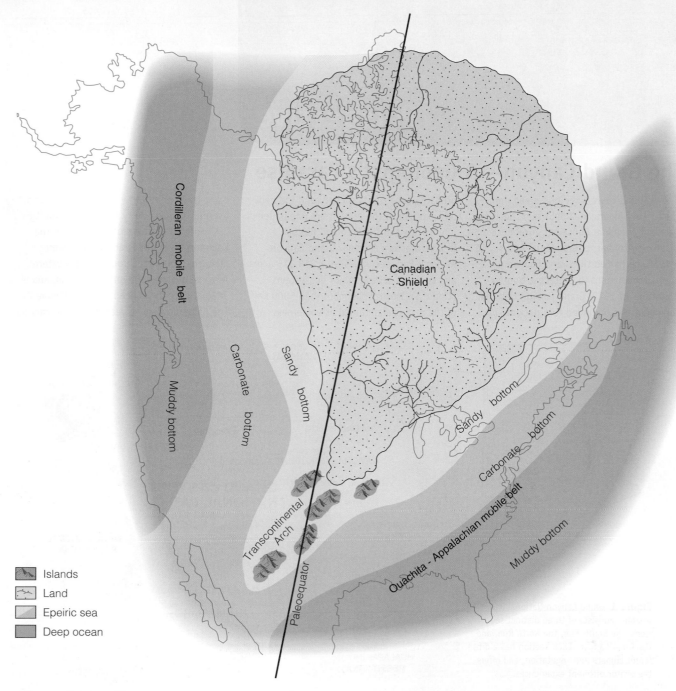

• Figure 10.4 **Cambrian Paleogeography of North America** Note the position of the Cambrian paleoequator. During this time, North America straddled the equator as indicated in Figure 10.2a.

from New Mexico to Minnesota and the Lake Superior region.

The sediments deposited both on the craton and along the shelf area of the craton margin show abundant evidence of shallow-water deposition. The only difference between the shelf and craton deposits is that the shelf deposits are thicker. In both areas, the sands are generally clean, well sorted, and commonly contain ripple marks and small-scale cross-bedding. Many of the carbonates are bioclastic (composed of fragments of organic remains), contain stromatolites, or have oolitic (small, spherical calcium carbonate grains) textures. Such sedimentary structures and textures indicate shallow-water deposition.

Perspective

The Grand Canyon—A Geologist's Paradise

"The Grand Canyon is the one great sight which every American should see," declared President Theodore Roosevelt. "We must do nothing to mar its grandeur." And so, in 1908, he named the Grand Canyon a national monument to protect it from exploitation. In 1919, Grand Canyon National Monument was upgraded to a national park primarily because its scenery and the geology exposed in the canyon are unparalleled.

Located in the Colorado Plateau in northwestern Arizona, Grand Canyon National Park encompasses 1,218,375 acres and contains several major ecosystems (Figure 1). Most of the more than 5000 km² of the park is maintained as wilderness, with many trails affording visitors both day and overnight backcountry hiking opportunities.

Formed by the erosive power of the Colorado River, the Grand Canyon winds more than 500 km through northwestern Arizona, averages a depth of 1220 m, and is 1830 m deep at its deepest point (Figure 2). The Colorado River is responsible for carving

Figure 1 Grand Canyon National Park, Arizona, consists of three distinct sections: the South Rim, the North Rim, and the Inner Canyon. Each section has a different climate and vegetation, and offers the visitor different experiences.

Figure 2 The Grand Canyon is world famous for its grandeur and beauty, and attracts approximately five million visitors a year. Mohave Point, on West Rim Drive, is a popular spot to view the South Rim of the Grand Canyon and the surrounding plateau area.

Figure 3 The South Rim of the Grand Canyon as viewed from Mohave Point. The Colorado River, which can be seen in the center of this view, is responsible for carving the Grand Canyon and other canyons in the area.

the Grand Canyon and other canyons during the past 30 million years or so in response to the general uplift of the Colorado Plateau (Figure 3).

Major John Wesley Powell, a Civil War veteran, led the first geologic expedition down the Colorado River through the Grand Canyon in 1869 (Figure 4). Without any maps or other information, Powell and his group ran the many rapids of the Colorado River in fragile wooden boats, hastily recording what they saw. Powell wrote in his diary that "all about me are interesting geologic records. The book is open and I read as I run." Following this first exploration, he led a second expedition in 1871, which included a photographer, a surveyor, and three topographers. This expedition made detailed topographic and geologic maps of the Grand Canyon area, as well as the first photographic record of the region.

The rocks of the Grand Canyon record more than one billion years of Earth history (Figure 5). The Vishnu Schist represents a major mountain-building episode that occurred during the Precambrian. Following erosion of this mountain range, sediments were deposited in a variety of marine, coastal, and terrestrial settings during the Paleozoic Era, and were lithified into the rock formations that are now exposed in the canyon walls (Figures 2 and 3).

Although best known for its geology, Grand Canyon National Park is home to 89 species of mammals, 56 species of reptiles and amphibians, more than 300 species of birds, 17 species of fish, and more than 1500 plant species (Figure 6). Its great biological diversity can be attributed to the fact that within its boundaries are five of the seven life zones and three of the four desert types in North America.

Figure 4 In 1869, Major John Wesley Powell, led the first geologic expedition down the Colorado River through the Grand Canyon in Arizona.

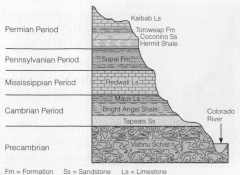

Permian Period — Kaibab Ls, Toroweap Fm, Coconino Ss, Hermit Shale
Pennsylvanian Period — Supai Fm
Mississippian Period — Redwall Ls
Cambrian Period — Mauv Ls, Bright Angel Shale, Tapeats Ss — Colorado River
Precambrian — Vishnu Schist

Fm = Formation   Ss = Sandstone   Ls = Limestone

Figure 5 The rocks of the Grand Canyon preserve more than one billion years of Earth history, ranging from periods of mountain building and erosion, as well as periods of transgressions and regressions of shallow seas over the area.

Figure 6 White aspen and pinyon pine, shown here in the DeMotte campground, 28 km north of the North Rim, are the typical trees found in the Grand Canyon.

The Cambrian of the Grand Canyon Region: A Transgressive Facies Model

Recall from Chapter 5 that sediments become increasingly finer the farther away from land one goes. Therefore, in a stable environment where sea level remains the same, coarse detrital sediments are typically deposited in the nearshore environment, and finer-grained sediments are deposited in the offshore environment. Carbonates form farthest from land in the area beyond the reach of detrital sediments. During a transgression, these facies (sediments that represent a particular environment) migrate in a landward direction (see Figure 5.8).

The Cambrian rocks of the Grand Canyon region (see Chapter 4 opening photo) provide an excellent model of the sedimentation patterns of a transgressing sea. The Grand Canyon region occupied the passive shelf and western margin of the craton during Sauk time. During the Neoproterozoic and Early Cambrian, most of the craton was above sea level and deposition of marine sediments was mainly restricted to the margins of the craton (continental shelves and slopes).

In the Grand Canyon region, the Tapeats Sandstone represents the basal transgressive shoreline deposits that accumulated as marine waters transgressed across the shelf and just onto the western margin of the craton during the

• Figure 10.5 **Cambrian Rocks of the Grand Canyon**

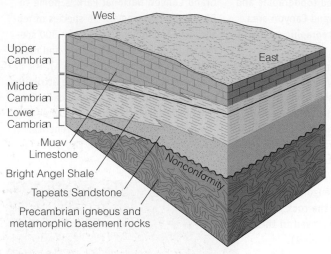

West

Upper Cambrian

Middle Cambrian

Lower Cambrian

East

Muav Limestone

Bright Angel Shale

Tapeats Sandstone

Nonconformity

Precambrian igneous and metamorphic basement rocks

a Block diagram of Cambrian strata exposed in the Grand Canyon illustrating the transgressive nature of the three formations.

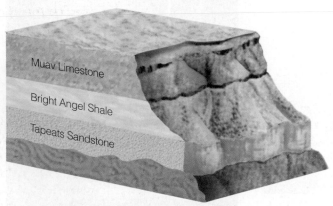

Muav Limestone

Bright Angel Shale

Tapeats Sandstone

b Block diagram of Cambrian rocks exposed along the Bright Angel Trail, Grand Canyon, Arizona.

Early Cambrian (• Figure 10.5). These sediments are clean, well-sorted sands of the type one would find on a beach today. As the transgression continued into the Middle Cambrian, muds of the Bright Angel Shale were deposited over the Tapeats Sandstone. By the Late Cambrian, the Sauk Sea had transgressed so far onto the craton that, in the Grand Canyon region, carbonates of the Muav Limestone were being deposited over the Bright Angel Shale.

This vertical succession of sandstone (Tapeats), shale (Bright Angel), and limestone (Muav) forms a typical transgressive sequence and represents a progressive migration of offshore facies toward the craton through time (Figure 10.5; see Figure 5.7).

Cambrian rocks of the Grand Canyon region also illustrate that many formations are time transgressive; that is, their age is not the same in every place they are found. Mapping and correlations based on faunal evidence indicate that deposition of the Muav Limestone had already started on the shelf before deposition of the Tapeats Sandstone was completed on the craton. Faunal analysis

of the Bright Angel Shale indicates that its age is Early Cambrian in California, and Middle Cambrian in the Grand Canyon region, thus illustrating the time-transgressive nature of formations and facies.

This same facies relationship also occurred elsewhere on the craton as the seas encroached from the Appalachian and Ouachita mobile belts onto the craton interior (• Figure 10.6). Carbonate deposition dominated on the craton as the Sauk transgression continued during the Early Ordovician, and the advancing Sauk Sea soon covered the islands of the Transcontinental Arch. By the end of Sauk time, much of the craton was submerged beneath a warm, equatorial epeiric sea (Figure 10.2a).

The Tippecanoe Sequence

As the Sauk Sea regressed from the craton during the Early Ordovician, a landscape of low relief emerged. The exposed rocks were predominantly limestones and dolostones deposited earlier as part of the Sauk transgression. Because North America was still located in a tropical environment when the seas regressed, these carbonates experienced extensive erosion at that time (• Figure 10.7). The resulting craton-wide unconformity thus marks the boundary between the Sauk and Tippecanoe sequences.

Like the Sauk Sequence, deposition of the **Tippecanoe Sequence** (Middle Ordovician–Early Devonian) began with a major transgression onto the craton. This transgressing sea deposited clean, well-sorted quartz sands over most of the craton. The best known of the Tippecanoe basal sandstones is the St. Peter Sandstone, an almost-pure quartz sandstone used in manufacturing glass. It occurs throughout much of the midcontinent and resulted from numerous cycles of weathering and erosion of Proterozoic and Cambrian sandstones deposited during the Sauk transgression (• Figure 10.8).

The Tippecanoe basal sandstones were followed by widespread carbonate deposition (Figure 10.7). The limestones were generally the result of deposition by calcium carbonate-secreting organisms such as corals, brachiopods, stromatoporoids, and bryozoans. Besides the limestones, there were also many dolostones. Most of the dolostones formed as a result of magnesium replacing calcium in calcite, thus converting limestones into dolostones.

In the eastern portion of the craton, the carbonates grade laterally into shales. These shales mark the farthest extent of detrital sediments derived from weathering and erosion of the Taconic Highlands, which resulted from a tectonic event taking place in the Appalachian mobile belt, and which we will discuss later.

Tippecanoe Reefs and Evaporites

Organic reefs are limestone structures constructed by living organisms, some of which contribute skeletal

• Figure 10.6 **Time Transgressive Cambrian Facies**

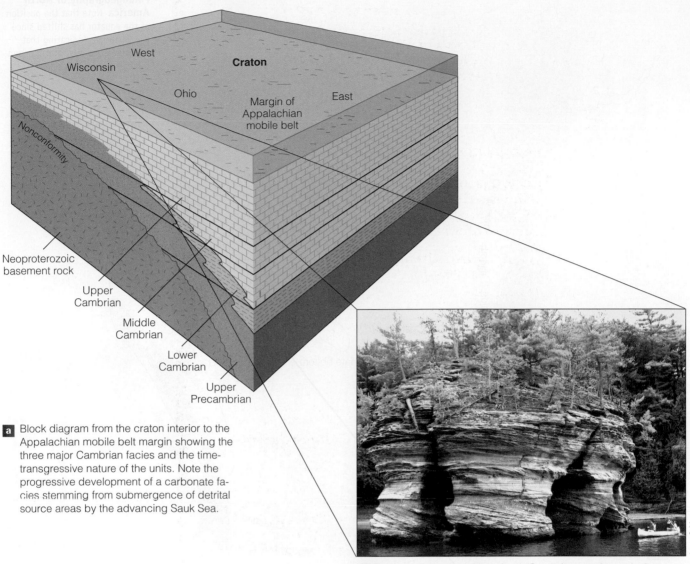

a Block diagram from the craton interior to the Appalachian mobile belt margin showing the three major Cambrian facies and the time-transgressive nature of the units. Note the progressive development of a carbonate facies stemming from submergence of detrital source areas by the advancing Sauk Sea.

b Outcrop of cross-bedded Upper Cambrian sandstone in the Dells area of Wisconsin.

James S. Monroe

materials to the reef framework (• Figure 10.9). Today, corals and calcareous algae are the most prominent reef builders, but in the geologic past other organisms played a major role.

Regardless of the organisms dominating reef communities, reefs appear to have occupied the same ecologic niche in the geologic past that they do today. Because of the ecologic requirements of reef-building organisms, present-day reefs are confined to a narrow latitudinal belt between approximately 30 degrees north and south of the equator. Corals, the major reef-building organisms today, require warm, clear, shallow water of normal salinity for optimal growth.

The size and shape of a reef are mostly the result of interactions among the reef-building organisms, the bottom topography, wind and wave action, and subsidence of the seafloor. Reefs also alter the area around them by forming barriers to water circulation or wave action.

Reefs typically are long, linear masses forming a barrier between a shallow platform on one side and a comparatively deep marine basin on the other side. Such reefs are known as *barrier reefs* (Figure 10.9). Reefs create and maintain a steep seaward front that absorbs incoming wave energy. As skeletal material breaks off from the reef front, it accumulates as talus along a fore-reef slope. The barrier reef itself is porous and composed of many different reef-building organisms. The lagoon area on the landward side of the reef is a low-energy, quiet-water zone where fragile, sediment-trapping organisms thrive. The lagoon area can also become the site of evaporite deposits when circulation to the open sea is cut off. Modern examples of barrier reefs are the Florida Keys, Bahama Islands, and the Great Barrier Reef of Australia.

Reefs have been common features in low latitudes since the Cambrian and have been built by a variety of organisms. The first skeletal builders of reef-like structures

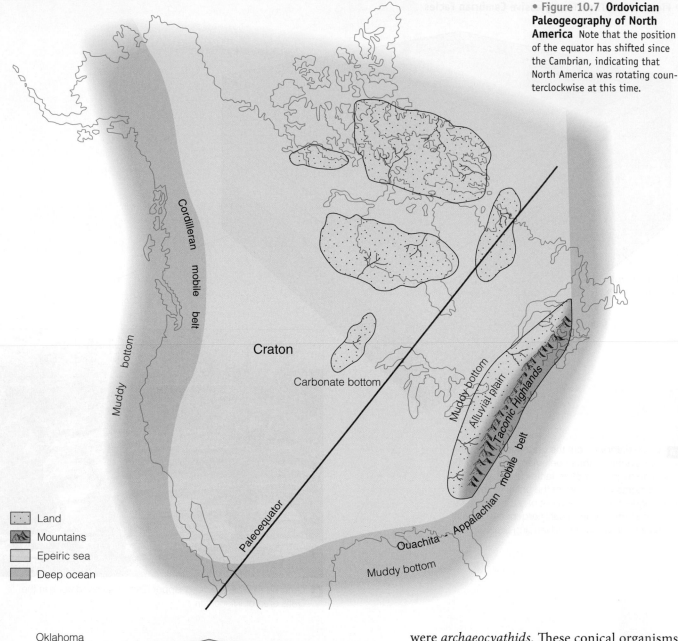

Cordilleran mobile belt

Muddy bottom

Craton

Carbonate bottom

Muddy bottom

Alluvial plain

Taconic Highlands

Appalachian mobile belt

Ouachita

Muddy bottom

Paleoequator

Land

Mountains

Epeiric sea

Deep ocean

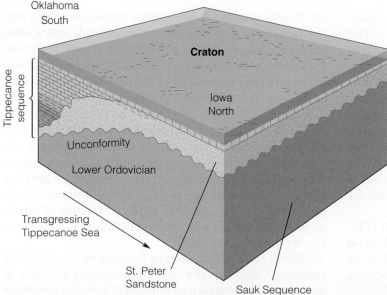

Oklahoma
South

Craton

Iowa
North

Tippecanoe sequence

Unconformity

Lower Ordovician

Transgressing Tippecanoe Sea

St. Peter Sandstone

Sauk Sequence

were *archaeocyathids*. These conical organisms lived during the Cambrian and had double, perforated, calcareous shell walls. Archaeocyathids built small mounds that have been found on all continents except South America (see Figure 12.6).

Beginning in the Middle Ordovician, stromatoporoid-coral reefs became common in the low latitudes, and similar reefs remained so throughout the rest of the Phanerozoic Eon. The burst of reef building seen in the Late Ordovician through Devonian probably occurred in response to evolutionary changes triggered by the

• Figure 10.8 **Transgressing Tippecanoe Sea** The transgression of the Tippecanoe Sea resulted in the deposition of the St. Peter Sandstone (Middle Ordovician) over a large area of the craton.

• Figure 10.9 **Organic Reefs**

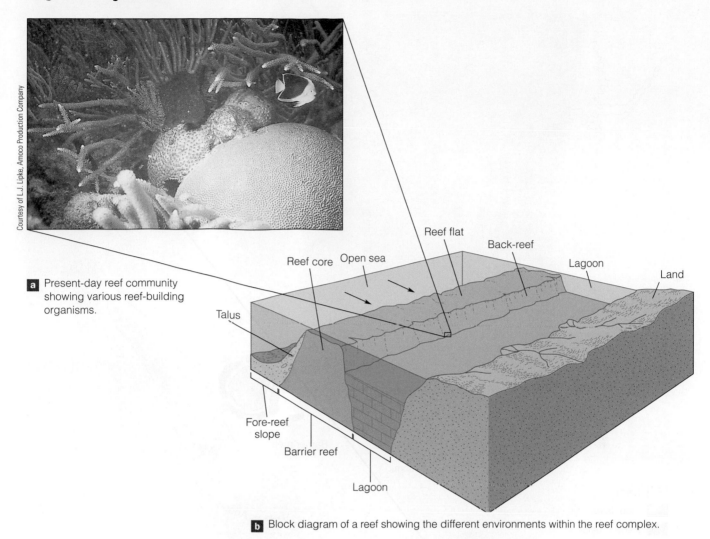

a Present-day reef community showing various reef-building organisms.

b Block diagram of a reef showing the different environments within the reef complex.

appearance of extensive carbonate seafloors and platforms beyond the influence of detrital sediments.

The Middle Silurian rocks (Tippecanoe Sequence) of the present-day Great Lakes region are world famous for their reef and evaporite deposits (• Figure 10.10). The best-known structure in the region, the Michigan Basin, is a broad, circular basin surrounded by large barrier reefs. No doubt these reefs contributed to increasingly restricted circulation and the precipitation of Upper Silurian evaporites inside the basin (• Figure 10.11).

Within the rapidly subsiding interior of the basin, other types of reefs are found. *Pinnacle reefs* are tall, spindly structures up to 100 m high. They reflect the rapid upward growth needed to maintain themselves near sea level during subsidence of the basin (Figure 10.11a). Besides the pinnacle reefs, bedded carbonates and thick sequences of rock salt and rock anhydrite are also found in the Michigan Basin (Figure 10.11).

As the Tippecanoe Sea gradually regressed from the craton during the Late Silurian, precipitation of evaporite minerals occurred in the Appalachian, Ohio, and Michigan basins (Figure 10.1). In the Michigan Basin alone, approximately 1500 m of sediments were deposited, nearly half of which are halite and anhydrite. How did such thick sequences of evaporites accumulate? One possibility is that when sea level dropped, the tops of the barrier reefs were as high as or above sea level, thus preventing the influx of new seawater into the basin. Evaporation of the basin seawater would result in the formation of brine, and as the brine became increasingly concentrated, the precipitation of salts would occur. A second possibility is that the reefs grew upward so close to sea level that they formed a sill or barrier that eliminated interior circulation and allowed for the evaporation of the seawater that produced a dense brine that eventually resulted in evaporite deposits (• Figure 10.12).

With North America still near the equator during the Silurian Period (Figure 10.2c), temperatures were probably high. As circulation to the Michigan Basin was restricted or ceased altogether, seawater within the basin evaporated

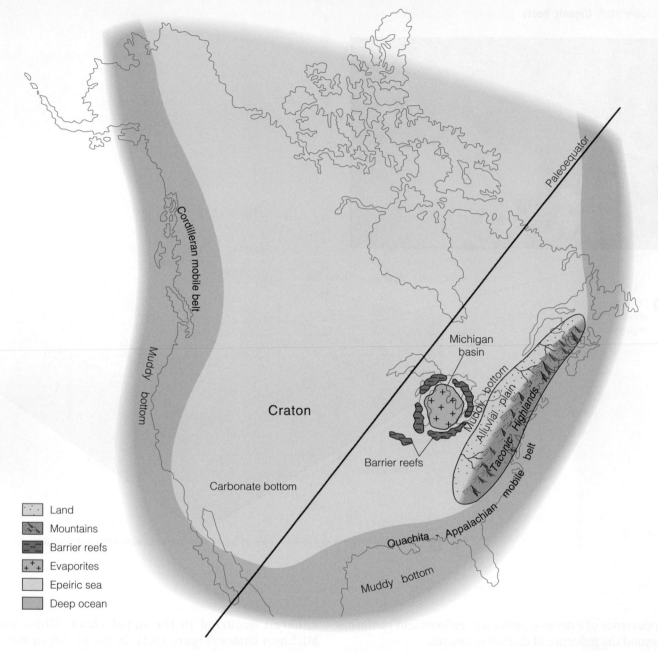

Land
Mountains
Barrier reefs
Evaporites
Epeiric sea
Deep ocean

• **Figure 10.10 Silurian Paleogeography of North America** Note the development of reefs in the Michigan, Ohio, and Indiana–Illinois–Kentucky areas.

and began forming brine. Because the brine was heavy, it concentrated near the bottom, with minerals precipitating on the basin floor and forming evaporite deposits. When seawater flowed back into the Michigan Basin either over the sill and through channels cut in the barrier reefs, this replenishment added new seawater, allowing the process of brine formation and precipitation of evaporites to repeat itself.

The order and type of salts precipitating from seawater depends on their solubility, the original concentration of seawater, and local conditions of the basin. In general, salts precipitate in a sequential order, beginning with the least soluble and ending with the most soluble. Therefore,

calcium carbonate usually precipitates out first, followed by gypsum,* and lastly halite. Many lateral shifts and interfingering of the limestone, anhydrite, and halite facies may occur, however, because of variations in the amount of seawater entering the basin and changing geologic conditions.

Thus, the periodic evaporation of seawater as just discussed could account for the observed vertical and lateral distribution of evaporites in the Michigan Basin. Associated

*Recall from Chapter 6 that gypsum ($CaSO_4 \cdot 2H_2O$) is the common sulfate precipitated from seawater, but when deeply buried, gypsum loses its water and is converted to anhydrite ($CaSO_4$).

• **Figure 10.11 The Michigan Basin**

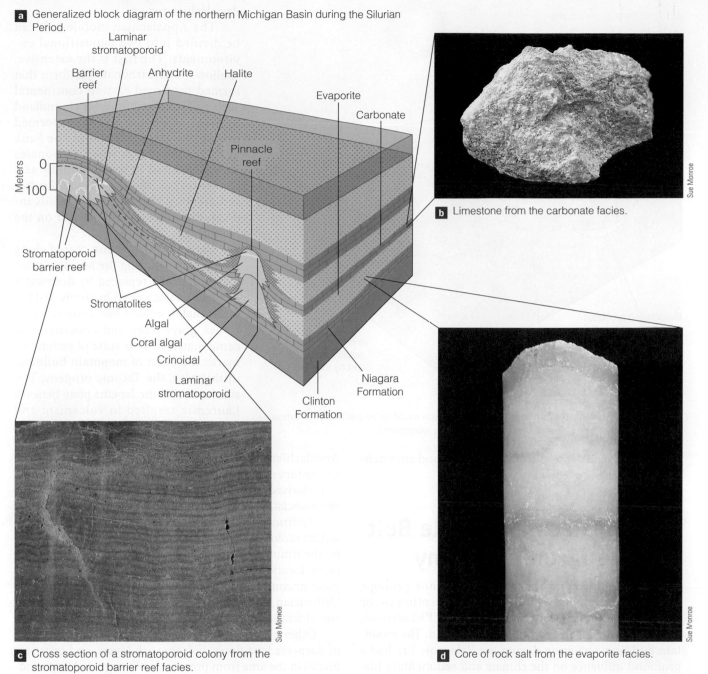

a Generalized block diagram of the northern Michigan Basin during the Silurian Period.

b Limestone from the carbonate facies.

c Cross section of a stromatoporoid colony from the stromatoporoid barrier reef facies.

d Core of rock salt from the evaporite facies.

with those evaporites, however, are pinnacle reefs, and the organisms constructing those reefs could not have lived in such a highly saline environment (Figure 10.11). How, then, can such contradictory features be explained? Numerous models have been proposed, ranging from cessation of reef growth followed by evaporite deposition, to alternation of reef growth and evaporite deposition. Although the Michigan Basin has been studied extensively for years, no model yet proposed completely explains the genesis and relationship of its various reef, carbonate, and evaporite facies.

The End of the Tippecanoe Sequence By the Early Devonian, the regressing Tippecanoe Sea had retreated to the craton margin, exposing an extensive lowland topography. During this regression, marine deposition was initially restricted to a few interconnected cratonic basins and, finally, by the end of the Tippecanoe, to only the mobile belts surrounding the craton.

As the Tippecanoe Sea regressed during the Early Devonian, the craton experienced mild deformation forming many domes, arches, and basins (Figure 10.1). These structures were mostly eroded during the time the craton

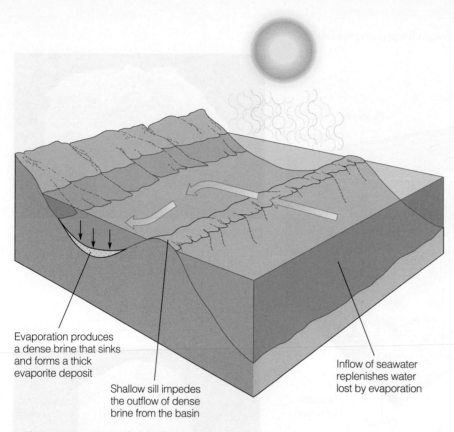

Evaporation produces
a dense brine that sinks
and forms a thick
evaporite deposit

Shallow sill impedes
the outflow of dense
brine from the basin

Inflow of seawater
replenishes water
lost by evaporation

• **Figure 10.12 Evaporite Sedimentation** Silled basin model for evaporite sedimentation by direct precipitation from seawater. Vertical scale is greatly exaggerated.

was exposed, and deposits from the ensuing and encroaching Kaskaskia Sea eventually covered them.

The Appalachian Mobile Belt and the Taconic Orogeny

Having examined the Sauk and Tippecanoe geologic history of the craton, we now turn our attention to the Appalachian mobile belt, where the first Phanerozoic orogeny began during the Middle Ordovician. The mountain building occurring during the Paleozoic Era had a profound influence on the climate and sedimentary history of the craton. In addition, it was part of the global tectonic regime that sutured the continents together, forming Pangaea by the end of the Paleozoic.

Throughout Sauk time (Neoproterozoic–Early Ordovician), the Appalachian region was a broad, passive, continental margin. Sedimentation was closely balanced by subsidence as extensive carbonate deposits succeeded thick, shallow marine sands. During this time, movement along a divergent plate boundary was widening the **Iapetus Ocean** (• Figure 10.13a). Beginning with the subduction of the Iapetus plate beneath Laurentia (an oceanic–continental convergent plate boundary), the Appalachian mobile belt was born (Figure 10.13b). The resulting **Taconic orogeny**—named after the present-day Taconic Mountains of eastern New York, central Massachusetts, and Vermont—was the

first of several orogenies to affect the Appalachian region.

The Appalachian mobile belt can be divided into two depositional environments. The first is the extensive, shallow-water carbonate platform that formed the broad eastern continental shelf and stretched from Newfoundland to Alabama (Figure 10.13a). It formed during the transgression of the Sauk Sea onto the craton when carbonates were deposited in a vast shallow sea. Stromatolites, mud cracks, and other sedimentary structures and fossils indicate the shallow water depth on the platform.

Carbonate deposition ceased along the east coast during the Middle Ordovician and was replaced by deepwater deposits characterized by thinly bedded black shales, graded beds, coarse sandstones, graywackes, and associated volcanic material. This suite of sediments marks the onset of mountain building, in this case, the Taconic orogeny. The subduction of the Iapetus plate beneath Laurentia resulted in volcanism and downwarping of the carbonate platform (Figure 10.13b). Throughout the Appalachian mobile belt, facies patterns, paleocurrents, and sedimentary structures all indicate that these deposits were derived from the east, where the Taconic Highlands and associated volcanoes were rising.

Additional structural, stratigraphic, petrologic, and sedimentologic evidence has provided much information on the timing and origin of this orogeny. For example, at many locations within the Taconic belt, pronounced angular unconformities occur where steeply dipping Lower Ordovician rocks are overlain by gently dipping or horizontal Silurian and younger rocks.

Other evidence includes volcanic activity in the form of deep-sea lava flows, volcanic ash layers, and intrusive bodies in the area from present-day Georgia to Newfoundland. These igneous rocks show a clustering of radiometric ages corresponding to the Middle to Late Ordovician. In addition, regional metamorphism coincides with the radiometric dates.

The final piece of evidence for the Taconic orogeny is the development of a large **clastic wedge,** an extensive accumulation of mostly detrital sediments deposited adjacent to an uplifted area. These deposits are thickest and coarsest nearest the highland area and become thinner and finer grained away from the source area, eventually grading into the carbonate cratonic facies (• Figure 10.14). The clastic wedge resulting from the erosion of the Taconic Highlands is referred to as the **Queenston Delta.** Careful mapping and correlation of these deposits indicate that more than 600,000 km³ of rock were eroded from the Taconic

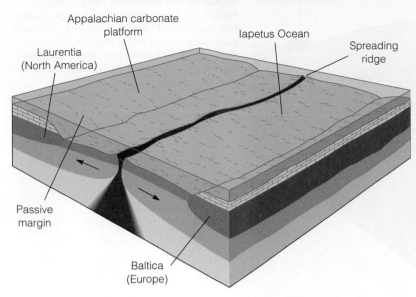

a During the Neoproterozoic to the Early Ordovician, the Iapetus Ocean was opening along a divergent plate boundary. Both the east coast of Laurentia and the west coast of Baltica were passive continental margins with large carbonate platforms.

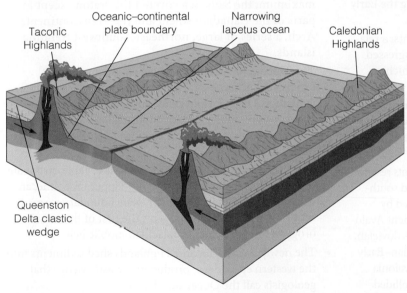

b Beginning in the Middle Ordovician, the passive margins of Laurentia and Baltica changed to active oceanic–continental plate boundaries, resulting in orogenic activity (Figure 10.2).

Highlands. Based on this figure, geologists estimate that the Taconic Highlands were at least 4000 m high.

The Taconic orogeny marked the first pulse of mountain building in the Appalachian mobile belt and was a response to the subduction taking place beneath the east coast of Laurentia. As the Iapetus Ocean narrowed and closed, another orogeny occurred in Europe during the Silurian. The *Caledonian orogeny* was essentially a mirror image of the Taconic orogeny and the Acadian orogeny (see Chapter 11) and was part of the global mountain building episode that occurred during the Paleozoic Era (Figure 10.2c). Even though the Caledonian orogeny occurred during Tippecanoe time, we discuss it in the next chapter because it was intimately related to the Devonian Acadian orogeny.

Early Paleozoic Mineral Resources

Early Paleozoic-age rocks contain a variety of important mineral resources, including sand and gravel for construction, building stone, and limestone used in the manufacture of cement. Important sources of industrial or silica sand are the Upper Cambrian Jordan Sandstone of Minnesota and Wisconsin, the Lower Silurian Tuscarora Sandstone in Pennsylvania and Virginia, and the Middle Ordovician St. Peter Sandstone. The latter, the basal sandstone of the Tippecanoe Sequence (Figure 10.8), occurs in several states, but the best-known area of production is in La Salle County, Illinois. Silica sand has a variety of uses, including the manufacture of glass, refractory bricks for blast furnaces, and molds for casting iron, aluminum, and copper alloys. Some silica sands, called hydraulic fracturing sands, are pumped into wells to fracture oil-or gas-bearing rocks and provide permeable passageways for the oil or gas to migrate to the well.

Thick deposits of Silurian evaporites, mostly rock salt (NaCl) and rock gypsum ($CaSO_4 \cdot 2H_2O$) altered to rock anhydrite ($CaSO_4$), underlie parts of Michigan, Ohio, New York, and adjacent areas in Ontario, Canada. These rocks are important sources of various salts. In addition, barrier and pinnacle reefs in carbonate rocks associated with these evaporites are the reservoirs for oil and gas in Michigan and Ohio.

The host rocks for deposits of lead and zinc in southeast Missouri are Cambrian dolostones, although some Ordovician rocks contain these metals as well. These deposits have been mined since 1720 but have been largely depleted. Now most lead and zinc mined in Missouri comes from Mississippian-age sedimentary rocks.

The Silurian Clinton Formation crops out from Alabama north to New York, and equivalent rocks are found in Newfoundland. This formation has been mined for iron in many places. In the United States, the richest ores and most extensive mining occurred near Birmingham, Alabama, but only a small amount of ore is currently produced in that area.

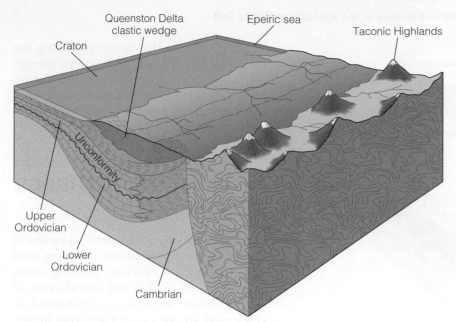

Craton

Queenston Delta
clastic wedge

Epeiric sea

Taconic Highlands

Unconformity

Upper
Ordovician

Lower
Ordovician

Cambrian

• **Figure 10.14 Reconstruction of the Taconic Highlands and Queenston Delta Clastic Wedge** The Queenston Delta clastic wedge, resulting from the erosion of the Taconic Highlands, consists of thick, coarse-grained detrital sediments nearest the highlands and thins laterally into finer-grained sediments in the epeiric seas covering the craton.

SUMMARY

Table 10.1 summarizes the geologic history of the North American craton and mobile belts as well as global events, sea level changes, and major evolutionary events during the Early Paleozoic.

- Most continents consist of two major components: a relatively stable craton over which epeiric seas transgressed and regressed, surrounded by mobile belts in which mountain building took place.

- Six major continents and numerous microcontinents and island arcs existed at the beginning of the Paleozoic Era; all of these were dispersed around the globe at low latitudes during the Cambrian.

- During the Ordovician and Silurian, plate movements resulted in a changing global geography. Gondwana moved southward and began to cross the South Pole as indicated by Upper Ordovician tillite deposits; the microcontinent Avalonia separated from Gondwana during the Early Ordovician, and collided with Baltica during the Late Ordovician–Early Silurian; Baltica, along with the newly attached Avalonia moved northwestward relative to Laurentia and collided with it to form Laurasia during the Silurian.

- Geologists divide the geologic history of North America into cratonic sequences that formed during craton-wide transgressions and regressions.

- The first major marine transgression onto the craton resulted in deposition of the Sauk Sequence. At its maximum, the Sauk Sea covered the craton except for parts of the Canadian shield and the Transcontinental Arch, a series of large, northeast–southwest trending islands.

- The Tippecanoe Sequence began with deposition of extensive sandstone over the exposed and eroded Sauk landscape. During Tippecanoe time, extensive carbonate deposition took place. In addition, large barrier reefs enclosed basins, resulting in evaporite deposition within these basins.

- The eastern edge of North America was a stable carbonate platform during Sauk time. During Tippecanoe time, an oceanic–continental convergent plate boundary formed, resulting in the Taconic orogeny, the first of three major orogenies to affect the Appalachian mobile belt.

- The newly formed Taconic Highlands shed sediments into the western epeiric sea, producing a clastic wedge that geologists call the Queenston Delta.

- Early Paleozoic-age rocks contain a variety of mineral resources, including building stone, limestone for cement, silica sand, hydrocarbons, evaporites, and iron ore.

IMPORTANT TERMS

Appalachian mobile belt,
  p. 198
Baltica, p. 199
China, p. 199
clastic wedge, p. 212
Cordilleran mobile belt, p. 198
craton, p. 198
cratonic sequence, p. 201

epeiric sea, p. 198
Gondwana, p. 199
Iapetus Ocean, p. 212
Kazakhstania, p. 199
Laurentia, p. 199
mobile belt, p. 198
organic reef, p. 206
Ouachita mobile belt, p. 198

Queenston Delta, p. 212
Sauk Sequence, p. 202
sequence stratigraphy,
  p. 201
Siberia, p. 199
Taconic orogeny, p. 212
Tippecanoe Sequence, p. 206
Transcontinental Arch, p. 202

REVIEW QUESTIONS

1. Which was the first major transgressive sequence onto the North American craton?
   a. _____ Absaroka; b. _____ Sauk; c. _____ Zuni; d. _____ Kaskaskia; e. _____ Tippecanoe.

2. What type of plate interaction produced the Taconic orogeny?
   a. _____ Divergent; b. _____ Transform; c. _____ Oceanic-oceanic convergent; d. _____ Oceanic–continental convergent; e. _____ Continental–continental convergent.

3. During which sequence did the eastern margin of Laurentia change from a passive plate margin to an active plate margin?
   a. _____ Zuni; b. _____ Tippecanoe; c. _____ Sauk; d. _____ Kaskaskia; e. _____ Absaroka.

4. A major transgressive–regressive cycle bounded by craton-wide unconformities is a(n)
   a. _____ biostratigraphic unit; b. _____ cratonic sequence; c. _____ orogeny; d. _____ shallow sea; e. _____ cyclothem.

5. An elongated area marking the site of mountain building is a(n)
   a. _____ cyclothem; b. _____ mobile belt; c. _____ platform; d. _____ shield; e. _____ craton.

6. During which sequence were evaporites and reef carbonates the predominant cratonic rocks?
   a. _____ Kaskaskia; b. _____ Zuni; c. _____ Sauk; d. _____ Absaroka; e. _____ Tippecanoe.

7. What Middle Ordovician formation is an important source of industrial silica sand?
   a. _____ St. Peter; b. _____ Tuscarora; c. _____ Jordan; d. _____ Oriskany; e. _____ Clinton.

8. The ocean separating Laurentia from Baltica is called the
   a. _____ Panthalassa; b. _____ Tethys; c. _____ Iapetus; d. _____ Atlantis; e. _____ Perunica.

9. Which mobile belt is located along the eastern side of North America?
   a. _____ Franklin; b. _____ Cordilleran; c. _____ Ouachita; d. _____ Appalachian; e. _____ answers a and b.

10. During deposition of the Sauk Sequence, the only area above sea level besides the Transcontinental Arch was the
    a. _____ Cratonic margin; b. _____ Canadian shield; c. _____ Queenston Delta; d. _____ Appalachian mobile belt; e. _____ Taconic Highlands.

11. Weathering of which highlands produced the Queenston Delta clastic wedge?
    a. _____ Transcontinental Arch; b. _____ Acadian; c. _____ Taconic; d. _____ Sevier; e. _____ Caledonian Highlands.

12. The vertical sequence of the Tapeats Sandstone, Bright Angel Shale, and Muav Limestone represents
    a. _____ a transgression; b. _____ time transgressive formations; c. _____ rocks of the Grand Canyon, Arizona; d. _____ sediments deposited by the Sauk Sea; e. _____ all of the previous answers.

13. At the beginning of the Cambrian, there were _____ major continents.
    a. _____ 3; b. _____ 4; c. _____ 5; d. _____ 6; e. _____ 7.

14. What are some methods geologists can use to determine the locations of continents during the Paleozoic Era?

15. Discuss why cratonic sequences are a convenient way to study the geologic history of the Paleozoic Era.

16. Discuss how the Cambrian rocks of the Grand Canyon illustrate the sedimentation patterns of a transgressive sea.

17. Discuss how sequence stratigraphy can be used to make global correlations and why it is so useful in reconstructing past events.

18. What evidence indicates that the Iapetus Ocean began closing during the Middle Ordovician?

19. Discuss how the evaporites of the Michigan Basin may have formed during the Silurian Period.

20. What evidence in the geologic record indicates that the Taconic orogeny occurred?

APPLY YOUR KNOWLEDGE

1. According to estimates made from mapping and correlation, the Queenston Delta contains more than 600,000 km³ of rock eroded from the Taconic Highlands. Based on this figure, geologists estimate the Taconic Highlands were at least 4000 m high. They also estimate that the Catskill Delta (see Chapter 11) contains three times as much sediment as the Queenston Delta. From what you know about the geographic distribution of the Taconic Highlands and the Acadian Highlands (see Chapter 11), can you estimate how high the Acadian Highlands might have been?

2. Paleogeographic maps of what the world looked like during the Paleozoic Era can be found in almost every Earth history book and in numerous scientific journals. What criteria are used to determine the location of ancient continents and ocean basins, and why are there minor differences in the location and size of these paleocontinents among the various books and articles?

3. You work for a travel agency and are putting together a raft trip down the Colorado River through the Grand Canyon. In addition to the usual information about such a trip, what kind of geologic information would you include in your brochure to make the trip appealing from an educational standpoint as well?

Table 10.1
Summary of Early Paleozoic Geologic and Evolutionary Events

Age (Millions of Years)	Geologic Period	Sequence	Relative Changes in Sea Level		Cordilleran Mobile Belt	Craton	Ouachita Mobile Belt	Appalachian Mobile Belt	Major Events Outside North America	Major Evolutionary Events
			Rising	Falling						
416	Silurian	Tippecanoe				Extensive barrier reefs and evaporites common		Acadian orogeny	Caledonian orogeny	First jawed fish evolve / Early land plants—seedless vascular plants
444	Ordovician	Tippecanoe		Present sea level		Queenston Delta clastic wedge		Taconic orogeny	Continental glaciation in Southern Hemisphere	Extinction of many marine invertebrates near end of Ordovician
	Ordovician					Transgression of Tippecanoe Sea				Plants move to land?
		Sauk				Regression exposing large areas to erosion				Major adaptive radiation of all invertebrate groups
488	Cambrian	Sauk				Canadian shield and Transcontinental Arch only areas above sea level				Many trilobites become extinct near end of Cambrian
	Cambrian					Transgression of Sauk Sea				Earliest vertebrates—jawless fish evolve
542										

CHAPTER **11**

LATE PALEOZOIC EARTH HISTORY

Tullimonstrum gregarium, also known as the Tully Monster, is Illinois's official state fossil. Left: Specimen from Pennsylvanian rocks, Mazon Creek locality, Illinois. Right: Reconstruction of the Tully Monster (about 30 cm long).

[OUTLINE]

Introduction

Late Paleozoic Paleogeography

The Devonian Period

The Carboniferous Period

The Permian Period

Late Paleozoic Evolution of North America

The Kaskaskia Sequence

Reef Development in Western Canada

Perspective *The Canning Basin, Australia—A Devonian Great Barrier Reef*

Black Shales

The Late Kaskaskia—A Return to Extensive Carbonate Deposition

The Absaroka Sequence

What Are Cyclothems, and Why Are They Important?

Cratonic Uplift—The Ancestral Rockies

The Middle Absaroka—More Evaporite Deposits and Reefs

History of the Late Paleozoic Mobile Belts

Cordilleran Mobile Belt

Ouachita Mobile Belt

Appalachian Mobile Belt

What Role Did Microplates and Terranes Play in the Formation of Pangaea?

Late Paleozoic Mineral Resources

Summary

At the end of this chapter, you will have learned that

- Movement of the six major continents during the Paleozoic Era resulted in the formation of the supercontinent Pangaea at the end of the Paleozoic.

- In addition to the large-scale plate interactions during the Paleozoic, microplate and terrane activity also played an important role in forming Pangaea.

- Most of the Kaskaskia Sequence is dominated by carbonates and associated evaporites.

- Transgressions and regressions over the low-lying craton during the Absaroka Sequence resulted in cyclothems and the formation of coals.

- During the Late Paleozoic Era, mountain-building activity took place in the Appalachian, Ouachita, and Cordilleran mobile belts.

- The Caledonian, Acadian, Hercynian, and Alleghenian orogenies were all part of the global tectonic activity resulting from the assembly of Pangaea.

- Late Paleozoic-age rocks contain a variety of mineral resources, including petroleum, coal, evaporites, and various metallic deposits.

Introduction

Approximately 300 million years ago in what is now Illinois, sluggish rivers flowed southwestward through swamps and built large deltas that extended outward into a subtropical shallow sea. These rivers deposited huge quantities of mud, which entombed the plants and animals living in the area. Rapid burial and the formation of ironstone concretions thus preserved many of them as fossils. Known as the Mazon Creek fossils, for the area in northeastern Illinois where most specimens are found, they provide us with significant insights about the soft-part anatomy of the region's biota. Because of the exceptional preservation of this ancient biota, Mazon Creek fossils are known throughout the world and many museums have extensive collections from the area.

During Pennsylvanian time, two major habitats existed in northeastern Illinois. One was a swampy, forested lowland of the subaerial delta, and the other was the shallow marine environment of the actively prograding delta.

In the warm, shallow waters of the delta front lived numerous cnidarians, mollusks, echinoderms, arthropods, worms, and fish. The swampy lowlands surrounding the delta were home to more than 400 plant species, numerous insects, spiders, and other animals such as scorpions and amphibians. In the ponds, lakes, and rivers were many fish, shrimp, and ostracods. Almost all of the plants were seedless vascular plants, typical of the kinds that flourished in the coal-forming swamps during the Pennsylvanian Period.

One of the more interesting Mazon Creek fossils is the Tully Monster, which is not only unique to Illinois but also is its official state fossil (see the chapter opening photo). Named for Francis Tully, who first discovered it in 1958, *Tullimonstrum gregarium* (its scientific name) was a small (up to 30 cm long), soft-bodied animal that lived in the warm, shallow seas covering Illinois about 300 million years ago.

The Tully Monster had a relatively long proboscis that contained a "claw" with small teeth in it. The round to oval-shaped body was segmented and contained a crossbar, whose swollen ends some interpreted as the animal's

sense organs. The tail had two horizontal fins. It probably swam like an eel, with most of the undulatory movement occurring behind the two sense organs. There presently is no consensus as to what phylum the Tully Monster belongs or to what animals it might be related.

The Late Paleozoic Era was a time not only of interesting evolutionary innovations and novelties such as the Tully Monster but also when the world's continents were colliding along convergent plate boundaries. These collisions profoundly influenced both Earth's geologic and its biologic history and eventually formed the supercontinent Pangaea by the end of the Permian Period.

Late Paleozoic Paleogeography

The Late Paleozoic was a time marked by continental collisions, mountain building, fluctuating sea levels, and varied climates. Coals, evaporites, and tillites testify to the variety of climatic conditions experienced by the different continents during the Late Paleozoic. Major glacial and interglacial episodes took place over much of Gondwana as it continued moving over the South Pole during the Late Mississippian to Early Permian. The growth and retreat of continental glaciers during this time profoundly affected the world's biota as well as contributed to global sea level changes. Collisions between continents not only led to the formation of the supercontinent Pangaea by the end of the Permian, but also resulted in mountain building that strongly influenced oceanic and atmospheric circulation patterns. By the end of the Paleozoic, widespread arid and semiarid conditions governed much of Pangaea.

The Devonian Period
Recall from Chapter 10 that Baltica, which had earlier united with Avalonia, collided

along a convergent plate boundary during the Silurian to form the larger continent of **Laurasia.** This collision, which closed the northern Iapetus Ocean, is marked by the **Caledonian orogeny.**

During the Devonian, as the southern Iapetus Ocean narrowed between Laurasia and Gondwana, mountain building continued along the eastern margin of Laurasia as a result of the **Acadian orogeny** (• Figure 11.1a).

• Figure 11.1 **Paleozoic Paleogeography** Paleogeography of the world during the **a** Late Devonian Period and **b** Early Carboniferous Period.

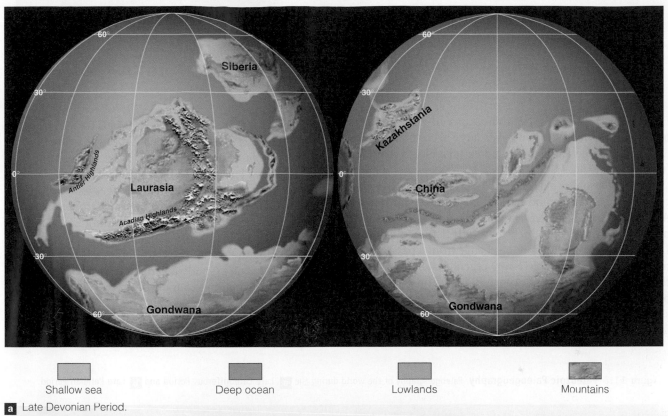

Shallow sea Deep ocean Lowlands Mountains

a Late Devonian Period.

Shallow sea Deep ocean Lowlands Mountains Glaciation

b Early Carboniferous Period.

		Phanerozoic Eon					
Archean Eon	Proterozoic Eon						
		Paleozoic Era					
Precambrian		Cambrian	Ordovician	Silurian	Devonian	Mississippian	Pennsylvanian	Permian
						Carboniferous	

2500 MYA   *542 MYA*   *251 MYA*

Erosion of the ensuing highlands spread vast amounts of reddish fluvial sediments over large areas of northern Europe (Old Red Sandstone) and eastern North America (the Catskill Delta).

Other Devonian tectonic events, probably related to the collision of Laurentia and Baltica, include the Cordilleran **Antler orogeny,** the *Ellesmere orogeny* along the northern margin of Laurentia (which may reflect the collision of Laurentia with Siberia), and the change from a passive continental margin to an active convergent plate boundary in the Uralian mobile belt of eastern Baltica. The distribution of reefs, evaporites, and red beds, as well as the existence of similar floras throughout the world, suggest a rather uniform global climate during the Devonian Period.

The Carboniferous Period During the Carboniferous Period, southern Gondwana moved over the South Pole, resulting in extensive continental glaciation (Figures 11.1b and • 11.2a). The advance and retreat of these glaciers produced global changes in sea level that affected sedimentation patterns on the cratons. As Gondwana moved northward, it began colliding with Laurasia during the Early Carboniferous and continued suturing with it during the rest of the Carboniferous (Figures 11.1b and 11.2a). Because Gondwana rotated clockwise relative to Laurasia, deformation generally progressed in a northeast-to-southwest direction along the Hercynian, Appalachian, and Ouachita mobile belts of the two continents. The final

• Figure 11.2 **Paleozoic Paleogeography** Paleogeography of the world during the **a** Late Carboniferous Period and **b** Late Permian Period.

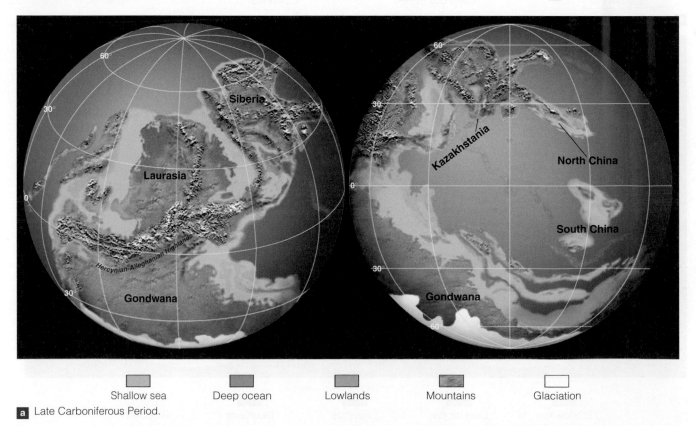

Shallow sea   Deep ocean   Lowlands   Mountains   Glaciation

a Late Carboniferous Period.

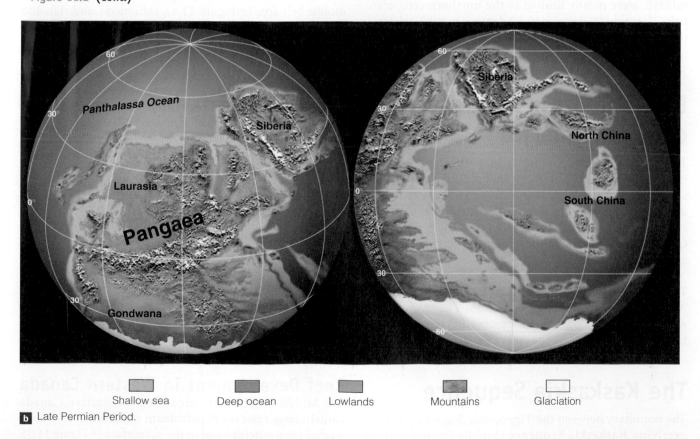

Phanerozoic Eon									
Mesozoic Era			Cenozoic Era						
Triassic	Jurassic	Cretaceous	Paleogene			Neogene		Quaternary	
			Paleocene	Eocene	Oligocene	Miocene	Pliocene	Pleistocene	Holocene

251 MYA

66 MYA

phase of collision between Gondwana and Laurasia is indicated by the Ouachita Mountains of Oklahoma, formed by thrusting during the Late Carboniferous and Early Permian.

Elsewhere, Siberia collided with Kazakhstania and moved toward the Uralian margin of Laurasia (Baltica), colliding with it during the Early Permian. By the end of the Carboniferous, the various continental landmasses were fairly close together as Pangaea began taking shape.

The Carboniferous coal basins of eastern North America, western Europe, and the Donets Basin of the Ukraine all lay in the equatorial zone, where rainfall was high and temperatures were consistently warm. The absence of strong seasonal growth rings in fossil plants from these coal basins indicates such a climate. The fossil plants found in the coals of Siberia, however, show well-developed growth rings, signifying seasonal growth with abundant rainfall and distinct seasons such as in the temperate zones (latitudes 40 degrees to 60 degrees north).

Glacial conditions and the movement of large continental ice sheets in the high southern latitudes are indicated by widespread tillites and glacial striations in southern Gondwana. These ice sheets spread toward the equator and, at their maximum growth, extended well into the middle temperate latitudes.

• Figure 11.2 (Cont.)

b Late Permian Period.

Shallow sea   Deep ocean   Lowlands   Mountains   Glaciation

The Permian Period

The assembly of Pangaea was essentially completed during the Permian as a result of the many continental collisions that began during the Carboniferous (Figure 11.2b). Although geologists generally agree on the configuration and location of the western half of the supercontinent, there is no consensus on the number or configuration of the various terranes and continental blocks that composed the eastern half of Pangaea. Regardless of the exact configuration of the eastern portion, geologists know that the supercontinent was surrounded by various subduction zones and moved steadily northward during the Permian. Furthermore, an enormous single ocean, the **Panthalassa,** surrounded Pangaea and spanned Earth from pole to pole (Figure 11.2b). Waters of this ocean probably circulated more freely than at present, resulting in more equable water temperatures.

The formation of a single large landmass had climatic consequences for the terrestrial environment as well. Terrestrial Permian sediments indicate that arid and semiarid conditions were widespread over Pangaea. The mountain ranges produced by the **Hercynian, Alleghenian,** and **Ouachita orogenies** were high enough to create rain shadows that blocked the moist, subtropical, easterly winds—much as the southern Andes Mountains do in western South America today. This produced very dry conditions in North America and Europe, as evident from the extensive Permian red beds and evaporites found in western North America, central Europe, and parts of Russia. Permian coals, indicating abundant rainfall, were mostly limited to the northern temperate belts (latitude 40 degrees to 60 degrees north), whereas the last remnants of the Carboniferous ice sheets continued their recession.

Late Paleozoic Evolution of North America

The Late Paleozoic cratonic history of North America included periods of extensive shallow-marine carbonate deposition and large coal-forming swamps as well as dry, evaporite-forming terrestrial conditions. Cratonic events largely resulted from sea level changes caused by Gondwanan glaciation and tectonic events related to the assemblage of Pangaea.

Mountain building that began with the Ordovician Taconic orogeny continued with the Caledonian, Acadian, Alleghenian, and Ouachita orogenies. These orogenies were part of the global tectonic process that resulted in the formation of Pangaea by the end of the Paleozoic Era.

The Kaskaskia Sequence

The boundary between the Tippecanoe Sequence and the overlying **Kaskaskia Sequence** (Middle Devonian–Late

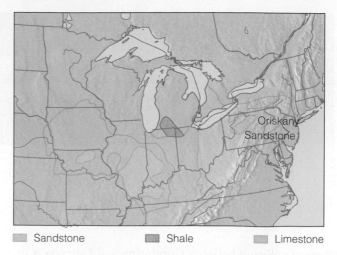

Sandstone   Shale   Limestone

• **Figure 11.3 Basal Rocks of the Kaskaskia Sequence** Extent of the basal rocks of the Kaskaskia Sequence in the eastern and north–central United States.

Mississippian) is marked by a major unconformity. As the Kaskaskia Sea transgressed over the low-relief landscape of the craton, most basal beds deposited consisted of clean, well-sorted quartz sandstones. A good example is the Oriskany Sandstone of New York and Pennsylvania and its lateral equivalents (• Figure 11.3). The Oriskany Sandstone, like the basal Tippecanoe St. Peter Sandstone, is an important glass sand as well as a good gas-reservoir rock.

The source areas for the basal Kaskaskia sandstones were primarily the eroding highlands of the Appalachian mobile belt area (• Figure 11.4), exhumed Cambrian and Ordovician sandstones cropping out along the flanks of the Ozark Dome, and exposures of the Canadian shield in the Wisconsin area. The lack of similar sands in the Silurian carbonate beds below the Tippecanoe–Kaskaskia unconformity indicates that the source areas of the basal Kaskaskia detrital rocks were submerged when the Tippecanoe Sequence was deposited. Stratigraphic studies indicate that these source areas were uplifted and the Tippecanoe carbonates removed by erosion before the Kaskaskia transgression.

Kaskaskian basal rocks elsewhere on the craton consist of carbonates that are frequently difficult to differentiate from the underlying Tippecanoe carbonates unless they are fossiliferous.

Except for widespread Upper Devonian and Lower Mississippian black shales, the majority of Kaskaskian rocks are carbonates, including reefs, and associated evaporite deposits. In many other parts of the world, such as southern England, Belgium, central Europe, Australia, and Russia, the Middle and early Late Devonian epochs were times of major reef building (see Perspective).

Reef Development in Western Canada

The Middle and Late Devonian reefs of western Canada contain large reserves of petroleum and have been widely studied from outcrops and in the subsurface (• Figure 11.5).

Perspective

The Canning Basin, Australia—A Devonian Great Barrier Reef

Rising majestically 50 to 100 meters above the surrounding plains, the Great Barrier Reef of the Canning Basin, Australia, is one of the largest and most spectacularly exposed fossil reef complexes in the world (Figure 1). This barrier reef complex developed during the Middle and Late Devonian Period, when a tropical epeiric sea covered the Canning Basin (Figure 11.1a).

The reefs themselves were constructed primarily by calcareous algae, stromatoporoids, and various corals, which also were the main components of other large reef complexes in the world at that time. Exposures along Windjana Gorge reveal the various features and facies of the Devonian Great Barrier Reef complex (Figure 2).

On the seaward side of the reef core is a steep fore-reef slope (see Figure 10.9), where such organisms as algae, sponges, and stromatoporoids lived. This facies contains considerable reef talus, an accumulation of debris eroded by waves from the reef front.

The reef core itself consists of unbedded limestones (see Figure 10.9) consisting largely of calcareous algae, stromatoporoids, and corals. The back-reef facies is bedded and is the major part of the total reef complex (see Figure 10.9). In this lagoonal environment lived a diverse and abundant assemblage of calcareous algae, stromatoporoids, corals, bivalves, gastropods, cephalopods, brachiopods, and crinoids.

Near the end of the Late Devonian, almost all the reef-building organisms—as well as much of the associated fauna of the Canning Basin Great Barrier Reef and other large barrier reef complexes—became extinct. As we will discuss in Chapter 12, few massive tabulate-rugose-stromatoporoid reefs are known from latest Devonian or younger rocks anywhere in the world.

Figure 1 Aerial view of Windjana Gorge showing the Devonian Great Barrier Reef.

Figure 2 Outcrop of the Devonian Great Barrier Reef along Windjana Gorge. The talus of the fore-reef can be seen on the left side of the picture sloping away from the reef core, which is unbedded. To the right of the reef core is the back-reef facies that is horizontally bedded.

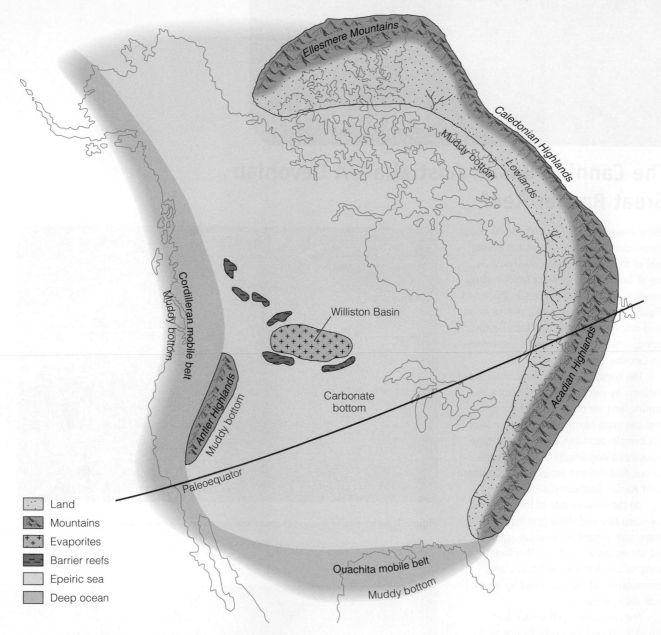

• Figure 11.4 **Devonian Paleogeography of North America**

These reefs began forming as the Kaskaskia Sea transgressed southward into western Canada. By the end of the Middle Devonian, they had coalesced into a large barrier reef system that restricted the flow of oceanic water into the back-reef platform, creating conditions for evaporite precipitation (Figure 11.5). In the back-reef area, up to 300 m of evaporites precipitated in much the same way as in the Michigan Basin during the Silurian (see Figure 10.11). More than half the world's potash, which is used in fertilizers, comes from these Devonian evaporites. By the middle of the Late Devonian, reef growth had stopped in the western Canada region, although nonreef carbonate deposition continued.

Black Shales

In North America, many areas of carbonate–evaporite deposition gave way to a greater proportion of shales and coarser detrital rocks beginning in the Middle Devonian and continuing into the Late Devonian. This change to detrital deposition resulted from the formation of new source areas brought about by the mountain-building activity associated with the Acadian orogeny in North America (Figure 11.4).

As the Devonian Period ended, a conspicuous change in sedimentation took place over the North American craton with the appearance of widespread black shales (• Figure 11.6a). In the eastern United States, these black shales are commonly called Chattanooga Shale, but they are known by a variety of local names elsewhere (for example, New Albany Shale and Antrim Shale). Although these black shales are best developed from the cratonic margins along the Appalachian mobile belt to the Mississippi

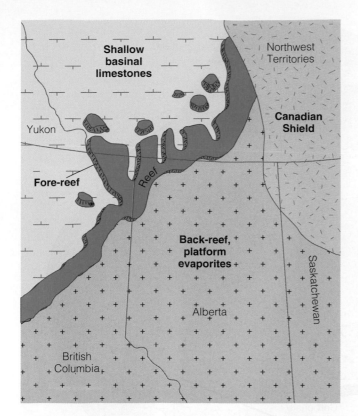

• **Figure 11.5 Devonian Reef Complex of Western Canada**
Reconstruction of the extensive Devonian Reef complex of western Canada. These extensive reefs controlled the regional facies of the Devonian epeiric seas.

Valley, correlative units can also be found in many western states and in western Canada (Figure 11.6a).

The Upper Devonian–Lower Mississippian black shales of North America are typically noncalcareous, thinly bedded, and less than 10 m thick (Figure 11.6b). Fossils are usually rare, but some Upper Devonian black shales do contain rich conodont (microscopic animals) faunas (see Figure 12.12a). Because most black shales lack body fossils, they are difficult to date and correlate. However, in places where conodonts, acritarchs (microscopic algae, see Figure 12.10), or plant spores are found, these fossils indicate that the lower beds are Late Devonian, and the upper beds are Early Mississippian in age.

Although the origin of these extensive black shales is still being debated, the essential features required to produce them include undisturbed anaerobic bottom water, a reduced supply of coarser detrital sediment, and high organic productivity in the overlying oxygenated waters. High productivity in the surface waters leads to a shower of organic material, which decomposes on the undisturbed seafloor and depletes the dissolved oxygen at the sediment–water interface.

The wide extent of such apparently shallow-water black shales in North America remains puzzling. Nonetheless, these shales are rich in uranium and are an important potential source rock for oil and gas in the Appalachian region.

The Late Kaskaskia—A Return to Extensive Carbonate Deposition
Following deposition of the widespread Upper Devonian–Lower Mississippian black shales, carbonate sedimentation on

• **Figure 11.6 Upper Devonian–Lower Mississippian Black Shales**

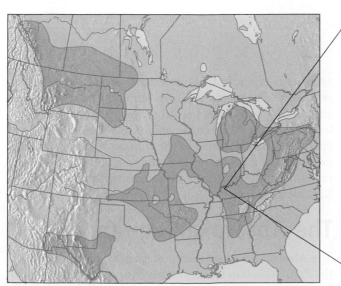

a The extent of the Upper Devonian to Lower Mississippian Chattanooga Shale and its equivalent units (such as the Antrim Shale and New Albany Shale) in North America.

b Upper Devonian New Albany Shale, Button Mold Knob Quarry, Kentucky.

Reed Wicander

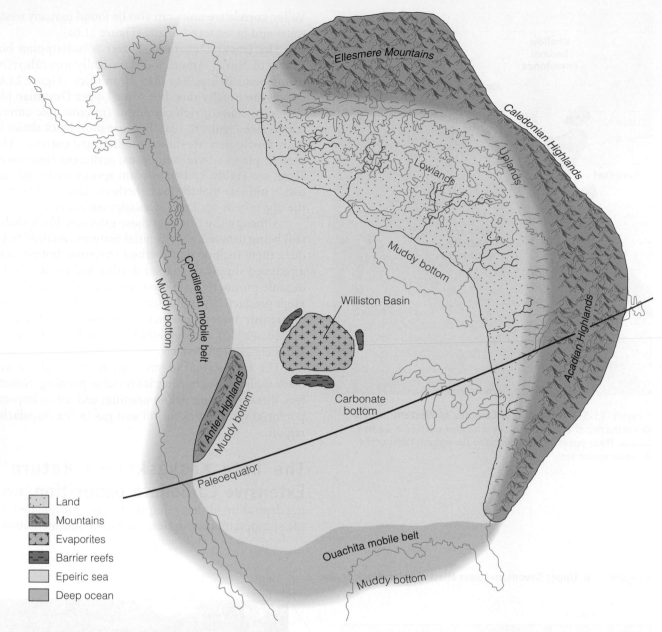

• Figure 11.7 Mississippian Paleogeography of North America

the craton dominated the remainder of the Mississippian Period (• Figure 11.7). During this time, a variety of carbonate sediments were deposited in the epeiric sea, as indicated by the extensive deposits of crinoidal limestones (rich in crinoid fragments), oolitic limestones, and various other limestones and dolostones. These Mississippian carbonates display cross-bedding, ripple marks, and well-sorted fossil fragments, all of which indicate a shallow-water environment. Analogous features can be observed on the present-day Bahama Banks. In addition, numerous small organic reefs occurred throughout the craton during the Mississippian. These were all much smaller than the large barrier reef complexes that dominated the earlier Paleozoic seas.

During the Late Mississippian regression of the Kaskaskia Sea from the craton, vast quantities of detrital

sediments replaced carbonate deposition. The resulting sandstones, particularly in the Illinois Basin, have been studied in great detail because they are excellent petroleum reservoirs. Before the end of the Mississippian, the epeiric sea had retreated to the craton margin, once again exposing the craton to widespread weathering and erosion resulting in a craton-wide unconformity at the end of the Kaskaskia Sequence.

The Absaroka Sequence

The **Absaroka Sequence** includes rocks deposited during the Pennsylvanian through Early Jurassic. In this chapter, however, we are concerned only with the Paleozoic rocks of the Absaroka Sequence. The extensive unconformity separating the Kaskaskia and Absaroka sequences

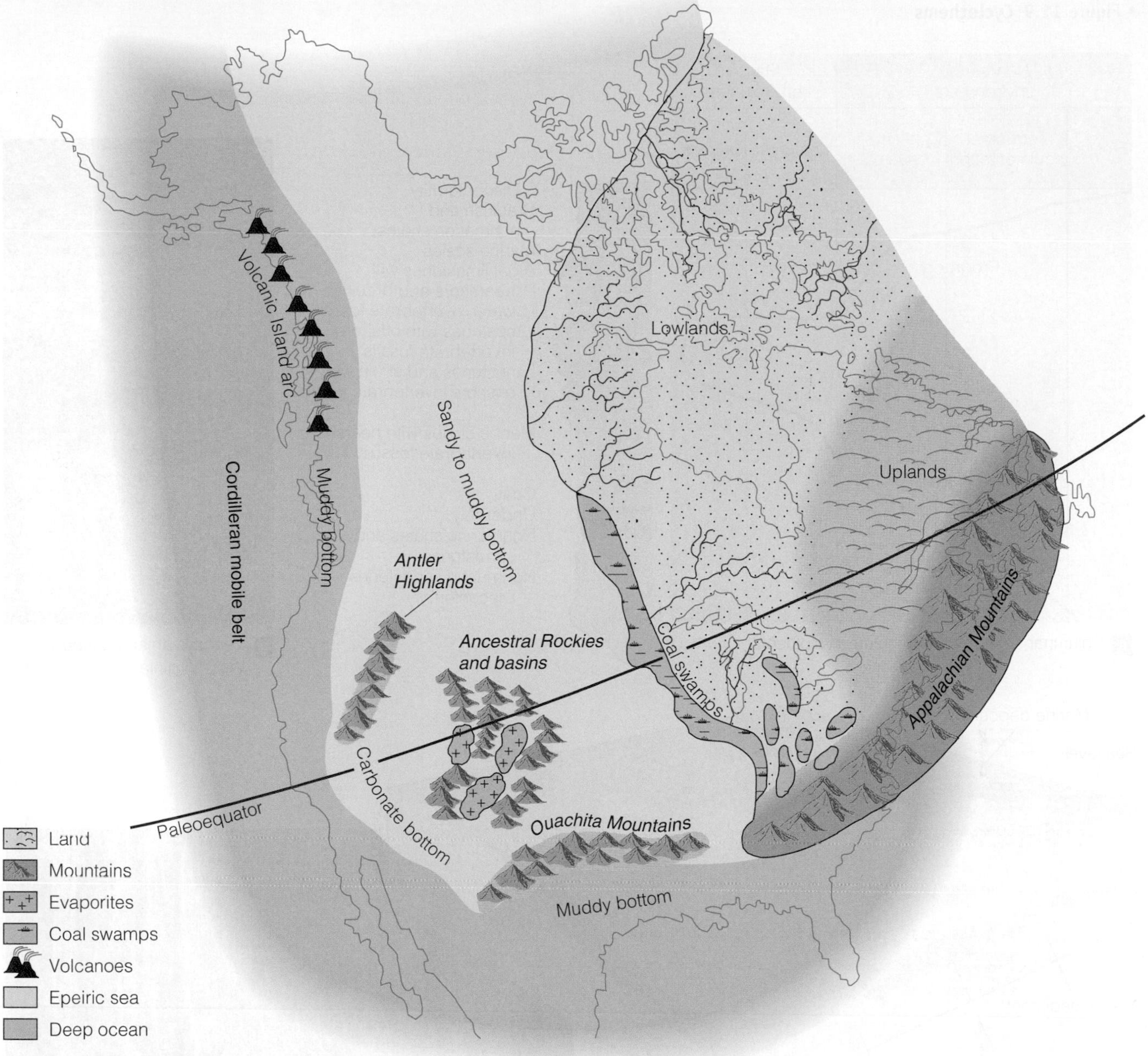
Volcanic Island arc
Cordilleran mobile belt
Muddy bottom
Sandy to muddy bottom
Antler Highlands
Ancestral Rockies and basins
Lowlands
Uplands
Coal swamps
Appalachian Mountains
Paleoequator
Carbonate bottom
Ouachita Mountains
Muddy bottom
Land
Mountains
Evaporites
Coal swamps
Volcanoes
Epeiric sea
Deep ocean

• **Figure 11.8 Pennsylvanian Paleogeography of North America**

essentially divides the strata into the North American Mississippian and Pennsylvanian systems. These two systems are closely equivalent to the European Lower and Upper Carboniferous systems, respectively. The rocks of the Absaroka Sequence not only differ from those of the Kaskaskia Sequence, but also result from different tectonic regimes.

The lowermost sediments of the Absaroka Sequence are confined to the margins of the craton. These deposits are generally thickest in the east and southeast, near the emerging highlands of the Appalachian and Ouachita mobile belts, and thin westward onto the craton. The rocks also reveal lateral changes from nonmarine detrital rocks and coals in the east, through transitional marine–nonmarine beds, to largely marine detrital rocks and limestones farther west (• Figure 11.8).

What Are Cyclothems, and Why Are They Important?

One characteristic feature of Pennsylvanian rocks is their repetitive pattern of alternating marine and nonmarine strata. Such rhythmically repetitive sedimentary sequences are called **cyclothems.** They result from repeated alternations of marine and nonmarine environments, usually in areas of low relief. Although seemingly simple, cyclothems reflect a delicate interplay between nonmarine deltaic and shallow-marine interdeltaic and shelf environments.

• Figure 11.9 **Cyclothems**

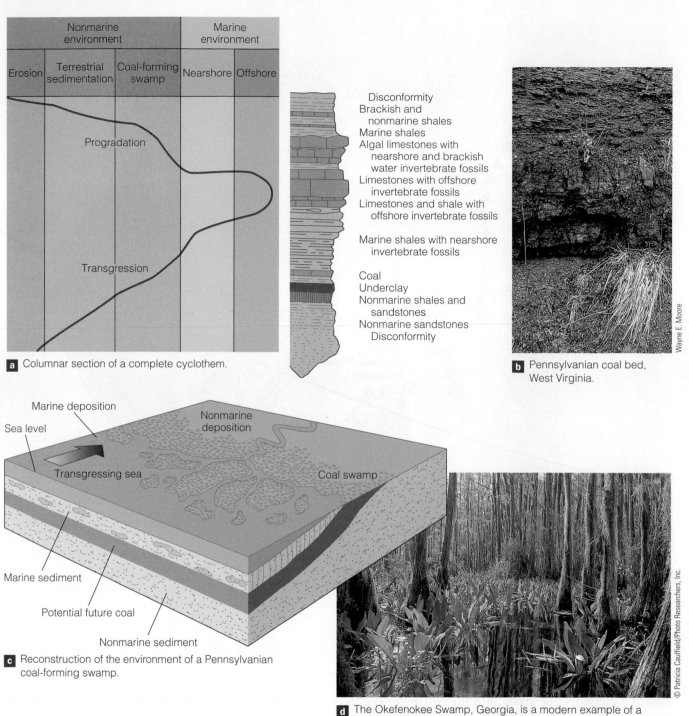

a Columnar section of a complete cyclothem.

b Pennsylvanian coal bed, West Virginia.

c Reconstruction of the environment of a Pennsylvanian coal-forming swamp.

d The Okefenokee Swamp, Georgia, is a modern example of a coal-forming environment, similar to those occurring during the Pennsylvanian Period.

For illustration, look at a typical coal-bearing cyclothem from the Illinois Basin (• Figure 11.9a). Such a cyclothem contains nonmarine units, capped by a coal unit and overlain by marine units. Figure 11.9a shows the depositional environments that produced the cyclothem. The initial units represent deltaic and fluvial deposits. Above them is an underclay that frequently contains root casts from the plants and trees that comprise the overlying coal. The coal bed results from accumulations of plant material and is overlain by marine units of alternating limestones and shales, usually with an abundant marine invertebrate fauna. The marine interval ends with an erosion surface. A new cyclothem begins with a nonmarine deltaic sandstone. All the beds illustrated in the idealized cyclothems are not always preserved because of abrupt changes from marine to nonmarine conditions or removal of some units by erosion.

Cyclothems represent transgressive and regressive sequences with an erosional surface separating one cyclothem from another. Thus, an idealized cyclothem passes upward from fluvial-deltaic deposits, through coals, to detrital shallow-water marine sediments, and finally to limestones typical of an open marine environment.

Such places as the Mississippi delta, the Florida Everglades, the Okefenokee Swamp, Georgia, and the Dutch lowlands represent modern coal-forming environments similar to those existing during the Pennsylvanian Period (Figure 11.9d). By studying these modern analogs, geologists can make reasonable deductions about conditions existing in the geologic past.

The Pennsylvanian coal swamps must have been widespread lowland areas with little topographic relief neighboring the sea (Figure 11.9c). In such cases, a very slight rise in sea level would have flooded these large lowland areas, whereas slight drops would have exposed large areas, resulting in alternating marine and nonmarine environments. The same result could also have been caused by a combination of rising sea level and progradation (the seaward extension of a delta by the accumulation of sediment) of a large delta, such as occurs today in Louisiana.

Such repetitive sedimentation over a widespread area requires an explanation. In most cases, local cyclothems of limited extent can be explained by rapid but slight changes in sea level in a swamp-delta complex of low relief near the sea such as by progradation or by localized crustal movement.

Explaining widespread cyclothems is more difficult. The hypothesis currently favored by many geologists is a rise and fall of sea level related to advances and retreats of Gondwanan continental glaciers. When the Gondwanan ice sheets advanced, sea level dropped; when they melted, sea level rose. Late Paleozoic cyclothem activity on all of the cratons closely corresponds to Gondwanan glacial–interglacial cycles.

Cratonic Uplift—The Ancestral Rockies

Recall that cratons are stable areas, and what deformation they do experience is usually mild. The Pennsylvanian Period, however, was a time of unusually severe cratonic deformation, resulting in uplifts of sufficient magnitude to expose Precambrian basement rocks. In addition to newly formed highlands and basins, many previously formed arches and domes, such as the Cincinnati Arch, Nashville Dome, and Ozark Dome, were also reactivated (see Figure 10.1).

During the Late Absaroka (Pennsylvanian), the area of greatest deformation was in the southwestern part of the North American craton, where a series of fault-bounded uplifted blocks formed the **Ancestral Rockies** (• Figure 11.10a). These mountain ranges had diverse

geologic histories and were not all elevated at the same time. Uplift of these mountains, some of which were elevated more than 2 km along near-vertical faults, resulted in erosion of overlying Paleozoic sediments and exposure of the Precambrian igneous and metamorphic basement rocks (Figure 11.10b). As the mountains eroded, tremendous quantities of coarse, red arkosic sand and conglomerate were deposited in the surrounding basins. These sediments are preserved in many areas, including the rocks of the Garden of the Gods near Colorado Springs (Figure 11.10c) and at the Red Rocks Amphitheatre near Morrison, Colorado.

Intracratonic mountain ranges are unusual, and their cause has long been debated. It is currently thought that the collision of Gondwana with Laurasia along the Ouachita mobile belt (Figure 11.2a) generated great stresses in the southwestern region of the North American craton. These crustal stresses were relieved by faulting. Movement along these faults produced uplifted cratonic blocks and down-warped adjacent basins, forming a series of related ranges and basins.

The Middle Absaroka—More Evaporite Deposits and Reefs

While the various intracratonic basins were filling with sediment during the Late Pennsylvanian, the epeiric sea slowly began retreating from the craton. During the Early Permian, the Absaroka Sea occupied a narrow region from Nebraska through west Texas (• Figure 11.11). By the Middle Permian, it had retreated to west Texas and southern New Mexico. The thick evaporite deposits in Kansas and Oklahoma show the restricted nature of the Absaroka Sea during the Early and Middle Permian and its southwestward retreat from the central craton.

During the Middle and Late Permian, the Absaroka Sea was restricted to west Texas and southern New Mexico, forming an interrelated complex of lagoonal, reef, and open-shelf environments (• Figure 11.12). Three basins separated by two submerged platforms developed in this area during the Permian. Massive reefs grew around the basin margins (• Figure 11.13), and limestones, evaporites, and red beds were deposited in the lagoonal areas behind the reefs. As the barrier reefs grew and the passageways between the basins became more restricted, Late Permian evaporites gradually filled the individual basins.

Spectacular deposits representing the geologic history of this region can be seen today in the Guadalupe Mountains of Texas and New Mexico where the Capitan Limestone forms the caprock of these mountains (• Figure 11.14). These reefs have been extensively studied because tremendous oil production comes from this region.

By the end of the Permian Period, the Absaroka Sea had retreated from the craton, exposing continental red beds that had been deposited over most of the southwestern and eastern region (Figure 11.2b).

• Figure 11.10 The Ancestral Rockies

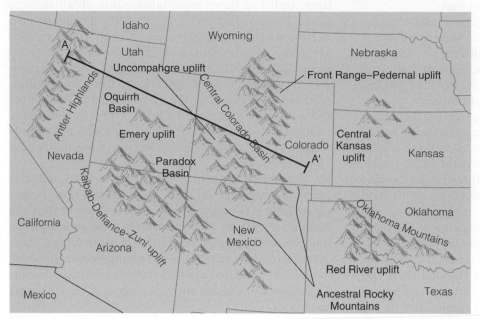

a Location of the principal Pennsylvanian highland areas and basins of the southwestern part of the craton.

b Block diagram of the Ancestral Rockies, elevated by faulting during the Pennsylvanian Period. Erosion of these mountains produced coarse, red sediments deposited in the basins adjacent to the Ancestral Rockies.

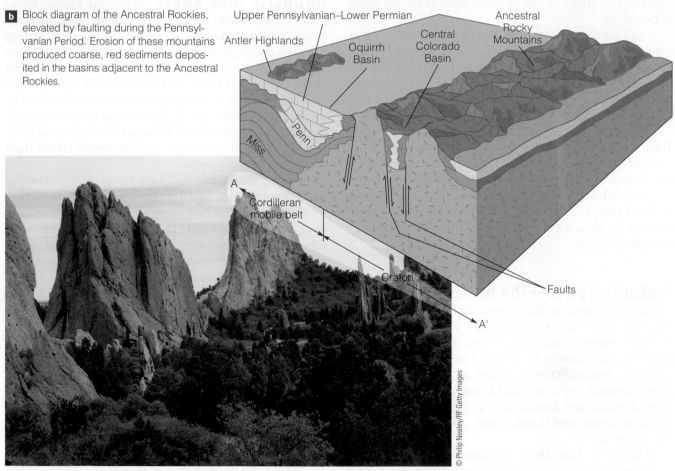

c Garden of the Gods, storm sky view from Near Hidden Inn, Colorado Springs, Colorado.

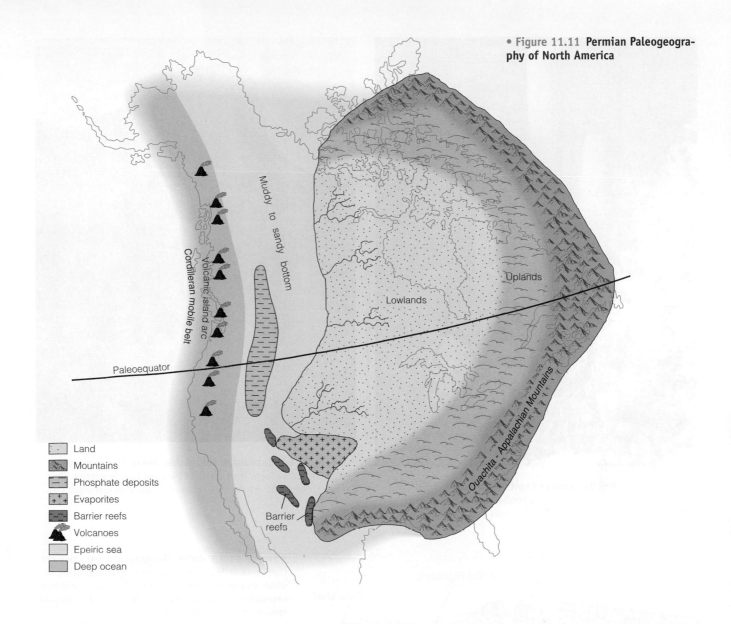

• Figure 11.11 **Permian Paleogeography of North America**

Land

Mountains

Phosphate deposits

Evaporites

Barrier reefs

Volcanoes

Epeiric sea

Deep ocean

Cordilleran mobile belt

Volcanic island arc

Muddy to sandy bottom

Paleoequator

Uplands

Lowlands

Ouachita - Appalachian Mountains

Barrier reefs

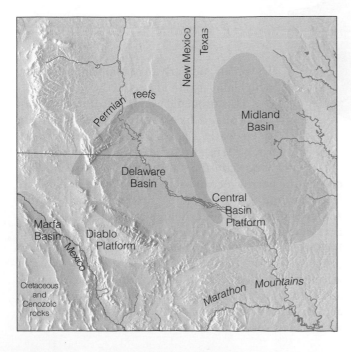

Texas

New Mexico

Permian reefs

Midland Basin

Delaware Basin

Central Basin Platform

Marfa Basin

Diablo Platform

Mexico

Cretaceous and Cenozoic rocks

Marathon Mountains

History of the Late Paleozoic Mobile Belts

Having examined the Kaskaskian and Absarokian history of the craton, we now turn our attention to the orogenic activity in the mobile belts. The mountain building occurring during this time had a profound influence on the climate and sedimentary history of the craton. In addition, it was part of the global tectonic regime that sutured the continents together, forming Pangaea by the end of the Paleozoic Era.

Cordilleran Mobile Belt During the Neoproterozoic and Early Paleozoic, the Cordilleran area

• Figure 11.12 **West Texas Permian Basins and Surrounding Reefs**
During the Middle and Late Permian, an interrelated complex of lagoonal, barrier reef, and open-shelf environments formed in the west Texas and southern New Mexico area. Much of the tremendous oil production in this region comes from these reefs.

• **Figure 11.13 Middle Permian Capitan Limestone Reef Environment** A reconstruction of the Middle Permian Capitan Limestone reef environment. Shown are brachiopods, corals, bryozoans, and large glass sponges.

• **Figure 11.14 Guadalupe Mountains, Texas** The prominent Capitan Limestone forms the caprock of the Guadalupe Mountains. The Capitan Limestone is rich in fossil corals and associated reef organisms.

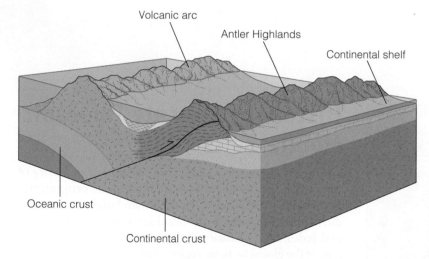

Volcanic arc

Antler Highlands

Continental shelf

Oceanic crust

Continental crust

• **Figure 11.15 Antler Orogeny** Reconstruction of the Cordilleran mobile belt during the Early Mississippian in which deep-water continental slope deposits were thrust eastward over shallow-water continental shelf carbonates, forming the Antler Highlands.

Devonian and Early Mississippian, producing a highland area (• Figure 11.15).

This orogenic event, the *Antler orogeny,* was caused by subduction, resulting in deep-water deposits and oceanic crustal rocks being thrust eastward over shallow-water continental shelf sediments, and thus closing the narrow ocean basin separating the island arc from the craton (Figure 11.15). Erosion of the resulting Antler Highlands produced large quantities of sediment that were deposited to the east in the epeiric sea covering the craton and to the west in the deep sea. The Antler orogeny was the first in a series of orogenic events to affect the Cordilleran mobile belt. During the Mesozoic and Cenozoic, this area was the site of major tectonic activity caused by oceanic–continental convergence and accretion of various terranes.

was a passive continental margin along which extensive continental shelf sediments were deposited (see Figures 9.7 and 10.5). Thick deposits of marine sediments graded laterally into thin cratonic units as the Sauk Sea transgressed onto the craton. Beginning in the Middle Paleozoic, an island arc formed off the western margin of the craton. A collision between this eastward-moving island arc and the western border of the craton took place during the Late

Ouachita Mobile Belt

The Ouachita mobile belt extends for approximately 2100 km from the subsurface of Mississippi to the Marathon region of Texas. Approximately 80% of the former mobile belt is buried beneath a Mesozoic and Cenozoic sedimentary cover. The two major exposed areas in this region are the Ouachita Mountains of Oklahoma and Arkansas and the Marathon Mountains of Texas.

During the Late Proterozoic to Early Mississippian, shallow-water detrital and carbonate sediments were deposited on a broad continental shelf, and in the deeper-water portion of the adjoining mobile belt, bedded cherts and shales were accumulating (• Figure 11.16a). Beginning in the Mississippian Period, the sedimentation rate increased dramatically as the region changed from a passive continental margin to an active convergent plate boundary, marking the beginning of the *Ouachita orogeny* (Figure 11.16b).

Thrusting of sediments continued throughout the Pennsylvanian and Early Permian, driven by the compressive forces generated along the zone of subduction as Gondwana collided with Laurasia (Figure 11.16c). The collision of Gondwana and Laurasia is marked by the formation of a large mountain range, most of which eroded during the Mesozoic Era. Only the rejuvenated Ouachita and Marathon Mountains remain of this once lofty mountain range.

The Ouachita deformation was part of the general worldwide tectonic activity that occurred when Gondwana united with Laurasia. The Hercynian, Appalachian, and Ouachita mobile belts were continuous and marked the southern boundary of Laurasia (Figure 11.2). The tectonic activity that uplifted the Ouachita mobile belt was very complex and involved not only the collision of Laurasia and Gondwana, but also several microplates and terranes between the continents that eventually became part of Central America. The compressive forces impinging on the Ouachita mobile belt also affected the craton by broadly uplifting the southwestern part of North America.

• **Figure 11.16** **Ouachita Mobile Belt** Plate tectonic model for deformation of the Ouachita mobile belt.

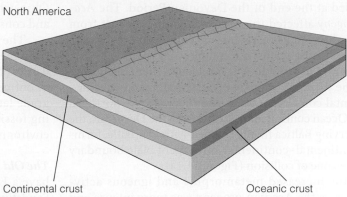

a Depositional environment prior to the beginning of orogenic activity.

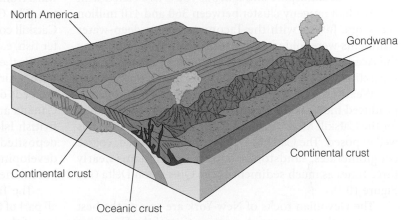

b Incipient continental collision between North America and Gondwana began during the Mississippian Period.

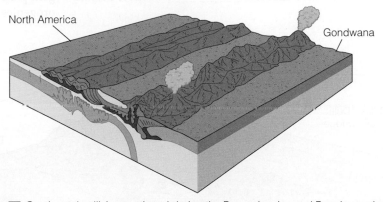

c Continental collision continued during the Pennsylvanian and Permian periods.

Appalachian Mobile Belt

Caledonian Orogeny The Caledonian mobile belt stretches along the western border of Baltica and includes the present-day countries of Scotland, Ireland, and Norway (see Figure 10.2c). During the Middle Ordovician, subduction along the boundary between the Iapetus plate and Baltica (Europe) began, forming a mirror image of the convergent plate boundary off the east coast of Laurentia (North America).

The culmination of the *Caledonian orogeny* (mentioned earlier) occurred during the Late Silurian and Early Devonian with the formation of a mountain range along the western margin of Baltica (see Figure 10.2c). Red-colored sediments deposited along the front of the Caledonian Highlands formed a large clastic wedge known as the *Old Red Sandstone*.

Acadian Orogeny The third Paleozoic orogeny to affect Laurentia and Baltica began during the Late Silurian and concluded at the end of the Devonian Period. The *Acadian orogeny* affected the Appalachian mobile belt from Newfoundland to Pennsylvania as sedimentary rocks were folded and thrust against the craton.

As with the preceding Taconic and Caledonian orogenies, the Acadian orogeny occurred along an oceanic–continental convergent plate boundary. As the northern Iapetus Ocean continued to close during the Devonian, the plate carrying Baltica finally collided with Laurentia, forming a continental–continental convergent plate boundary along the zone of collision (Figure 11.1a).

As the increased metamorphic and igneous activity indicates, the Acadian orogeny was more intense and lasted longer than the Taconic orogeny. Radiometric dates from the metamorphic and igneous rocks associated with the Acadian orogeny cluster between 360 and 410 million years ago. Just as with the Taconic orogeny, deep-water sediments were folded and thrust northwestward during the Acadian orogeny, producing angular unconformities separating Upper Silurian from Mississippian rocks.

Weathering and erosion of the Acadian Highlands produced the **Catskill Delta,** a thick clastic wedge named for the Catskill Mountains in upstate New York, where it is well exposed. The Catskill Delta, composed of red, coarse conglomerates, sandstones, and shales, contains nearly three times as much sediment as the Queenston Delta (see Figure 10.14).

The Devonian rocks of New York are among the best studied on the continent. A cross section of the Devonian strata clearly reflects an eastern source (Acadian Highlands) for the Catskill facies (• Figure 11.17). These detrital rocks can be traced from eastern Pennsylvania, where the coarse-grained deposits are approximately 3 km thick, to Ohio, where the deltaic facies are only about 100 m thick and consist of cratonic shales and carbonates.

The red beds of the Catskill Delta derive their color from the hematite in the sediments. Plant fossils and oxidation of the hematite indicate the beds were deposited in a continental environment. Toward the west, the red beds grade laterally into gray sandstones and shales containing fossil tree trunks, which indicate a swamp or marsh environment.

The Old Red Sandstone The red beds of the Catskill Delta have a European counterpart in the Devonian Old Red Sandstone of the British Isles (Figure 11.17). The Old Red Sandstone was a Devonian clastic wedge that grew eastward from the Caledonian Highlands onto the Baltica craton. The Old Red Sandstone, just like its North American Catskill counterpart, contains numerous fossils of freshwater fish, early amphibians, and land plants.

By the end of the Devonian Period, Baltica and Laurentia were sutured together, forming Laurasia (Figure 11.17). The red beds of the Catskill Delta can be traced north, through Canada and Greenland, to the Old Red Sandstone of the British Isles and into northern Europe. These beds were deposited in similar environments along the flanks of developing mountain chains formed at convergent plate boundaries.

The Taconic, Caledonian, and Acadian orogenies were all part of the same major orogenic event related to the closing of the Iapetus Ocean (see Figures 10.13b and 11.17). This event began with paired oceanic–continental convergent plate boundaries during the Taconic and Caledonian orogenies and culminated along a continental–continental convergent plate boundary during the Acadian orogeny as Laurentia and Baltica became sutured. After this, the Hercynian–Alleghenian orogeny began, followed by orogenic activity in the Ouachita mobile belt.

Hercynian–Alleghenian Orogeny The Hercynian mobile belt of southern Europe and the Appalachian and Ouachita mobile belts of North America mark the zone along which Europe (part of Laurasia) collided with Gondwana (Figure 11.1). While Gondwana and southern Laurasia

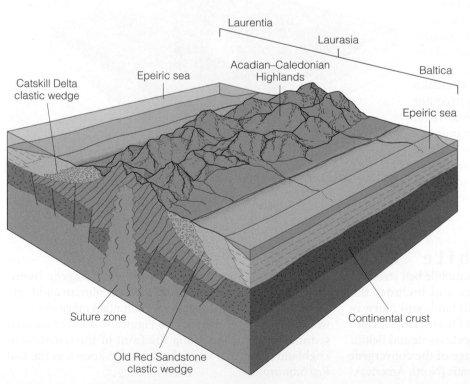

Catskill Delta clastic wedge

Epeiric sea

Laurentia

Laurasia

Acadian–Caledonian Highlands

Baltica

Epeiric sea

Suture zone

Old Red Sandstone clastic wedge

Continental crust

• **Figure 11.17 Formation of Laurasia**
Block diagram showing the area of collision between Laurentia and Baltica. Note the bilateral symmetry of the Catskill Delta clastic wedge and the Old Red Sandstone and their relationship to the Acadian and Caledonian Highlands.

collided during the Pennsylvanian and Permian in the area of the Ouachita mobile belt, eastern Laurasia (Europe and southeastern North America) joined together with Gondwana (Africa) as part of the *Hercynian–Alleghenian orogeny* (Figure 11.2).

Initial contact between eastern Laurasia and Gondwana began during the Mississippian Period along the Hercynian mobile belt. The greatest deformation occurred during the Pennsylvanian and Permian periods and is referred to as the *Hercynian orogeny*. The central and southern parts of the Appalachian mobile belt (from New York to Alabama) were folded and thrust toward the craton as eastern Laurasia and Gondwana were sutured. This event in North America is referred to as the *Alleghenian orogeny*.

These three Late Paleozoic orogenies (Hercynian, Alleghenian, and Ouachita) represent the final joining of Laurasia and Gondwana into the supercontinent Pangaea during the Permian.

What Role Did Microplates and Terranes Play in the Formation of Pangaea?

We have presented the geologic history of the mobile belts bordering the Paleozoic continents in terms of subduction along convergent plate boundaries. It is becoming increasingly clear, however, that accretion along the continental margins is more complicated than the somewhat simple, large-scale plate interactions we have described. Geologists now recognize that numerous terranes or microplates existed during the Paleozoic and were involved in the orogenic events that occurred during that time.

In this chapter and the previous one, we have been concerned only with the six major Paleozoic continents. However, terranes and microplates of varying size were present during the Paleozoic and participated in the formation of Pangaea. For example, as we mentioned in the previous chapter, the microcontinent of *Avalonia* consisted of some coastal parts of New England, southern New Brunswick, much of Nova Scotia, the Avalon Peninsula of eastern Newfoundland, southeastern Ireland, Wales, England, and parts of Belgium and northern France. This microcontinent separated from Gondwana in the Early Ordovician and existed as a separate continent until it collided with Baltica during the Late Ordovician–Early Silurian and then with Laurentia (as part of Baltica) during the Silurian (see Figures 10.2b, 10.2c, and 11.1a).

Other terranes and microplates include *Iberia-Armorica* (a portion of southern France, Sardinia, and most of the Iberian peninsula), *Perunica* (Bohemia), numerous Alpine fragments (especially in Austria), as well as many other bits and pieces of island arcs and suture zones. Not only did these terranes and microplates separate and move away from the larger continental landmasses during the

Paleozoic, but they usually developed their own unique faunal and floral assemblages.

Thus, although the basic history of the formation of Pangaea during the Paleozoic remains essentially the same, geologists now realize that microplates and terranes also played an important role in the formation of Pangaea. Furthermore, they help explain some previously anomalous geologic and paleontologic situations.

Late Paleozoic Mineral Resources

Late Paleozoic-age rocks contain a variety of important mineral resources including energy resources and metallic and nonmetallic mineral deposits. Petroleum and natural gas are recovered in commercial quantities from rocks ranging in age from the Devonian through Permian. For example, Devonian rocks in the Michigan Basin, Illinois Basin, and the Williston Basin of Montana, South Dakota, and adjacent parts of Alberta, Canada, have yielded considerable amounts of hydrocarbons. Permian reefs and other strata in the western United States, particularly Texas, have also been prolific producers.

Although Permian coal beds are known from several areas, including Asia, Africa, and Australia, much of the coal in North America and Europe comes from Pennsylvanian (Upper Carboniferous) deposits. Large areas in the Appalachian region and the midwestern United States are underlain by vast coal deposits (Figure 11.18). These coal deposits formed from the lush vegetation that flourished in Pennsylvanian coal-forming swamps (Figure 11.9).

Much of this coal is bituminous coal, which contains about 80% carbon. It is a dense, black coal that has been so thoroughly altered that plant remains can be seen only rarely. Bituminous coal is used to make *coke,* a hard, gray substance made up of the fused ash of bituminous coal. Coke is used to fire blast furnaces for steel production.

Some Pennsylvanian coal from North America is *anthracite,* a metamorphic type of coal containing up to 98% carbon. Most anthracite is in the Appalachian region (Figure 11.18). It is especially desirable because it burns with a smokeless flame and yields more heat per unit volume than other types of coal. Unfortunately, it is the least common type—much of the coal used in the United States is bituminous.

A variety of Late Paleozoic-age evaporite deposits are important nonmetallic mineral resources. The Zechstein evaporites of Europe extend from Great Britain across the North Sea and into Denmark, the Netherlands, Germany, eastern Poland, and Lithuania. Besides the evaporites themselves, Zechstein deposits form the caprock for the large reservoirs of the gas fields of the Netherlands and part of the North Sea region.

Other important evaporite mineral resources include those of the Permian Delaware Basin of West Texas and New Mexico and Devonian evaporites in the Elk Point

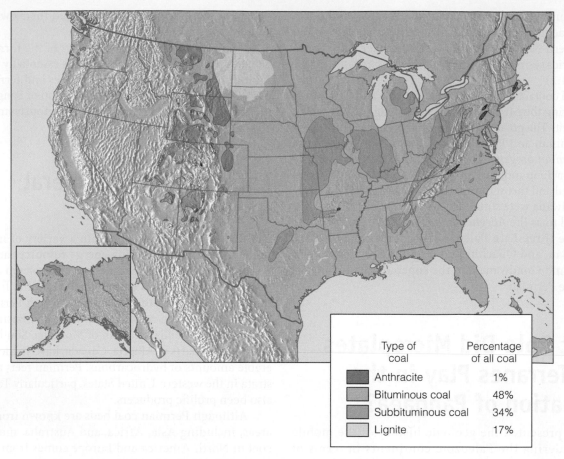

Type of coal	Percentage of all coal
Anthracite	1%
Bituminous coal	48%
Subbituminous coal	34%
Lignite	17%

● Figure 11.18 **Distribution of Coal Deposits in the United States** The age of the coals in the midwestern states and the Appalachian region are mostly Pennsylvanian, whereas those in the West are mostly Cretaceous and Cenozoic.

Basin of Canada. In Michigan, gypsum is mined and used in the construction of sheetrock. Late Paleozoic-age limestones from many areas in North America are used in manufacturing cement. Limestone is also mined and used in blast furnaces for steel production.

Most silica sand mined in the United States comes from east of the Mississippi River, and much of this comes from Late Paleozoic-age rocks. For example, the Devonian Ridgeley Formation is mined in West Virginia, Maryland, and Pennsylvania, and the Devonian Sylvania Sandstone is mined near Toledo, Ohio. Recall from Chapter 10 that

silica sand is used in manufacturing glass; for refractory bricks in blast furnaces; for molds for casting aluminum, iron, and copper alloys; and for a variety of other uses.

Metallic mineral resources, including tin, copper, gold, and silver are also known from Late Paleozoic-age rocks, especially those deformed during mountain building. Although the precise origin of the Missouri lead and zinc deposits remains unresolved, much of the ores of these metals come from Mississippian-age rocks. In fact, mines in Missouri account for a substantial amount of all domestic production of lead ores.

SUMMARY

Table 11.1 summarizes the geologic history of the North American craton and mobile belts as well as global events, sea level changes, and evolutionary events during the Late Paleozoic.

- During the Late Paleozoic, Baltica and Laurentia collided, forming Laurasia. Siberia and Kazakhstania collided and finally were sutured to Laurasia. Gondwana moved over the South Pole and experienced several glacial–interglacial periods, resulting in global sea level changes and transgressions and regressions along the low-lying craton margins.

- Laurasia and Gondwana underwent a series of collisions beginning in the Carboniferous. During the

Permian, the formation of Pangaea was completed. Surrounding the supercontinent was the global ocean, Panthalassa.

- The Late Paleozoic history of the North American craton can be deciphered from the rocks of the Kaskaskia and Absaroka sequences.

- The basal beds of the Kaskaskia Sequence that were deposited on the exposed Tippecanoe surface consisted either of sandstones derived from the eroding Taconic Highlands, or of carbonate rocks.

- Most of the Kaskaskia Sequence is dominated by carbonates and associated evaporites. The Devonian Period was a time of major reef building in western Canada, southern England, Belgium, Australia, and Russia.

- Widespread black shales were deposited over large areas of the craton during the Late Devonian and Early Mississippian.

- The Mississippian Period was dominated, for the most part, by carbonate deposition.

- Transgressions and regressions, probably caused by advancing and retreating Gondwanan ice sheets, over the low-lying North American craton, resulted in cyclothems and the formation of coals during the Pennsylvanian Period.

- Cratonic mountain building, specifically the Ancestral Rockies, occurred during the Pennsylvanian Period and resulted in thick nonmarine detrital sediments and evaporites being deposited in the intervening basins.

- By the Early Permian, the Absaroka Sea occupied a narrow zone of the south–central craton. Here, several large reefs and associated evaporites developed. By the end of the Permian Period, this epeiric sea had retreated from the craton.

- The Cordilleran mobile belt was the site of the Antler orogeny, a minor Devonian orogeny during which deep-water sediments were thrust eastward over shallow-water sediments.

- During the Pennsylvanian and Early Permian, mountain building occurred in the Ouachita mobile belt. This tectonic activity was partly responsible for the cratonic uplift in the southwest, resulting in the Ancestral Rockies.

- The Caledonian, Acadian, Hercynian, and Alleghenian orogenies were all part of the global tectonic activity that resulted from the assembly of Pangaea.

- During the Paleozoic Era, numerous microplates and terranes, such as Avalonia, Iberia–Armorica, and Perunica, existed and played an important role in forming Pangaea.

- Late Paleozoic-age rocks contain a variety of mineral resources, including petroleum, coal, evaporites, silica sand, lead, zinc, and other metallic deposits.

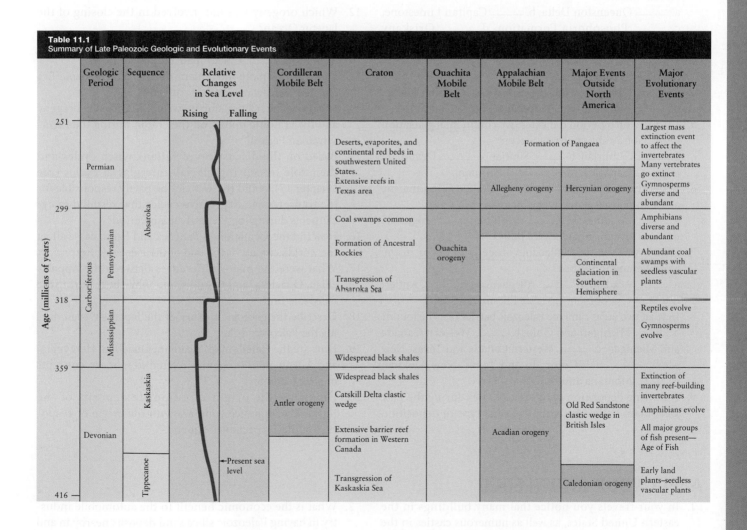

Table 11.1
Summary of Late Paleozoic Geologic and Evolutionary Events

IMPORTANT TERMS

Absaroka Sequence, p. 226
Acadian orogeny, p. 219
Alleghenian orogeny, p. 222
Ancestral Rockies, p. 229
Antler orogeny, p. 220

Caledonian orogeny, p. 219
Catskill Delta, p. 234
cyclothem, p. 227
Hercynian orogeny, p. 222
Kaskaskia Sequence, p. 222

Laurasia, p. 219
Ouachita orogeny, p. 222
Panthalassa Ocean, p. 222

REVIEW QUESTIONS

1. Which of the following resulted from intracratonic deformation?
   a. _____ Antler Highlands; b. _____ Ancestral Rockies; c. _____ Acadian Highlands; d. _____ Caledonian Highlands; e. _____ Taconic Highlands.

2. The Catskill Delta clastic wedge resulted from weathering and erosion of the _____ highlands.
   a. _____ Taconic; b. _____ Nevadan; c. _____ Transcontinental Arch; d. _____ Acadian; e. _____ Sevier.

3. The European Old Red Sandstone is the equivalent of the North American
   a. _____ Queenston Delta; b. _____ Capitan Limestone; c. _____ Phosphoria Formation; d. _____ Oriskany Sandstone; e. _____ Catskill Delta.

4. During which Paleozoic cratonic sequence were cyclothems common?
   a. _____ Sauk; b. _____ Absaroka; c. _____ Kaskaskia; d. _____ Zuni; e. _____ Tippecanoe.

5. During which period did extensive continental glaciation of the Gondwana continent occur?
   a. _____ Cambrian; b. _____ Silurian; c. _____ Devonian; d. _____ Carboniferous; e. _____ Permian.

6. Repetitive sedimentary sequences of alternating marine and nonmarine sedimentary rocks are
   a. _____ cyclothems; b. _____ reefs; c. _____ orogenies; d. _____ evaporites; e. _____ tillites.

7. Which was the first Paleozoic orogeny to occur in the Cordilleran mobile belt?
   a. _____ Acadian; b. _____ Alleghenian; c. _____ Antler; d. _____ Caledonian; e. _____ Ellesmere.

8. In what two areas can Late Paleozoic barrier reefs be found?
   a. _____ Michigan and Ohio; b. _____ Western Canada and Michigan; c. _____ Western Canada and Texas–New Mexico; d. _____ Colorado and Texas–New Mexico; e. _____ Montana and Utah.

9. Following deposition of the black shales during the Late Devonian–Early Mississippian, what type of deposition predominated on the craton during the remainder of the Mississippian Period?
   a. _____ Carbonates; b. _____ Clastics; c. _____ Evaporites; d. _____ Volcanics; e. _____ Cherts and graywackes.

10. The Ancestral Rockies formed during which geologic period?
    a. _____ Permian; b. _____ Pennsylvanian; c. _____ Mississippian; d. _____ Devonian; e. _____ Silurian.

11. The economically valuable deposit in a cyclothem is
    a. _____ gravel; b. _____ metallic ore; c. _____ coal; d. _____ carbonates; e. _____ evaporites.

12. Which orogeny was *not* involved in the closing of the Iapetus Ocean?
    a. _____ Alleghenian; b. _____ Acadian; c. _____ Taconic; d. _____ Caledonian; e. _____ Antler.

13. Discuss how plate movement during the Paleozoic Era affected worldwide weather patterns.

14. What was the relationship between the Ouachita orogeny and the cratonic uplifts on the craton during the Pennsylvanian Period?

15. Based on the discussion of Milankovitch cycles and their role in causing glacial–interglacial cycles (see Chapter 17), could these cycles be partly responsible for the transgressive–regressive cycles that resulted in cyclothems during the Pennsylvanian Period?

16. How did the formation of Pangaea and Panthalassa affect the world's climate at the end of the Paleozoic Era?

17. What were the major differences between the Appalachian, Ouachita, and Cordilleran mobile belts during the Paleozoic Era?

18. Describe the geologic history of the Iapetus Ocean during the Paleozoic Era.

19. How are the Caledonian, Acadian, Ouachita, Hercynian, and Alleghenian orogenies related to modern concepts of plate tectonics?

20. How does the origin of evaporite deposits of the Kaskaskia Sequence compare with the origin of evaporites of the Tippecanoe Sequence?

APPLY YOUR KNOWLEDGE

1. In your travels you notice that many buildings in the eastern United States, as well as numerous castles in the United Kingdom, seem to be constructed of the same coarse-grained red sandstones and conglomerates. How would you account for such a coincidence? Or is it really a coincidence? Explain.

2. What is the economic benefit to the automobile industry in having Paleozoic silica sand deposits nearby in and around Toledo, Ohio?

3. You are the geology team leader of an international mining company. Your company holds the mineral rights on large blocks of acreage in various countries along the

west coast of Africa. The leases on these mineral rights will shortly expire, and you've been given the task of evaluating which leases are the most promising. How do you think your knowledge of Paleozoic plate tectonics can help you in these evaluations?

4. This close-up of a Devonian red rock from a building in Glasgow, Scotland, shows a distinctive sedimentary structure. Identify the sedimentary structure, indicate what type of environment you think it was deposited in, and give the name of the formation this rock comes from.

James S. Monroe

Smithsonian Institution, Transparency No. 86-13471A

Diorama of the environment
and biota of the Phyllopod
bed of the Middle Cambrian
Burgess Shale, British Columbia,
Canada. In the background is
a vertical wall of a submarine
escarpment with algae growing on it. The large, cylindrical ribbed organisms on the
muddy bottom in the fore-
ground are sponges.

[OUTLINE]

Introduction

What Was the Cambrian Explosion?

The Emergence of a Shelly Fauna

Paleozoic Invertebrate Marine Life

    The Present Marine Ecosystem

    Cambrian Marine Community

    Perspective *Trilobites–Paleozoic
Arthropods*

    The Burgess Shale Biota
Ordovician Marine Community

Silurian and Devonian Marine
Communities

Carboniferous and Permian Marine
Communities

Mass Extinctions

    The Permian Mass Extinction

Summary

- Animals with skeletons appeared abruptly at the beginning of the Paleozoic Era and experienced a short period of rapid evolutionary diversification.

- The present marine ecosystem is a complex organization of organisms that interrelate and interact not only with each other, but also with the physical environment.

- The Cambrian Period was a time of many evolutionary innovations during which almost all the major invertebrate phyla evolved.

- The Ordovician Period witnessed striking changes in the marine community, resulting in a dramatic increase in diversity of the shelly fauna, followed by a mass extinction at the end of the Ordovician.

- The Silurian and Devonian periods were a time of rediversification and recovery for many of the invertebrate phyla as well as a time of major reef building.

- Following the Late Devonian extinctions, the marine community again experienced renewed adaptive radiation and diversification during the Carboniferous and Permian periods.

- Mass extinctions occur when anomalously high numbers of species go extinct in a short period of time. The greatest recorded mass extinction in Earth's history occurred at the end of the Permian Period.

Introduction

On August 30 and 31, 1909, near the end of the summer field season, Charles D. Walcott, geologist and head of the Smithsonian Institution, was searching for fossils along a trail on Burgess Ridge between Mount Field and Mount Wapta, near Field, British Columbia, Canada. On the west slope of this ridge, he discovered the first soft-bodied fossils from the Burgess Shale, a discovery of immense importance in deciphering the early history of life. During the following week, Walcott and his collecting party split open numerous blocks of shale, many of which yielded carbonized impressions of a number of soft-bodied organisms beautifully preserved on bedding planes. Walcott returned to the site the following summer and located the shale stratum that was the source of his fossil-bearing rocks in the steep slope above the trail. He quarried the site and shipped back thousands of fossil specimens to the United States National Museum of Natural History, where he later catalogued and studied them.

The importance of Walcott's discovery is not that it was another collection of well-preserved Cambrian fossils, but rather that it allowed geologists a rare glimpse into a world previously almost unknown: that of the soft-bodied animals that lived some 530 million years ago. The beautifully preserved fossils from the Burgess Shale present a much more complete picture of a Middle Cambrian community than do deposits containing only fossils of the hard parts of organisms. In fact, 60% of the total fossil assemblage of more than 100 genera is composed of soft-bodied animals, a percentage comparable to present-day marine communities.

What conditions led to the remarkable preservation of the Burgess Shale fauna? The depositional site of the Burgess Shale lay at the base of a steep submarine escarpment. The animals whose exquisitely preserved fossil remains are found in the Burgess Shale lived in and on mud banks that formed along the top of this escarpment. Periodically, this unstable area would slump and slide down the escarpment as a turbidity current. At the base, the mud and animals carried with it were deposited in a deep-water anoxic environment devoid of life. In such an environment, bacterial degradation did not destroy the buried animals, and thus they were compressed by the weight of the overlying sediments and eventually preserved as carbonaceous films.

In the following two chapters, we examine the history of Paleozoic life as a system in which its parts consist of a series of interconnected biologic and geologic events. The underlying processes of evolution and plate tectonics are the forces that drove this system. The opening and closing of ocean basins, transgressions and regressions of epeiric seas, the formation of mountain ranges, and the changing positions of the continents profoundly affected the evolution of the marine and terrestrial communities.

A time of tremendous biologic change began with the appearance of skeletonized animals near the Precambrian–Cambrian boundary. Following this event, marine invertebrates began a period of adaptive radiation and evolution, during which the Paleozoic marine invertebrate community greatly diversified. Indeed, the history of the Paleozoic marine invertebrate community was one of diversification and extinction, culminating at the end of the Paleozoic Era in the greatest mass extinction in Earth history.

What Was the Cambrian Explosion?

At the beginning of the Paleozoic Era, animals with skeletons appeared rather abruptly in the fossil record. In fact, their appearance is described as an explosive development of new types of animals and is referred to as the "Cambrian explosion" by most scientists. This sudden appearance of new animals in the fossil record is rapid, however, only in the context of geologic time, having taken place over millions of years during the Early Cambrian Period.

Archean Eon	Proterozoic Eon	Phanerozoic Eon							
Precambrian		Paleozoic Era							
		Cambrian	Ordovician	Silurian	Devonian	Mississippian	Pennsylvanian	Permian	
							Carboniferous		

2500 MYA — 542 MYA — 251 MYA

This seemingly sudden appearance of animals in the fossil record is not a recent discovery. Early geologists observed that the remains of skeletonized animals appeared rather abruptly in the fossil record. Charles Darwin addressed this problem in *On the Origin of Species by Means of Natural Selection* (1859) and observed that, without a convincing explanation, such an event was difficult to reconcile with his newly expounded evolutionary theory.

The sudden appearance of shelled animals during the Early Cambrian contrasts sharply with the biota living during the preceding Proterozoic Eon. Up until the evolution of the Ediacaran fauna, Earth was populated primarily by single-celled organisms. Recall from Chapter 9 that the Ediacaran fauna, which is found on all continents except Antarctica, consists primarily of multicelled soft-bodied organisms. Microscopic calcareous tubes, presumably housing wormlike suspension-feeding organisms, have also been found at some localities (see Figure 9.21b). In addition, trails and burrows, which represent the activities of worms and other sluglike animals, are also found associated with Ediacaran faunas throughout the world. The trails and burrows are similar to those made by present-day soft-bodied organisms.

Until recently, it appeared that there was a fairly long period of time between the extinction of the Ediacaran fauna and the first Cambrian fossils. That gap has been considerably narrowed in recent years with the discovery of new Proterozoic fossiliferous localities. Now, Proterozoic fossil assemblages continue right to the base of the Cambrian.

Nonetheless, the cause of the sudden appearance of so many different animal phyla during the Early Cambrian is still a hotly debated topic. Newly developed molecular techniques that allow evolutionary biologists to compare molecular sequences of the same gene from different species are being applied to the phylogeny, or evolutionary history, of many organisms. In addition, new fossil sites and detailed stratigraphic studies are shedding light on the early history and ancestry of the various invertebrate phyla.

The Cambrian explosion probably had its roots firmly planted in the Proterozoic. However, the mechanism

or mechanisms that triggered this event are still being investigated. Although some would argue for a single causal event, it is more likely that the Cambrian explosion was a combination of factors, both biological and geological. For example, geologic evidence indicates that Earth was glaciated one or more times during the Proterozoic, followed by global warming during the Cambrian. These global environmental changes may have stimulated evolution and contributed to the Cambrian explosion.

Others would argue that a change in the chemistry of the oceans favored the evolution of a mineralized skeleton. In this scenario, an increase in the concentration of calcium from the Neoproterozoic through the Early Cambrian allowed for the precipitation of calcium carbonate and calcium phosphate, compounds that comprise the shells of most invertebrates.

Another hypothesis gaining favor is that the rapid evolution of a skeletonized fauna was a response to the evolution of predators. A shell or mineralized covering would provide protection against predation by the various predators evolving during this time.

An interesting line of research related to the Cambrian explosion involves *Hox* genes, which are sequences of genes that control the development of individual regions of the body. Studies indicate that the basic body plans for all animals were apparently established by the end of the Cambrian explosion, and only minor modifications have occurred since then. Whatever the ultimate cause of the Cambrian explosion, the appearance of a skeletonized fauna and the rapid diversification of that fauna during the Early Cambrian were major events in life history.

The Emergence of a Shelly Fauna

The earliest organisms with hard parts are Proterozoic calcareous tubes found associated with Ediacaran faunas from several locations throughout the world. These were followed by other microscopic skeletonized fossils from the Early Cambrian (• Figure 12.1) and the appearance

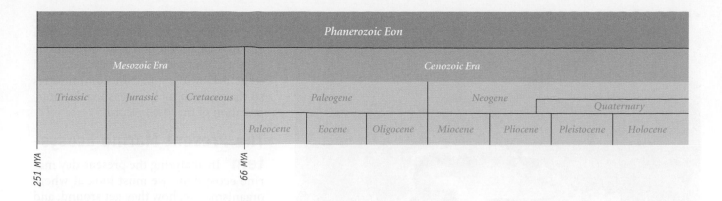

Phanerozoic Eon											
Mesozoic Era			Cenozoic Era								
Triassic	Jurassic	Cretaceous	Paleogene			Neogene			Quaternary		
			Paleocene	Eocene	Oligocene	Miocene	Pliocene		Pleistocene	Holocene	

251 MYA

66 MYA

• **Figure 12.1** **Lower Cambrian Shelly Fossils** Two small (several millimeters in size) Lower Cambrian shelly fossils.

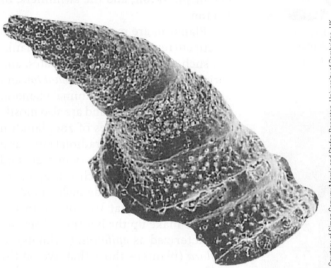

a *Lapworthella,* a conical sclerite (a piece of armor covering) from Australia.

Courtesy of Simon Conway Morris and Stefan Bengston, University of Cambridge, UK

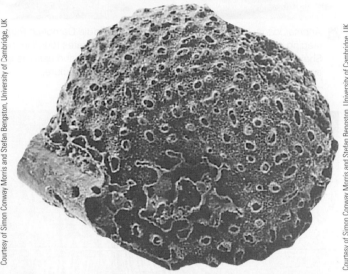

b *Archaeooides,* an enigmatic spherical fossil from the Mackenzie Mountains, Northwest Territories, Canada.

Courtesy of Simon Conway Morris and Stefan Bengston, University of Cambridge, UK

of large skeletonized animals during the Cambrian explosion. Along with the question of why animals appeared so suddenly in the fossil record is the equally intriguing question of why they initially acquired skeletons and what selective advantage skeletons provided. A variety of explanations about why marine organisms evolved skeletons have been proposed, but none are completely satisfactory or universally accepted.

The formation of an exoskeleton, or shell, confers many advantages on an organism: (1) It provides protection against ultraviolet radiation, allowing animals to move into shallower waters; (2) it helps prevent drying out in an intertidal environment; (3) a supporting skeleton, whether an exo- or endoskeleton, allows animals to increase their size and provides attachment sites for muscles; and (4) it provides protection against predators. Evidence of actual fossils of predators and specimens

of damaged prey, as well as antipredatory adaptations in some animals, indicates that the impact of predation during the Cambrian was great, leading some scientists to hypothesize that the rapid evolution of a shelly invertebrate fauna was a response to the rise of predators (• Figure 12.2). With predators playing an important role in the Cambrian marine ecosystem, any mechanism or feature that protected an animal would certainly be an adaptive advantage to the organism.

Scientists currently have no clear answer as to why marine organisms evolved mineralized skeletons during the Cambrian explosion and shortly thereafter. They undoubtedly evolved because of a variety of biologic and environmental factors. Whatever the reason, the acquisition of a mineralized skeleton was a major evolutionary innovation that allowed invertebrates to successfully occupy a wide variety of marine habitats.

• Figure 12.2 **Cambrian Predation**

© DEA PICTURE LIBRARY/Getty Images

a *Anomalocaris,* a predator from the Early and Middle Cambrian Period, is shown feeding on *Opabinia,* one of many extinct invertebrates found in the Burgess Shale. *Anomalocaris* was approximately 45 cm long and used its gripping appendages to bring its prey to its circular mouth structure.

Courtesy of the Geological Survey of Canada. Photo 202109-5

b Wounds (area just above the ruler) to the body of the trilobite *Olenellus robsonensis.* The wounds have healed, demonstrating that they occurred when the animal was alive and were not inflicted on an empty shell.

Paleozoic Invertebrate Marine Life

Having considered the origin, differentiation, and evolution of the Precambrian–Cambrian marine biota, we now examine the changes that occurred in the marine invertebrate community during the Paleozoic Era. Rather than focusing on the history of each invertebrate phylum (Table 12.1), we will survey the evolution of the marine invertebrate communities through time, concentrating on the major features and changes that took place. To do that, we need to briefly examine the nature and structure of living marine communities so that we can make a reasonable interpretation of the fossil record.

The Present Marine Ecosystem In analyzing the present-day marine ecosystem, we must look at where organisms live, how they get around, and how they feed (• Figure 12.3). Organisms that live in the water column above the seafloor are called *pelagic.* They can be divided into two main groups: the floaters, or **plankton,** and the swimmers, or **nekton.**

Plankton are mostly passive and go where currents carry them. Plant plankton such as diatoms, dinoflagellates, and various algae are called *phytoplankton* and are mostly microscopic. Animal plankton are called *zooplankton* and are also mostly microscopic. Examples of zooplankton include foraminifera, radiolarians, and jellyfish. The nekton are swimmers and are mainly vertebrates such as fish; the invertebrate nekton include cephalopods.

Organisms that live on or in the seafloor make up the **benthos.** They are characterized as *epifauna* (animals) or *epiflora* (plants)—those that live on the seafloor—or as *infauna*—animals that live in and move through the sediments. The benthos are further divided into those organisms that stay in one place, called *sessile,* and those that move around on or in the seafloor, called *mobile.*

The feeding strategies of organisms are also important in terms of their relationships with other organisms in the marine ecosystem. There are basically four feeding groups: **Suspension-feeding** animals remove or consume microscopic plants and animals as well as dissolved nutrients from the water, **herbivores** are plant eaters, **carnivore-scavengers** are meat eaters, and **sediment-deposit feeders** ingest sediment and extract the nutrients from it.

We can define an organism's place in the marine ecosystem by where it lives and how it eats. For example, an articulate brachiopod is a benthic, epifaunal suspension feeder, whereas a cephalopod is a nektonic carnivore.

An ecosystem includes several *trophic levels,* which are tiers of food production and consumption within a feeding hierarchy. The feeding hierarchy, and hence energy flow, in

TABLE **12.1** | The Major Invertebrate Groups and Their Stratigraphic Ranges

Phylum Protozoa	Cambrian—Recent	**Phylum Mollusca**	Cambrian—Recent
Class Sarcodina	Cambrian—Recent	Class Monoplacophora	Cambrian—Recent
Order Foraminifera	Cambrian—Recent	Class Gastropoda	Cambrian—Recent
Order Radiolaria	Cambrian—Recent	Class Bivalvia	Cambrian—Recent
Phylum Porifera	Cambrian—Recent	Class Cephalopoda	Cambrian—Recent
Class Demospongea	Cambrian—Recent	**Phylum Annelida**	Precambrian—Recent
Order Stromatoporida	Cambrian—Oligocene	**Phylum Arthropoda**	Cambrian—Recent
Phylum Archaeocyatha	Cambrian	Class Trilobita	Cambrian—Permian
Phylum Cnidaria	Cambrian—Recent	Class Crustacea	Cambrian—Recent
Class Anthozoa	Ordovician—Recent	Class Insecta	Silurian—Recent
Order Tabulata	Ordovician—Permian	**Phylum Enchinodermata**	Cambrian—Recent
Order Rugosa	Ordovician—Permian	Class Blastoidea	Ordovician—Permian
Order Scleractinia	Triassic—Recent	Class Crinoidea	Cambrian—Recent
Phylum Bryozoa	Ordovician—Recent	Class Echinoidea	Ordovician—Recent
Phylum Brachiopoda	Cambrian—Recent	Class Asteroidea	Ordovician—Recent
Class Inarticulata	Cambrian—Recent	**Phylum Hemichordata**	Cambrian—Recent
Class Articulata	Cambrian—Recent	Class Graptolithina	Cambrian—Mississippian

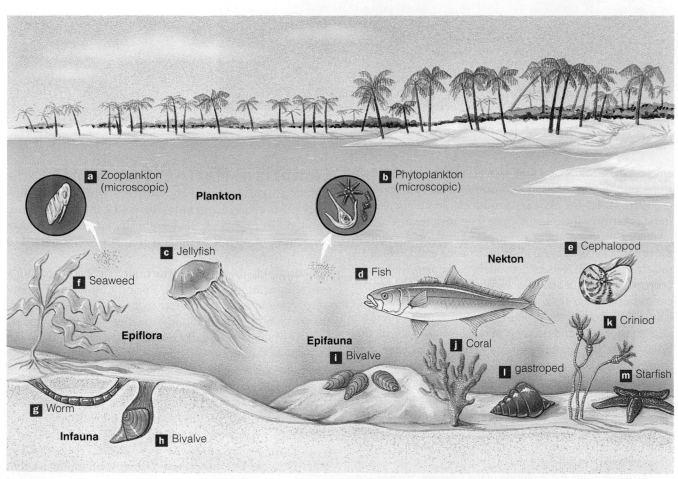

• **Figure 12.3 Marine Ecosystem** Where and how animals and plants live in the marine ecosystem. Plankton: **a** through **c**. Nekton: **d** and **e**. Benthos: **f** through **m**. Sessile epiflora: **f**. Sessile epifauna: **i**, **j**, and **k**. Mobile epifauna: **l** and **m**. Infauna: **g** and **h**. Suspension feeders: **i**, **j**, and **k**. Herbivores: **l**. Carnivore-scavengers: **m**. Sediment-deposit feeders: **g**.

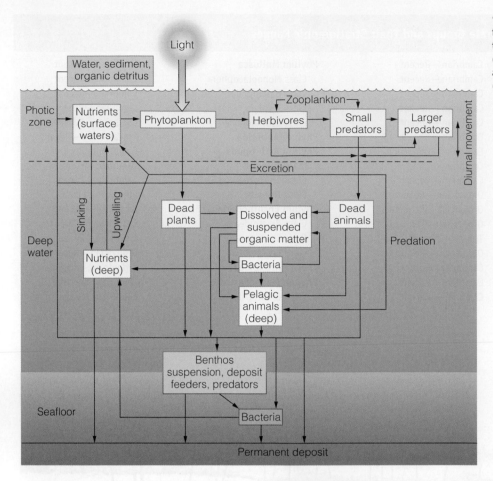

• Figure 12.4 **Marine Food Web** Marine food web showing the relationship between the producers (phytoplankton), consumers (herbivores, small and large predators, pelagic animals, and benthos), and decomposers (bacteria).

an ecosystem comprise a food web of complex interrelationships among the producers, consumers, and decomposers (• Figure 12.4). The **primary producers,** or *autotrophs,* are those organisms that manufacture their own food. Virtually all marine primary producers are phytoplankton. Feeding on the primary producers are the primary consumers, which are mostly suspension feeders. Secondary consumers feed on the primary consumers and thus are predators, whereas tertiary consumers, which are also predators, feed on the secondary consumers. Besides the producers and consumers, there are also transformers and decomposers. These are bacteria that break down the dead organisms that have not been consumed into organic compounds, which are then recycled.

When we look at the marine realm today, we see a complex organization of organisms interrelated by trophic interactions and affected by changes in the physical environment. When one part of the system changes, the whole structure changes, sometimes almost insignificantly, other times catastrophically.

As we examine the evolution of the Paleozoic marine ecosystem, keep in mind how geologic and evolutionary changes have had a significant impact on its composition and structure. For example, the major transgressions onto the craton opened up vast areas of shallow seas that could be inhabited. The movement of continents affected oceanic circulation patterns as well as causing environmental changes.

Cambrian Marine Community

The Cambrian Period was a time during which many new body plans evolved and animals moved into new niches. As might be expected, the Cambrian witnessed a higher percentage of such experiments than any other period of geologic history.

Although almost all of the major invertebrate phyla evolved during the Cambrian Period (Table 12.1), many were represented by only a few species. Whereas trace fossils are common and echinoderms diverse, trilobites, inarticulate brachiopods, and archaeocyathids comprised the majority of Cambrian skeletonized life (• Figure 12.5).

Trilobites were by far the most conspicuous element of the Cambrian marine invertebrate community and made up approximately half of the total fauna. Most trilobites were benthic, mobile, sediment-deposit feeders that crawled or swam along the seafloor (see Perspective).

Cambrian *brachiopods* were mostly types called inarticulates. They secreted a chitinophosphate shell, composed of the organic compound chitin combined with calcium phosphate. Inarticulate brachiopods also lacked a tooth-and-socket arrangement along the hinge line where the two shells pivot. The articulate brachiopods, which have a tooth-and-socket arrangement, were also present, but did not become abundant until the Ordovician Period.

• Figure 12.5 **Cambrian Marine Community** Reconstruction of a Cambrian marine community showing floating jellyfish, swimming arthropods, benthonic sponges, and scavenging trilobites.

• Figure 12.6 **Archaeocyathids** Restoration of a Cambrian reef-like structure built by archaeocyathids.

The third major group of Cambrian organisms was the *archaeocyathids* (• Figure 12.6). These organisms were benthic, sessile, presumably suspension feeders that constructed reef-like structures beginning in the Early Cambrian. Archaeocyathids went extinct at the end of the Cambrian.

The rest of the Cambrian fauna consisted of representatives of most of the other major phyla, including many organisms that were short-lived evolutionary experiments (• Figure 12.7). As might be expected during times of adaptive radiation and evolutionary experimentation, many of the invertebrates that evolved during the Cambrian soon became extinct.

• Figure 12.7 **The Primitive Echinoderm** *Helicoplacus Helicoplacus,* a primitive echinoderm that became extinct 20 million years after it first evolved about 510 million years ago. Such an organism (a representative of one of several short-lived echinoderm classes) illustrates the "experimental" nature of the Cambrian invertebrate fauna.

Perspective

Trilobites–Paleozoic Arthropods

Trilobites, an extinct class of arthropods, are probably the favorite and most sought after of invertebrate fossils. They lived from the Early Cambrian until the end of the Permian and were most diverse during the Late Cambrian. More than 15,000 species of trilobites have been described, and they are currently grouped into nine orders.

Trilobites had a worldwide distribution throughout the Paleozoic and they lived in all marine environments, from shallow, nearshore waters to deep oceanic settings. They occupied a wide variety of habitats. Most were bottom dwellers, crawling around the seafloor and scavenging organic detritus or feeding on microorganisms or algae. Others were free-swimming predators or filter-feeders living throughout the water column.

The trilobite body is divided into three parts (Figure 1); a cephalon (head), containing the eyes, mouth, and sensory organs; the thorax (body), composed of individual segments; and the pygidium (tail). Interestingly, the name trilobite does not refer to its three main body parts, but means *three-lobed*, which corresponds to the three longitudinal lobes of the thorax–the axial lobe, and the two flanking pleural lobes.

The appendages of trilobites are rarely preserved. However, from specimens in which the soft part anatomy is preserved as an impression, paleontologist know that beneath each thoracic segment was a two-part appendage consisting of a gill-bearing outer branch used for respiration and an inner branch or walking leg composed of articulating limb segments (Figure 2).

Trilobites range in size from a few millimeters in length, as found in many of the

Figure 1 *Cedaria minor,* from the Cambrian Weeks Formation, Utah, illustrates the major body parts of a trilobite.

Figure 2 Model of dorsal **a** and ventral **b** anatomy of *Triarthrus eatoni,* a Late Ordovician trilobite from the Frankfort Shale, New York.

a Dorsal view.

b Ventral view.

agnostid trilobites (Figure 3), to more than 70 cm, as recorded by the world's largest trilobite (Figure 4). Most trilobites, however, range between 3 and 10 cm long.

Trilobites were either blind, as in the agnostids (Figure 3), or possessed paired eyes. Holochroal eyes are characterized by generally hexagonal-shaped biconvex lenses that are packed closely together and cov-

ered by a single corneal layer that covers all of the lenses (Figure 5). This type of eye is similar to a modern compound eye and results in mosaic vision.

Schizochroal eyes are composed of relatively large, individual lenses, each of which is covered by its own cornea and separated from adjacent lenses by a thick dividing wall (Figure 6). Schizochroal lenses are arranged in rows and columns, and most likely produced a visual field consisting of shadows.

Although trilobites had a hard dorsal exoskeleton, useful, in part, as a protection against predators, some trilobites also had the ability to enroll, presumably to protect their antennae, limbs, and soft ventral appendages. By doing so, the trilobite could protect its soft ventral anatomy and still view its surroundings (Figure 7).

As mentioned above, trilobites were a very successful group. They first appeared in the Early Cambrian, rapidly diversified, reached their maximum diversity in the Late Cambrian, and then suffered major reductions in diversity near the end of the Cambrian, at the end of the Ordovician, and again near the end of the Devonian.

As yet, no consensus exists on what caused the trilobite extinctions, but a combination of factors was likely involved, possibly including a reduction of shelf space, increased competition, and a rise in predators. Some scientists have also suggested that a cooling of the seas may have played a role, particularly for the extinctions at the end of the Ordovician Period. The trilobites, like most of their Permian marine invertebrate contemporaries, finally fell victim to the Permian mass extinction event.

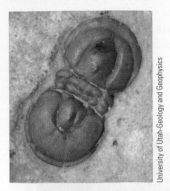

Figure 3 *Hypagnostus parvifrons,* a small agnostid trilobite from the Cambrian Marjum Formation, Utah.

University of Utah-Geology and Geophysics

Courtesy of E. N. K. Clarkson, used with permission

Figure 5 Holochroal eyes from the trilobite *Scutellum campaniferum.*

© The Manitoba Museum/Dr. Graham Young

Figure 4 *Isotellus rex,* the world's largest trilobite (more than 70 cm long), from the Ordovician Churchill River Group, Manitoba, Canada.

Courtesy of E. N. K. Clarkson, used with permission

Figure 6 Schizochroal eyes from the trilobite *Eophacops trapeziceps.*

© Sinclair Stammers/SPL/Photo Researchers, Inc.

Figure 7 An enrolled specimen of the Ordovician trilobite *Pilomera fisheri* from Putilowa, Poland.

The Burgess Shale Biota No discussion of Cambrian life would be complete without mentioning one of the best examples of a preserved soft-bodied fauna and flora: the Burgess Shale biota. As the Sauk Sea transgressed from the Cordilleran shelf onto the western edge of the craton, Early Cambrian-age sands were covered by Middle Cambrian-age black muds that allowed a diverse soft-bodied benthic community to be preserved. As we discussed in the Introduction, Charles Walcott discovered these fossils near Field, British Columbia, Canada, in 1909. They represent one of the most significant fossil finds of the 20th century because they consist of carbonized impressions of soft-bodied animals and plants that are rarely preserved in the fossil record (• Figure 12.8). This discovery, therefore, provides us with a valuable glimpse of rarely preserved organisms as well as the soft-part anatomy of many extinct groups.

In recent years, the reconstruction, classification, and interpretation of many of the Burgess Shale fossils have undergone a major change that has led to new theories and explanations of the Cambrian explosion of life. Recall that during the Neoproterozoic, multicelled organisms evolved, and shortly thereafter animals with hard parts made their first appearance. These were followed by an explosion of invertebrate groups during the Cambrian, many of which are now extinct. These Cambrian organisms represent the rootstock and basic body plans from which all present-day invertebrates evolved.

• **Figure 12.8 Fossils from the Burgess Shale** Some of the fossil animals preserved in the Burgess Shale.

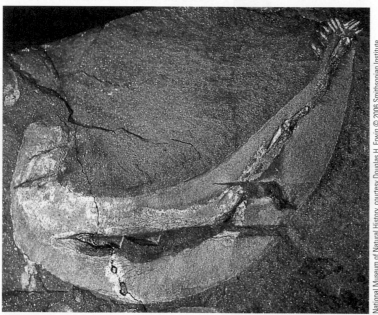

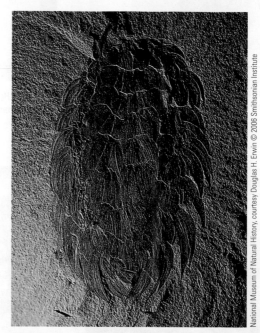

a *Ottoia*, a carnivorous worm.

b *Wiwaxia*, a scaly armored sluglike creature whose affinities remain controversial.

c *Hallucigenia*, a velvet worm.

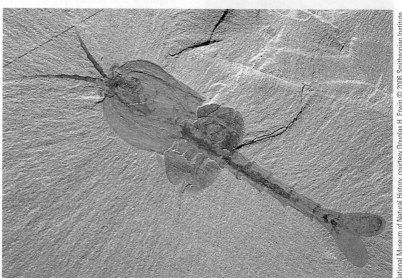

d *Waptia*, an arthropod.

The question that paleontologists are still debating is how many phyla arose during the Cambrian, and at the center of that debate are the Burgess Shale fossils. For years, most paleontologists placed the bulk of the Burgess Shale organisms into existing phyla, with only a few assigned to phyla that are now extinct. Thus, the phyla of the Cambrian world were viewed as being essentially the same in number as the phyla of the present-day world but with fewer species in each phylum. According to this view, the history of life has been simply a gradual increase in the diversity of species within each phylum through time. The number of basic body plans has, therefore, remained more or less constant since the initial radiation of multicelled organisms.

This view, however, has been challenged by other paleontologists who think the initial explosion of varied life-forms in the Cambrian was promptly followed by a short period of experimentation and then extinction of many phyla. The richness and diversity of modern life-forms are the result of repeated variations of the basic body plans that survived the Cambrian extinctions. In other words, life was much more diverse in terms of phyla during the Cambrian than it is today. The reason why members of the Burgess Shale biota look so strange to us is that no living organisms possess their basic body plan and, therefore, many of them have been reassigned into new phyla.

Discoveries of new Cambrian fossils at localities such as Sirius Passet, Greenland, and Yunnan, China, have resulted in reassignment of some Burgess Shale specimens back into extant phyla. If these reassignments to known phyla prove to be correct, then no massive extinction event followed the Cambrian explosion, and life has gradually increased in diversity through time. Currently, there is no clear answer to this debate, and the outcome will probably be decided as more fossil discoveries are made.

Ordovician Marine Community

A major transgression that began during the Middle Ordovician (Tippecanoe Sequence) resulted in widespread inundation of the craton. This vast epeiric sea, which had a uniformly warm climate during this time, opened many new marine habitats that were soon filled by a variety of organisms.

Not only did sedimentation patterns change dramatically from the Cambrian to the Ordovician, but the fauna underwent equally striking changes. Whereas trilobites, inarticulate brachiopods, and archaeocyathids dominated the Cambrian invertebrate community, the Ordovician was characterized by the adaptive radiation of many other animal phyla (such as articulate brachiopods, bryozoans, and corals), with a consequent dramatic increase in the diversity of the total shelly fauna (• Figure 12.9). The Ordovician was also a time of increased diversity and abundance of the *acritarchs* (organic-walled phytoplankton of unknown affinity), which were the major phytoplankton group of the Paleozoic Era and the primary food source of the suspension feeders (• Figure 12.10).

• **Figure 12.9 Middle Ordovician Marine Community** Reconstruction of a Middle Ordovician seafloor fauna. Cephalopods, crinoids, colonial corals, bryozoans, trilobites, and brachiopods are shown.

• Figure 12.10 **Late Ordovician Acritarchs** Acritarchs from the Upper Ordovician Sylvan Shale, Oklahoma. Acritarchs are organic-walled phytoplankton and were the primary food source for suspension feeders during the Paleozoic Era.

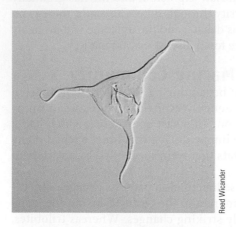

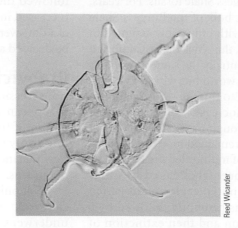

• Figure 12.11 **Representative Brachiopods and Graptolites**

a Brachiopods are benthic, sessile, suspension feeders.

b Graptolites are planktonic suspension feeders. Shown is *Phyllograptus angustifolius* from Norway.

During the Cambrian, archaeocyathids were the main builders of reef-like structures, but bryozoans, stromatoporoids, and tabulate and rugose corals assumed that role beginning in the Middle Ordovician. Many of these reefs were small patch reefs similar in size to those of the Cambrian but of a different composition, whereas others were quite large. As with present-day reefs, Ordovician reefs showed a high diversity of organisms and were dominated by suspension feeders.

Three Ordovician fossil groups have proved particularly useful for biostratigraphic correlation—the *articulate brachiopods, graptolites,* and *conodonts.* The articulate brachiopods, present since the Cambrian, began a period of major diversification in the shallow-water marine environment during the Ordovician (• Figure 12.11a). They became a conspicuous element of the invertebrate fauna during the Ordovician and in succeeding Paleozoic periods.

Because the vast majority of graptolites were planktonic and thus carried about by ocean currents, and because most species existed for less than a million years, graptolites are excellent guide fossils. They were especially abundant during the Ordovician and Silurian periods. Graptolites are most commonly found in black shales where they are preserved

as carbonaceous impressions (Figure 12.11b).

Conodonts are a group of well-known, small, tooth-like fossils composed of the mineral apatite (calcium phosphate), the same mineral that composes bone (• Figure 12.12a). Although conodonts have been known for more than 150 years, their affinity was debated until the discovery of the conodont animal in 1983 (Figure 12.12b). Several specimens of the carbonized impression of the conodont animal from Lower Carboniferous rocks of Scotland show it is a member of a group of primitive jawless animals assigned to the phylum Chordata. Study of the specimens indicates that the conodont animal was probably an elongate swimming organism. The wide distribution and short stratigraphic range of individual conodont species make them excellent fossils for biostratigraphic zonation and correlation.

The end of the Ordovician was a time of mass extinctions in the marine realm. More than 100 families of marine invertebrates became extinct, and in North America alone, about half of the brachiopods and bryozoans died out. What caused such an event? Many geologists think these extinctions were the result of extensive glaciation in Gondwana at the end of the Ordovician Period (see Chapter 10).

Silurian and Devonian Marine Communities

The mass extinction at the end of the Ordovician was followed by rediversification and recovery of many of the decimated groups. Brachiopods, bryozoans, gastropods, bivalves, corals, crinoids, and graptolites were just some of the groups that rediversified during the Silurian.

As we discussed in Chapters 10 and 11, the Silurian and Devonian were times of major reef building. Whereas most of the Silurian radiations of invertebrates represented the repopulation of niches, organic reef builders diversified in new ways, building massive reefs larger than any produced during the Cambrian or Ordovician. This repopulation was probably caused, in part, by renewed transgressions over the craton, and although a major drop in sea level occurred at the end of the Silurian, the Middle Paleozoic sea level was generally high (see Table 10.1).

• Figure 12.12 **Conodonts and the Conodont Animal**

a Conodonts are microscopic toothlike fossils. *Cahabagnathus sweeti*, Copenhagen Formation (Middle Ordovician), Monitor Range, Nevada (*left*); *Scolopodus*, sp., Shingle Limestone, Single Pass, Nevada (*right*).

Courtesy of Stig M. Bergstrom, Ohio State University

Photo by J. K. Ingham, supplied courtesy of R. J. Aldridge, University of Leicester, Leicester, England

b The conodont animal preserved as a carbonized impression in the Lower Carboniferous Granton Shrimp Bed in Edinburgh, Scotland. The animal measures about 40 mm long and 2 mm wide.

The Silurian and Devonian reefs were dominated by tabulate and colonial rugose corals and stromatoporoids (• Figure 12.13). Although the fauna of these Silurian and Devonian reefs was somewhat different from that of earlier reefs and reef-like structures, the general composition and structure are the same as in present-day reefs.

The Silurian and Devonian periods were also the time when *eurypterids* (arthropods with scorpion-like bodies and impressive pincers) were abundant, and unlike many other marine invertebrates, eurypterids expanded into brackish and freshwater habitats (• Figure 12.14). *Ammonoids,* a subclass of cephalopods, evolved from nautiloids during the Early Devonian and rapidly diversified. With their distinctive suture patterns, short stratigraphic ranges, and widespread distribution, ammonoids are excellent guide fossils for the Devonian through Cretaceous periods (• Figure 12.15).

Near the end of the Devonian, another mass extinction occurred that resulted in a worldwide near-total collapse of the massive marine reef communities. On land, however, the seedless vascular plants were seemingly unaffected. Thus, extinctions at this time were most extensive among marine life, particularly in the reef and pelagic communities.

The demise of the Middle Paleozoic reef communities highlights the geographic aspects of the Late Devonian mass extinction. The tropical groups were most severely affected; in contrast, the higher latitude communities were seemingly little affected. Apparently, an episode of global cooling was largely responsible for the extinctions

© The Field Museum, Chicago, Geo80821c

• Figure 12.13 **Middle Devonian Marine Reef Community** Reconstruction of a Middle Devonian reef from the Great Lakes area of North America. Shown are corals, bryozoans, cephalopods, trilobites, crinoids, and brachiopods.

• Figure 12.14 **Silurian Brackish Water Community** Restoration of a Silurian brackish water scene near Buffalo, New York. Shown are algae, eurypterids, gastropods, worms, and shrimp.

• Figure 12.15 **Ammonoid Cephalopod** A Late Devonian-age ammonoid cephalopod from Erfoud, Morocco. The distinctive suture pattern, short stratigraphic range, and wide geographic distribution make ammonoids excellent guide fossils.

near the end of the Devonian. During such a cooling, the disappearance of tropical conditions would have had a severe effect on reef and other warm-water organisms. Cool-water species, in contrast, could have simply migrated toward the equator. Although cooling temperatures certainly played an important role in the Late Devonian extinctions, the closing of the Iapetus Ocean and the orogenic events of this time (see Figure 11.1a) undoubtedly

also played a role by reducing the area of shallow shelf environments where many marine invertebrates lived.

Carboniferous and Permian Marine Communities
The Carboniferous invertebrate marine community responded to the Late Devonian extinctions in much the same way the Silurian invertebrate marine community responded to the Late Ordovician extinctions—that is, by renewed adaptive radiation and rediversification. The brachiopods and ammonoids quickly recovered and again assumed important ecologic roles. Other groups, such as the lacy bryozoans and crinoids, reached their greatest diversity during the Carboniferous. With the decline of the stromatoporoids and the tabulate and rugose corals, large organic reefs such as those existing earlier in the Paleozoic virtually disappeared and were replaced by small patch reefs. These reefs were dominated by crinoids, blastoids, lacy bryozoans, brachiopods, and calcareous algae and flourished during the Late Paleozoic (• Figure 12.16). In addition, bryozoans and crinoids contributed large amounts of skeletal debris to the formation of the vast bedded limestones that constitute the majority of Mississippian sedimentary rocks.

The Permian invertebrate marine faunas resembled those of the Carboniferous but were not as widely distributed because of the restricted size of the shallow seas on the cratons and the reduced shelf space along the continental margins (see Figure 11.11). The spiny and odd-shaped productids dominated the brachiopod assemblage and constituted

• **Figure 12.16 Late Mississippian Marine Community** Reconstruction of marine life during the Mississippian, based on an Upper Mississippian fossil site at Crawfordville, Indiana. Invertebrate animals shown include crinoids, blastoids, lacy bryozoans, brachiopods, and small corals.

an important part of the reef complexes that formed in the Texas region during the Permian (• Figure 12.17). The fusulinids (spindle-shaped foraminifera), which first evolved during the Late Mississippian and greatly diversified during the Pennsylvanian (• Figure 12.18), experienced a further diversification during the Permian. Because of their abundance, diversity, and worldwide occurrence, fusulinids are important guide fossils for Pennsylvanian and Permian strata. Bryozoans, sponges, and some types of calcareous algae also were common elements of the Permian invertebrate fauna.

Mass Extinctions

Throughout geologic history, various plant and animal species have become extinct. In fact, extinction is a common feature of the fossil record, and the rate of extinction through time has fluctuated only slightly. Just as new species evolve, others become extinct. There have, however, been brief intervals in the geologic past during which mass extinctions have eliminated large numbers of species. Extinctions of this magnitude could only occur due to radical changes in the environment on a regional or global scale.

When we look at the different mass extinctions that have occurred during the geologic past, several common themes stand out. The first is that mass extinctions typically have affected life both in the sea and on land. Second, tropical organisms, particularly in the marine realm, apparently are more affected than organisms from the temperate and high-latitude regions. Third, some animal groups repeatedly experience mass extinctions.

When we examine the mass extinctions for the past 650 million years, we see that the first major extinction involved only the acritarchs. Several extinction events occurred during the Cambrian, and these affected only marine invertebrates, particularly trilobites. Three other marine mass extinctions took place during the Paleozoic Era: one at the end of the Ordovician, involving many invertebrates; one near the end of the Devonian, affecting the major barrier reef–building organisms as well as the primitive armored fish; and the most severe at the end of the Permian, when about 90% of all marine invertebrate species and more than 65% of all land animals became extinct.

The Mesozoic Era experienced several mass extinctions, the most devastating occurring at the end of the Cretaceous, when almost all large animals, including dinosaurs, flying reptiles, and seagoing animals such as plesiosaurs and ichthyosaurs, became extinct. Many scientists think the terminal Cretaceous mass extinction was caused by a meteorite impact (see Chapter 15).

Several mass extinctions also occurred during the Cenozoic Era (see Chapter 18). The most severe was near the end of the Eocene Epoch and is correlated with global cooling and climatic change. The most recent extinction occurred near the end of the Pleistocene Epoch.

Although many scientists think of the marine mass extinctions as sudden events from a geologic perspective, they were rather gradual from a human perspective, occurring

• **Figure 12.17 Permian Patch-Reef Marine Community** Reconstruction of a Permian patch-reef community from the Glass Mountains of west Texas. Shown are algae, productid brachiopods, cephalopods, sponges, and corals.

The American Museum of Natural History, #K10269

Lower Permian Owens Valley Group, California

• **Figure 12.18 Fusulinids** Fusulinids are spindle-shaped, microscopic benthonic foraminifera that are excellent guide fossils for the Pennsylvanian and Permian periods. Shown here are three natural cross-sections of fusulinids from the Lower Permian Owens Valley Group, Inyo County, California. The two elongated specimens are *Parafusulina* sp. and the circular specimen in the lower right (view is a cross-section perpendicular to the long axis of the specimen) is an unidentified fusulinid.

over hundreds of thousands and even millions of years. Furthermore, many geologists think that climatic changes, rather than a single catastrophic event, were primarily responsible for the extinctions, particularly in the marine realm. Evidence of glacial episodes or other signs of climatic change, such as global warming, have been correlated with the extinctions recorded in the fossil record.

The Permian Mass Extinction The greatest recorded mass extinction to affect Earth occurred at the end of the Permian Period (• Figure 12.19). By the time the Permian ended, roughly 50% of all marine invertebrate families and about 90% of all marine invertebrate species became extinct. Fusulinids, rugose and tabulate corals, and several bryozoan and brachiopod orders, as well as trilobites and blastoids, did not survive the end of the Permian. All of these groups had been very successful during the Paleozoic Era. In addition, more than 65% of all amphibians and reptiles, as well as nearly 33% of insects on land, also became extinct.

What caused such a crisis for both marine and land-dwelling organisms? Various hypotheses have been proposed, but no completely satisfactory answer has yet been found. Because the extinction event extended over many millions of years at the end of the Permian, a meteorite impact such as occurred at the end of the Cretaceous Period (see Chapter 15) can be reasonably discounted.

A reduction in the habitable marine shelf area caused by the formation of Pangaea and a widespread marine regression resulting from glaciation can also be rejected as the primary cause of the Permian extinctions. By the end of

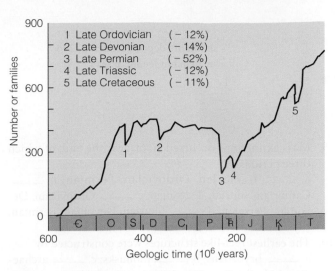

• Figure 12.19 Phanerozoic Marine Diversity Phanerozoic diversity for marine invertebrate and vertebrate families. Note the three episodes of Paleozoic mass extinctions, with the greatest occurring at the end of the Permian Period.

the Permian, most collisions of the continents had already taken place, such that the reduction in the shelf area had already occurred before the mass extinctions began in earnest. Furthermore, the widespread glaciation that took place during the Carboniferous was now waning in the Permian.

Currently, many scientists think an episode of deep-sea anoxia and increased oceanic CO_2 levels, resulted in a highly stratified ocean during the Late Permian. In other words, there was very little, if any, circulation of oxygen-rich surface waters into the deep ocean. During this time, stagnant waters also covered the shelf regions, thus affecting the shallow marine fauna.

In addition, there is also evidence of increased global warming during the Late Permian. This would also contribute to a stratified global ocean because warming of the high latitudes would significantly reduce or eliminate the downwelling of cold, dense, oxygenated waters from the polar areas into the deep oceans at lower latitudes, as occurs today. This would result in stagnant, stratified oceans, rather than a well-mixed, oxygenated oceanic system.

During the Late Permian, widespread volcanic and continental fissure eruptions were also taking place, releasing additional carbon dioxide into the atmosphere and contributing to increased climatic instability and ecologic collapse. By the end of the Permian, a near collapse of both the marine and terrestrial ecosystem had occurred. Although the ultimate cause of such devastation is still being debated and investigated, it is safe to say that it was probably a combination of interconnected and related geologic and biologic events.

SUMMARY

Table 12.2 summarizes the major evolutionary and geologic events of the Paleozoic Era and shows their relationships to each other.

- Multicelled organisms presumably had a long Precambrian history, during which they lacked hard parts.

- Invertebrates with hard parts suddenly appeared during the Early Cambrian in what is called the Cambrian explosion. Skeletons provided such advantages as protection against predators and support for muscles, enabling organisms to grow large and increase locomotor efficiency. Hard parts probably evolved as a result of various geologic and biologic factors rather than a single cause.

- Marine organisms are classified as plankton if they are floaters, nekton if they swim, and benthos if they live on or in the seafloor.

- Marine organisms are divided into four basic feeding groups: suspension feeders, which consume microscopic plants and animals as well as dissolved nutrients from water; herbivores, which are plant eaters; carnivore-scavengers, which are meat eaters; and sediment-deposit feeders, which ingest sediment and extract nutrients from it.

- The marine ecosystem consists of various trophic levels of food production and consumption. At the base are primary producers, on which all other organisms are dependent. Feeding on the primary producers are the primary consumers, which in turn are fed on by higher levels of consumers. The decomposers are bacteria that break down the complex organic compounds of dead organisms and recycle them within the ecosystem.

- The Cambrian invertebrate community was dominated by three major groups: the trilobites, inarticulate brachiopods, and archaeocyathids. Little specialization existed among the invertebrates, and most phyla were represented by only a few species.

- The Middle Cambrian Burgess Shale contains one of the finest examples of a well-preserved, soft-bodied biota in the world.

- The Ordovician marine invertebrate community marked the beginning of the dominance by the shelly fauna and the start of large-scale reef building. The end of the Ordovician Period was a time of major extinction for many invertebrate phyla.

- The Silurian and Devonian periods were times of diverse faunas dominated by reef-building animals, whereas the Carboniferous and Permian periods saw a great decline in invertebrate diversity.

- Mass extinctions are times when anomalously high numbers of organisms go extinct in a short period of time. Such events have occurred several times during the past 650 million years.

- A major extinction occurred at the end of the Paleozoic Era, affecting the invertebrates as well as the vertebrates. Its cause is still the subject of debate.

IMPORTANT TERMS

benthos, p. 244

carnivore-scavenger, p. 244

herbivore, p. 244

nekton, p. 244

plankton, p. 244

primary producer, p. 246

sediment-deposit feeder, p. 244

suspension feeder, p. 244

REVIEW QUESTIONS

1. Organisms that manufacture their own food are
   a. _____ autotrophs; b. _____ herbivores; c. _____ benthos; d. _____ epifaunal; e. _____ none of the previous answers.

2. The major organic-walled phytoplankton group of the Paleozoic Era was
   a. _____ acritarchs; b. _____ coccolithophorids; c. _____ diatoms; d. _____ dinoflagellates; e. _____ graptolites.

3. Which group of planktonic invertebrates that were especially abundant during the Ordovician and Silurian periods are excellent guide fossils?
   a. _____ Brachiopods; b. _____ Cephalopods; c. _____ Fusulinids; d. _____ Graptolites; e. _____ Trilobites.

4. Organisms living in the water column above the seafloor are
   a. _____ benthic; b. _____ epifaunal; c. _____ infaunal; d. _____ epifloral; e. _____ pelagic.

5. The Burgess Shale fauna is significant because it contains the
   a. _____ first shelled animals; b. _____ carbonized impressions of many extinct soft-bodied animals; c. _____ fossils of rare marine plants; d. _____ earliest known benthic community; e. _____ conodont animal.

6. Brachiopods are
   a. _____ benthic mobile carnivores; b. _____ benthic mobile scavengers; c. _____ benthic suspension feeders; d. _____ nektonic carnivore-scavengers; e. _____ planktonic primary producers.

7. What type of invertebrates dominated the Ordovician invertebrate community?
   a. _____ Epifloral planktonic primary producers; b. _____ Infaunal nektonic carnivores; c. _____ Infaunal benthic sessile suspension feeders; d. _____ Epifaunal benthic mobile suspension feeders; e. _____ Epifaunal benthic sessile suspension feeders.

8. An exoskeleton is advantageous because it
   a. _____ prevents drying out in an intertidal environment; b. _____ provides protection against ultraviolet radiation; c. _____ provides protection against predators; d. _____ provides attachment sites for development of strong muscles; e. _____ all of the previous answers.

9. Mass extinctions occurred at, or near the end, of which three periods?
   a. _____ Cambrian, Ordovician, Permian; b. _____ Cambrian, Silurian, Devonian; c. _____ Ordovician, Devonian, Permian; d. _____ Silurian, Devonian, Permian; e. _____ Cambrian, Devonian, Permian.

10. The earliest reef-like structures were constructed by
    a. _____ bryozoans; b. _____ mollusks; c. _____ archaeocyathids; d. _____ sponges; e. _____ corals.

11. The age of the Burgess Shale is
    a. _____ Cambrian; b. _____ Ordovician; c. _____ Silurian; d. _____ Devonian; e. _____ Mississippian.

12. The greatest recorded mass extinction in Earth history took place at the end of which period?
    a. _____ Cambrian; b. _____ Ordovician; c. _____ Devonian; d. _____ Permian; e. _____ Cretaceous.

13. The three invertebrate groups that comprised the majority of Cambrian skeletonized life were
    a. _____ trilobites, archaeocyathids, brachiopods; b. _____ echinoderms, corals, bryozoans; c. _____ brachiopods, archaeocyathids, corals; d. _____ trilobites, echinoderms, corals; e. _____ trilobites, brachiopods, corals.

14. The _____ and _____ were times of major reef building.
    a. _____ Cambrian, Ordovician; b. _____ Ordovician, Silurian; c. _____ Silurian, Devonian; d. _____ Devonian, Mississippian; e. _____ Mississippian, Pennsylvanian.

15. Discuss how changing geologic conditions affected the evolution of invertebrate life during the Paleozoic Era.

16. If the Cambrian explosion of life was partly the result of filling unoccupied niches, why don't we see such rapid evolution following mass extinctions such as those that occurred at the end of the Permian and Cretaceous periods?

17. Discuss the significance of the appearance of the first shelled animals and possible causes for the acquisition of a mineralized exoskeleton.

18. Discuss how the incompleteness of the fossil record may play a role in what is known as the Cambrian explosion.

19. What are the major differences between the Cambrian marine community and the Ordovician marine community?

20. Discuss some of the possible causes for the Permian mass extinction.

APPLY YOUR KNOWLEDGE

1. Draw a marine food web that shows the relationships among the producers, consumers, and decomposers.

2. One concern of environmentalists is that environmental degradation is leading to vast reductions in global biodiversity. As a paleontologist, you are aware mass extinctions have taken place throughout Earth history.

What facts and information can you provide from your geological perspective that will help focus the debate as to whether or not Earth's biota is being adversely affected by such human activities as industrialization, and what the possible outcome(s) might be if global biodiversity is severely reduced?

Table 12.2
Major Evolutionary and Geologic Events of the Paleozoic Era

Age (millions of years)	Geologic Period		Invertebrates	Vertebrates	Plants	Major Geologic Events
251 —	Permian		Largest mass extinction event to affect the invertebrates	Acanthodians, placoderms, and pelycosaurs become extinct Therapsids and pelycosaurs the most abundant reptiles	Gymnosperms diverse and abundant	Formation of Pangaea Alleghenian orogeny Hercynian orogeny
299 —	Carboniferous	Pennsylvanian	Fusulinids diversify	Amphibians abundant and diverse	Coal swamps with flora of seedless vascular plants and gymnosperms	Coal-forming swamps common Formation of Ancestral Rockies Continental glaciation in Gondwana
318 —		Mississippian	Crinoids, lacy bryozoans, blastoids become abundant Renewed adaptive radiation following extinctions of many reef-builders	Reptiles evolve	Gymnosperms appear (may have evolved during Late Devonian)	Ouachita orogeny Widespread deposition of black shale
359 —	Devonian		Extinctions of many reef-building invertebrates near end of Devonian Reef building continues Eurypterids abundant	Amphibians evolve All major groups of fish present—Age of Fish	First seeds evolve Seedless vascular plants diversify	Widespread deposition of black shale Acadian orogeny Antler orogeny
416 —	Silurian		Major reef building Diversity of invertebrates remains high	Ostracoderms common Acanthodians, the first jawed fish, fish evolve	Early land plants— seedless vascular plants	Caledonian orogeny Extensive barrier reefs and evaporites
444 —	Ordovician		Extinctions of a variety of marine invertebrates near end of Ordovician Major adaptive radiation of all invertebrate groups Suspension feeders dominant	Ostracoderms diversify	Plants move to land?	Continental glaciation in Gondwana Taconic orogeny
488 —	Cambrian		Many trilobites become extinct near end of Cambrian Trilobites, brachiopods, and archaeocyathids are most abundant	Earliest vertebrates— jawless fish called ostracoderms		First Phanerozoic transgression (Sauk) onto North American craton
542 —						

Courtesy of Ken Higgs, Department of Geology, University College, Cork, Ireland

CHAPTER 13

PALEOZOIC LIFE HISTORY: VERTEBRATES AND PLANTS

Tetrapod trackway at Valentia Island, Ireland. These fossilized footprints, which are more than 365 million years old, are evidence of one of the earliest four-legged animals on land.

[OUTLINE]

Introduction

Vertebrate Evolution

Fish

Amphibians—Vertebrates Invade the Land

Evolution of the Reptiles—The Land Is Conquered

Plant Evolution

Perspective *Palynology: A Link Between Geology and Biology*

Silurian and Devonian Floras

Late Carboniferous and Permian Floras

Summary

At the end of this chapter, you will have learned that

- Vertebrates first evolved during the Cambrian Period, and fish diversified rapidly during the Paleozoic Era.

- Amphibians first appear in the fossil record during the Late Devonian, having made the transition from water to land, and became extremely abundant during the Pennsylvanian Period when coal-forming swamps were widespread.

- The evolution of the amniote egg allowed reptiles to colonize all parts of the land beginning in the Late Mississippian.

- The pelycosaurs or finback reptiles were the dominant reptiles during the Permian and were the ancestors to the therapsids or mammal-like reptiles.

- The earliest land plants are known from the Ordovician Period, whereas the oldest known vascular land plants first appear in the Middle Silurian.

- Seedless vascular plants, such as ferns, were very abundant during the Pennsylvanian Period.

- With the onset of arid conditions during the Permian Period, the gymnosperms became the dominant element of the world's flora.

Introduction

The discovery in 1992 of fossilized tetrapod footprints more than 365 million years old has forced paleontologists to rethink how and when animals emerged onto land. The Late Devonian trackway that Swiss geologist Iwan Stössel discovered that year on Valentia Island, off the southwest coast of Ireland, has helped shed light on the early evolution of tetrapods (from the Greek *tetra,* meaning "four," and *podos,* meaning "foot"). Given these footprints, geologists estimate that the creature was longer than three feet and had fairly large back legs. Furthermore, instead of walking on dry land, this animal was probably walking or wading around in a shallow, tropical stream, filled with aquatic vegetation and predatory fish. This hypothesis is based on the fact that the trackway showed no evidence of a tail being dragged behind it. Unfortunately, no bones are associated with the tracks to help reconstruct what this primitive tetrapod looked like.

One of the intriguing questions paleontologists ask is, Why did limbs evolve in the first place? It was probably not for walking on land. In fact, many scientists think aquatic limbs made it easier to move around in streams, lakes, or swamps that were choked with water plants or other debris. The transition from water to land and the role limbs played in this adaptation to a new environment is further discussed in this chapter.

Presently, there are many more questions about the evolution of the earliest tetrapods than there are answers. During the 1990s, only a few Devonian tetrapods were known, and the evolutionary transition from water to land involved the well-known phylogeny of *Eusthenopteron* (crossopterygian lobe-finned fish) to *Ichthyostega* (amphibian). Today, however, paleontologists have a more detailed knowledge of both of these groups, and with the recent discovery of additional fossils, are currently able to fill in the gaps between the fish and amphibians. Such discoveries now make it possible to infer the evolution of various features, leading to a more complete fish–amphibian phylogeny. In addition, as more paleoenvironmental and paleoecologic data and analysis from a variety of sites are

made available, a better understanding of the linkage between morphological changes and the environment is fast emerging. Furthermore, new technologies now provide the means to extract more and more detailed information from the fossils in ways that weren't possible previously.

In Chapter 12, we examined the Paleozoic history of invertebrates, beginning with the acquisition of hard parts and concluding with the massive Permian extinctions that claimed about 90% of all invertebrates and more than 65% of all amphibians and reptiles. In this chapter, we examine the Paleozoic evolutionary history of vertebrates and plants.

One of the striking parallels between plants and animals is that in making the transition from water to land, both plants and animals had to solve the same basic problems. For both groups, the method of reproduction proved the major barrier to expansion into the various terrestrial environments. With the evolution of the seed in plants and the amniote egg in animals, this limitation was removed, and both groups expanded into all terrestrial habitats.

Vertebrate Evolution

A **chordate** (phylum Chordata) is an animal that has, at least during part of its life cycle, a notochord, a dorsal hollow nerve cord, and gill slits (• Figure 13.1). **Vertebrates,** which are animals with backbones, are simply a subphylum of chordates.

The ancestors and early members of the phylum Chordata were soft-bodied organisms that left few fossils (• Figure 13.2). Consequently, we know little about the early evolutionary history of the chordates or vertebrates. Surprisingly, a close relationship exists between echinoderms (see Table 12.1) and chordates, and they may even have shared a common ancestor. This is because the development of the embryos of echinoderms and chordates are the same in both groups, that is, the cells divide by radial cleavage so that the cells are aligned directly above

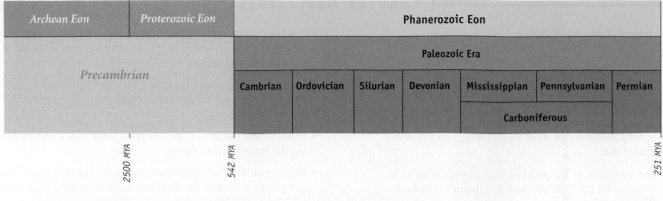

Archean Eon	Proterozoic Eon	Phanerozoic Eon							
		Paleozoic Era							
Precambrian		Cambrian	Ordovician	Silurian	Devonian	Mississippian	Pennsylvanian	Permian
						Carboniferous		

2500 MYA

542 MYA

251 MYA

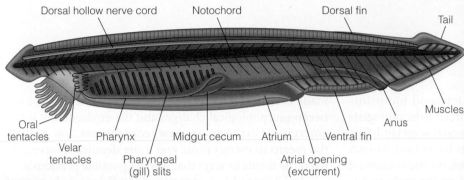

Dorsal hollow nerve cord   Notochord   Dorsal fin   Tail

Oral tentacles   Velar tentacles   Pharynx   Pharyngeal (gill) slits   Midgut cecum   Atrium   Ventral fin   Atrial opening (excurrent)   Anus   Muscles

• **Figure 13.1 Three Characteristics of a Chordate** The structure of the lancelet *Amphioxus* (approximately 5 cm. long) illustrates the three characteristics of a chordate: a notochord, a dorsal hollow nerve cord, and gill slits.

each other (• Figure 13.3a). In all other invertebrates, cells undergo spiral cleavage, which results in having cells nested between each other in successive rows (Figure 13.3b). Furthermore, the biochemistry of muscle activity and blood proteins, and the larval stages are similar in both echinoderms and chordates.

The evolutionary pathway to vertebrates thus appears to have taken place much earlier and more rapidly than many scientists have long thought. Based on fossil evidence and recent advances in molecular biology, one

scenario suggests that vertebrates evolved shortly after an ancestral chordate, probably resembling *Yunnanozoon*, and acquired a second set of genes. According to this hypothesis, a random mutation produced a duplicate set of genes, letting the ancestral vertebrate animal evolve entirely new body structures that proved to be evolutionarily advantageous. Not all scientists accept this hypothesis, and the origin of vertebrates is still hotly debated.

Fish

The most primitive vertebrates are fish, and some of the oldest fish remains are found in the Upper Cambrian Deadwood Formation in northeastern Wyoming (• Figure 13.4). Here, phosphatic scales and plates of *Anatolepis*, a primitive member of the class Agnatha (jawless fish), have been recovered from marine sediments. All known Cambrian and Ordovician fossil fish have been found in

• **Figure 13.3 Cell Cleavage**

• **Figure 13.2 *Yunnanozoon lividum*** Found in 525-million-year-old rocks in Yunnan province, China, *Yunnanozoon lividum*, a 5-cm-long animal, is one of the oldest known chordates.

Dr. Lars Ramskold

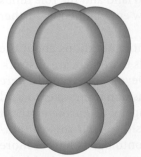

Radial Cleavage

a Arrangement of cells resulting from radial cleavage is characteristic of chordates and echinoderms. In this configuration, cells are directly above each other.

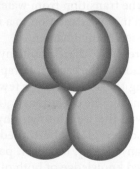

Spiral Cleavage

b Arrangement of cells resulting from spiral cleavage. In this arrangement, cells in successive rows are nested between each other. Spiral cleavage is characteristic of all invertebrates except echinoderms.

Phanerozoic Eon									
Mesozoic Era			Cenozoic Era						
Triassic	Jurassic	Cretaceous	Paleogene			Neogene		Quaternary	
			Paleocene	Eocene	Oligocene	Miocene	Pliocene	Pleistocene	Holocene

251 MYA

66 MYA

shallow, nearshore marine deposits, whereas the earliest nonmarine (freshwater) fish remains have been found in Silurian strata. This does not prove that fish originated in the oceans, but it does lend strong support to the idea.

John E. Repetski/USGS

• **Figure 13.4 Ostracoderm Fish Plate** A fragment of a plate from *Anatolepis* cf. *A. heintzi* from the Upper Cambrian Deadwood Formation of Wyoming. *Anatolepis* is one of the oldest known fish.

As a group, fish range from the Late Cambrian to the present (• Figure 13.5). The oldest and most primitive of the class Agnatha are the **ostracoderms,** whose name means "bony skin" (Table 13.1). These are armored, jawless fish that first evolved during the Late Cambrian, reached their zenith during the Silurian and Devonian, and then became extinct.

The majority of ostracoderms lived on the seafloor. *Hemicyclaspis* is a good example of a bottom-dwelling ostracoderm (• Figure 13.6a). Vertical scales allowed *Hemicyclaspis* to wiggle sideways, propelling itself along the seafloor, and the eyes on the top of its head allowed it to see such predators as cephalopods and jawed fish approaching from above. While moving along the sea bottom, it probably sucked up small bits of food and sediments through its jawless mouth. Another type of ostracoderm, represented by *Pteraspis*, was more elongated and probably an active

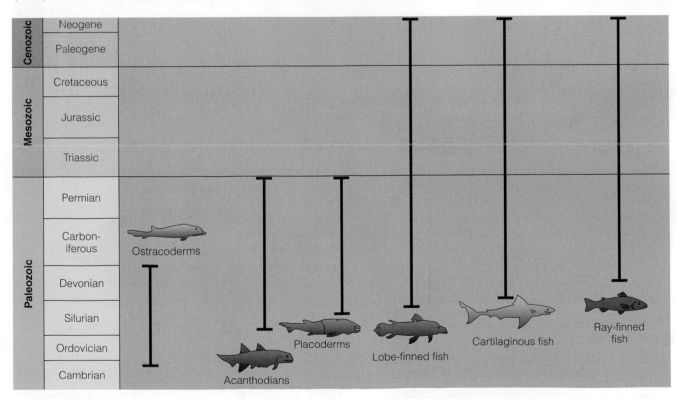

• **Figure 13.5 Geologic Ranges of the Major Fish Groups** Ostracoderms are early members of the class Agnatha (jawless fish). Acanthodians (class Acanthodii) are the first fish with jaws. Placoderms (class Placodermii) are armored, jawed fish. Lobe-finned (subclass Sarcopterygii) and ray-finned (subclass Actinopterygii) fish are members of the class Osteichthyes (bony fish), whereas cartilaginous fish belong to the class Chondrichthyes.

TABLE **13.1**   Brief Classification of Fish Groups Referred to in the Text

Classification	Geologic Range	Living Example
Class Agnatha (jawless fish)	Late Cambrian–Recent	Lamprey, hagfish
Early members of the class are called ostracoderms		No living ostracoderms
Class Acanthodii (the first fish with jaws)	Early Silurian–Permian	None
Class Placodermii (armored jawed fish)	Late Silurian–Permian	None
Class Chondrichthyes (cartilagenous fish)	Devonian–Recent	Sharks, rays, skates
Class Osteichthyes (bony fish)	Devonian–Recent	Tuna, perch, bass, pike, catfish, trout, salmon, lungfish, *Latimeria*
Subclass Actinopterygii (ray-finned fish)	Devonian–Recent	Tuna, perch, bass, pike, catfish, trout, salmon
Subclass Sarcopterygii (lobe-finned fish)	Devonian–Recent	Lungfish, *Latimeria*
Order Coelacanthimorpha	Devonian–Recent	*Latimeria*
Order Dipnoi	Devonian–Recent	Lungfish
Order Crossopterygii	Devonian–Permian	None
Suborder Rhipidistia	Devonian–Permian	None

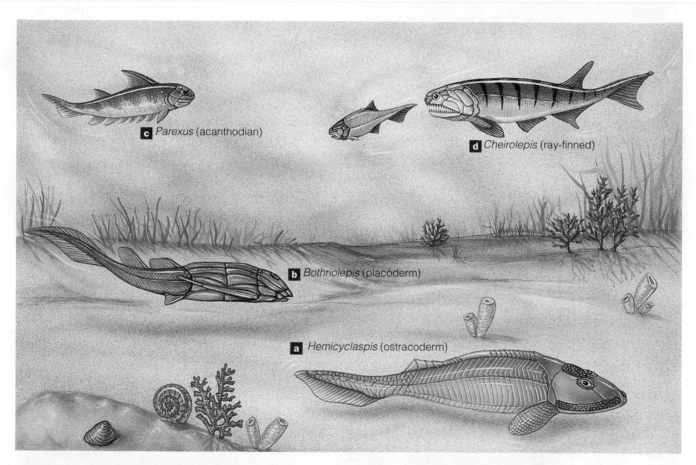

• Figure 13.6 **Devonian Seafloor** Recreation of a Devonian seafloor showing **a** an ostracoderm, **b** a placoderm, **c** an acanthodian, and **d** a ray-finned fish.

swimmer, although it also seemingly fed on small pieces of food that it was able to suck up.

The evolution of jaws was a major evolutionary advance among primitive vertebrates. Although their jawless ancestors could only feed on detritus, jawed fish could chew food and become active predators, thus opening many new ecologic niches.

The vertebrate jaw is an excellent example of evolutionary opportunism. Various studies suggest that the jaw originally evolved from the first two or three anterior gill

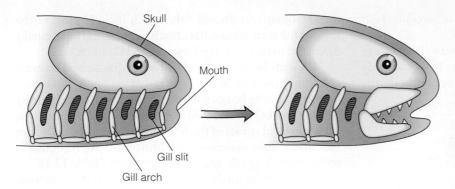

Skull

Mouth

Gill slit

Gill arch

• **Figure 13.7 Evolution of the Vertebrate Jaw** The evolution of the vertebrate jaw is thought to have begun from the modification of the first two or three anterior gill arches. This theory is based on the comparative anatomy of living vertebrates.

arches of jawless fish. Because the gills are soft, they are supported by gill arches of bone or cartilage. The evolution of the jaw may thus have been related to respiration rather than to feeding (• Figure 13.7). By evolving joints in the forward gill arches, jawless fish could open their mouths wider. Every time a fish opened and closed its mouth, it would pump more water past the gills, thereby increasing the oxygen intake. The modification from rigid to hinged forward gill arches let fish increase both their food consumption and oxygen intake, and the evolution of the jaw as a feeding structure rapidly followed.

The fossil remains of the first jawed fish are found in Lower Silurian rocks and belong to the *acanthodians* (class Acanthodii), a group of small, enigmatic fish characterized by large spines, paired fins, scales covering much of the body, jaws, teeth, and greatly reduced body armor (Figure 13.6c, Table 13.1). Although their relationship to other fish is not yet well established, many scientists think the acanthodians included the probable ancestors of the present-day bony and cartilaginous fish groups. The acanthodians were most abundant during the Devonian, declined in importance through the Carboniferous, and became extinct during the Permian.

The other jawed fish, the **placoderms** (class Placodermii), whose name means "plate-skinned," evolved during the Silurian. Placoderms were heavily armored, jawed fish that lived in both freshwater and the ocean, and, like the acanthodians, reached their peak of abundance and diversity during the Devonian.

The placoderms showed considerable variety, including small bottom dwellers (Figure 13.6b), as well as large major predators such as *Dunkleosteus,* a Late Devonian fish that lived in the midcontinental North American epeiric seas (• Figure 13.8a). It was by far the largest

fish of the time, reaching a length of more than 12 m. It had a heavily armored head and shoulder region, a huge jaw lined with razor-sharp bony teeth, and a flexible tail, all features consistent with its status as a ferocious predator.

Besides the abundant acanthodians, placoderms, and ostracoderms, other fish groups, such as the cartilaginous and bony fish, also evolved during the Devonian Period. Small wonder, then, that the Devonian is informally called the "Age of Fish," because all major fish groups were present during this time period.

The **cartilaginous fish** (class Chrondrichthyes) (Table 13.1), represented today by sharks, rays, and skates, first evolved during the Early Devonian, and by the Late Devonian, primitive marine sharks such as *Cladoselache* were abundant (Figure 13.8b). Cartilaginous fish have never been as numerous or as diverse as their cousins, the bony fish, but they were, and still are, important members of the marine vertebrate fauna.

Along with cartilaginous fish, the **bony fish** (class Osteichthyes) (Table 13.1) also first evolved during the Devonian. Because bony fish are the most varied and numerous of all the fishes, and because the amphibians evolved from them, their evolutionary history is particularly important. There are two groups of bony fish: the common *ray-finned fish* (subclass Actinopterygii) (Figure 13.8d) and

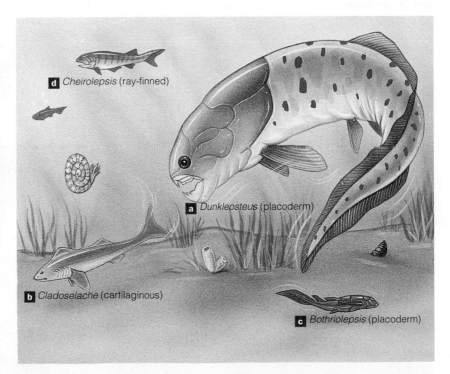

d *Cheirolepis* (ray-finned)

a *Dunkleosteus* (placoderm)

b *Cladoselache* (cartilaginous)

c *Bothriolepis* (placoderm)

• **Figure 13.8 Late Devonian Seascape** A Late Devonian marine scene from the midcontinent of North America. **a** The giant placoderm *Dunkleosteus* (length more than 12 m) is pursuing **b** the shark *Cladoselache* (length up to 1.2 m), a cartilaginous fish. Also shown are **c** the bottom-dwelling placoderm *Bothriolepis* and **d** the swimming ray-finned fish *Cheirolepis*, both of which reached a length of 40–50 cm.

the less familiar **lobe-finned fish** (subclass Sarcopterygii) (Table 13.1).

The term *ray-finned* refers to the way the fins are supported by thin bones that spread away from the body (• Figure 13.9a). From a modest freshwater beginning during

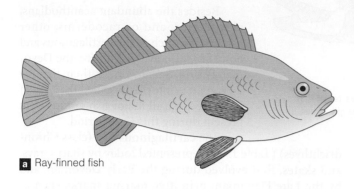

a Ray-finned fish

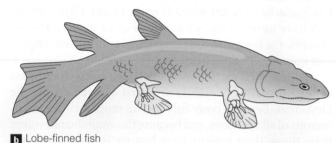

b Lobe-finned fish

• **Figure 13.9 Ray-finned and Lobe-finned Fish** Arrangement of fin bones for **a** a typical ray-finned fish and **b** a lobe-finned fish. The muscles extend into the fin of the lobe-finned fish, allowing greater flexibility of movement than that of the ray-finned fish.

• **Figure 13.10 Rhipidistian Crossopterygian** The crossopterygians are the group from which the amphibians are thought to have evolved. *Eusthenopteron*, a member of the rhipidistian crossopterygians, had an elongated body and paired fins that could be used for moving about on land.

the Devonian, ray-finned fish, which include most of the familiar fish such as trout, bass, perch, salmon, and tuna, rapidly diversified to dominate the Mesozoic and Cenozoic seas.

Present-day lobe-finned fish are characterized by muscular fins. The fins do not have radiating bones but rather have articulating bones with the fin attached to the body by a fleshy shaft (Figure 13.9b). Such an arrangement allows for a powerful stroke of the fin, making the fish an effective swimmer. Three orders of lobe-finned fish are recognized: *coelacanths, lungfish,* and *crossopterygians* (Table 13.1).

Coelacanths (order Coelacanthimorpha) are marine lobe-finned fish that evolved during the Middle Devonian and were thought to have gone extinct at the end of the Cretaceous. In 1938, however, a fisherman caught a coelacanth in the deep waters off Madagascar (see Figure 7.15a), and since then, several dozen more have been caught, both there and in Indonesia.

Lungfish (order Dipnoi) were fairly abundant during the Devonian, but today only three freshwater genera exist, one each in South America, Africa, and Australia. Their present-day distribution presumably reflects the Mesozoic breakup of Gondwana (see Chapter 14).

The "lung" of a modern-day lungfish is actually a modified swim bladder that most fish use for buoyancy in swimming. In lungfish, this structure absorbs oxygen, allowing them to breath air when the lakes or streams in which they live become stagnant and dry up. During such times, they burrow into the sediment to prevent dehydration and breath through their swim bladder until the stream begins flowing or the lake they were living in fills with water. When they are back in the water, lungfish then rely on gill respiration.

The **crossopterygians** (order Crossopterygii) are an important group of lobe-finned fish, because it is probably from them that amphibians evolved. However, the transition between crossopterygians and true amphibians is not as simple as it was once portrayed. Among the crossopterygians, the *rhipidistians* appear to be the ancestral group (Table 13.1). These fish, reaching lengths of over 2 m, were the dominant freshwater predators during the Late Paleozoic. *Eusthenopteron,* a good example of a rhipidistian crossopterygian and the classic example of the transitional form between fish and amphibians, had an elongated body that helped it move swiftly through the water, and paired, muscular fins that many scientists thought could be used for moving on land (• Figure 13.10). The structural

• Figure 13.11 **Similarities between Crossopterygians and Labyrinthodonts** Similarities between the crossopterygian lobe-finned fish and the labyrinthodont amphibians.

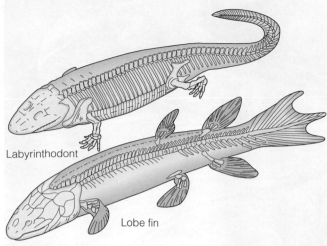

Labyrinthodont

Lobe fin

a Skeletal similarity.

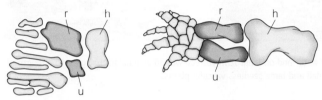

b Comparison of the limb bones of a crossopterygian *(left)* and amphibian *(right);* color identifies the bones (u = ulna, shown in blue, r = radius, mauve, h = humerus, gold) that the two groups have in common.

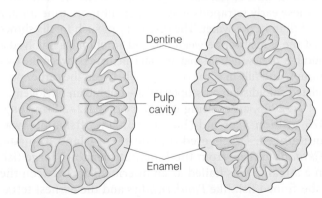

Dentine

Pulp cavity

Enamel

c Comparison of tooth cross sections shows the complex and distinctive structure found in both the crossopterygians *(left)* and labyrinthodont amphibians *(right).*

similarity between crossopterygian fish and the earliest amphibians is striking and one of the most widely cited examples of a transition from one major group to another (• Figure 13.11). However, recent discoveries of older lobe-finned fish and newly published findings of tetrapod-like fish are filling the gaps in the evolution from fish to tetrapods.

Before discussing this transition and the evolution of amphibians, it is useful to place the evolutionary history of Paleozoic fish in the larger context of Paleozoic

evolutionary events. Certainly, the evolution and diversification of jawed fish as well as eurypterids and ammonoids had a profound effect on the marine ecosystem. Previously defenseless organisms either evolved defensive mechanisms or suffered great losses, possibly even extinction.

Recall from Chapter 12 that trilobites experienced extinctions at the end of the Cambrian, recovered slightly during the Ordovician, and then declined greatly from the end of the Ordovician to final extinction at the end of the Permian. Perhaps their lightly calcified external covering made them easy prey for the rapidly evolving jawed fish and cephalopods.

Ostracoderms, although armored, would also have been easy prey for the swifter jawed fishes. Ostracoderms became extinct by the end of the Devonian, a time that coincides with the rapid evolution of jawed fish. Placoderms, like acanthodians, greatly decreased in abundance after the Devonian and became extinct by the end of the Paleozoic Era. In contrast, cartilaginous and ray-finned bony fish expanded during the Late Paleozoic, as did the ammonoid cephalopods (see Figure 12.15), the other major predators of the Late Paleozoic seas.

Amphibians—Vertebrates Invade the Land

Although amphibians were the first vertebrates to live on land, they were not the first land-living organisms. Land plants, which probably evolved from green algae, first evolved during the Ordovician. Furthermore, insects, millipedes, spiders, and even snails invaded the land before amphibians. Fossil evidence indicates that such land-dwelling arthropods as scorpions and flightless insects had evolved by at least the Devonian.

The transition from water to land required animals to surmount several barriers. The most critical were desiccation, reproduction, the effects of gravity, and the extraction of oxygen from the atmosphere by lungs rather than from water by gills. Up until the 1990s, the traditional evolutionary sequence had a rhipidistian crossopterygian, like *Eusthenopteron,* evolving into a primitive amphibian such as *Ichthyostega.* At that time, fossils of those two genera were about all paleontologists had to work with, and although there were gaps in morphology, the link between crossopterygians and these earliest amphibians was easy to see (Figure 13.11).

Crossopterygians already had a backbone and limbs that could be used for walking and lungs that could extract oxygen (Figure 13.11). The oldest known amphibian fossils, on the other hand, found in the Upper Devonian Old Red Sandstone of eastern Greenland and belonging to the genus *Ichthyostega,* had streamlined bodies, long tails, and fins along their backs, in addition to four legs, a strong backbone, a rib cage, and pelvic and pectoral girdles, all of which were structural adaptations for walking on land (• Figure 13.12). These earliest amphibians thus appear to

• Figure 13.12 **Late Devonian Landscape** Shown is *Ichthyostega*, both in the water and on land. *Ichthyostega* was an amphibian that grew to a length of approximately 1 m. The flora at this time was diverse, and consisted of a variety of small and large seedless vascular plants.

have inherited many characteristics from the crossopterygians with little modification (Figure 13.11).

Nevertheless, the question still remained as to when animals made the transition from water to land, and perhaps even more intriguing is the question of why limbs evolved in the first place. The answer to that question, as we mentioned in the Introduction, is that they probably were not for walking on land. In fact, the current thinking of many scientists is that aquatic limbs made it easier for animals to move around in streams, lakes, or swamps that were choked with water plants or other debris. The fossil evidence that began to emerge in the 1990s also seems to support this hypothesis.

Panderichthys, a large (up to 1.3 m long), Late Devonian (~380 million years ago) lobe-finned fish from Latvia, was essentially a contemporary of *Eusthenopteron*. It had a large tetrapod-like head with a pointed snout, dorsally located eyes, and modifications to that part of the skull related to the ear region. From paleoenvironmental evidence, *Panderichthys* lived in shallow tidal flats or estuaries, using its lobed fins to maneuver around in the shallow waters.

Fossils of *Acanthostega,* a tetrapod found in 360-million-year-old rocks from Greenland, reveal an animal that had limbs, but was clearly unable to walk on land. Paleontologist Jennifer Clack, who has recovered and analyzed hundreds of specimens of *Acanthostega,* points out that *Acanthostega*'s limbs were not strong enough to support its weight on land, and its rib cage was too small for the necessary muscles needed to hold its body off the ground. In addition, *Acanthostega* had gills and lungs, meaning that it could survive on land, but it was more suited for the water. Clack thinks *Acanthostega* used its limbs to maneuver around in

swampy, plant-filled waters, where swimming would be difficult and limbs would be an advantage.

Fragmentary fossils from other tetrapods living at about the same time as *Acanthostega* suggest, however, that some of these early tetrapods may have spent more time on dry land than in the water. The discovery of such fossils shows that the transition between fish and amphibians involved a number of new genera that are intermediate between the two groups, and fills in some of the gaps between the earlier postulated rhipidistian crossopterygian–amphibian phylogeny.

In 2006, an exciting discovery of a 1.2–2.8-m-long, 375-million-year-old (Late Devonian) "fishapod" was announced. Discovered on Ellesmere Island, Canada, *Tiktaalik roseae*, from the Inuktitut meaning "large fish in a stream," was hailed as an intermediary between the lobe-finned fish like *Panderichthys* and the earliest tetrapod, *Acanthostega* (• Figure 13.13a).

Tiktaalik roseae is truly a "fishapod" in that it has a mixture of both fish and tetrapod characteristics (Figure 13.13b). For example, it has gills and fish scales but also a broad skull, eyes on top of its head, a flexible neck and large rib cage that could support its body on land or in shallow water, and lungs, all of which are tetrapod features. What really excited scientists, however, was that *Tiktaalik roseae* has the beginnings of a true tetrapod forelimb, complete with functional wrist bones and five digits, as well as a modified ear region. Sedimentological evidence suggests *Tiktaalik roseae* lived in a shallow water habitat associated with the Late Devonian floodplains of Laurasia.

As previously mentioned, the oldest known amphibian, *Ichthyostega*, had skeletal features that

• **Figure 13.13** ***Tiktaalik roseae*** *Tiktaalik roseae*, a "fishapod," has been hailed as an intermediary between lobe-finned fish and tetrapods, because it has characteristics of both fish and tetrapods.

a Skeleton of *Taktaalik roseae*.

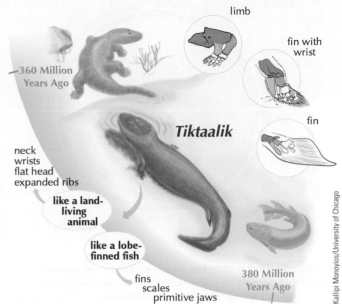

b Diagram illustrating how *Tiktaalik roseae* is a transitional species between lobe-finned fish and tetrapods.

• **Figure 13.14 Carboniferous Coal Swamp** The amphibian fauna during the Carboniferous was varied. Shown is the serpentlike *Dolichosoma* (foreground) and the large labyrinthodonts amphibian *Eryops*.

allowed it to spend its life on land. Because amphibians did not evolve until the Late Devonian, they were a minor element of the Devonian terrestrial ecosystem. Like other groups that moved into new and previously unoccupied niches, amphibians underwent rapid adaptive radiation and became abundant during the Carboniferous and Early Permian.

The Late Paleozoic amphibians did not at all resemble the familiar frogs, toads, newts, and salamanders that make up the modern amphibian fauna. Rather, they displayed a broad spectrum of sizes, shapes, and modes of life (• Figure 13.14). One group of amphibians was the **labyrinthodonts,** so named for the labyrinthine wrinkling and folding of the chewing surface of their teeth (Figure 13.11c). Most labyrinthodonts were large animals, as much as 2 m in length. These typically sluggish creatures lived in swamps and streams, eating fish, vegetation, insects, and other small amphibians (Figure 13.14).

Labyrinthodonts were abundant during the Carboniferous when swampy conditions were widespread (see Chapter 11) but soon declined in abundance during the Permian, perhaps in response to changing climatic conditions. Only a few species survived into the Triassic.

Evolution of the Reptiles— The Land Is Conquered

Amphibians were limited in colonizing the land because they had to return to water to lay their gelatinous eggs. The evolution of the **amniote egg** (• Figure 13.15) freed reptiles from this constraint. In such an egg, the developing embryo is surrounded by a liquid-filled sac called the *amnion* and

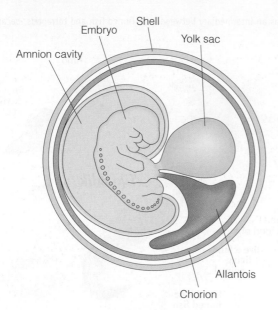

• Figure 13.15 **Amniote Egg** In an amniote egg, the embryo is surrounded by a liquid sac (amnion cavity) and provided with a food source (yolk sac) and waste sac (allantois). The evolution of the amniote egg freed reptiles from having to return to the water for reproduction and let them inhabit all parts of the land.

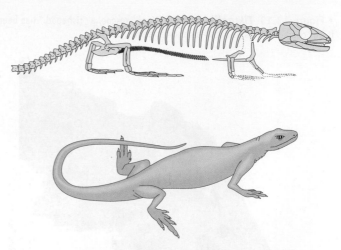

• Figure 13.16 **Hylonomus lyelli** Reconstruction and skeleton of one of the oldest known reptiles, *Hylonomus lyelli,* from the Pennsylvanian Period. Fossils of this animal have been collected from sediments that filled tree stumps. *Hylonomus lyelli* was about 30 cm long.

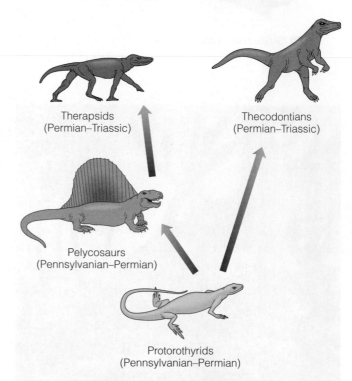

• Figure 13.17 **Evolutionary Relationship among the Paleozoic Reptiles**

provided with both a yolk, or food sac, and an allantois, or waste sac. In this way, the emerging reptile is in essence a miniature adult, bypassing the need for a larval stage in the water. The evolution of the amniote egg allowed vertebrates to colonize all parts of the land, because they no longer had to return to the water as part of their reproductive cycle.

Many of the differences between amphibians and reptiles are physiologic and are not preserved in the fossil record. Nevertheless, amphibians and reptiles differ sufficiently in skull structure, jawbones, ear location, and limb and vertebral construction to suggest that reptiles evolved from labyrinthodont ancestors by the Late Mississippian. This assessment is based on the discovery of a well-preserved fossil skeleton of the oldest known reptile, *Westlothiana,* and other fossil reptile skeletons from Late Mississippian-age rocks in Scotland.

Other early reptile fossils occur in the Lower Pennsylvanian Joggins Formation in Nova Scotia, Canada. Here, remains of *Hylonomus* are found in the sediments filling in tree trunks (• Figure 13.16). These earliest reptiles from Scotland and Canada were small and agile and fed largely on grubs and insects. They are loosely grouped together as **protorothyrids,** whose members include the earliest known reptiles (• Figure 13.17). During the Permian Period, reptiles diversified and began displacing many amphibians. The reptiles succeeded partly because of their advanced method of reproduction and their more advanced jaws and teeth, as well as their tough skin and scales to prevent desiccation, and their ability to move rapidly on land.

The **pelycosaurs,** or finback reptiles, evolved from the protorothyrids during the Pennsylvanian and were the dominant reptile group by the Early Permian. They evolved into a diverse assemblage of herbivores, exemplified by the herbivore *Edaphosaurus* and carnivores such as *Dimetrodon* (• Figure 13.18). An interesting feature of the

pelycosaurs is their sail. It was formed by vertebral spines that, in life, were covered with skin. The sail has been variously explained as a type of sexual display, a means of protection, and a display to look more ferocious. The current consensus seems to be that the sail served as some type of rudimentary thermoregulatory device, raising the reptile's temperature by catching the sun's rays or cooling it by facing the wind. Because pelycosaurs are considered to be the group from which therapsids evolved, it is interesting that they may have had some sort of body-temperature control.

The pelycosaurs became extinct during the Permian and were succeeded by the **therapsids,** mammal-like reptiles that evolved from the carnivorous pelycosaur ancestry

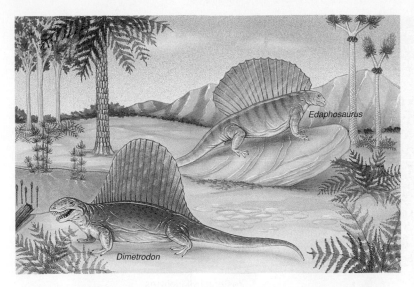

• Figure 13.18 **Pelycosaurs** Most pelycosaurs or finback reptiles have a characteristic sail on their back. One hypothesis explains the sail as a rudimentary thermoregulatory device. Other hypotheses are that it was a type of sexual display or a device to make the reptile look more intimidating. Shown here are the carnivore *Dimetrodon* and the herbivore *Edaphosaurus*.

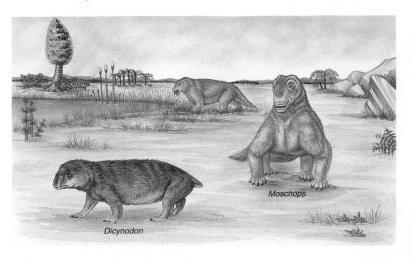

• Figure 13.19 **Therapsids** A Late Permian scene in southern Africa showing various therapsids including *Dicynodon* and *Moschops*. Many paleontologists think therapsids were endothermic and may have had a covering of fur, as shown here.

and rapidly diversified into herbivorous and carnivorous lineages (• Figure 13.19). Therapsids were small- to medium-sized animals that displayed the beginnings of many mammalian features: fewer bones in the skull, because many of the small skull bones were fused; enlarged lower jawbone; differentiation of teeth for various functions such as nipping, tearing, and chewing food; and more vertically placed legs for greater flexibility, as opposed to the way the legs sprawled out to the side in primitive reptiles.

Furthermore, many paleontologists think therapsids were *endothermic,* or warm-blooded, enabling them to maintain a constant internal body temperature. This characteristic would have let them expand into a variety of habitats, and indeed, the Permian rocks in which their fossil remains are found are distributed not only in low latitudes but also in middle and high latitudes as well.

As the Paleozoic Era came to an end, the therapsids constituted about 90% of the known reptile genera and occupied a wide range of ecologic niches. The mass extinctions that decimated the marine fauna at the close of the Paleozoic had an equally great effect on the terrestrial population (see Chapter 12). By the end of the Permian, about 90% of all marine invertebrate species were extinct, compared with more than two-thirds of all amphibians and reptiles. Plants, in contrast, apparently did not experience as great a turnover as animals.

Plant Evolution

When plants made the transition from water to land, they had to solve most of the same problems that animals did: desiccation, support, and the effects of gravity. Plants did so by evolving a variety of structural adaptations that were fundamental to the subsequent radiations and diversification that occurred during the Silurian, Devonian, and later periods (see Perspective) (Table 13.2). Most experts agree that the ancestors of land plants first evolved in a marine environment, then moved into a freshwater environment before finally progressing onto land. In this way, the differences in osmotic pressures between saltwater and freshwater were overcome while the plant was still in the water.

The higher land plants are divided into two major groups: nonvascular and vascular. Most land plants are **vascular,** meaning they have a tissue system of specialized cells for the movement of water and nutrients. The *nonvascular* plants, such as bryophytes (liverworts, hornworts, and mosses) and fungi, do not have these specialized cells and are typically small and usually live in low, moist areas.

The earliest land plants from the Middle to Late Ordovician were probably small and bryophyte-like in their overall organization (but not necessarily related to bryophytes). The evolution of vascular tissue in plants was an important step because it allowed transport of nutrients and water.

Discoveries of probable vascular plant megafossils and characteristic spores indicate to many paleontologists that vascular plants evolved well before the Middle Silurian. Sheets of cuticle-like cells—that is, the cells that cover the surface of present-day land plants—and tetrahedral clusters that closely resemble the spore tetrahedrals of primitive land plants have been reported from Middle to Upper Ordovician rocks from western Libya and elsewhere (• Figure 13.20).

The ancestor of terrestrial vascular plants was probably some type of green alga. Although no fossil record of the transition from green algae to terrestrial vascular plants has

Major Events in the Evolution of Land Plants. The Devonian Period was a time of rapid evolution for the land plants. Major events were the appearance of leaves, heterospory, secondary growth and the emergence of seeds

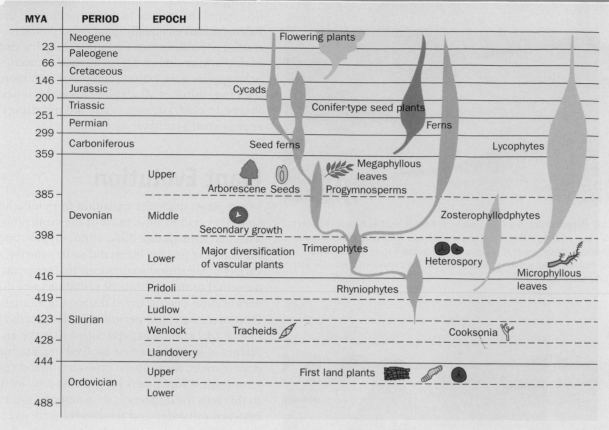

MYA	PERIOD	EPOCH
23	Neogene	
66	Paleogene	
146	Cretaceous	
200	Jurassic	
251	Triassic	
299	Permian	
359	Carboniferous	
385	Devonian	Upper
398	Devonian	Middle
416	Devonian	Lower
419		Pridoli
423	Silurian	Ludlow
428	Silurian	Wenlock
444		Llandovery
488	Ordovician	Upper / Lower

Flowering plants

Cycads

Conifer-type seed plants

Ferns

Seed ferns

Lycophytes

Arborescene   Seeds   Megaphyllous leaves

Progymnosperms

Secondary growth

Zosterophyllodphytes

Major diversification of vascular plants   Trimerophytes

Heterospory

Microphyllous leaves

Rhyniophytes

Tracheids

Cooksonia

First land plants

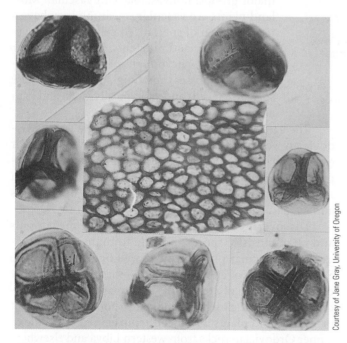

Courtesy of Jane Gray, University of Oregon

• Figure 13.20 **Upper Ordovician Plant Spores and Cells** Fossils that closely resemble the spore tetrahedrals of primitive land plants. The sheet of cuticle-like cells *(center)* is from the Upper Ordovician Melez Chograne Formation of Libya. The others are from the Upper Ordovician Djeffara Formation of Libya.

been found, comparison of their physiology reveals a strong link. Primitive *seedless vascular plants* (discussed later in this chapter), such as ferns, resemble green algae in their pigmentation, important metabolic enzymes, and type of rapid reproductive cycle. Furthermore, green algae are one of the few plant groups to have made the transition from saltwater to freshwater.

The evolution of terrestrial vascular plants from an aquatic, probably green algal ancestry was accompanied by various modifications that let them occupy this new, harsh environment. Besides the primary function of transporting water and nutrients throughout a plant, vascular tissue also provides some support for the plant body. Additional strength is derived from the organic compounds *lignin* and *cellulose,* found throughout a plant's walls.

The problem of desiccation was circumvented by the evolution of *cutin,* an organic compound found in the outer-wall layers of plants. Cutin also provides additional resistance to oxidation, the effects of ultraviolet light, and the entry of parasites.

Roots evolved in response to the need to collect water and nutrients from the soil and to help anchor the plant in the ground. The evolution of *leaves* from tiny outgrowths on the stem or from branch systems provided plants with an efficient light-gathering system for photosynthesis.

Perspective

Palynology: A Link between Geology and Biology

Palynology is the study of organic micro-fossils called palynomorphs. These include such familiar items as spores and pollen (both of which cause allergies for many people) (see Figure 17.7), but also such unfamiliar organisms as acritarchs (see Figure 12.10), dinoflagellates (marine and freshwater single-celled phytoplankton, some species of which in high concentrations make shellfish toxic to humans) (see Figure 15.4c), chitinozoans (vase-shaped microfossils of unknown origin), scolecodonts (jaws of marine annelid worms), and microscopic colonial algae. Fossil palynomorphs are extremely resistant to decay and are extracted from sedimentary rocks by dissolving the rocks in various acids.

A specialty of palynology that attracts many biologists and geologists is the study of spores and pollen. By examining the fossil spores and pollen preserved in sedimentary rocks, palynologists can tell when plants colonized Earth's surface (Figure 13.20), which in turn influenced weathering and erosion rates, soil formation, and changes in the composition of atmospheric gases. Furthermore, because plants are not particularly common as fossils, the study of spores and pollen can frequently reveal the time and region for the origin and extinction of various plant groups.

Analysis of fossil spores and pollen is used to solve many geologic and biologic problems. One of the more important uses of fossil spores and pollen is determining the geologic age of sedimentary rocks. Because spores and pollen are microscopic,

resistant to decay, deposited in both marine and terrestrial environments, extremely abundant, and are part of the life cycle of plants (Figure 13.23), they are very useful for determining age. Many spore and pollen species have narrow time ranges that make them excellent guide fossils.

Rocks considered lacking in fossils by paleontologists who were looking only for megafossils often actually contain thousands, even millions, of fossil spores or pollen grains that allow palynologists to date these so-called unfossiliferous rocks. Fossil spores and pollen are also useful in determining the environment and climate in the past. Their presence in sedimentary rocks helps palynologists determine what plants and trees were living at the time, even if the fossils of those plants

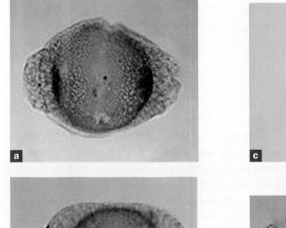
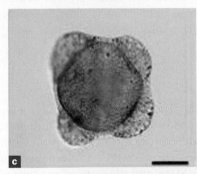
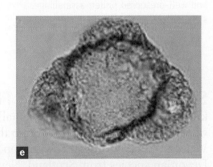
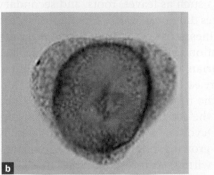
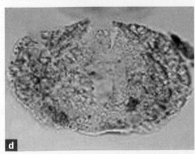
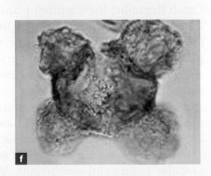

Figure 1 Normal and abnormal pollen grains of *Klausipollenites schaubergeri* from Nedubrovo, Russia **a–c** and the Junggar Basin, China **d–f**. Normal pollen grains of *Klausipollenites schaubergeri* are bisaccate, that is they have a central body with two air sacs **a** and **d**, whereas abnormal forms have three or four air sacs **b–c**; **e–f**. The high percentage of abnormal pollen grains of *Klausipollenites schaubergeri* recovered from sedimentary rocks spanning the Permian-Trissic boundary in Russia and China have been used as evidence of increased atmospheric pollution related to the Permian mass extinctions. The scale bar in **c** represents 25 μm (25/1000 mm). Photos courtesy of Clinton B. Foster, Petroleum and Marine Division, Geoscience Australia, and Sergey A. Afonin, formerly of the Palaeontological Institute, Moscow, Russia.

and trees are not preserved. Plants are very sensitive to climatic changes, and by plotting the abundance and types of vegetation present, based on their preserved spores and pollen, palynologists can determine past climates and changes in climates through time (see Figure 17.7).

An interesting study by C. B. Foster and S. A. Afonin published in 2005 related morphologic abnormalities in gymnosperm pollen grains to deteriorating atmospheric conditions around the Permian–Triassic boundary—that is, at the time of the global Permian extinction event. One cause of morphologic abnormalities in living gymnosperms and angiosperms is environmental stress on the parent plant. Plants are sensitive indicators of environmental change, and studies have shown that pollen wall abnormalities are caused by atmospheric pollution, ultraviolet-B (UV-B) radiation, or a combination of both.

Processing of samples from nonmarine sedimentary rock sequences that span the Permian–Triassic boundary from the Junggar Basin, Xinjiang Province, China, and Nedubrovo, Russia, yielded a diverse, abundant, and well-preserved pollen assemblage. Examination of the assemblage revealed

that greater than 3% of the pollen showed morphologic abnormalities (Figure 1). Based on this finding, the authors concluded these abnormalities were the result of atmospheric pollution, including increased UV-B radiation, caused by extensive volcanism. Other studies show similar pollen morphologic abnormalities in end-Permian sediments ranging from Greenland to Australia.

Following these studies, a 2007 publication discussed the role that the Siberian Traps flood basalts may have played in causing widespread ozone depletion, thus allowing an increase in UV-B radiation at the end of the Permian. The authors of this study used a two-dimensional atmospheric chemistry–transport model to assess the impact of the Siberian Traps eruption in altering the end-Permian stratospheric ozone layer. The authors also noted the increase in pollen morphologic abnormalities reported earlier in the literature.

A recent study (2008) using microspectroscopy examined the biochemistry of the outer wall of the spores of *Lycopodium magellanicum* and *Lycopodium annotinum* from present-day high-northern- and southern-latitude localities. Their study indicates that the concentrations of two

ultraviolet-B-absorbing compounds reflects variations in ultraviolet-B radiation. As such, these compounds may prove valuable for reconstructing past variations in stratospheric ozone, which screens the surface biota from harmful UV-B radiation.

Recall from Chapter 12 that one of the possible causes of the Permian extinction was increased global warming caused by higher atmospheric carbon dioxide levels and volcanic activity. One way of increasing atmospheric carbon dioxide levels is from volcanic and continental fissure eruptions. Although the cause of the end-Permian mass extinctions is still unresolved, the results of these three studies point out some of the exciting research being conducted on climate change and its impact on Earth's biota.

From this short survey of palynology, it can be seen that the study of spores and pollen provides a tremendous amount of information about the vegetation in the past, its evolution, the type of climate, and changes in climate. In addition, spores and pollen are very useful for relative dating of rocks and correlating marine and terrestrial rocks, both regionally and globally.

Silurian and Devonian Floras

The earliest known vascular land plants are small Y-shaped stems assigned to the genus *Cooksonia* from the Middle Silurian of Wales and Ireland. Together with Upper Silurian and Lower Devonian species from Scotland, New York State, and the Czech Republic, these earliest plants were small, simple, leafless stalks with a spore-producing structure at the tip (• Figure 13.21). They are known as **seedless vascular plants** because they did not produce seeds. They also did not have a true root system. A *rhizome,* the underground part of the stem, transferred water from the soil to the plant and anchored the plant in the ground. The sedimentary rocks in which these plant fossils are found indicate that they lived in low, wet, marshy, freshwater environments.

An interesting parallel can be seen between seedless vascular plants and amphibians. When they made the transition from water to land, both plants and animals had to overcome the same problems such a transition involved. Both groups, while successful, nevertheless required a source of water in order to reproduce. In the case of amphibians, their gelatinous egg had to remain moist, and the

seedless vascular plants required water for the sperm to travel through to reach the egg.

From this simple beginning, the seedless vascular plants evolved many of the major structural features characteristic of modern plants such as leaves, roots, and secondary growth. These features did not all evolve simultaneously but rather at different times, a pattern known as *mosaic evolution.* This diversification and adaptive radiation took place during the Late Silurian and Early Devonian and resulted in a tremendous increase in diversity (• Figure 13.22). During the Devonian, the number of plant genera remained about the same, yet the composition of the flora changed. Whereas the Early Devonian landscape was dominated by relatively small, low-growing, bog-dwelling types of plants, the Late Devonian witnessed forests of large, tree-sized plants up to 10 m tall.

In addition to the diverse seedless vascular plant flora of the Late Devonian, another significant floral event took place. The evolution of the seed at this time liberated land plants from their dependence on moist conditions and allowed them to spread over all parts of the land.

• **Figure 13.21** *Cooksonia* The earliest known fertile land plant was *Cooksonia*, seen in this fossil from the Upper Silurian of South Wales. *Cooksonia* consisted of upright, branched stems terminating in sporangia (spore-producing structures). It also had a resistant cuticle and produced spores typical of a vascular plant. These plants probably lived in moist environments such as mud flats. This specimen is 1.49 cm long.

Seedless vascular plants require moisture for successful fertilization because the sperm must travel to the egg on the surface of the gamete-bearing plant (gametophyte) to produce a successful spore-generating plant (sporophyte). Without moisture, the sperm would dry out before reaching the egg (• Figure 13.23a). In the seed method of reproduction, the spores are not released to the environment, as they are in the seedless vascular plants, but are retained on the spore-bearing plant, where they grow into the male and female forms of the gamete-bearing generation.

In the case of the **gymnosperms,** or flowerless seed plants, these are the male and female cones (Figure 13.23b). The male cone produces pollen, which contains the sperm and has a waxy coating to prevent desiccation, and the egg, or embryonic seed, is contained in the female cone. After fertilization, the seed then develops into a mature, cone-bearing plant. In this way, the need for a moist environment for the gametophyte generation is solved. The significance of this development is that seed plants, like reptiles, were no longer restricted to wet areas but were free to migrate into previously unoccupied dry environments.

Before seed plants evolved, an intermediate evolutionary step was necessary. This was the development of *heterospory*, whereby a species produces two types of spores: a large one (megaspore) that gives rise to the female gamete-bearing plant and a small one (microspore) that produces the male gamete-bearing plant (Table 13.2). The earliest evidence of heterospory is found in the Early Devonian plant *Chaleuria cirrosa*, which produced spores of two distinct sizes (• Figure 13.24). The appearance of heterospory was followed several million years later by

• **Figure 13.22 Early Devonian Landscape** Reconstruction of an Early Devonian landscape showing some of the earliest land plants.

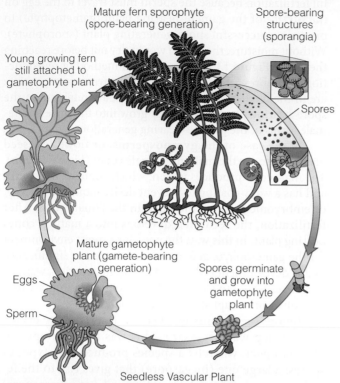

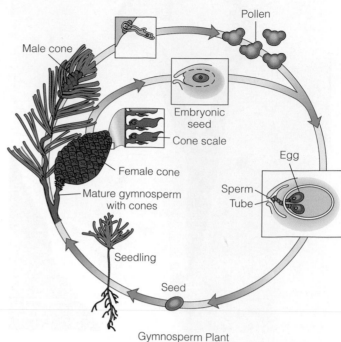

Gymnosperm Plant

a Generalized life history of a seedless vascular plant. The mature sporophyte plant produces spores, which, upon germination, grow into small gametophyte plants that produce sperm and eggs. The fertilized eggs grow into the spore-producing mature plant, and the sporophyte–gametophyte life cycle begins again.

b Generalized life history of a gymnosperm plant. The mature plant bears both male cones that produce sperm-bearing pollen grains and female cones that contain embryonic seeds. Pollen grains are transported to the female cones by the wind. Fertilization occurs when the sperm moves through a moist tube growing from the pollen grain and unites with the embryonic seed, which then grows into a cone-bearing mature plant.

Courtesy of Patricia G. Gensel, University of North Carolina

• Figure 13.24 *Chaleuria cirrosa* Specimen of the Early Devonian plant *Chaleuria cirrosa* from New Brunswick, Canada. This plant was heterosporous, meaning that it produced two sizes of spores.

the emergence of progymnosperms—Middle and Late Devonian plants with fernlike reproductive habits and a gymnosperm anatomy—which gave rise in the Late Devonian to such other gymnosperm groups as the seed ferns and conifer-type seed plants (Table 13.2).

Although the seedless vascular plants dominated the flora of the Carboniferous coal-forming swamps, the gymnosperms made up an important element of the Late Paleozoic flora, particularly in the nonswampy areas.

Late Carboniferous and Permian Floras

As discussed earlier, the rocks of the Pennsylvanian Period (Late Carboniferous) are the major source of the world's coal. Coal results from the alteration of plant remains accumulating in low, swampy areas. The geologic and geographic conditions of the Pennsylvanian were ideal for the growth of seedless vascular plants, and consequently these coal swamps had a very diverse flora (• Figure 13.25).

It is evident from the fossil record that whereas the Early Carboniferous flora was similar to its Late Devonian counterpart, a great deal of evolutionary experimentation was taking place that would lead to the highly successful Late Paleozoic flora of the coal swamps and adjacent habitats. Among the seedless vascular plants, the *lycopsids* and *sphenopsids* were the most important coal-forming groups of the Pennsylvanian Period.

The lycopsids were present during the Devonian, chiefly as small plants, but by the Pennsylvanian, they were the dominant element of the coal swamps, achieving heights up to 30 m in such genera as *Lepidodendron* and *Sigillaria*. The Pennsylvanian lycopsid trees are interesting because they lacked branches except at their top, which had elongate leaves similar to the individual palm leaf of today. As the trees grew, the leaves were replaced from the top, leaving prominent and characteristic rows or spirals of scars on the trunk.

• Figure 13.25 **Pennsylvanian Coal Swamp** Reconstruction of a Pennsylvanian coal swamp with its characteristic vegetation of seedless vascular plants.

The sphenopsids, the other important coal-forming plant group, are characterized by being jointed and having horizontal underground stem-bearing roots. Many of these plants, such as *Calamites,* average 5 to 6 m tall. Small, seedless vascular plants and seed ferns formed a thick undergrowth or ground cover beneath these treelike plants. Living sphenopsids include the horsetail *(Equisetum)* or scouring rushes (• Figure 13.26).

Not all plants were restricted to the coal-forming swamps. Among those plants that occupied higher and drier ground were some of the *cordaites,* a group of tall gymnosperm trees that grew up to 50 m high and probably formed vast forests (• Figure 13.27). Another important nonswamp dweller was

Glossopteris, the famous plant so abundant in Gondwana (see Figure 3.1), whose distribution is cited as critical evidence that the continents have moved through time.

The floras that were abundant during the Pennsylvanian persisted into the Permian, but because of climatic and geologic changes resulting from tectonic events (see Chapter 11), they declined in abundance and importance. By the end of the Permian, the cordaites became extinct, and the lycopsids and sphenopsids were reduced to mostly small, creeping forms. Gymnosperms, with lifestyles more suited to the warmer and drier Permian climates, diversified and came to dominate the Permian, Triassic, and Jurassic landscapes.

• Figure 13.26 **Horsetail *Equisetum*** Living sphenopsids include the horsetail *Equisetum*.

• Figure 13.27 **Late Carboniferous Cordaite Forest** Cordaites were a group of gymnosperm trees that grew up to 50 m tall.

SUMMARY

Table 13.3 summarizes the major evolutionary and geologic events of the Paleozoic Era and shows their relationships to each other.

- Chordates are characterized by a notochord, dorsal hollow nerve cord, and gill slits. The earliest chordates were soft-bodied organisms that were rarely fossilized. Vertebrates are a subphylum of the chordates.

- Fish are the earliest known vertebrates with their first fossil occurrence in Upper Cambrian rocks. They have had a long and varied history, including jawless and jawed armored forms (ostracoderms and placoderms), cartilaginous forms, and bony forms. It is from the lobe-finned fish that amphibians evolved.

- The link between crossopterygian lobe-finned fish and the earliest amphibians is convincing and includes a close similarity of bone and tooth structures. However, new fossil discoveries show that the transition between the two groups is more complicated than originally hypothesized, and includes several intermediate forms.

- Amphibians evolved during the Late Devonian, with labyrinthodont amphibians becoming the dominant terrestrial vertebrate animals during the Carboniferous.

- The Late Mississippian marks the earliest fossil record of reptiles. The evolution of an amniote egg was the critical factor that allowed reptiles to completely colonize the land.

- Pelycosaurs were the dominant reptiles during the Early Permian, whereas therapsids dominated the landscape for the rest of the Permian Period.

- In making the transition from water to land, plants had to overcome the same basic problems as animals—namely, desiccation, reproduction, and gravity.

- The earliest fossil record of land plants is from Middle to Upper Ordovician rocks. These plants were probably small and bryophyte-like in their overall organization.

- The evolution of vascular tissue was an important event in plant evolution as it allowed nutrients and water to be transported throughout the plant and provided the plant with additional support.

- The ancestor of terrestrial vascular plants was probably some type of green alga based on such similarities as pigmentation, metabolic enzymes, and the same type of reproductive cycle.

- The earliest seedless vascular plants were small, leafless stalks with spore-producing structures on their tips. From this simple beginning, plants evolved many of the major structural features characteristic of today's plants.

- By the end of the Devonian Period, forests with tree-sized plants up to 10 m had evolved. The Late Devonian also witnessed the evolution of the flowerless seed plants (gymnosperms), whose reproductive style freed them from having to stay near water.

- The Carboniferous Period was a time of vast coal swamps, where conditions were ideal for the seedless vascular plants. With the onset of more arid conditions during the Permian, the gymnosperms became the dominant element of the world's flora.

IMPORTANT TERMS

amniote egg, p. 269
bony fish, p. 265
cartilaginous fish, p. 265
chordate, p. 261
crossopterygian, p. 266
gymnosperm, p. 275

labyrinthodont, p. 269
lobe-finned fish, p. 266
ostracoderm, p. 263
pelycosaur, p. 270
placoderm, p. 265
protorothyrid, p. 270

seedless vascular plant, p. 274
therapsid, p. 270
vascular plant, p. 271
vertebrate, p. 261

REVIEW QUESTIONS

1. Labyrinthodonts are
   a. _____ plants; b. _____ fish; c. _____ amphibians; d. _____ reptiles; e. _____ none of the previous answers.

2. Based on similarity of embryo cell division, which invertebrate phylum is most closely allied with the chordates?
   a. _____ Mollusca; b. _____ Echinodermata; c. _____ Porifera; d. _____ Annelida; e. _____ Arthropoda.

3. Which of the following groups did amphibians evolve from?
   a. _____ Coelacanths; b. _____ Ray-finned fish; c. _____ Lobe-finned fish; d. _____ Pelycosaurs; e. _____ Therapsids.

4. Which was the first plant group that did not require a wet area for the reproductive part of its life cycle?
   a. _____ Seedless vascular; b. _____ Naked seedless; c. _____ Gymnosperms; d. _____ Angiosperms; e. _____ Flowering.

5. Which plant group first successfully invaded land?
   a. _____ Seedless vascular; b. _____ Gymnosperms;
   c. _____ Naked seed bearing; d. _____ Angiosperms;
   e. _____ Flowering.

6. An organism must possess which of the following during at least part of its life cycle to be classified a chordate?
   a. _____ Notochord, dorsal solid nerve cord, lungs;
   b. _____ Vertebrae, dorsal hollow nerve cord, gill slits;
   c. _____ Vertebrae, dorsal hollow nerve cord, lungs;
   d. _____ Notochord, ventral solid nerve cord, lungs;
   e. _____ Notochord, dorsal hollow nerve cord, gill slits.

7. Which of the following is thought by many scientists to be endothermic?
   a. _____ Crossopterygians; b. _____ Therapsids;
   c. _____ Amphibians; d. _____ Rhipidistians; e. _____ Labyrinthodonts.

8. Which reptile group gave rise to the mammals?
   a. _____ Labyrinthodonts; b. _____ Acanthodians;
   c. _____ Pelycosaurs; d. _____ Protothyrids; e. _____ Therapsids.

9. The Age of Fish is which period?
   a. _____ Cambrian; b. _____ Silurian; c. _____ Devonian; d. _____ Pennsylvanian; e. _____ Permian.

10. Which evolutionary innovation allowed reptiles to colonize all of the land?
    a. _____ Tear ducts; b. _____ Additional bones in the jaw; c. _____ The middle-ear bones; d. _____ An egg that contained a food-and-waste sac and surrounded the embryo in a fluid-filled sac; e. _____ Limbs and a backbone capable of supporting the animals on land.

11. Pelycosaurs are
    a. _____ jawless fish; b. _____ jawed armored fish;
    c. _____ reptiles; d. _____ amphibians; e. _____ plants.

12. Which algal group was the probable ancestor to vascular plants?
    a. _____ Yellow; b. _____ Blue-green; c. _____ Red;
    d. _____ Brown; e. _____ Green.

13. In which period were amphibians and seedless vascular plants most abundant?
    a. _____ Permian; b. _____ Pennsylvanian; c. _____ Mississippian; d. _____ Silurian; e. _____ Cambrian.

14. The discovery of *Tiktaalik roseae* is significant because it is
    a. _____ the ancestor of modern reptiles; b. _____ an intermediate between lobe-finned fish and amphibians; c. _____ the first vascular land plant; d. _____ the "missing-link" between amphibians and reptiles; e. _____ the oldest known fish.

15. What are the major differences between the seedless vascular plants and the gymnosperms, and why are these differences significant in terms of exploiting the terrestrial environment?

16. Outline the evolutionary history of fish.

17. Describe the problems that had to be overcome before organisms could inhabit and completely colonize the land.

18. Discuss the significance and possible advantages of the pelycosaur sail.

19. Why were the reptiles so much more successful at extending their habitat than the amphibians?

20. Discuss how changing geologic conditions affected the evolution of plants and vertebrates.

APPLY YOUR KNOWLEDGE

1. Based on what you know about Carboniferous geology (Chapter 12), why was this time period so advantageous to the evolution of both plants and amphibians?

2. Because of the recent controversy concerning the teaching of evolution in the public schools, your local school board has asked you to make a 30-minute presentation on the history of life and how such a history is evidence that evolution is a valid scientific theory. With so much material to cover and so little time, you decide to focus on the Paleozoic evolutionary history of vertebrates. You have chosen the Paleozoic Era because during this time the stage was set, so to speak, for the later evolution of dinosaurs, birds, and mammals, groups with which most citizens are familiar. What features of the Paleozoic vertebrate fossil record would you emphasize, and how would you go about convincing the school board that evolution has taken place? How would the recent discovery of *Tiktaalik roseae* help your argument?

Table 13.3
Major Evolutionary and Geologic Events of the Paleozoic Era

Age (millions of years)	Geologic Period		Invertebrates	Vertebrates	Plants	Major Geologic Events
251	Permian		Largest mass extinction event to affect the invertebrates	Acanthodians, placoderms, and pelycosaurs become extinct — Therapsids and pelycosaurs the most abundant reptiles	Gymnosperms diverse and abundant	Formation of Pangaea — Alleghenian orogeny — Hercynian orogeny
299	Carboniferous	Pennsylvanian	Fusulinids diversify	Amphibians abundant and diverse	Coal swamps with flora of seedless vascular plants and gymnosperms	Coal-forming swamps common — Formation of Ancestral Rockies — Continental glaciation in Gondwana
318		Mississippian	Crinoids, lacy bryozoans, blastoids become abundant — Renewed adaptive radiation following extinctions of many reef-builders	Reptiles evolve	Gymnosperms appear (may have evolved during Late Devonian)	Ouachita orogeny — Widespread deposition of black shale
359	Devonian		Extinctions of many reef-building invertebrates near end of Devonian — Reef building continues — Eurypterids abundant	Amphibians evolve — All major groups of fish present—Age of Fish	First seeds evolve — Seedless vascular plants diversify	Widespread deposition of black shale — Acadian orogeny — Antler orogeny
416	Silurian		Major reef building — Diversity of invertebrates remains high	Ostracoderms common — Acanthodians, the first jawed fish, evolve	Early land plants— seedless vascular plants	Caledonian orogeny — Extensive barrier reefs and evaporites
444	Ordovician		Extinctions of a variety of marine invertebrates near end of Ordovician — Major adaptive radiation of all invertebrate groups — Suspension feeders dominant	Ostracoderms diversify	Plants move to land?	Continental glaciation in Gondwana — Taconic orogeny
488	Cambrian		Many trilobites become extinct near end of Cambrian — Trilobites, brachiopods, and archaeocyathids are most abundant	Earliest vertebrates— jawless fish called ostracoderms		First Phanerozoic transgression (Sauk) onto North American craton
542						

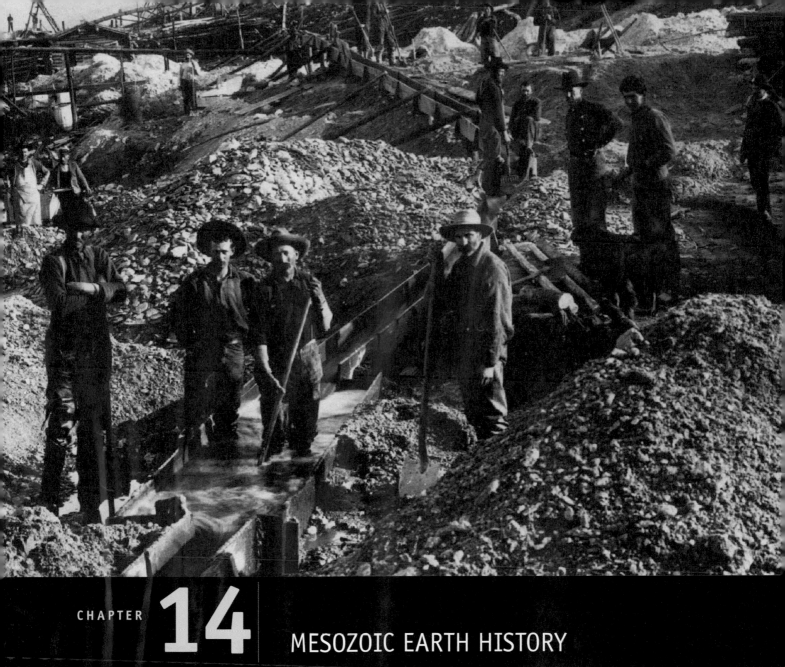

© Bettmann/Corbis

CHAPTER 14

MESOZOIC EARTH HISTORY

By 1852, during the California gold rush, mining operations were well under way on the American River near Sacramento.

[OUTLINE]

Introduction

The Breakup of Pangaea

    The Effects of the Breakup of Pangaea on Global Climates and Ocean Circulation Patterns

Mesozoic History of North America

Continental Interior

Eastern Coastal Region

Gulf Coastal Region

Western Region

    Mesozoic Tectonics

    Mesozoic Sedimentation

    Perspective Petrified Forest National Park

What Role Did Accretion of Terranes Play in the Growth of Western North America?

Mesozoic Mineral Resources

Summary

At the end of this chapter, you will have learned that

- The Mesozoic breakup of Pangaea profoundly affected geologic and biologic events.

- During the Triassic Period most of North America was above sea level.

- During the Jurassic Period a seaway flooded the interior of western North America.

- A global rise in sea level during the Cretaceous Period resulted in an enormous interior seaway that extended from the Gulf of

Mexico to the Arctic Ocean and divided North America into two large landmasses.

- Western North America was affected by four interrelated orogenies that took place along an oceanic–continental plate boundary.

- Terrane accretion played an important role in the Mesozoic geologic history of western North America.

- Coal, petroleum, uranium, and copper deposits are major Mesozoic mineral resources.

Introduction

Approximately 150 to 210 million years after the emplacement of massive plutons created the Sierra Nevada (Nevadan orogeny), gold was discovered at Sutter's Mill on the South Fork of the American River at Coloma, California. On January 24, 1848, James Marshall, a carpenter building a sawmill for John Sutter, found bits of the glittering metal in the mill's tailrace. Soon settlements throughout the state were completely abandoned as word of the chance for instant riches spread throughout California.

Within a year after the news of the gold discovery reached the East Coast, the Sutter's Mill area was swarming with more than 80,000 prospectors, all hoping to make their fortune. At least 250,000 gold seekers worked the Sutter's Mill area, and although most were Americans, prospectors came from all over the world, even as far away as China. Most thought the gold was simply waiting to be taken and didn't realize prospecting was hard work.

No one thought much about the consequences of so many people converging on the Sutter's Mill area, all intent on making easy money. In fact, life in the mining camps was extremely hard and expensive. The shop owners and traders frequently made more money than the prospectors. In reality, only a few prospectors ever hit it big or were even moderately successful. The rest barely eked out a living until they eventually abandoned their dream and went home.

Although many prospectors searched for the source of the gold, or the mother lode, the gold that they recovered, however, was mostly in the form of placer deposits (deposits of sand and gravel containing gold particles large enough to be recovered by panning). Weathering of gold-bearing igneous rocks and mechanical separation of minerals by density during stream transport forms placer deposits. Panning, a common method for recovering placer deposits, is performed by dipping a shallow pan into a streamed to capture sediment. The material is then swirled around and the lighter material is poured off. Gold, being about six times heavier than most sand grains and rock chips, concentrates on the bottom of the pan and can then be picked out.

Whereas some prospectors dug $30,000 worth of gold dust a week out of a single claim and some gold was found practically sitting on the surface of the ground, most of

this easy gold was recovered very early during the gold rush. Most prospectors barely made a living wage working their claims. Nevertheless, during the five years from 1848 to 1853 that constituted the gold rush proper, more than $200 million in gold was extracted.

The Mesozoic Era (251 to 66 million years ago) was an important time in Earth history. The major geologic event was the breakup of Pangaea, which affected oceanic and climatic circulation patterns and influenced the evolution of the terrestrial and marine biotas. Other important Mesozoic geologic events resulting from plate movement include the origin of the Atlantic Ocean basin and the Rocky Mountains, the accumulation of vast salt deposits that eventually formed salt domes adjacent to which oil and natural gas were trapped, and the emplacement of huge batholiths that account for the origin of various mineral resources, including the gold that fueled the California gold rush of the mid-1800s.

The Breakup of Pangaea

Just as the formation of Pangaea influenced geologic and biologic events during the Paleozoic, the breakup of this supercontinent had profound geologic and biologic effects during the Mesozoic. The movement of continents affected the global climatic and oceanic regimes as well as the climates of the individual continents. Populations became isolated or were brought into contact with other populations, leading to evolutionary changes in the biota. So great was the effect of this breakup on the world that it forms the central theme of this chapter.

Because of the magnetic anomalies preserved in the oceanic crust (see Figure 3.12), geologists have a very good record of the history of Pangaea's breakup, and the direction of movement of the various continents during the Mesozoic and Cenozoic eras.

Pangaea's breakup began with rifting between Laurasia and Gondwana during the Triassic (• Figure 14.1a). By the end of the Triassic, the newly formed and expanding Atlantic Ocean separated North America from Africa. This was

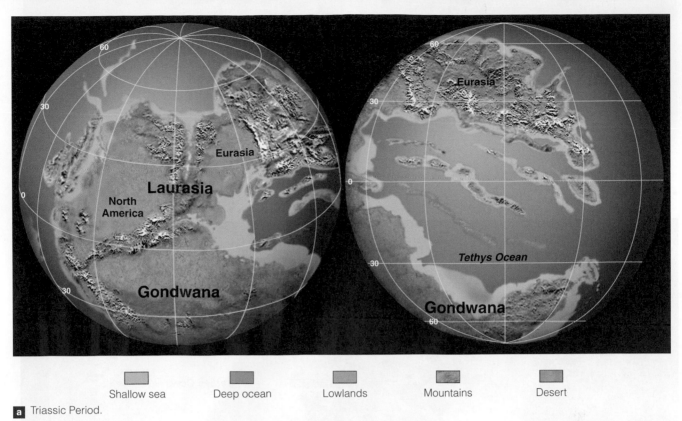

ⓐ Triassic Period.

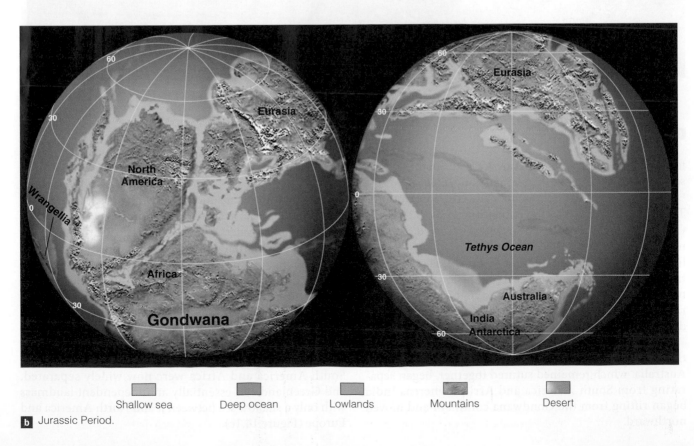

ⓑ Jurassic Period.

Archean Eon	Proterozoic Eon	Phanerozoic Eon							
Precambrian		Paleozoic Era							
		Cambrian	Ordovician	Silurian	Devonian	Mississippian	Pennsylvanian	Permian	
						Carboniferous			

2500 MYA 542 MYA 251 MYA

• Figure 14.1 (cont.)

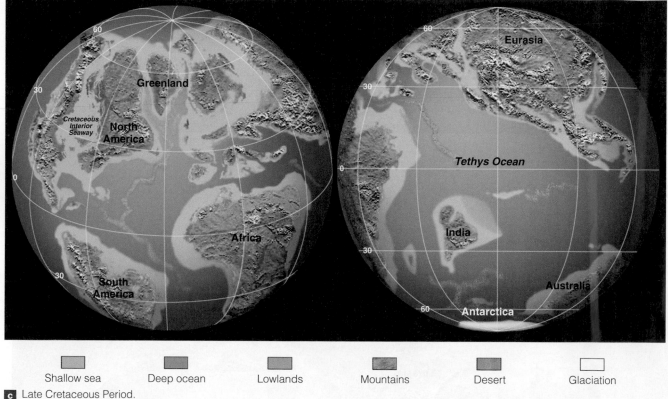

Shallow sea Deep ocean Lowlands Mountains Desert Glaciation

c Late Cretaceous Period.

followed by the rifting of North America from South America, sometime during the Late Triassic and Early Jurassic.

Separation of the continents allowed water from the Tethys Sea to flow into the expanding central Atlantic Ocean, whereas Pacific Ocean waters flowed into the newly formed Gulf of Mexico, which at that time was little more than a restricted bay (• Figure 14.2). During that time, these areas were located in the low tropical latitudes, where high temperatures and high rates of evaporation were ideal for the formation of thick evaporite deposits.

The initial breakup of Gondwana took place during the Late Triassic and Jurassic periods. Antarctica and Australia, which remained sutured together, began separating from South America and Africa, whereas India began rifting from the Gondwana continent and moved northward.

South America and Africa began rifting apart during the Jurassic (Figure 14.1b) and the subsequent separation of these two continents formed a narrow basin where thick evaporite deposits accumulated from the evaporation of southern ocean waters (Figure 14.2). During this time, the eastern end of the Tethys Sea began closing as a result of the clockwise rotation of Laurasia and the northward movement of Africa. This narrow Late Jurassic and Cretaceous seaway between Africa and Europe was the forerunner of the present Mediterranean Sea.

By the end of the Cretaceous, Australia and Antarctica had detached from each other, and India had moved into the low southern latitudes and was nearly to the equator. South America and Africa were now widely separated, and Greenland was essentially an independent landmass with only a shallow sea between it and North America and Europe (Figure 14.1c).

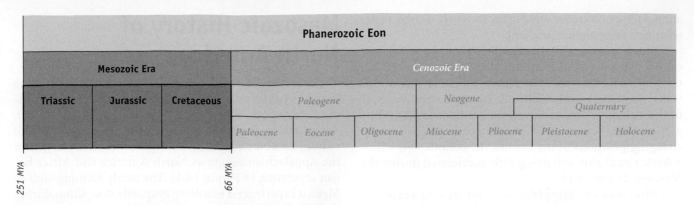

Phanerozoic Eon										
Mesozoic Era			Cenozoic Era							
Triassic	Jurassic	Cretaceous	Paleogene			Neogene		Quaternary		
			Paleocene	Eocene	Oligocene	Miocene	Pliocene	Pleistocene		Holocene

251 MYA

99 MYA

the Appalachians of eastern North America and Africa began separating (Figure 14.1b). The north-trending Mid-Atlantic Ridge could have evolved from this rift. During the Late Triassic and Jurassic, North America separated

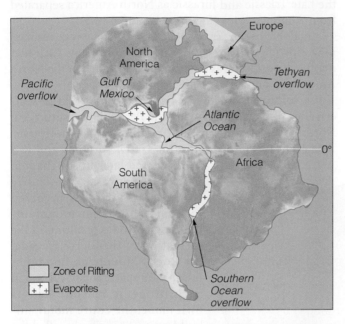

• Figure 14.2 **Evaporite Formation** Evaporites accumulated in shallow basins as Pangaea broke apart during the Early Mesozoic. Water from the Tethys Sea flowed into the central Atlantic Ocean, and water from the Pacific Ocean flowed into the newly formed Gulf of Mexico. Marine water from the south flowed into the southern Atlantic Ocean.

A global rise in sea level during the Cretaceous resulted in worldwide transgressions onto the continents. Higher heat flow and rapid expansion of oceanic ridges were responsible for these transgressions. By the Middle Cretaceous, sea level was probably as high as at any time since the Ordovician, and about one-third of the present land area was inundated by epeiric seas (Figure 14.1c).

The final stage in Pangaea's breakup occurred during the Cenozoic. During this time, Australia continued moving northward, and Greenland was completely separated from Europe and North America and formed a separate landmass.

The Effects of the Breakup of Pangaea on Global Climates and Ocean Circulation Patterns By the end of the Permian Period, Pangaea extended from pole to pole, covered about one-fourth of Earth's surface, and was surrounded by Panthalassa, a

global ocean that encompassed approximately 300 degrees of longitude (see Figure 11.2b). Such a configuration exerted tremendous influence on the world's climate and resulted in generally arid conditions over large parts of Pangaea's interior.

The world's climates result from the complex interaction between wind and ocean currents and the location and topography of the continents. In general, dry climates occur on large landmasses in areas remote from sources of moisture and where barriers to moist air exist, such as mountain ranges. Wet climates occur near large bodies of water or where winds can carry moist air over land.

Past climatic conditions can be inferred from the distribution of climate-sensitive deposits. Evaporite deposits result when evaporation exceeds precipitation. Although sand dunes and red beds may form locally in humid regions, they are characteristic of arid regions. Coal forms in both warm and cool humid climates. Vegetation that is eventually converted into coal requires at least a good seasonal water supply; thus, coal deposits are indicative of humid conditions.

Widespread Triassic evaporites, red beds, and desert dunes in the low and middle latitudes of North and South America, Europe, and Africa indicate dry climates in those regions, whereas coal deposits are found mainly in the high latitudes, indicating humid conditions (Figure 14.1a). These high latitude coals are analogous to today's Scottish peat bog or Canadian muskeg. The lands bordering the Tethys Sea were probably dominated by seasonal monsoon rains resulting from the warm, moist winds and warm oceanic currents impinging against the east-facing coast of Pangaea.

The temperature gradient between the tropics and the poles also affects oceanic and atmospheric circulation. The greater the temperature difference between the tropics and the poles, the steeper the temperature gradient, and thus, the faster the circulation of the oceans and atmosphere. Oceans absorb about 90% of the solar radiation they receive, whereas continents absorb only about 50%, even less if they are snow covered. The rest of the solar radiation is reflected back into space. Areas dominated by seas are warmer than those dominated by continents. By knowing the distribution of continents and ocean basins, geologists can generally estimate the average annual temperature for any region on Earth, as well as determine a temperature gradient.

The breakup of Pangaea during the Late Triassic caused the global temperature gradient to increase because the Northern Hemisphere continents moved farther northward, displacing higher-latitude ocean waters. Because of the steeper global temperature gradient produced by a decrease in temperature in the high latitudes and the changing positions of the continents, oceanic and atmospheric circulation patterns greatly accelerated during the Mesozoic (• Figure 14.3).

Although the temperature gradient and seasonality on land were increasing during the Jurassic and Cretaceous, the middle-and higher-latitude oceans were still quite warm because warm waters from the Tethys Sea were circulating to the higher latitudes. The result was a relatively equable worldwide climate through the end of the Cretaceous.

Mesozoic History of North America

The beginning of the Mesozoic Era was essentially the same in terms of tectonism and sedimentation as the preceding Permian Period in North America (see Figure 11.11). Terrestrial sedimentation continued over much of the craton, and block faulting and igneous activity began in the Appalachian region as North America and Africa began separating (• Figure 14.4). The newly forming Gulf of Mexico experienced extensive evaporite deposition during the Late Triassic and Jurassic as North America separated from South America (Figures 14.2 and • 14.5).

A global rise in sea level during the Cretaceous resulted in worldwide transgressions onto the continents such that marine deposition was continuous over much of the North American Cordilleran (• Figure 14.6).

A volcanic island arc system that formed off the western edge of the craton during the Permian was sutured to North America sometime during the Permian or Triassic. This event is referred to as the *Sonoma orogeny* and will be discussed later in the chapter. During the Jurassic, the entire Cordilleran area was involved in a series of major mountain-building episodes resulting in the formation of the Sierra Nevada, the Rocky Mountains, and other lesser mountain ranges. Although each orogenic episode has its own name, the entire mountain-building event is simply called the *Cordilleran orogeny* (also discussed later in this chapter). With this simplified overview of the Mesozoic history of North America in mind, we will now examine the specific regions of the continent.

Continental Interior

Recall that the history of the North American craton is divided into unconformity-bound sequences reflecting advances and retreats of epeiric seas over the craton (see Figure 10.3). Although these

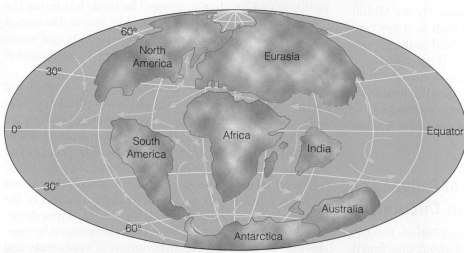

a Triassic Period

b Cretaceous Period

• Figure 14.3 **Mesozoic Oceanic Circulation Patterns** Oceanic circulation evolved from a a simple pattern in a single ocean (Panthalassa) with a single continent (Pangaea) to b a more complex pattern in the newly formed oceans of the Cretaceous Period.

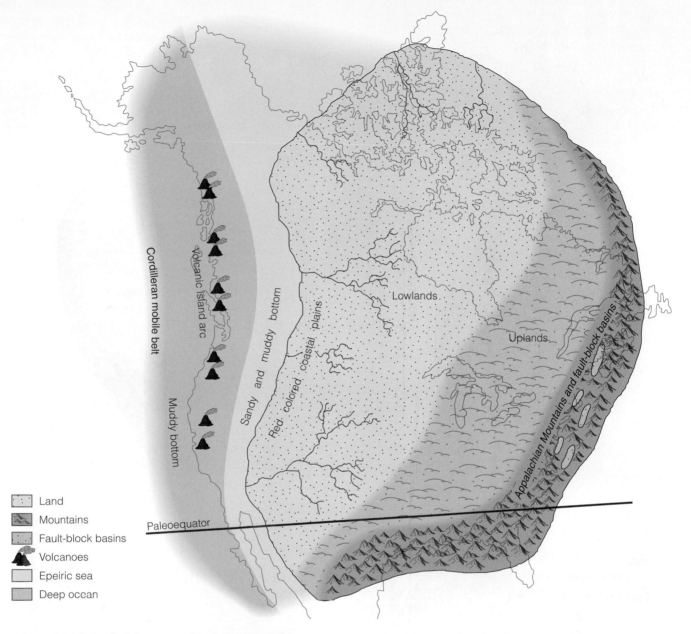

Land
Mountains
Fault-block basins
Volcanoes
Epeiric sea
Deep ocean

Cordilleran mobile belt
Volcanic island arc
Muddy bottom
Sandy and muddy bottom
Red colored coastal plains
Lowlands
Uplands
Appalachian Mountains and fault-block basins
Paleoequator

• **Figure 14.4 Triassic Paleogeography of North America**

transgressions and regressions played a major role in the Paleozoic geologic history of the continent, they were not as important during the Mesozoic. Most of the continental interior during the Mesozoic was well above sea level and was not inundated by epeiric seas. Consequently, the two Mesozoic cratonic sequences, the *Absaroka Sequence* (Late Mississippian to Early Jurassic) and *Zuni Sequence* (Early Jurassic to Early Paleocene) (see Figure 10.3), are not treated separately here; instead, we will examine the Mesozoic history of the three continental margin regions of North America.

Eastern Coastal Region

During the Early and Middle Triassic, coarse detrital sediments derived from erosion of the recently uplifted

Appalachians (Alleghenian orogeny) filled the various intermontane basins and spread over the surrounding areas. As weathering and erosion continued during the Mesozoic, this once lofty mountain system was reduced to a low-lying plain.

During the Late Triassic, the first stage in the breakup of Pangaea began with North America separating from Africa. Fault-block basins developed in response to upwelling magma beneath Pangaea in a zone stretching from present-day Nova Scotia to North Carolina (• Figure 14.7). Erosion of the fault-block mountains filled the adjacent basins with great quantities (up to 6000 m) of poorly sorted red nonmarine detrital sediments known as the *Newark Group*.

Reptiles roamed along the margins of the various lakes and streams that formed in these basins, leaving their footprints and trackways in the soft sediments (• Figure 14.8). Although the Newark Group rocks contain numerous

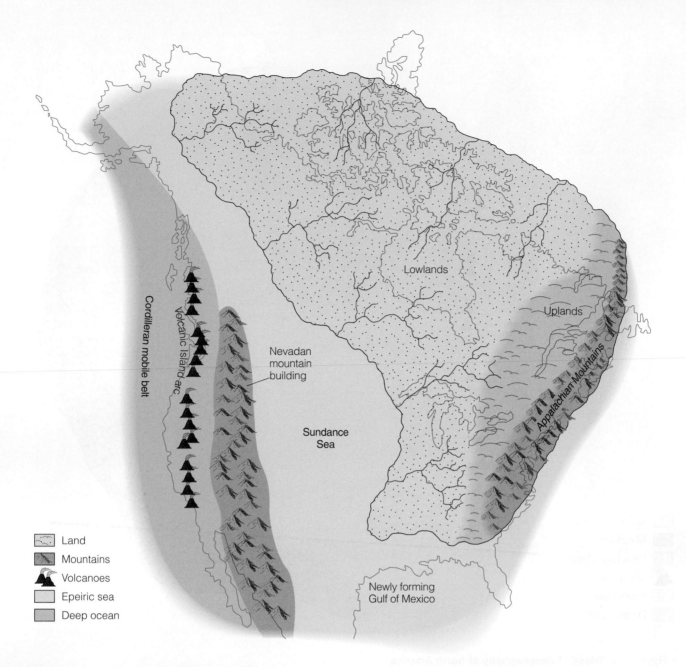

Cordilleran mobile belt

Volcanic Island arc

Nevadan mountain building

Sundance Sea

Lowlands

Uplands

Appalachian Mountains

Newly forming Gulf of Mexico

• **Figure 14.5 Jurassic Paleogeography of North America**

dinosaur footprints, they are almost completely devoid of dinosaur bones! The Newark Group is mostly Late Triassic in age, but in some areas deposition did not begin until the Early Jurassic.

Concurrent with sedimentation in the fault-block basins were extensive lava flows that blanketed the basin floors, as well as intrusions of numerous dikes and sills. The most famous intrusion is the prominent Palisades sill along the Hudson River in the New York–New Jersey area (Figure 14.7d).

As the Atlantic Ocean grew, rifting ceased along the eastern margin of North America, and this once active convergent plate margin became a passive, trailing continental margin. The fault-block mountains produced by this rifting continued to erode during the Jurassic and Early Cretaceous until all that was left was an area of low-relief.

The sediments produced by this erosion contributed to the growing eastern continental shelf. During the Cretaceous Period, the Appalachian region was reelevated and once again shed sediments onto the continental shelf, forming a gently dipping, seaward-thickening wedge of rocks up to 3000 m thick. These rocks are currently exposed in a belt extending from Long Island, New York, to Georgia.

Gulf Coastal Region

The Gulf Coastal region was above sea level until the Late Triassic (Figure 14.4). As North America separated from South America during the Late Triassic and Early Jurassic, the Gulf of Mexico began to form (Figure 14.5).

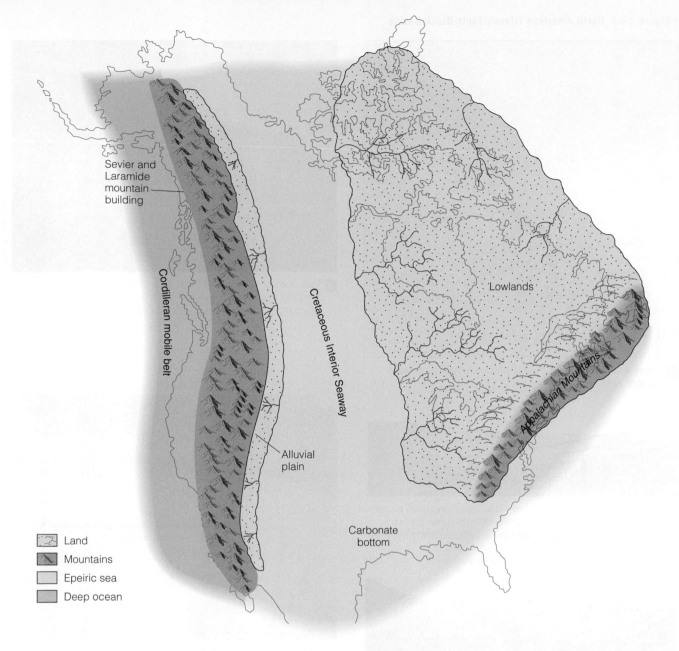

Sevier and
Laramide
mountain
building

Cordilleran mobile belt

Cretaceous Interior Seaway

Lowlands

Appalachian Mountains

Alluvial
plain

Carbonate
bottom

Land

Mountains

Epeiric sea

Deep ocean

• **Figure 14.6 Cretaceous Paleogeography of North America**

With oceanic waters flowing into this newly formed, shallow, restricted basin, conditions were ideal for evaporite formation. More than 1000 m of evaporites were precipitated at this time, and most geologists think that these Jurassic evaporites are the source for the Cenozoic salt domes found today in the Gulf of Mexico and southern Louisiana. The history of these salt domes and their associated petroleum accumulations is discussed in Chapter 16.

By the Late Jurassic, circulation in the Gulf of Mexico was less restricted, and evaporite deposition ended. Normal marine conditions returned to the area with alternating transgressing and regressing seas. The resulting sediments were covered and buried by thousands of meters of Cretaceous and Cenozoic sediments.

During the Cretaceous, the Gulf Coastal region, like the rest of the continental margin, was flooded by northward-transgressing seas (Figure 14.6). As a result, nearshore sandstones are overlain by finer sediments characteristic of deeper waters. Following an extensive regression at the end of the Early Cretaceous, a major transgression began, during which a wide seaway extended from the Arctic Ocean to the Gulf of Mexico (Figure 14.6). Sediments deposited in the Gulf Coastal region during the Cretaceous formed a seaward-thickening wedge.

Reefs were also widespread in the Gulf Coastal region during the Cretaceous and were composed primarily of bivalves called *rudists* (see Chapter 15). Because of their high porosity and permeability, rudistoid reefs make excellent petroleum reservoirs. A good example of a Cretaceous

• Figure 14.7 **North American Triassic Fault-Block Basins**

a Areas where Triassic fault-block basin deposits crop out in eastern North America.

d Palisades of the Hudson River. This sill was one of many intruded into the Newark sediments during the Late Triassic rifting that marked the separation of North America from Africa.

Courtesy of John Favre

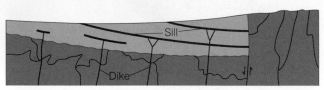

b After the Appalachians were eroded to a low-lying plain by the Middle Triassic, fault-block basins such as this one (shown in cross section) formed as a result of Late Triassic rifting between North America and Africa.

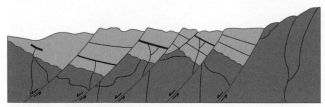

c These valleys accumulated tremendous thickness of sediments and were themselves broken by a complex of normal faults during rifting.

Courtesy of Dinosaur State Park

• Figure 14.8 **Triassic Newark Group Reptile Footprints** Reptile tracks in the Triassic Newark Group were uncovered during the excavation for a new state building in Hartford, Connecticut. Because the tracks were so spectacular, the building site was moved, and the excavation was designated as a state park.

reef complex occurs in Texas where the reef trend strongly influenced the carbonate platform deposition of the region (• Figure 14.9). The facies patterns of these Cretaceous carbonate rocks are as complex as those in the major barrier-reef systems of the Paleozoic Era.

Western Region

Mesozoic Tectonics The Mesozoic geologic history of the North American Cordilleran mobile belt is very complex, involving the eastward subduction of the oceanic

Farallon plate under the continental North American plate. Activity along this oceanic–continental convergent plate boundary resulted in an eastward movement of deformation. This orogenic activity progressively affected the trench and continental slope, the continental shelf, and the cratonic margin, causing a thickening of the continental crust. In addition, the accretion of terranes and microplates along the western margin of North America also played a significant role in the Mesozoic tectonic history of this area.

Except for the Late Devonian–Early Mississippian Antler orogeny (see Figure 11.15), the Cordilleran region

• Figure 14.9 Early Cretaceous Shelf-Margin Sedimentary Facies

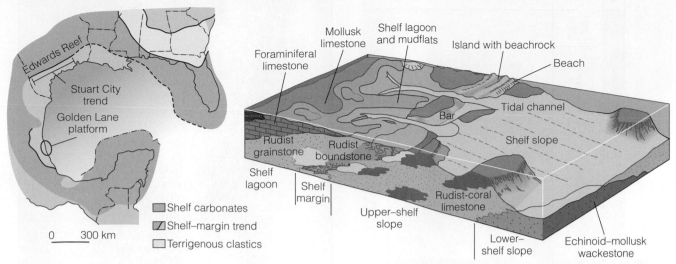

Shelf carbonates
Shelf–margin trend
Terrigenous clastics

0   300 km

a Early Cretaceous shelf-margin facies around the Gulf of Mexico Basin. The reef trend is shown as a black line.

b Reconstruction of the depositional environment and facies changes across the Stuart City reef trend, south Texas.

of North America experienced little tectonism during the Paleozoic. However, an island arc and ocean basin formed off the western North American craton during the Permian (Figure 14.4), followed by subduction of an oceanic plate beneath the island arc and the thrusting of oceanic and island arc rocks eastward against the craton margin (• Figure 14.10). This event, similar to the preceding Antler orogeny, and known as the **Sonoma orogeny,** occurred at or near the Permian–Triassic boundary. Like the Antler orogeny, it resulted in the suturing of island-arc terranes along the western edge of North America and the formation of a landmass called *Sonomia.*

Following the Late Paleozoic–Early Mesozoic destruction of the volcanic island arc during the Sonoma orogeny, the western margin of North America became an oceanic–continental convergent plate boundary. During the Late Triassic, a steeply dipping subduction zone developed along the western margin of North America in response to the westward movement of North America over the Farallon plate. This newly created oceanic–continental plate boundary controlled Cordilleran tectonics for the rest of the Mesozoic and for most of the Cenozoic Era; this subduction zone marks the beginning of the modern circum-Pacific orogenic system.

The general term **Cordilleran orogeny** is applied to the mountain-building activity that began during the Jurassic and continued into the Cenozoic (• Figure 14.11). The Cordilleran orogeny consisted of a series of individually named, but interrelated, mountain-building events, or pulses, that occurred in different regions at different times but overlapped to some extent. Most of this Cordilleran orogenic activity is related to the continued westward movement of the North American plate as it overrode the Farallon plate and its history is highly complex.

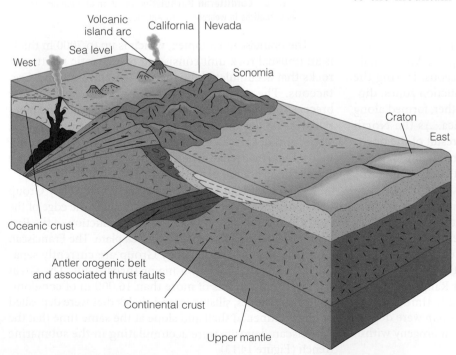

• Figure 14.10 **Sonoma Orogeny** Tectonic activity that culminated in the Permian–Triassic Sonoma orogeny in western North America. The Sonoma orogeny was the result of a collision between the southwestern margin of North America and an island arc system.

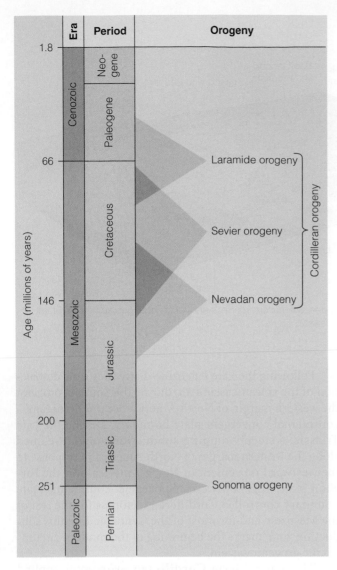

• Figure 14.11 **Mesozoic Cordilleran Orogenies** Mesozoic orogenies occurring in the Cordilleran mobile belt.

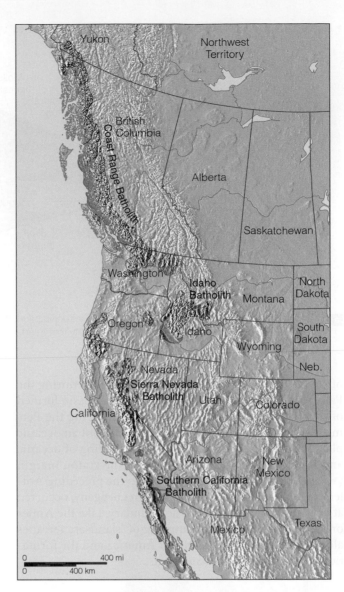

• Figure 14.12 **Cordilleran Batholiths** Location of Jurassic and Cretaceous batholiths in western North America.

The first pulse of the Cordilleran orogeny, the **Nevadan orogeny** (Figure 14.11), began during the Mid to Late Jurassic and continued into the Cretaceous. During the Middle to early Late Jurassic, two subduction zones, dipping in opposite directions from each other, formed along the western margin of North America. As the North American plate moved westward, as a result of the opening of the Atlantic Ocean, it soon overrode the westerly dipping subduction zone, destroying it, and leaving only the easterly dipping subduction zone along its western periphery. As the easterly dipping ocean crust continued to be subducted, large volumes of granitic magma were generated at depth beneath the western edge of North America. These granitic masses were emplaced as huge batholiths that are now recognized as the Sierra Nevada, Southern California, Idaho, and Coast Range batholiths (• Figure 14.12). It was also during this time that the *Franciscan Complex* and *Great Valley Group* were deposited and deformed as part of the Nevadan orogeny within the Cordilleran mobile belt.

The Franciscan Complex, which is up to 7000 m thick, is an unusual rock unit consisting of a chaotic mixture of rocks that accumulated during the Late Jurassic and Cretaceous. The various rock types—graywacke, volcanic breccia, siltstone, black shale, chert, pillow basalt, and blueschist metamorphic rocks—suggest that continental shelf, slope, and deep-sea environments were brought together in a submarine trench when North America overrode the subducting Farallon plate (• Figure 14.13).

The Franciscan Complex and the Great Valley Group that lies east of it were both squeezed against the edge of the North American craton as a result of subduction of the Farallon plate beneath the North America plate. The Franciscan Complex and the Great Valley Group are currently separated from each other by a major thrust fault. The Great Valley Group consists of more than 16,000 m of conglomerates, sandstones, siltstones, and shales that were deposited on the continental shelf and slope at the same time that the Franciscan deposits were accumulating in the submarine trench (Figure 14.13).

• Figure 14.13 **The Franciscan Complex**

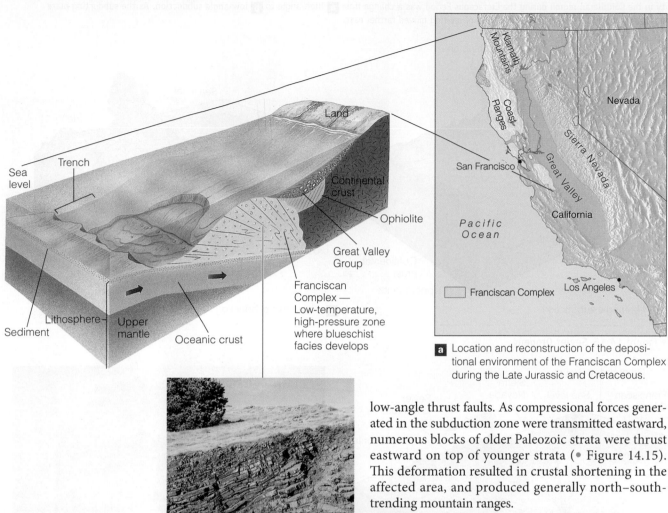

a Location and reconstruction of the depositional environment of the Franciscan Complex during the Late Jurassic and Cretaceous.

b Bedded chert exposed in Marin County, California. Most of the layers are about 5 cm thick.

James S. Monroe

low-angle thrust faults. As compressional forces generated in the subduction zone were transmitted eastward, numerous blocks of older Paleozoic strata were thrust eastward on top of younger strata (• Figure 14.15). This deformation resulted in crustal shortening in the affected area, and produced generally north–south-trending mountain ranges.

During the Late Cretaceous to Early Cenozoic, the final pulse of the Cordilleran orogeny took place (Figure 14.11). The **Laramide orogeny** developed east of the Sevier orogenic belt in the present-day Rocky Mountain areas of New Mexico, Colorado, and Wyoming. Most features of the present-day Rocky Mountains resulted from the Cenozoic phase of the Laramide orogeny, and for that reason, it will be discussed in Chapter 16.

By the Late Cretaceous, most of the volcanic and plutonic activity had migrated eastward into Nevada and Idaho. This migration was probably caused by a change from high-angle to low-angle subduction, resulting in the subducting oceanic plate reaching its melting depth farther east (• Figure 14.14). Thrusting occurred progressively farther east so that by the Late Cretaceous it extended all the way to the Idaho–Washington border.

The second pulse of the Cordilleran orogeny, the **Sevier orogeny,** affected western North America from Alaska to Mexico, and was mostly a Cretaceous event, even though it began in the Late Jurassic and is associated with the tectonic activity of the earlier Nevadan orogeny (Figure 14.11). Subduction of the Farallon plate beneath the North American plate during this time resulted in numerous overlapping,

Mesozoic Sedimentation Concurrent with the tectonism in the Cordilleran mobile belt, Early Triassic sedimentation on the western continental shelf consisted of shallow-water marine sandstones, shales, and limestones. During the Middle and Late Triassic, the western shallow seas regressed farther west, exposing large areas of former seafloor to erosion. Marginal marine and nonmarine Triassic rocks, particularly red beds, contribute to the spectacular and colorful scenery of the region.

The Lower Triassic *Moenkopi Formation* of the southwestern United States consists of a succession of brick red and chocolate-colored mudstones (• Figure 14.16). Such sedimentary structures as desiccation cracks and ripple marks, as well as fossil amphibians and reptiles and their tracks, indicate deposition in a variety of continental

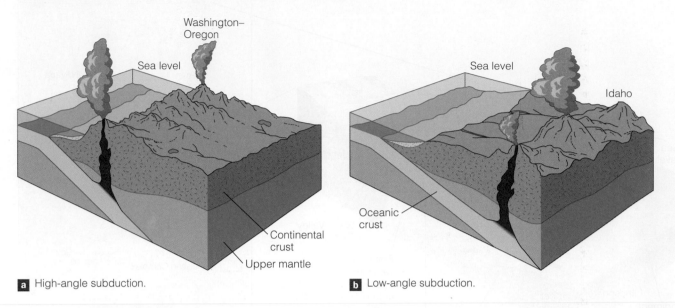

a High-angle subduction.

b Low-angle subduction.

• **Figure 14.15 Sevier Orogeny**

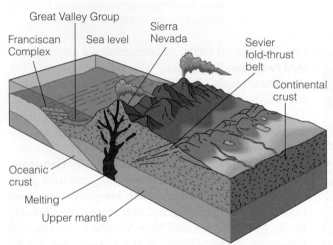

a Restoration showing the associated tectonic features of the Late Cretaceous Sevier orogeny caused by subduction of the Farallon plate under the North American plate.

b The Keystone thrust fault, a major fault in the Sevier overthrust belt, is exposed west of Las Vegas, Nevada. The sharp boundary between the light-colored Mesozoic rocks and the overlying dark-colored Paleozoic rocks marks the trace of the Keystone thrust fault.

environments, including stream channels, floodplains, and both freshwater and brackish ponds. Thin tongues of marine limestones indicate brief incursions of the sea, whereas local beds with gypsum and halite crystal casts attest to a rather arid climate. Unconformably overlying the Moenkopi is the Upper Triassic *Shinarump Conglomerate,* a widespread unit generally less than 50 m thick.

Above the Shinarump are the multicolored shales, siltstones, and sandstones of the Upper Triassic *Chinle Formation* (Figure 14.16a). This formation is widely exposed throughout the Colorado Plateau and is probably most famous for its petrified wood, spectacularly exposed in Petrified Forest National Park, Arizona (see Perspective). Whereas fossil ferns are found here, the park is best known

for its abundant and beautifully preserved logs of gymnosperms, especially conifers and plants called *cycads* (see Figure 15.6). Fossilization resulted from the silicification of the plant tissues. Weathering of volcanic ash beds interbedded with fluvial and deltaic Chinle sediments provided most of the silica for silicification. Some trees were preserved in place, but most were transported during floods and deposited on sandbars and on floodplains, where fossilization took place. After burial, silica-rich groundwater percolated through the sediments and silicified the wood.

Although best known for its petrified wood, the Chinle Formation has also yielded fossils of labyrinthodont amphibians, phytosaurs, and small dinosaurs (see Chapter 15 for a discussion of the latter two animal groups). In addition,

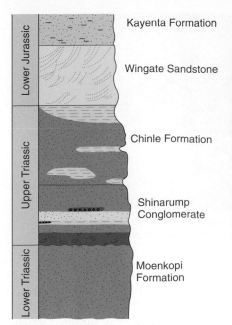

a Stratigraphic column of Triassic and Lower Jurassic formations in the western United States.

b View of the East Entrance of Zion Canyon, Zion National Park, Utah. The massive light-colored rocks are the Jurassic Navajo Sandstone, and the slope-forming rocks below the Navajo are the Lower Jurassic Kayenta Formation.

palynologic studies show a similar assemblage of pollen from the Chinle Formation and the Lower Newark Group on the east coast, indicating that the Chinle Formation and Lower Newark Group units are the same age.

Early Jurassic-age deposits in a large part of the western region consist mostly of clean, cross-bedded sandstones indicative of wind-blown deposits. The lowermost unit is the *Wingate Sandstone,* a desert dune deposit, which is overlain by the *Kayenta Formation,* a stream and lake deposit (Figure 14.16a). These two formations are well exposed in southwestern Utah. The thickest and most prominent of the Jurassic cross-bedded sandstones is the *Navajo Sandstone* (Figure 14.16b), a widespread formation that accumulated in a coastal dune environment along the southwestern margin of the craton. The sandstone's most distinguishing feature is its large-scale cross-beds, some of which are more than 25 m high (• Figure 14.17). The upper part of the Navajo contains smaller cross-beds as well as dinosaur and crocodilian fossils.

Marine conditions returned to the region during the Middle Jurassic when a seaway called the **Sundance Sea** twice flooded the interior of western North America (Figure 14.5). The resulting deposits, the *Sundance Formation,* were produced from erosion of tectonic highlands to the west that paralleled the shoreline. These highlands resulted from intrusive igneous activity and associated volcanism that began during the Triassic.

During the Late Jurassic, a mountain chain formed in Nevada, Utah, and Idaho as a result of the deformation produced by the Nevadan orogeny. As the mountain chain grew and shed sediments eastward, the Sundance Sea began retreating northward. A large part of the area formerly occupied by the Sundance Sea was then covered by multicolored sandstones, mudstones, shales, and occasional lenses of conglomerates that comprise the world-famous *Morrison Formation* (• Figure 14.18a).

The Morrison Formation contains one of the world's richest assemblages of Jurassic dinosaur remains. Although most of the skeletons are broken up, as many as 50 individuals have been found together in a small area. Such a concentration indicates that the skeletons were brought together during times of flooding and deposited on sandbars in stream channels. Soils in the Morrison indicate that the climate was seasonally dry.

Although most major museums have either complete dinosaur skeletons or at least bones from the Morrison Formation, the best place to see the bones still embedded in the rocks is the visitors' center at Dinosaur National Monument near Vernal, Utah (Figure 14.18b).

• Figure 14.17 **Navajo Sandstone** Large cross-beds of the Jurassic Navajo Sandstone exposed in Zion National Park, Utah.

• Figure 14.18 **Morrison Formation**

a Panoramic view of the Jurassic Morrison Formation as seen from the visitors' center at Dinosaur National Monument, near Vernal, Utah.

b North wall of visitors' center showing dinosaur bones in bas relief, just as they were deposited 140 million years ago.

Shortly before the end of the Early Cretaceous, Arctic waters spread southward over the craton, forming a large inland sea in the Cordilleran region. Mid-Cretaceous transgressions also occurred on other continents, and all were part of the global mid-Cretaceous rise in sea level that resulted from accelerated seafloor spreading as Pangaea continued to fragment. These Middle Cretaceous transgressions are marked by widespread black shale deposition within the oceanic areas, the shallow sea shelf areas, and the continental regions that were inundated by the transgressions.

By the beginning of the Late Cretaceous, this incursion joined the northward-transgressing waters from the Gulf area to create an enormous **Cretaceous Interior Seaway** that occupied the area east of the Sevier orogenic belt. Extending from the Gulf of Mexico to the Arctic Ocean, and more than 1500 km wide at its maximum extent, this seaway effectively divided North America into two large landmasses until just before the end of the Late Cretaceous (Figure 14.6).

Deposition in this seaway and the resulting sedimentary rock sequences are a result of complex interactions involving sea level changes, sediment supply from the adjoining landmasses, tectonics, and climate. Cretaceous deposits less than 100 m thick indicate that the eastern margin of the Cretaceous Interior Seaway subsided slowly and received little sediment from the emergent, low-relief craton to the east. The western shoreline, however, shifted back and forth, primarily in response to fluctuations in the supply of sediment from the Cordilleran Sevier orogenic belt to the west. The facies relationships show lateral changes from conglomerate and coarse sandstone adjacent to the mountain belt through finer sandstones, siltstones, shales, and even limestones and chalks in the east (• Figure 14.19). During times of particularly active mountain building, these coarse clastic wedges of gravel and sand prograded even further east.

As the Mesozoic Era ended, the Cretaceous Interior Seaway withdrew from the craton. During this regression, marine waters retreated to the north and south, and marginal marine and continental deposition formed widespread coal-bearing deposits on the coastal plain.

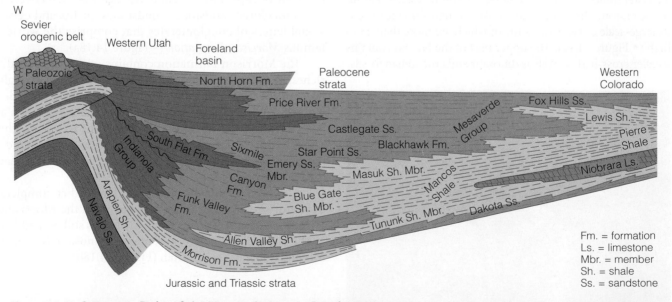

• Figure 14.19 **Cretaceous Facies of the Western Cretaceous Interior Seaway** This restored west–east cross section of Cretaceous facies of the western Cretaceous Interior Seaway shows their relationship to the Sevier orogenic belt.

Perspective

Petrified Forest National Park

Petrified Forest National Park is located in eastern Arizona about 42 km east of Holbrook (Figure 1). The park consists of two sections: the Painted Desert, which is north of Interstate 40, and the Petrified Forest, which is south of the Interstate.

The Painted Desert is a brilliantly colored landscape whose colors and hues change constantly throughout the day. The multicolored rocks of the Triassic Chinle Formation have been weathered and eroded to form a badlands topography of numerous gullies, valleys, ridges, mounds, and mesas. The Chinle Formation is composed predominantly of various-colored shale beds. These shales and associated volcanic ash layers are easily weathered and eroded. Interbedded locally with the shales are lenses of conglomerates, sandstones, and limestones, which are more resistant to weathering and erosion than the shales and form resistant ledges.

The Petrified Forest was originally set aside as a national monument to protect the large number of petrified logs that lay exposed in what is now the southern part of the park (Figure 2). When the transcontinental railroad constructed a coaling and watering stop in Adamana, Arizona, passengers were encouraged to take excursions to "Chalcedony Park," as the area was then called, to see the petrified forests. In a short time, collectors and souvenir hunters hauled off tons of petrified wood, quartz crystals, and Native American relics. Not until a huge rock crusher was built to crush the logs for the manufacture of abrasives was the area declared a national monument and the petrified forests preserved and protected.

During the Triassic Period, the climate of the area was much wetter than today, with many rivers, streams, and lakes. About 40 fossil plant species have been identified from the Chinle Formation. These include numerous seedless vascular plants such as rushes and ferns, as well as gymnosperms such as cycads and conifers. Such plants thrive in floodplains and marshes. Most logs are conifers and belong to the genus *Araucarioxylon* (Figure 2). Some of these trees were more than 60 m tall and up to 4 m in diameter. Apparently, most of the conifers grew on higher ground or riverbanks. Although many trees were buried in place, most seem to have been uprooted and transported by raging streams during times of flooding. Burial of the logs was rapid, and groundwater saturated with silica from the ash of nearby volcanic eruptions quickly permineralized the trees.

Deposition continued in the Colorado Plateau region during the Jurassic and Cretaceous, further burying the Chinle Formation. During the Laramide orogeny, the Colorado Plateau area was uplifted and eroded, exposing the Chinle Formation. Because the Chinle is mostly shales, it was easily eroded, leaving the more resistant petrified logs and log fragments exposed on the surface—much as we see them today (Figure 2).

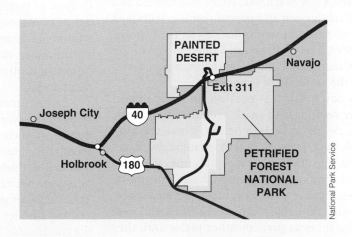

Figure 1 Petrified Forest National Park, Arizona, consists of two parts: the Painted Desert, and the Petrified Forest. The Painted Desert is located north of Interstate 40, whereas the Petrified Forest stretches north of Highway 180 and south of Interstate 40.

Figure 2 Petrified Forest National Park, Arizona. All of the logs here are *Araucarioxylon*, the most abundant tree in the park. The petrified logs have been weathered from the Chinle Formation, and are mostly in the position in which they were buried some 200 million years ago.

© Stephen J. Krasemeon/Photo Researchers, Inc.

What Role Did Accretion of Terranes Play in the Growth of Western North America?

In the preceding sections, we have discussed orogenies along convergent plate boundaries resulting in continental accretion. Much of the material accreted to continents during such events is simply eroded older continental crust; however, a significant amount of new material is added to continents as well, such as igneous rocks that formed as a consequence of subduction and partial melting. Although subduction is the predominant influence on the tectonic history in many regions of orogenesis, other processes are also involved in mountain building and continental accretion, especially the accretion of terranes.

Geologists now know that portions of many mountain systems are composed of small accreted lithospheric blocks that are clearly of foreign origin. These **terranes** differ completely in their fossil content, stratigraphy, structural trends, and paleomagnetic properties from the rocks of the surrounding mountain system and adjacent craton. In fact, these terranes are so different from adjacent rocks that most geologists think they formed elsewhere and were carried great distances as parts of other plates until they collided with other terranes or continents.

Geologic evidence indicates that more than 25% of the entire Pacific Coast from Alaska to Baja California consists of accreted terranes. These accreted terranes are composed of volcanic island arcs, oceanic ridges, seamounts, volcanic plateaus, hot spot tracks, and small fragments of continents that were scraped off and accreted to the continent's margin

as the oceanic plate on which they were carried was subducted under the continent.

Geologists estimate that more than 100 different-sized terranes have been added to the western margin of North America during the last 200 million years (• Figure 14.20). The Wrangellian terranes (Figure 14.1b) are a good example of terranes that have been accreted to North America's western margin (Figure 14.20).

The basic plate tectonic reconstruction of orogenies and continental accretion remains unchanged, but the details of such reconstructions are decidedly different in view of terrane tectonics. For example, growth along active continental margins is faster than along passive continental margins because of the accretion of terranes. Furthermore, these accreted terranes are often new additions to a continent, rather than reworked older continental material.

So far, most terranes have been identified in mountains of the North American Pacific Coast region, but a number of such plates are suspected to be present in other mountain systems as well. They are more difficult to recognize in older mountain systems, such as the Appalachians, however, because of greater deformation and erosion. Thus, terranes provide another way of viewing Earth and gaining a better understanding of the geologic history of the continents.

Mesozoic Mineral Resources

Although much of the coal in North America is Pennsylvanian or Paleogene in age, important Mesozoic coals occur in the Rocky Mountains states. These are mostly lignite and

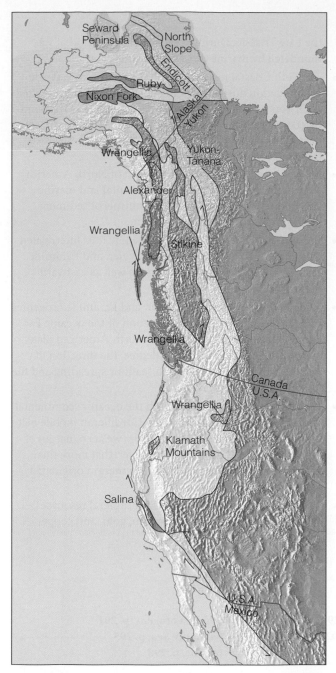

• **Figure 14.20 Mesozoic Terranes** Some of the accreted lithospheric blocks, called terranes, that form the western margin of the North American craton. The dark brown blocks probably originated as terranes and were accreted to North America. The light green blocks are possibly displaced parts of North America. The North American craton is shown in dark green. See Figure 14.1b for the position of the Wrangellian terranes during the Jurassic.

bituminous coals, but some local anthracites are present as well. Particularly widespread in western North America are coals of Cretaceous age. Mesozoic coals are also known from Australia, Russia, and China.

Large concentrations of petroleum occur in many areas of the world, but more than 50% of all proven reserves are in the Persian Gulf region. During the Mesozoic Era, what is now the Gulf region was a broad passive continental margin conducive for the formation of petroleum. Similar conditions existed in what is now the Gulf Coast

region of the United States and Central America. Here, petroleum and natural gas also formed on a broad shelf over which transgressions and regressions occurred. In this region, the hydrocarbons are largely in reservoir rocks that were deposited as distributary channels on deltas and as barrier-island and beach sands. Some of these hydrocarbons are associated with structures formed adjacent to rising salt domes. The salt, called the *Louann Salt,* initially formed in a long, narrow sea when North America separated from Europe and North Africa during the fragmentation of Pangaea (Figure 14.2).

The richest uranium ores in the United States are widespread in Mesozoic rocks of the Colorado Plateau area of Colorado and adjoining parts of Wyoming, Utah, Arizona, and New Mexico. These ores, consisting of fairly pure masses of a complex potassium-, uranium-, vanadium-bearing mineral called *carnotite,* are associated with plant remains in sandstones that were deposited in ancient stream channels.

As noted in Chapter 9, Proterozoic banded iron formations are the main sources of iron ores. There are, however, some important exceptions. For example, the Jurassic-age "Minette" iron ores of western Europe, composed of oolitic limonite and hematite, are important ores in France, Germany, Belgium, and Luxembourg. In Great Britain, low-grade iron ores of Jurassic age consist of oolitic siderite, which is an iron carbonate. And in Spain, Cretaceous rocks are the host rocks for iron minerals.

South Africa, the world's leading producer of gemquality diamonds and among the leaders in industrial diamond production, mines these minerals from kimberlite pipes, conical igneous intrusions of dark gray or blue igneous rock. Diamonds, which form at great depth where pressure and temperature are high, are brought to the surface during the explosive volcanism that forms kimberlite pipes. Although kimberlite pipes have formed throughout geologic time, the most intense episode of such activity in South Africa and adjacent countries was during the Cretaceous Period. Emplacement of Triassic and Jurassic diamond-bearing kimberlites also occurred in Siberia.

In the Introduction we noted that the mother lode or source for the placer deposits mined during the California gold rush is in Jurassic-age intrusive rocks of the Sierra Nevada. Gold placers are also known in Cretaceous-age conglomerates of the Klamath Mountains of California and Oregon.

Porphyry copper was originally named for copper deposits in the western United States mined from porphyritic granodiorite; however, the term now applies to large, low-grade copper deposits disseminated in a variety of rocks. These porphyry copper deposits are an excellent example of the relationship between convergent plate boundaries and the distribution, concentration, and exploitation of valuable metallic ores. Magma generated by partial melting of a subducting plate rises toward the surface, and as it cools, it precipitates and concentrates various metallic ores. The world's largest copper deposits were formed during the Mesozoic and Cenozoic in a belt along the western margins of North and South America (see Figure 3.30).

SUMMARY

Table 14.1 summarizes the geologic history of North America, as well as global events and sea level changes during the Mesozoic Era.

- We can summarize the breakup of Pangaea as follows:
  1. During the Late Triassic, North America began separating from Africa. This was followed by the rifting of North America from South America.
  2. During the Late Triassic and Jurassic periods, Antarctica and Australia—which remained sutured together—began separating from South America and Africa, and India began rifting from Gondwana.
  3. South America and Africa began separating during the Jurassic, and Europe and Africa began converging during this time.
  4. The final stage in Pangaea's breakup occurred during the Cenozoic, when Greenland completely separated from Europe and North America.

- The breakup of Pangaea influenced global climatic and atmospheric circulation patterns. Although the temperature gradient from the tropics to the poles gradually increased during the Mesozoic, overall global temperatures remained equable.

- An increased rate of seafloor spreading during the Cretaceous Period caused sea level to rise and transgressions to occur.

- Except for incursions along the continental margin and two major transgressions (the Sundance Sea and the Cretaceous Interior Seaway), the North American craton was above sea level during the Mesozoic Era.

- The Eastern Coastal Plain was the initial site of the separation of North America from Africa that began during the Late Triassic. During the Cretaceous Period, it was inundated by marine transgressions.

- The Gulf Coastal region was the site of major evaporite accumulation during the Jurassic as North America rifted from South America. During the Cretaceous, it was inundated by a transgressing sea, which, at its maximum, connected with a sea transgressing from the north to create the Cretaceous Interior Seaway.

- Mesozoic rocks of the western region of North America were deposited in a variety of continental and marine environments. One of the major controls of sediment distribution patterns was tectonism.

- Western North America was affected by four interrelated orogenies: the Sonoma, Nevadan, Sevier, and Laramide. Each involved igneous intrusions, as well as eastward thrust faulting and folding.

- The cause of the Nevadan, Sevier, and Laramide orogenies was the changing angle of subduction of the oceanic Farallon plate under the continental North American plate. The timing, rate, and, to some degree, the direction of plate movement were related to seafloor spreading and the opening of the Atlantic Ocean.

- Orogenic activity associated with the oceanic–continental convergent plate boundary in the Cordilleran mobile belt explains the structural features of the western margin of North America. It is thought, however, that more than 25% of the North American western margin originated from the accretion of terranes.

- Mesozoic rocks contain a variety of mineral resources, including coal, petroleum, uranium, gold, and copper.

IMPORTANT TERMS

Cordilleran orogeny, p. 291

Cretaceous Interior Seaway, p. 296

Laramide orogeny, p. 293

Nevadan orogeny, p. 292

Sevier orogeny, p. 293

Sonoma orogeny, p. 291

Sundance Sea, p. 295

terrane, p. 298

REVIEW QUESTIONS

1. The formation or complex responsible for the spectacular scenery of the Painted Desert and Petrified Forest is the
   a. _____ Franciscan; b. _____ Morrison; c. _____ Chinle; d. _____ Wingate; e. _____ Navajo.

2. A possible cause for the eastward migration of igneous activity in the Cordilleran region during the Cretaceous was a change from
   a. _____ high-angle to low-angle subduction; b. _____ divergent plate margin activity to subduction; c. _____ subduction to divergent plate margin activity; d. _____ oceanic–oceanic convergence to oceanic–continental convergence; e. _____ divergent to convergent plate margin activity.

3. The first Mesozoic orogeny in the Cordilleran region was the
   a. _____ Sevier; b. _____ Laramide; c. _____ Sonoma; d. _____ Antler; e. _____ Nevadan.

4. During the Jurassic, the newly forming Gulf of Mexico was the site of primarily what type of deposition?
   a. _____ Evaporites; b. _____ Siliciclastics; c. _____ Volcaniclastics; d. _____ Detrital; e. _____ Answers b and c.

5. Triassic rifting between which two continental landmasses initiated the breakup of Pangaea?
   a. _____ India and Australia; b. _____ Antarctica and India; c. _____ South America and Africa; d. _____ North America and Eurasia; e. _____ Laurasia and Gondwana.

6. The first major seaway to flood North America was the
   a. _____ Cretaceous Interior Seaway; b. _____ Sundance;
   c. _____ Cordilleran; d. _____ Zuni; e. _____ Newark.
7. The orogeny responsible for the present-day Rocky Mountains is the
   a. _____ Sevier; b. _____ Nevadan; c. _____ Antler;
   d. _____ Sonoma; e. _____ Laramide.
8. The time of greatest post-Paleozoic inundation of the craton occurred during which geologic period?
   a. _____ Triassic; b. _____ Jurassic; c. _____ Cretaceous;
   d. _____ Paleogene; e. _____ Neogene.
9. Which orogeny produced the Sierra Nevada, Southern California, Idaho, and Coast Range batholiths?
   a. _____ Laramide; b. _____ Sonoma; c. _____ Nevadan;
   d. _____ Sevier; e. _____ None of the previous answers.
10. Which formation or group filled the Late Triassic fault-block basins of the east coast of North America with red nonmarine sediments?
    a. _____ Morrison; b. _____ Chinle; c. _____ Navajo;
    d. _____ Franciscan; e. _____ Newark.
11. Why is the paleogeography of the Mesozoic Era in some ways easier to reconstruct and more accurate than for the Paleozoic Era?
12. Briefly explain some of the evidence geologists use to interpret climatic conditions during the Mesozoic Era.
13. Why are Mesozoic-age coals mostly lignite and bituminous, whereas Paleozoic-age coals tend to be high-grade bituminous and anthracite?
14. From a plate tectonic perspective, how does the orogenic activity that occurred in the Cordilleran mobile belt during the Mesozoic Era differ from that which took place in the Appalachian mobile belt during the Paleozoic?
15. How did the Mesozoic rifting that took place on the East Coast of North America affect the tectonics in the Cordilleran mobile belt?
16. Compare the tectonic setting and depositional environment of the Gulf of Mexico evaporites with the evaporite sequences of the Paleozoic Era.
17. What effect did the breakup of Pangaea have on oceanic and climatic circulation patterns? Compare the oceanic circulation pattern during the Triassic with that during the Cretaceous.
18. How does terrane accretion relate to the Mesozoic orogenies that took place on the western margins of North America?
19. Compare the tectonics of the Sonoma and Antler orogenies.
20. Explain and diagram how increased seafloor spreading can cause a rise in sea level along the continental margins.

APPLY YOUR KNOWLEDGE

1. The breakup of Pangaea influenced the distribution of continental landmasses, ocean basins, and oceanic and atmospheric circulation patterns, which in turn affected the distribution of natural resources, landforms, and the evolution of the world's biota. Reconstruct a hypothetical history of the world for a different breakup of Pangaea— one in which the continents separate in a different order or rift apart in a different configuration. How would such a scenario affect the distribution of natural resources? Would the distribution of coal and petroleum reserves be the same? How might evolution be affected? Would human history be different?

2. The gold discovered at Sutter's Mill, California, in 1848 sparked the California gold rush. This gold is widely disseminated throughout the granitic rocks of the Sierra Nevada batholith and was concentrated in placer deposits. During the Nevadan orogeny, several other large granitic batholiths were intruded in the Cordillera region of North America. Have there been any gold discoveries associated with these intrusions? Why?

3. Because of political events in the Middle East, the oil-producing nations of this region have reduced the amount of petroleum they export, resulting in shortages in the United States. To alleviate U.S. dependence on overseas oil, the major oil companies want Congress to let them explore for oil in many of our national parks and some environmentally sensitive offshore areas. As director of the National Park system, you have been called to testify at the congressional hearing addressing this possibility. What arguments would you use to discourage such exploration? Would a knowledge of the geology of the area be helpful in your testimony? Explain.

Table 14.1
Summary of Mesozoic Geologic Events

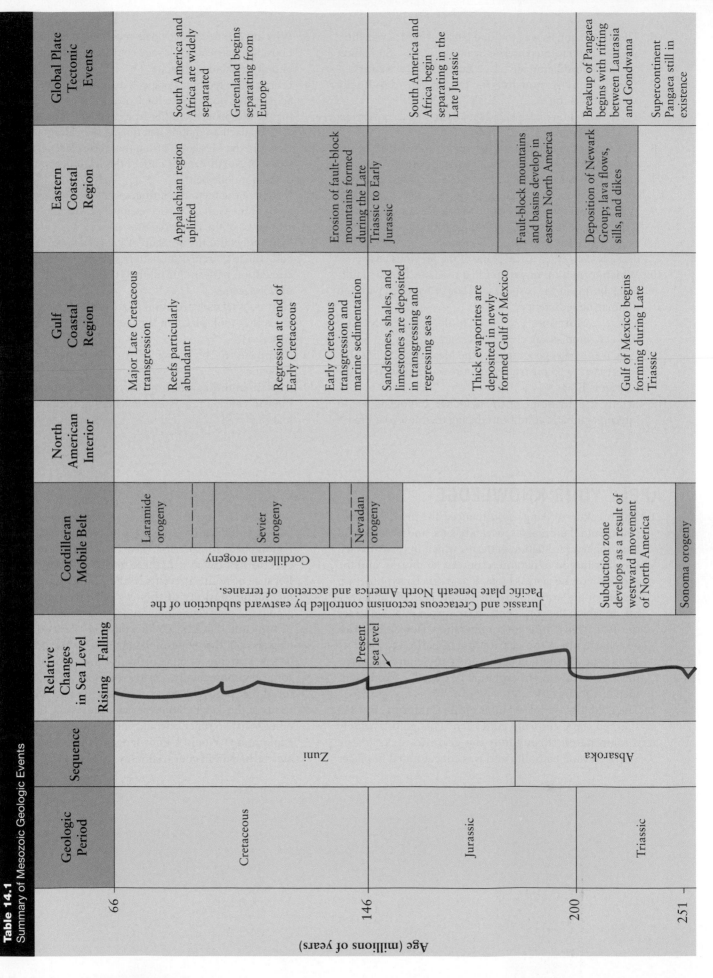

Age (millions of years)	Geologic Period	Sequence	Relative Changes in Sea Level (Rising — Falling)	Cordilleran Mobile Belt	North American Interior	Gulf Coastal Region	Eastern Coastal Region	Global Plate Tectonic Events
66	Cretaceous	Zuni	*Present sea level*	Laramide orogeny / Sevier orogeny / Nevadan orogeny (Cordilleran orogeny)		Major Late Cretaceous transgression; Reefs particularly abundant; Regression at end of Early Cretaceous; Early Cretaceous transgression and marine sedimentation; Sandstones, shales, and limestones are deposited in transgressing and regressing seas; Thick evaporites are deposited in newly formed Gulf of Mexico	Appalachian region uplifted; Erosion of fault-block mountains formed during the Late Triassic to Early Jurassic	South America and Africa are widely separated; Greenland begins separating from Europe; South America and Africa begin separating in the Late Jurassic
146	Jurassic		Jurassic and Cretaceous tectonism controlled by eastward subduction of the Pacific plate beneath North America and accretion of terranes.				Fault-block mountains and basins develop in eastern North America	
200	Triassic	Absaroka		Subduction zone develops as a result of westward movement of North America; Sonoma orogeny		Gulf of Mexico begins forming during Late Triassic	Deposition of Newark Group; lava flows, sills, and dikes	Breakup of Pangaea begins with rifting between Laurasia and Gondwana; Supercontinent Pangaea still in existence
251								

© Eivind Bovor

In this scene from the Late Cretaceous, *Ankylosaurus* is defending itself from the large predator *Tyrannosaurus*. *Ankylosaurus* was the most heavily armored dinosaur, and it had a large club at the end of its tail that was almost certainly used for defense. This animal measured 8 to 10 m long and weighed about 4.5 metric tons. At 13 m long and weighing perhaps 5 metric tons, *Tyrannosaurus* was one of the largest carnivorous dinosaurs.

[OUTLINE]

Introduction

Marine Invertebrates and Phytoplankton

Aquatic and Semiaquatic Vertebrates

   The Fishes

   Amphibians

Plants—Primary Producers on Land

The Diversification of Reptiles

   Archosaurs and the Origin of Dinosaurs

   Dinosaurs

   Warm-Blooded Dinosaurs?

   Flying Reptiles

   Mesozoic Marine Reptiles

   Crocodiles, Turtles, Lizards, and Snakes

From Reptiles to Birds

   Perspective *Mary Anning and Her Contributions to Paleontology*

Origin and Evolution of Mammals

   Cynodonts and the Origin of Mammals

   Mesozoic Mammals

Mesozoic Climates and Paleobiogeography

Mass Extinctions—A Crisis in Life History

Summary

At the end of this chapter, you will have learned that

- Marine invertebrates that survived the Paleozoic extinctions diversified and repopulated the seas.

- Land plant communities changed considerably when flowering plants evolved during the Cretaceous.

- Reptile diversification began during the Mississippian and continued throughout the Mesozoic Era.

- Among the Mesozoic reptiles, dinosaurs had evolved by the Late Triassic and soon became the dominant land-dwelling vertebrate animals.

- In addition to dinosaurs, the Mesozoic was also the time of flying reptiles and marine reptiles, as well as turtles, lizards, snakes, and crocodiles.

- Birds evolved from reptiles, probably from some small carnivorous dinosaur.

- Mammals evolved from reptiles only distantly related to dinosaurs, and they existed as contemporaries with dinosaurs.

- The transition from reptiles to mammals is very well supported by fossil evidence.

- Several varieties of Mesozoic mammals are known, all of which were small, and their diversity remained low.

- The proximity of continents and generally mild Mesozoic climates allowed many plants and animals to spread over extensive geographic areas.

- Extinctions at the end of the Mesozoic Era were second in magnitude only to the Paleozoic extinctions. These extinctions have received more attention than any others because dinosaurs were among the victims.

Introduction

Ever since 1842 when Sir Richard Owen first used the term *dinosaur*, these animals have been the object of intense curiosity, the subject matter for countless articles and books, TV specials, and movies. The current interest in dinosaurs was certainly fueled by the release of the movie *Jurassic Park* (1993) and its sequels *The Lost World* (1997) and *Jurassic Park* III (2001), as well as several others that featured dinosaurs. No other group of animals living or extinct has so thoroughly captured the public imagination (see the chapter opening image), but dinosaurs are only one of several groups of remarkable Mesozoic reptiles and other animals.

In addition to dinosaurs, the Mesozoic Era was also the time when flying reptiles (pterosaurs), several types of marine reptiles (ichthyosaurs, plesiosaurs, and mosasaurs), and huge crocodiles proliferated. Also present were turtles, lizards, and snakes, although the fossil record for the last two groups is not so good. All in all, the Mesozoic land fauna, the skies, and the seas were populated by many types of reptiles. In fact, reptiles were so common that geologist informally refer to the Mesozoic as "The Age of Reptiles." Keep in mind, though, that these "Age of" designations simply reflect our personal preferences; there were far more species of insects and fishes at this time.

Certainly the Mesozoic was an important time in the evolution of reptiles, and recent discoveries have added much to our knowledge of these extinct creatures. For example, remarkable discoveries of feathered dinosaurs in China have important implications about the warm-blooded-cold-blooded dinosaur debate, and the relationships of dinosaurs to birds. And speaking of birds, they first appeared during the Jurassic, probably having evolved from small carnivorous dinosaurs. Mammals evolved from mammal-like reptiles during the Triassic. In fact, mammals were contemporaries with dinosaurs, although they were not nearly as diverse and all were small.

Important changes also took place in Cretaceous land plant communities when flowering plants evolved and soon became the most numerous and diverse of all land plants. Among the invertebrate animals that survived the Paleozoic mass extinctions the survivors diversified during the Triassic and repopulated the seas, accounting for the success of several types of cephalopods, bivalves, and others. In short, biotic diversity once again increased in all realms of the organic world, only to decrease again at the end of the Mesozoic. This Mesozoic extinction was second in magnitude to the one at the end of the Paleozoic, but because dinosaurs are among its victims, it is more widely known.

One of the main emphases in this book has been the systems approach to Earth and life history. Remember, that the distribution of land and sea has a profound influence on oceanic circulation, which in turn partly controls climate, and that the proximity of continents partly determines the geographic distribution of organisms. Pangaea began fragmenting during the Triassic and continues to do so, and as a result, intercontinental interchange among faunas became increasingly difficult for most organisms. In fact, South America and Australia were isolated from the other continents and their faunas, evolving in isolation, became increasingly different from those elsewhere.

Marine Invertebrates and Phytoplankton

Following the Permian mass extinctions, the Mesozoic was a time when marine invertebrates repopulated the seas.

The Early Triassic invertebrate fauna was not very diverse, but by the Late Triassic, the seas were once again swarming with invertebrates—from planktonic foraminifera to cephalopods. The brachiopods that had been so abundant during the Paleozoic never completely recovered from their near extinction, and although they still exist today, the bivalves have largely taken over their ecologic niche.

Mollusks such as cephalopods, bivalves, and gastropods were the most important elements of the Mesozoic marine invertebrate fauna. Their rapid evolution and the fact that many cephalopods were nektonic make them excellent guide fossils (• Figure 15.1). The Ammonoidea, cephalopods with wrinkled sutures, constitute three groups: the goniatites, ceratites, and ammonites. The latter, though present during the entire Mesozoic, were most prolific during the Jurassic and Cretaceous. Most ammonites were coiled, with some attaining diameters of 2 m, whereas others were uncoiled and led a near benthonic existence (Figure 15.1). Ammonites went extinct at the end of the Cretaceous, but, two related groups of cephalopods survived into the Cenozoic: the *nautiloids,* including the living pearly nautilus, and the *coleoids,* represented by the extinct

belemnoids (• Figure 15.2) which are good Jurassic and Cretaceous guide fossils, as well as the living squid and octopus.

Mesozoic bivalves diversified to inhabit many epifaunal and infaunal niches. Oysters and clams (epifaunal suspension feeders) became particularly diverse and abundant, and despite a reduction in diversity at the end of the Cretaceous, remain important animals in the marine fauna today.

As is true now, where shallow marine waters were warm and clear, coral reefs proliferated. However, reefs did not rebound from the Permian extinctions until the Middle Triassic. An important reef-builder throughout the Mesozoic was a group of bivalves known as *rudists.* Rudists are important because they displaced corals as the main reef-builders during the later Mesozoic and are good guide fossils for the Late Jurassic and Cretaceous.

A new and familiar type of coral also appeared during the Triassic, the *scleractinians.* Whether scleractinians evolved from rugose corals or from an as yet unknown soft-bodied ancestor with no known fossil record is still unresolved. In addition to the familiar present-day reef-building colonial scleractinian corals, solitary or individual scleractinian corals also inhabited relatively deep waters during the Mesozoic.

Another invertebrate group that prospered during the Mesozoic was the echinoids. Echinoids were exclusively epifaunal during the Paleozoic but branched out into the infaunal habitat during the Mesozoic. Both groups began a major adaptive radiation during the Late Triassic that continued throughout the remainder of the Mesozoic and into the Cenozoic.

A major difference between Paleozoic and Mesozoic marine invertebrate faunas was the increased abundance and diversity of burrowing organisms. With few exceptions,

• **Figure 15.1 Cretaceous Seascape** Cephalopods such as the Late Cretaceous ammonoids *Baculites* (foreground) and *Helioceros* (background) were present throughout the Mesozoic, but they were most abundant during the Jurassic and Cretaceous. They were important predators, and they are excellent guide fossils.

• **Figure 15.2 Belemnoids** Belemnoids are extinct squidlike cephalopods that were particularly abundant during the Cretaceous. They are excellent guide fossils for the Jurassic and Cretaceous. Shown here are several belemnoids swimming in a Cretaceous sea.

Archean Eon	Proterozoic Eon	Phanerozoic Eon							
Precambrian			Paleozoic Era						
			Cambrian	Ordovician	Silurian	Devonian	Mississippian	Pennsylvanian	Permian
							Carboniferous		

2500 MYA   542 MYA   251 MYA

Paleozoic burrowers were soft-bodied animals such as worms. The bivalves and echinoids, which were epifaunal animals during the Paleozoic, evolved various means of entering infaunal habitats. This trend toward an infaunal existence may have been an adaptive response to increasing predation from the rapidly evolving fish and cephalopods. Bivalves, for instance, expanded into the infaunal niche during the Mesozoic, and by burrowing, they escaped predators.

The foraminifera (single-celled consumers) diversified rapidly during the Jurassic and Cretaceous and continued to be diverse and abundant to the present. The planktonic forms (• Figure 15.3), in particular, diversified rapidly, but most genera became extinct at the end of the Cretaceous. The planktonic foraminifera are excellent guide fossils for the Cretaceous.

The primary producers in the Mesozoic seas were various types of microorganisms. *Coccolithophores* are an important group of calcareous phytoplankton (• Figure 15.4a) that first evolved during the Jurassic and became extremely common during the Cretaceous. *Diatoms* (Figure 15.4b), which build skeletons of silica, made their appearance during the Cretaceous, but they are more important as

primary producers during the Cenozoic. Diatoms are presently most abundant in cooler oceanic waters, and some species inhabit freshwater lakes. *Dinoflagellates*, which are organic-walled phytoplankton, were common during the Mesozoic and today are the major primary producers in warm water (Figure 15.4c).

In general terms, we can think of the Mesozoic as a time of increasing complexity among the marine invertebrate fauna. At the beginning of the Triassic, diversity was low and food chains were short. Near the end of the Cretaceous, though, the marine invertebrate fauna was highly complex, with interrelated food chains.

• **Figure 15.4** **Primary Producers** Coccolithophores, diatoms, and dinoflagellates were primary producers in the Mesozoic and Cenozoic oceans. **a** A Miocene coccolith from the Gulf of Mexico (left): a Pliocene-Miocene coccolith from the Gulf of Mexico (right). **b** Upper Miocene diatoms from Java (left and right). **c** Eocene dinoflagellates from Alabama (left) and the Gulf of Mexico (right).

Dept. of Paleobiology. Smithsonian Institution

• **Figure 15.3** **Planktonic Foraminifera** Planktonic foraminifera diversified and became abundant during the Jurassic and Cretaceous, and continued to be diverse and abundant throughout the Cenozoic. Many planktonic foraminifera are excellent guide fossils for the Cretaceous, such as species of the genus *Globotruncana,* which is restricted to the Upper Cretaceous. Shown here are three views of the holotype (the specimen that defines the species) of *Globotruncana loeblichi,* a Late Cretaceous planktonic foraminifera.

Phanerozoic Eon										
Mesozoic Era			Cenozoic Era							
Triassic	Jurassic	Cretaceous	Paleogene			Neogene		Quaternary		
			Paleocene	Eocene	Oligocene	Miocene	Pliocene	Pleistocene	Holocene	

251 MYA

66 MYA

This evolutionary history reflects changing geologic conditions influenced by plate tectonic activity, as discussed in Chapter 3.

Aquatic and Semiaquatic Vertebrates

Remember that the fishes evolved by Cambrian time and then diversified, especially during the Devonian Period when several major groups appeared (see Chapter 13). Amphibians also evolved during the Devonian and persist to the present, but their greatest abundance and diversity occurred during the Pennsylvanian.

The Fishes Today, Earth's oceans, lakes, rivers, and streams are populated by an estimated 24,000 species of bony fishes, whereas only about 930 species of cartilaginous fishes exist, and nearly all of them are confined to the seas. We know that sharks and other cartilaginous fishes became more abundant during the Mesozoic, but they never came close to matching the diversity of the bony fishes. Nevertheless, sharks, an evolutionarily conservative group, were and remain important in the marine fauna, especially as predators, although a few strain plankton from seawater.

Only a few species of lungfishes and crossopterygians persisted into the Mesozoic, and the latter declined and were nearly extinct at the end of the era. In fact, only one species of crossopterygian exists now (see Figure 7.15a), and the group has no known Cenozoic fossil record. The lungfishes have not fared much better—only three species exist, one each in Africa, South America, and Australia.

All bony fishes, except for lungfishes and crossopterygians, belong to three groups, which for convenience we call *primitive*, *intermediate*, and *advanced*. Superficially, they resemble one another, but important changes took place as one group replaced another. For example, the internal skeleton of the primitive and intermediate varieties was partly cartilaginous, but in the advanced group it was completely bony. The primitive group existed mostly

© Dmitry Bogdanov

• Figure 15.5 **The Mesozoic Fish *Leedsichthys* (background) and the Short-Necked Plesiosaur *Liopluerodon*** This fish, from the intermediate group of fishes (Holostei), was one of the largest ever, but because some of its spine is missing its total length is not known; estimates range from 9 to 30 m. It was probably a plankton feeder.

during the Paleozoic, but by the Jurassic the intermediate group predominated, and among the latter group was the largest fish known (• Figure 15.5). The advanced group, formally known as *teleosts*, became the most diverse of all bony fishes by the Cretaceous in both fresh and saltwater habitats, and now are the most varied and numerous of all vertebrate animals.

Amphibians The labyrinthodont amphibians were common during the latter part of the Paleozoic, but the few surviving Mesozoic species died out by the end of the Triassic. Since their greatest diversity and abundance during the Pennsylvanian, amphibians have made up only a small part of the total vertebrate fauna. A recently analyzed fossil dubbed a *frogmander* from the Permian of Texas coupled with molecular evidence from living amphibians leads some investigators to conclude that frogs and salamanders diverged during the Late Permian or Early Triassic. Frogs and salamanders were certainly present by the Mesozoic, but their fossil records are poor.

Plants—Primary Producers on Land

Just as during the Late Paleozoic, seedless vascular plants and gymnosperms dominated Triassic and Jurassic land plant communities, and, in fact, representatives of both groups are still common (• Figure 15.6). Among the gymnosperms, the large seed ferns became extinct by the end of the Triassic, but *ginkgos* (see Figure 7.15b) remained abundant and still exist in isolated regions, and *conifers* continued to diversify and are now widespread, particularly at high elevations and high latitudes. A new group of gymnosperms known as *cycads* made their appearance during the Triassic. They became widespread and now exist in tropical and semitropical areas (Figure 15.5b).

The long dominance of seedless vascular plants and gymnosperms ended during the Early Cretaceous, perhaps the Late Jurassic, when many of them were replaced by the flowering plants or **angiosperms** (• Figure 15.7).

• Figure 15.6 **Mesozoic Vegetation**

a This Jurassic landscape was dominated by seedless vascular plants, especially ferns, as well as gymnosperms including the conifers and cycads.

b These living cycads look much like some of the vegetation shown in **a**.

• Figure 15.7 **The Angiosperms or Flowering Plants**

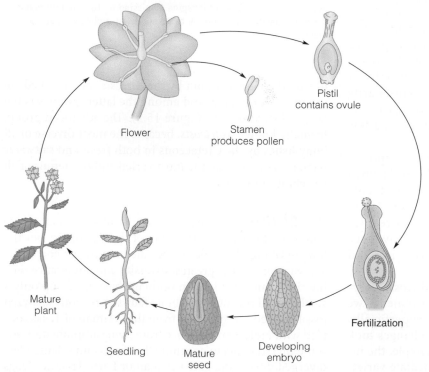

Flower
Stamen produces pollen
Pistil contains ovule
Fertilization
Developing embryo
Mature seed
Seedling
Mature plant

a The reproductive cycle in angiosperms.

b *Archaefructus sinensis* from Lower Cretaceous rocks in China is among the oldest known angiosperms.

c Restoration of *Archaefructus sinensis*.

Unfortunately, the fossil record of the earliest angiosperms is sparse so their precise ancestors remain obscure. Nevertheless, studies of living plants and the fossils that are available indicate close relationships to gymnosperms.

Since they first evolved, angiosperms have adapted to nearly every terrestrial habitat from high mountains to deserts, marshes and swamps, and some have even adapted to shallow coastal waters. Several features account for their phenomenal evolutionary success, including enclosed seeds, and above all the origin of flowers, which attract animal pollinators, especially insects. All organisms interact in some ways and influence the evolution others, but the interrelationships among flowering plants and insects are so close that biologists refer to changes in one induced by the other as *coevolution*.

The 250,000 to 300,000 species of angiosperms that now exist, accounting for more than 90% of all land plant species, and the fact that they inhabit some environments hostile to other plants, is a testimony to their success. Nevertheless, seedless vascular plants and gymnosperms remain important in the worlds' flora.

The Diversification of Reptiles

We already mentioned that because of the diversity of reptiles the Mesozoic is informally called "The Age of Rep-

tiles." Remember, though, that reptiles actually first appeared during the Mississippian, and when they first evolved from amphibians they did not look very different from their ancestors. Nevertheless, this group of so-called *stem reptiles* gave rise to all other reptiles, birds, and mammals (• Figure 15.8). All living reptiles, crocodiles, lizards, snakes, the tuatara, and turtles, are cold-blooded, lay amniotic eggs, practice internal fertilization, and have a tough, scaly skin. In addition, dinosaurs, the extinct pterosaurs, as well as all living reptiles, with the exception of turtles, have two openings on the side of the skull in the temporal area.

Dinosaurs are included among the reptiles but they possess several characteristics that set them apart. They had teeth set in individual sockets, a reduced lower leg bone (fibula), a pelvis anchored to the vertebral column by three or more vertebrae, a ball-like head on the upper leg bone (femur), and elongate bones in the palate. In addition, dinosaurs had a fully upright posture with the limbs directly beneath their bodies, rather than the sprawling stance or semi-erect posture as in other reptiles. In fact, their upright posture and other limb modifications may have been partly responsible for their incredible success.

Contrary to popular belief, there were no flying dinosaurs or fully aquatic ones, although there were Mesozoic reptiles that filled these niches. Nor were all dinosaurs large, even though some certainly were. Also, dinosaurs

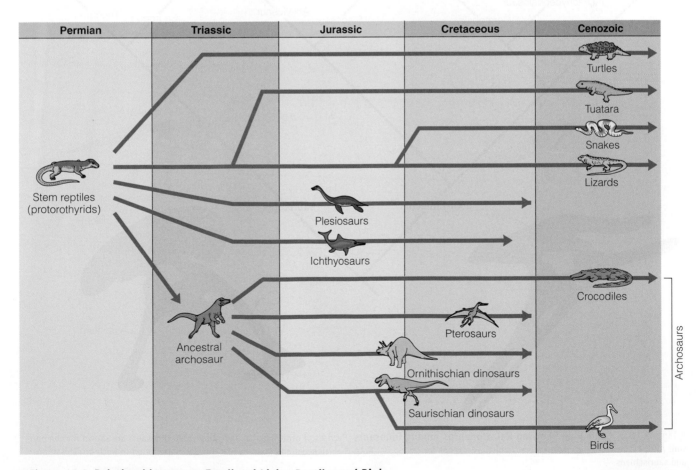

• Figure 15.8 **Relationships among Fossil and Living Reptiles and Birds**

lived only during the Mesozoic Era, unless we consider the birds, which evolved from one specific group of dinosaurs.

Archosaurs and the Origin of Dinosaurs

Reptiles known as **archosaurs** (*archo* meaning "ruling," and "*sauros*" meaning "lizard") include crocodiles, pterosaurs (flying reptiles), dinosaurs, and birds. Including such diverse animals in a single group implies that they share a common ancestor, and indeed they possess several characteristics that unite them. All archosaurs have teeth set in individual sockets, except today's birds, but even the earliest birds had this feature. In addition, these animals have a single skull opening in front of the eye that is not found in other reptiles.

Dinosaurs share many characteristics and yet they differ enough for paleontologists to recognize two distinct orders based primarily on their type of pelvis: the **Saurischia** and the **Ornithischia.** Saurischian dinosaurs have a lizard-like pelvis and are thus informally called lizard-hipped dinosaurs, whereas ornithischians have a bird-like pelvis and are called bird-hipped dinosaurs (• Figure 15.9). For decades, paleontologists thought that each order evolved independently during the Late Triassic, but it is now clear that they had a common ancestor much like the archosaurs known from Middle Triassic rocks in Argentina. These long-legged, small (less than 1 m long) dinosaur ancestors walked and ran on their hind limbs, so they were **bipedal,** as opposed to **quadrupedal** animals that move on all four limbs.

Dinosaurs

The term **dinosaur** was proposed in 1842 by Sir Richard Owen to mean "fearfully great lizard," although now "fearfully" has come to mean "terrible," thus the characterization of dinosaurs as "terrible lizards." Of course, we have no reason to think they were any more terrible than animals living today, and they were not

• **Figure 15.9 Cladogram Showing Relationships among Dinosaurs** Peivises of ornithischian and saurischian dinosaurs are shown for comparison. All the dinosaurs shown here were herbivores, except for theropods. Note that bidpedal and quadrupedal dinosaurs are found in both ornithischians and saurischians.

lizards. Nevertheless, these ideas persist and even their popularization in cartoons and movies has commonly been inaccurate and contributed to misunderstandings. For instance, many people think that all dinosaurs were large and that they were poorly adapted because they went extinct. Many were large—in fact among them were the largest animals ever to live on land. However, they varied from giants weighing several tens of metric tons to those that weighed only 2 or 3 kilograms. And to consider them poorly adapted is to ignore the fact that as a group they were extremely diverse and widespread for more then 140 million years!

Although various media now portray dinosaurs as more active animals, the mistaken belief that they were dim-witted lethargic beasts persists. Evidence now available indicates that some were brainy animals, at least by reptile standards, and more active than formerly thought. Perhaps some dinosaurs were even warm blooded. It also seems that some species cared for their young long after they hatched, a behavior that is found mostly in birds and mammals. Although many questions about dinosaurs remain unanswered, their fossils and the rocks containing them are revealing more and more about their evolution and behavior.

Saurischian Dinosaurs Paleontologists recognize two groups of saurischians known as *theropods* and *sauropods* (Figure 15.9). Theropods were bipedal carnivores that varied from tiny *Compsognathuis* to giants, such as *Tyrannosaurus* and similar, but even larger, genera (• Figure 15.10, and Table 15.1). Beginning in 1996, Chinese scientists have found several genera of theropods with feathers.* No one doubts that these dinosaurs had feathers, and molecular evidence indicates that they were composed of the same material as bird feathers.

The movie *Jurassic Park* popularized some of the smaller theropods, especially *Velociraptor*, a 1.8-m-long predator with large sickle-like claws on its back feet. This carnivore and its somewhat larger relative, *Deinonychus*, likely used these claws in a slashing type of attack (Figure 15.10b). Despite what you might see in movies and on TV, theropods, like predators today, probably avoided large, dangerous prey and went for the easy kill, preying on the young, old, or disabled, or they dined on carrion. No doubt the larger theropods simply chased smaller predators away from their kill. From the evidence available, some theropods such as diminutive *Coelophysis* and medium-sized *Deinonychus* hunted in packs.

The second group of saurischians, the sauropods, includes the truly giant, quadrupedal, herbivorous dinosaurs such as *Apatosaurus*, *Diplodocus*, and *Brachiosaurus*. Among the sauropods were the largest land animals ever (Table 15.1), and *Brachiosaurus* was a giant even by sauropod standards—it may have weighed 75 metric tons, and partial remains indicate that even

• Figure 15.10 **Theropod Dinosaurs Were Bipedal Carnivores** Theropods varied in size from tiny *Compsognathus* to giants such as *Tyrannosaurus* which was 13 m long and weighed several metric tons (see the chapter opening image).

a *Compsognathus* was about 60 cm long and weighed only 2 or 3 kg. Bones found in its ribcage indicate that it ate lizards.

b Life-like restoration of *Deinonychus*. It was 3 m long and may have weighed 80 kg. This dinosaur had a huge curved claw on each back foot.

Sue Monroe

---

*One feathered specimen purchased in Utah that was reported to have come from China turned out to be a forgery. It even appeared in *National Geographic* before scientists exposed it as a fraud.

Order	Suborder	Familiar Genera	Comments
TABLE 15.1			**Summary Chart for the Orders and Suborders of Dinosaurs (lengths and weights approximate, from several sources)**

Order	Suborder	Familiar Genera	Comments
Saurischia	Theropoda	*Allosaurus, Coelophysis, Compsognathus, Deinonychus, Tyrannosaurus,* * *Velociraptor*	Bipedal carnivores. Late Triassic to end of the Cretaceous. Size from 0.6 to 15 m long, 2 or 3 kg to 7.3 metric tons. Some smaller genera may have hunted in packs.
	Sauropoda	*Apatosaurus, Brachiosaurus, Camarasaurus, Diplodocus, Titanosaurus*	Giant quadrupedal herbivores. Late Triassic to Cretaceous, but most common during Jurassic. Size up to 27 m long, 75 metric tons.** Trackways indicate sauropods lived in herds. Preceded in fossil record by the smaller prosauropods.
Ornithischia	Ornithopoda	*Anatosaurus, Camptosaurus, Hypsilophodon, Iguanodon, Parasaurolophus*	Some ornithopods, such as *Anatosaurus*, had a flattened bill-like mouth and are called duck-billed dinosaurs. Size from a few meters long up to 13 m and 3.6 metric tons. Especially diverse and common during the Cretaceous. Primarily bipedal herbivores, but could also walk on all fours.
	Pachycepha-losauria	*Stegoceras*	*Stegoceras* only 2 m long and 55 kg, but larger species known. Thick bones of skull cap might have aided in butting contests for dominance and mates. Bipedal herbivores of the Cretaceous.
	Ankylosauria	*Ankylosaurus*	*Ankylosaurus* more than 7 m long and about 4.5 metric tons. Heavily armored with bony plates on top of head, back, and sides. Quadrupedal herbivore.
	Stegosauria	*Stegosaurus*	A variety of stegosaurs are known, but *Stegosaurus*, with bony plates on its back and a spiked tail, is best known. Plates probably were for absorbing and dissipating heat. Quadrupedal herbivores that were most common during the Jurassic. *Stegosaurus* 9 m long, 1.8 metric tons.
	Ceratopsia	*Triceratops*	Numerous genera known. Some early ones bipedal, but later large animals were quadrupedal herbivores. Much variation in size; *Triceratops* to 7.6 m long and 5.4 metric tons, with large bony frill over top of neck, three horns on skull, and beaklike mouth. Especially common during the Cretaceous.

*Tyrannosaurus at 4.5 metric tons was the largest known theropod, but now similar and larger animals are known from Argentina and Africa.

**Partial remains indicate even larger brachiosaurs existed, perhaps measuring 30 m long and weighing 100 metric tons.

larger ones existed. Trackways show that sauropods moved in herds.

Sauropods were preceded in the fossil record by the smaller, Late Triassic to Early Jurassic prosauropods, which were undoubtedly related to sauropods, but were probably not their ancestors. Sauropods were most common during the Jurassic; only a few genera existed during the Cretaceous.

Ornithischian Dinosaurs Recall that the distinguishing features of ornithischians is their bird-like pelvis. However, they differ from saurischians in other ways, too. For instance

ornithischians have no teeth in the front of the mouth, whereas saurischians do, and they also have ossified (bone-like) tendons in the back region. Scientists identify five groups of ornithischians: ornithopdods, pachycephalosaurs, ankylosaurs, stegosaurs, and ceratopsians (Figure 15.9, Table 15.1). Although you might not know the names of these groups, you have probably seen examples of them all.

In 1822, Gideon Mantell and his wife Mary Ann discovered some teeth that he later named *Iguanadon*, which proved to be a member of the ornithischian subgroup known as the ornithopods. Another ornithopod known as *Hadrosaurus* was discovered in North America in 1858, and it was the first dinosaur to be assembled and displayed in a museum. Among the several varieties of ornithopods, the duck-billed dinosaurs or hadrosaurs, were especially numerous during the Cretaceous, and several had crests on their heads that may have been used to amplify bellowing, for sexual display, or for species recognition (• Figure 15.11). All ornithopods were herbivores and primarily bipedal, but they had well developed forelimbs and could also walk in a quadrupedal manner.

Another duck-billed dinosaur of some interest is *Maisaura* ("good mother dinosaur") that nested in colonies and used the same nesting area repeatedly. Furthermore, their 2-m-diameter nests were spaced 7 m apart or about the length of an adult. Some nests contain the remains of juveniles up to 1 m long, which is much larger than when they hatched, so they must have stayed in the nest area where adults protected and perhaps fed them. Another interesting fact about these dinosaurs is that a bone bed in Montana has the remains of an estimated 10,000 individuals. The evidence indicates that they were overcome by volcanic gases and later buried by flood deposits.

The most distinctive feature of the bipedal, herbivorous pachycephalosaurs is their thick-boned domed skulls, although the earliest ones were flat headed. In any case, they varied from 1.0 to 4.5 m long and have been found only in Cretaceous-age rocks and only on the Northern Hemisphere continents. Although not accepted by all paleontologists, the traditional view is that these animals butted heads for dominance or competition for mates.

• **Figure 15.11 Hadrosaurs or Duck-billed Dinosaurs** This group of ornithopods had flattened bill-like mouths, and some had crests or other ornamentation on their heads. Images **a** and **b** are of fossils from China on temporary display at the Oregon Museum of Science and Industry in Portland.

a *Shantungosaurus* had the typical flattened bill-like mouth of hadrosaurs but it had no crest.

b *Corythosaurus* had a crest which was a bony extension of the skull.

c *Parasaurolophus* was also a crested hadrosaur.

d *Tsintaosaurus* had no crest, but it did have a bony projection from its skull, and thus is called the unicorn dinosaur.

The fossil record of ceratopsians (horned dinosaurs) indicates that small Late Jurassic bipeds were the ancestors of large Late Cretaceous quadrupeds such as *Triceratops* (• Figure 15.12a). *Triceratops* with three horns on its skull and related genera had a huge head and a bony frill over the top of the neck; they were especially common in North America during the Cretaceous. Fossil trackways show that these large herbivores moved in herds. Indeed, bone beds with dozens of individuals of a single species indicate that large numbers of animals perished quickly, probably during river crossings.

Everyone is familiar with *Stegosaurus* as a representative of the Stegosauria. It had the well-known plates along its back and spikes at the end of its tail for defense. The arrangement of the plates is not precisely known, but most restorations show two rows with the plates on one side offset from those on the other. Regardless of their arrangement, most paleontologists think the plates functioned to absorb and dissipate heat. These medium-size, quadrupedal, herbivores lived during the Jurassic. In addition to *Stegosaurus* there were several other genera that did not have broad plates on their back but rather had spikes.

Ankylosaurs were quadrupedal herbivores that were more heavily armored than any other dinosaur (Table 15.1), and as a result were not very fast; one estimate has their top speed at 10 km/hr. The animal's back, flanks, and top of its head were protected by bony armor, and the tail of some species, such as *Ankylosaurus*, ended in a huge bony club that could no doubt deliver a crippling blow to an attacking predator (see the chapter opening image). If the tail proved inadequate, the animal probably simply hunkered down; it would have been difficult even for *Tyrannosaurus* to flip over a 9-m-long, 4.5 metric ton *Ankylosaurus*.

Typically, dinosaurs have been depicted as aggressive, dangerous beasts, but we have every reason to think that they behaved much as land animals do now. Certainly, some lived in herds and no doubt interacted by bellowing, snorting, grunting, and foot stomping in defense, territorial disputes, and attempts to attract mates.

Warm-Blooded Dinosaurs?

Were dinosaurs **endotherms** (warm–blooded) like today's mammals and birds, or were they **ectotherms** (cold–blooded) like all of today's reptiles? Almost everyone now agrees some compelling evidence exists for dinosaur endothermy, but opinion is divided among (1) those holding that all dinosaurs were endotherms, (2) those who think only some were endotherms, and (3) those proposing that dinosaur metabolism, and thus the ability to regulate body temperature, changed as they matured.

Bones of endotherms typically have numerous passage-ways that, when the animals are alive, contain blood vessels, but the bones of ectotherms have considerably fewer passageways. Proponents of dinosaur

Sue Monroe

a Skeleton of the rhinoceros-sized, Late Cretaceous ceratopsian *Triceratops*.

b *Stegosaurus* from the Late Jurassic was about 9 m long. Notice the large plates along the back and the spikes on the tail.

endothermy note that dinosaur bones are more similar to the bones of living endotherms. However, crocodiles and turtles have this so-called endothermic bone, but they are ectotherms, and some small mammals have bone more typical of ectotherms. Perhaps bone structure is related more to body size and growth patterns than to endothermy, so this evidence is not conclusive.

Endotherms must eat more than comparably sized ectotherms because their metabolic rates are so much higher. Consequently, endothermic predators require large prey populations and thus constitute a much smaller proportion of the total animal population than their prey, usually only a few percent. In contrast, the proportion of ectothermic predators to prey may be as high as 50%. Where data are sufficient to allow an estimate, dinosaur predators made up 3% to 5% of the total population. Nevertheless, uncertainties in the data make this argument less than convincing for many paleontologists.

A large brain in comparison to body size requires a rather constant body temperature and thus implies endothermy. And some dinosaurs were indeed brainy, especially the small- and medium-sized theropods.

So brain size might be a convincing argument for these dinosaurs. Even more compelling evidence for theropod endothermy comes from their relationship to birds and from the recent discoveries in China of dinosaurs with feathers or a featherlike covering, (• Figure 15.13). Today, only endotherms have hair, fur, or feathers for insulation.

Some scientists point out that certain duck-billed dinosaurs grew and reached maturity much more quickly than would be expected for ectotherms and conclude that they must have been warm blooded. Furthermore, a fossil ornithopod discovered in 1993 has a preserved four-chambered heart much like that of living mammals and birds. Three-dimensional imaging of this structure, now on display at the North Carolina Museum of Natural Sciences, has convinced many scientists that this animal was an endotherm.

Good arguments for endothermy exist for several types of dinosaurs, particularly theropods, although the large sauropods were probably not endothermic, but nevertheless were capable of maintaining a rather constant body temperature. Large animals heat up and cool down more slowly than smaller ones because they have a small surface area compared to their volume. With their comparatively smaller surface area for heat loss, sauropods probably retained heat more effectively than their smaller relatives.

Flying Reptiles

Paleozoic insects were the first animals to achieve flight, but the first among vertebrates were **pterosaurs,** or flying reptiles, which were common in the skies from the Late Triassic until their extinction at the end of the Cretaceous (• Figure 15.14). Adaptations for flight include a wing membrane supported by an elongated fourth finger (Figure 15.14c), light, hollow bones; and development of those parts of the brain that controlled muscular coordination and sight.

Pterosaurs are generally depicted in movies as large, aggressive creatures, but some were no bigger than today's sparrows, robins, and crows. However, a few species had wingspans of several meters, and the wingspan of one Cretaceous pterosaur was at least 12 m! Nevertheless, even the very largest species probably weighed no more than a few tens of kilograms.

Experiments and studies of fossils indicate that the wing bones of large pterosaurs such as *Pteranodon* (Figure 15.14b) were too weak for sustained flapping. These comparatively large animals probably took advantage of rising air currents to stay airborne, mostly by soaring, but occasionally flapping their wings for maneuvering. In contrast, smaller pterosaurs probably stayed aloft by vigorously flapping their wings just as present-day small birds do.

At least one small pterosaur called *Sordes pilosus* (hairy devil) found in 1971 in what is now Kazahkstan had a coat of hair, or hairlike feathers. This outer covering and the fact that wing flapping requires a high metabolic rate and efficient respiratory and circulatory systems as in present-day birds, indicates that some, or perhaps all, pterosaurs were warm blooded.

Mesozoic Marine Reptiles

Several types of Mesozoic reptiles adapted to a marine environment, including turtles and some crocodiles, as well as the Triassic mollusk-crushing placodonts. Here, though, we concentrate on the ichthyosaurs and plesiosaurs and the less familiar mosasaurs. All were thoroughly aquatic marine predators, but other than all being reptiles, they were not closely related to one another. Furthermore, none were dinosaurs, although some popular media depict them as such.

The streamlined, rather porpoise-like **ichthyosaurs** varied from species measuring only 0.7 m long to giants more than 15 m long (• Figure 15.15a). Details of their ancestry are still not clear, but fossil ichthyosaurs from Japan prompted researcher Ryosuke Motani to say," I knew *Utatsusaurus* was exactly what paleontologists had been

• **Figure 15.13 Feathered Dinosaur** Restoration of the Early Cretaceous feathered dinosaur *Caudipteryx* from China. The fact that *Caudipteryx* had short forelimbs, symmetric feathers, and was larger than the oldest known bird indicate that it was flightless.

• Figure 15.14 **The Pterosaurs (Flying Reptiles)**

a *Pterodactylus* is a well known Late Jurassic long-tailed ptero-saur. Among the several species of this genus, wingspans var-ied from 50 cm to 2.5 m.

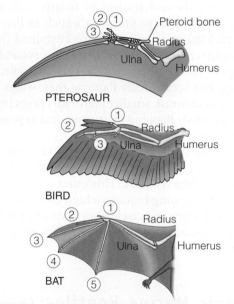

c In all flying vertebrates, the wing is a modified forelimb. A long fourth finger supports the pterosaur wing, but in birds the second and third are fused, and in bats fingers 2 through 5 support the wing.

b The short-tailed pterosaur known as *Pteranodon* was a large, Cretaceous animal with a wingspan of more than 6 m.

expecting to find for years: an ichthyosaur that looked like a lizard with legs."*

Ichthyosaurs used their powerful tail for propul-sion and maneuvered with their flipperlike forelimbs. They had numerous sharp teeth, and preserved stomach contents reveal a diet of fish, cephalopods, and other ma-rine organisms. It is doubtful that ichthyosaurs could come

onto land, so females must have retained eggs within their bodies and given birth to live young. A few fossils with small ichthyosaurs in the appropriate part of the body cav-ity support this interpretation.

An interesting side note in the history of paleontology is the story of Mary Anning (see Prespective), who, when she was only about 11 years old, discovered and directed the excavation of a nearly complete ichthyosaur in southern England.

The **plesiosaurs** belonged to one of two subgroups: short-necked and long-necked (Figure 15.5 and 15.15b). Most were modest-sized animals 3.6 to 6 m long, but one species found in Antarctica measures 15 m. Short-necked plesiosaurs might have been bottom feeders, but their long-necked cousins may have used their necks in a snakelike fashion, and their numerous sharp teeth, to capture fish. These animals probably came ashore to lay their eggs.

Mosasaurs were Late Cretaceous marine lizards related to the present-day Komodo dragon or monitor lizard. Some species measured no more than 2.5 m long, but a few such as *Tylosaurus* were large, measuring up to 9 m. Mosasaur limbs resemble paddles and were used mostly for maneuvering, whereas the long tail provided propulsion. All were predators, and preserved stomach contents indicate they ate fish, birds, smaller mosasaurs, and a variety of invertebrates, including ammonoids.

A possible mosasaur ancestor was found in Texas in 1989, although it was several years before it was examined

*Ryosuke Motani. 2004. Rulers of the Jurassic Seas. Scientific American. v. 14, no. 2. p. 7.

a The ichthyosaurs looked and probably lived much like today's porpoises. Most were a few meters long, but some exceeded 15 m long.

b Plesiosaurs were also aquatic, but their flipper-like forelimbs probably allowed them to come out on land. There were long- and short-necked plesiosaurs (see Figure 15.5).

and the results published. In any case, *Dallasaurus* from Upper Cretaceous rocks has a mosasaur skeleton, but rather than paddle-like flippers it retained complete limbs for walking on land.

Crocodiles, Turtles, Lizards, and Snakes

All crocodiles are amphibious, spending much of their time in water, but they are well equipped for walking on land. By Jurassic time, crocodiles had become the most common freshwater predators. Overall, crocodile evolution has been conservative, involving changes mostly in size from a meter or so in Jurassic forms to 15 m in some Cretaceous species.

Turtles, too, have been evolutionarily conservative since their appearance during the Triassic. The most remarkable feature of turtles is their heavy, bony armor; they are more thoroughly armored than any other vertebrate animal, living or fossil. Turtle ancestry is uncertain. One Permian animal had eight broadly expanded ribs, which may represent the first stages in the development of turtle armor.

Lizards and snakes are closely related, and lizards were in fact ancestral to snakes. The limbless condition in snakes (some lizards are limbless, too) and skull modifications that allow snakes to open their mouths very wide are the main difference between these two groups. Lizards are known from Upper Permian strata, but they were not abundant until the Late Cretaceous. Snakes first appear during the Cretaceous, but the families to which most living snakes belong differentiated since the Early Miocene. One Early Cretaceous genus from Israel shows characteristics intermediate between snakes and their lizard ancestors.

From Reptiles to Birds

Birds have feathers, whereas reptiles have scales or a tough beaded skin, and birds do not closely resemble any living reptile. So why do scientists think that birds evolved from reptiles? Long ago, scientists were aware of the probable relationships between these two groups of animals. Birds and reptiles both lay shelled, yolked eggs, and both share several skeletal characteristics, such as the way the jaw attaches to the skull. Furthermore, since 1860, approximately 10 fossils have been recovered from the Solnhofen Limestone of Germany that provide evidence for reptile–bird relationships. The fossils definitely have feathers and a wishbone, consisting of the fused clavicle bones so typical of birds, and yet, in most other physical characteristics, they most closely resemble small theropod dinosaurs.

These remarkable fossils, known as *Archaeopteryx* (from Greek "archaios," ancient and "pteryx", feather) are birds by definition, but their numerous reptilian features convince scientists that their ancestors were among theropods (• Figure 15.16).

Even the fused clavicles (wishbone) are found in several theropods, and during the last several years, paleontologists in China have discovered theropods with feathers, providing more evidence for this relationship. The few that oppose the theropod–bird view note that theropods are found in Cretaceous-age rocks, but *Archaeopteryx* is Jurassic. However, some of the fossils from China are about the same age as *Archaeopteryx* narrowing the gap between presumed ancestor and descendant.

Perspective

Mary Anning and Her Contributions to Paleontology

Men and women from many countries contribute to our understanding of prehistoric life, but this has not always been the case. The early history of paleontology was dominated by Western European males, but there was a notable exception: Mary Anning (1799–1847), who began a remarkable career as a fossil collector when she was only 11 years old (Figure 1).

Mary Anning was born in Lyme Regis on England's south coast. When only 15 months old, she survived a lightning strike that, according to one report, killed three girls, and according to another, killed a nurse tending her. Of the ten or so children in the Anning family, only Mary and her brother Joseph reached maturity.

In 1810 Mary's father, a cabinetmaker who also sold fossils part time, died leaving the family nearly destitute. Mary Anning expanded the fossil business and became a professional fossil collector known to the paleontologists of her time, some of whom visited her shop to buy fossils or gather information. She collected fossils from the Dorset coast near Lyme Regis and is reported to have been the inspiration for the tongue twister "She sells seashells by the sea shore."

Soon after her father's death, she made her first important discovery: a nearly complete skeleton of a Jurassic ichthyosaur, which was described in 1814 by Sir Everard Home. The sale of this fossil specimen provided considerable financial relief for her family. In 1821 she made a second major discovery and excavated the remains of another Mesozoic marine reptile, a plesiosaur. And in 1818 she excavated the remains of the first Mesozoic flying reptile (pterosaur) found in England, which was sent to the eminent geologist William Buckland of Oxford University.

By 1830 Mary Anning's fortunes began declining as collectors and museums had less money to spend on fossils. Indeed, she may once again have become destitute if it had not been for her friend Henry Thomas de la Beche, also a resident of Lyme Regis. De la Beche drew a fanciful scene called *Duria antiquior*, meaning "An earlier Dorset," in which he brought to life the fossils Mary Anning had collected. The scene was printed and sold widely, the proceeds of which went directly to Mary Anning.

Mary Anning died of cancer in 1847, and although only 48 years old, she had a fossil-collecting career that spanned 36 years. Her contributions to paleontology are now widely recognized, but, unfortunately, soon after her death she was mostly forgotten. Apparently, people who purchased her fossils were credited with finding them. So even though Mary Anning became a respected fossil collector, many scientists of that time could not accept that an untutored girl could possess such knowledge and skill. "It didn't occur to them to credit a woman from the lower classes with such astonishing work. So an uneducated little girl, with a quick mind and an accurate eye, played a key role in setting the course of the 19th-century geologic revolution. Then we simply forgot about her."[*]

Although Mary Anning's contributions to paleontology were mostly forgotten, beginning in 2002, The Palaeontological Association has presented The Mary Anning Award to someone who is "not professionally employed within palaeontology but who made an outstanding contribution to the subject[**]."

The Natural History Museum, London

Figure 1 Mary Anning made several important fossil discoveries in England during the early 1800s, but she was largely forgotten after her death in 1847.

---

[*]John Lienhard, University of Houston.
[**]The Palaeontological Association (http://www.palass.org/modules.php?name=palaeo&sec=Awards&page=122)

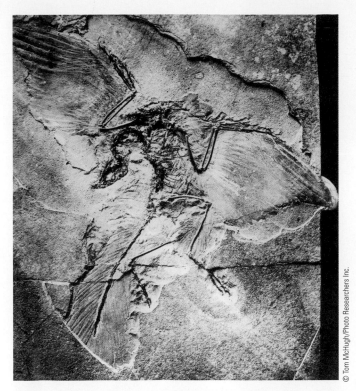

© Tom McHugh/Photo Researchers Inc.

• Figure 15.16 *Archaeopteryx* **From the Jurassic-age Solnhofen Limestone of Germany** Fossil showing the impressions of wing feathers. This animal had feathers and a wishbone, so it is a bird, but in most details of its anatomy it more closely resembles small theropod dinosaurs—it had reptile-like teeth, claws on its wings, and a long tail.

Another fossil bird from China that is slightly younger than *Archaeopteryx* retains ribs in the abdominal region just as *Archaeopteryx* and small theropods, but it has a reduced tail more like present-day birds. More fossils found in China in 2004 and 2005 of five specimens of an Early Cretaceous bird indicate that today's birds may have had an aquatic ancestor. With few exceptions, the bones of these birds, known as *Gansus yumenesis*, are much like those of living birds.

The fossils of *Archaeopteryx* are significant, but there are not enough of them or of other early birds to resolve whether *Archaeopteryx* was the ancestor of today's birds or an early bird that died out without leaving descendants. Of course, this does not diminish the fact that *Archaeopteryx* had both reptile and bird features (recall the concept of *mosaic evolution* from Chapter 7). However, there is another candidate for earliest bird. Some claim that fossils of two crow-sized individuals, known as *Protoavis*, from Upper Triassic rocks in Texas, are birds. Protoavis does have hollow bones and a wishbone as today's birds do, but because the specimens are fragmentary and no feather impressions were found, most paleontologists think that they are small theropods.

One hypothesis for the origin of bird flight—*from the ground up*—holds that the ancestors of birds were bipedal, fleet-footed ground dwellers that used their wings to leap into the air, at least for short distances, to catch insects or to escape predators. The *from-the-trees-down* hypothesis holds that bird ancestors were bipeds that climbed trees and used their wings for gliding or parachuting. The from-the-ground-up hypothesis is probably better supported in that a bipedal theropod ancestor is reasonable because small theropods had forelimbs much like those of *Archaeopteryx*. However, from-the-trees-down has an advantage because takeoff from an elevated position is easier, although landing is a challenge.

Origin and Evolution of Mammals

Recall from Chapter 13 that mammal-like reptiles called **therapsids** diversified into many species of herbivores and carnivores during the Permian Period. In fact, they were the most numerous and diverse land-dwelling vertebrates at that time. Among the therapsids one group known as **cynodonts** was the most mammal-like of all, and by Late Triassic time true mammals evolved from them.

Cynodonts and the Origin of Mammals

We can easily recognize living mammals as warm-blooded animals that have hair or fur and mammary glands and, except for the platypus and spiny anteater, give birth to live young. However, these criteria are not sufficient for recognizing fossil mammals; for them, we must rely on skeletal structure only. Several skeletal modifications took place during the transition from mammal-like reptiles to mammals, but distinctions between the two groups are based mostly on details of the middle ear, the lower jaw, and the teeth (Table 15.2). Fortunately, the evolution of mammals from cynodonts is so well documented by fossils that classification of some fossils as reptile or mammal is difficult.

Reptiles have one small bone in the middle ear (the stapes), whereas mammals have three: the incus, the malleus, and the stapes. Also, the lower jaw of a mammal is composed of a single bone called the dentary, but a reptile's jaw is composed of several bones (• Figure 15.17). In addition, a reptile's jaw is hinged to the skull at a contact between the articular and quadrate bones, whereas in mammals the dentary contacts the squamosal bone of the skull (Figure 15.17).

During the transition from cynodonts to mammals, the quadrate and articular bones that had formed the joint between the jaw and skull in reptiles were modified into the incus and malleus of the mammalian middle ear (Figure 15.17, Table 15.2). Fossils document the progressive enlargement of the dentary until it became the only element in the mammalian jaw. Likewise, a progressive change from the reptile to mammal jaw joint is documented by fossil evidence. In fact, some of the most advanced cynodonts were truly transitional, because they had a compound jaw joint consisting of (1) the articular and quadrate bones typical of reptiles and (2) the dentary and squamosal bones as in mammals (Table 15.2).

Several other features of cynodonts also indicate they were ancestors of mammals. Their teeth were becoming double-rooted as they are in mammals, and they were somewhat differentiated into distinct types that performed specific functions. In mammals the teeth are fully differentiated into incisors, canines, and chewing teeth, but typical reptiles do not have differentiated teeth (• Figure 15.18). In addition, mammals have only two sets of teeth during their lifetimes—a set of baby teeth and the permanent adult teeth. Reptiles, with the exception of some cynodonts, have teeth replaced continuously throughout their lives, but cynodonts in mammal fashion had only two sets of teeth. Another important feature of mammal teeth is occlusion; that is, the chewing teeth meet surface to surface to allow grinding. Thus, mammals chew their food, but reptiles, amphibians, and fish do not. However, tooth occlusion is known in some advanced cynodonts (Table 15.2).

Reptiles and mammals have a bony protuberance from the skull that fits into a socket in the first vertebra: the atlas. This structure, called the *occipital condyle*, is a single feature in typical reptiles, but in cynodonts it is partly divided into a double structure typical of mammals (Table 15.2). Another mammalian feature, the secondary palate, was partially developed in advanced cynodonts. This bony shelf separating the nasal passages from the mouth cavity is an adaptation for eating and breathing at the same time, a necessary requirement for endotherms with their high demands for oxygen.

In 1837, the German scientist Karl Reichert discovered that the embryos of mammals have an extra bone, the articular, in the lower jaw whereas adult mammals have one bone, the dentary. He also found an extra bone in the upper jaw called the quadrate. Furthermore, these two bones formed the jaw skull joint just as they do in reptiles. However, as the embryo matured, the articular and quadrate moved to the middle ear where they became the incus and malleus, and the jaw-skull joint typical of mammals developed. In fact, when opossums are born they have this reptile-type jaw-skull joint between the articular and quadrate, but as they

• **Figure 15.17 Evolution of the Mammal Jaw and Middle Ear** Note also that the teeth of mammals are fully differentiated into incisors, canines, and chewing teeth (premolars and molars).

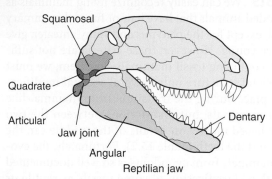

a Cynodonts had several bones in the jaw and a jaw-skull joint between the articular and quadrate bones.

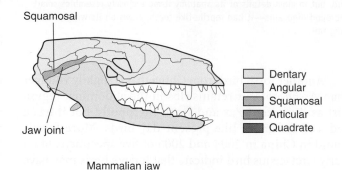

b Mammals have one bone in the jaw, the dentary, and a jaw-skull joint between the dentary and squamosal bones.

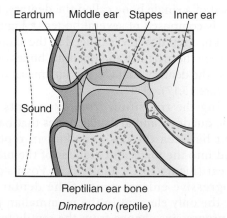

Dimetrodon (reptile)

c In reptiles, sound is carried from the ear drum through the stapes, the only bone in the middle ear.

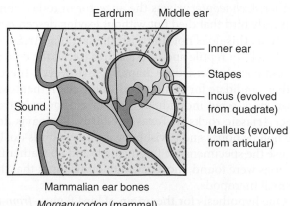

Morganucodon (mammal)

d Mammals have three middle ear bones: the stapes, malleus, and incus. The malleus and incus were derived from the articular and quadrate bones.

Features	Typical Reptile	Cynodont	Mammal
Lower Jaw	Dentary and several other bones	Dentary enlarged, other bones reduced	Dentary bone only, except in earliest mammals
Jaw-Skull Joint	Articular-quadrate	Articular-quadrate; some advanced cynodonts had both the reptile jaw-skull joint and the mammal jaw-skull joint	Dentary-squamosal
Middle-Ear Bones	Stapes	Stapes	Stapes, incus, malleus
Secondary Palate	Absent	Partly developed	Well developed
Teeth	No differentiation; chewing teeth single rooted	Some differentiation; chewing teeth partly double rooted	Fully differentiated into incisors, canines, and chewing teeth; chewing teeth double rooted
Tooth Replacement	Teeth replaced continuously	Only two sets of teeth in some advanced cynodonts	Two sets of teeth
Occipital Condyle	Single	Partly divided	Double
Occlusion (chewing teeth meet surface to surface to allow grinding)	No occlusion	Occlusion in some advanced cynodonts	Occlusion
Endothermic vs. Ectothermic	Ectothermic	Probably endothermic	Endothermic
Body Covering	Scales	One fossil shows it had skin similar to that of mammals	Skin with hair or fur

• Figure 15.18 **Comparison of the Teeth of a Mammal and a Reptile**

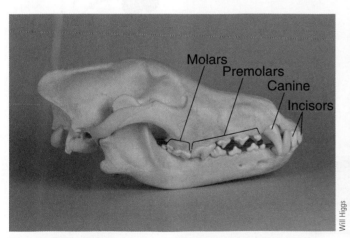

a This wolf skull shows that mammal teeth are differentiated into incisors, canines, premolars, and molars.

b Reptiles, represented here by a crocodile, may have teeth that vary somewhat in size, but otherwise they all look the same. The only exception is among some mammal-like reptiles.

develop further, these bones move to the middle ear and the a typical mammal-type jaw-skull joint forms.

Mesozoic Mammals

Mammals evolved during the Late Triassic not long after the first dinosaurs appeared, but for the rest of the Mesozoic Era most of them were small. There were, however, a few exceptions. One, a Middle Jurassic-age aquatic mammal found in China, measures about 50 cm long,

and it also has the distinction of being the oldest known fossil with fur. The other is an Early Cretaceous-age mammal called *Repenomamus giganticus*, also from China, that was about 1 m long, weighed 12 to 14 kg, and had the remains of a juvenile dinosaur in its stomach. Most other Mesozoic mammals were about the size of mice and rats, and they were not very diverse—certainly not as diverse as they were during the Cenozoic Era. Furthermore, they retained reptile characteristics but had mammalian features, too. The Triassic triconodonts, for instance, had the fully differentiated teeth typical of mammals, but they also had both the reptile and the mammal types of jaw joints. In short, some mammal features appeared sooner than others (remember the concept of *mosaic evolution* from Chapter 7).

The early mammals diverged into two distinct branches. One branch includes the triconodonts (• Figure 15.19) and their probable descendants, the **monotremes,** or egg-laying mammals such as the spiny anteater and platypus of the Australian region. The other branch includes the **marsupial (pouched) mammals** and the **placental mammals** and their ancestors, the eupantotheres (Figure 15.19).

Although the history of the monotremes is uncertain, fossils of several Mesozoic animals are relevant to the evolution of marsupials and placentals. In fact, the divergence of marsupials and placental mammals from a common ancestor took place during the Early Cretaceous (• Figure 15.20).

Mesozoic Climates and Paleobiogeography

Fragmentation of the supercontinent Pangaea began by the Late Triassic, but during much of the Mesozoic, close connections existed between the various landmasses. The proximity of these landmasses alone, however, is not enough to explain Mesozoic biogeographic distributions, because climates are also effective barriers to wide dispersal. During much of the Mesozoic, though, climates were more equable and lacked the strong north and south zonation characteristic of the present. In short, Mesozoic plants and animals had greater opportunities to occupy much more extensive geographic ranges.

Pangaea persisted as a supercontinent through most of the Triassic (see Figure 14.1a), and the Triassic climate was warm temperate to tropical, although some areas, such as the present southwestern United States,

• Figure 15.20 **Restorations of the Oldest Known Marsupial and Placental Mammals** Both of these restorations are based on fossils that were found in Lower Cretaceous rocks in China.

a *Sinodelphys*, the oldest known marsupial, was only about 15 cm long.

b This restoration shows *Eomaia*, which was only 12 or 13 cm long. It is the oldest known placental mammal.

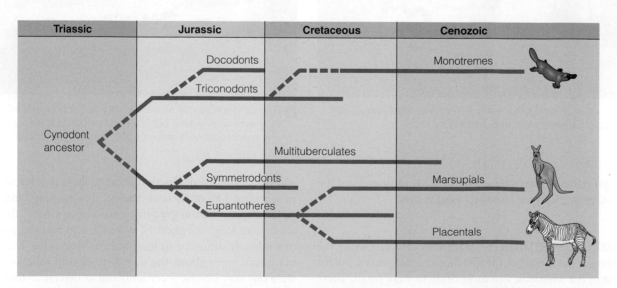

• Figure 15.19 **Relationships among the Early Mammals and their Descendants** Mammal evolution proceeded along two branches, one leading to today's monotremes, or egg-laying mammals, and the other to marsupial and placental animals.

were arid. Mild temperatures extended 50 degrees north and south of the equator, and even the polar regions may have been temperate. The fauna had a truly worldwide distribution. Some dinosaurs had continuous ranges across Laurasia and Gondwana, the peculiar gliding lizards lived in New Jersey and England, and reptiles known as phytosaurs lived in North America, Europe, and Madagascar.

By the Late Jurassic, Laurasia had become partly fragmented by the opening North Atlantic, but a connection still existed. The South Atlantic had begun to open so that a long, narrow sea separated the southern parts of Africa and South America. Otherwise the southern continents were still close together.

The mild Triassic climate persisted into the Jurassic. Ferns whose living relatives are now restricted to the tropics of southeast Asia lived as far as 63 degrees south and 75 degrees north. Dinosaurs roamed widely across Laurasia and Gondwana. For example, the giant sauropod *Brachiosaurus* is found in western North America and eastern Africa. Stegosaurs and some families of carnivorous dinosaurs lived throughout Laurasia and in Africa.

By the Late Cretaceous, the North Atlantic had opened further, and Africa and South America were completely separated (see Figure 14.1c). South America remained an island continent until late in the Cenozoic, and its fauna became increasingly different from faunas of the other continents (see Chapter 18). Marsupial mammals reached Australia from South America via Antarctica, but the South American connection was eventually severed. Placentals, other than bats and a few rodents, never reached Australia, explaining why marsupials continue to dominate the continent's fauna even today.

Cretaceous climates were more strongly zoned by latitude, but they remained warm and equable until the close of that period. Climates then became more seasonal and cooler, a trend that persisted into the Cenozoic. Dinosaur and mammal fossils demonstrate that interchange was still possible, especially between the various components of Laurasia.

Mass Extinctions—A Crisis in Life History

There are too few Precambrian fossils to know if mass extinctions occurred then, but we know that during the Phanerozoic there were at least five of these events when Earth's biotic diversity was drastically reduced and several others of lesser impact (see Figure 12.19). The greatest of these mass extinctions took place at the end of the Paleozoic Era, but the one of interest here was at the end of the Mesozoic. It has certainly attracted more attention than any other mass extinction because dinosaurs and many of their relatives died out, but it was equally devastating for several types of marine invertebrates, including ammonites, rudist bivalves, and some planktonic organisms.

Many hypotheses have been proposed to account for Mesozoic extinctions, but most have been dismissed as improbable or inconsistent with the available data. In 1980, however, a proposal was made that has gained wide acceptance. It was based on a discovery at the Cretaceous–Paleogene boundary in Italy of a 2.5-cm-thick clay layer with a notable concentration of the platinum-group element iridium. High iridium concentrations are now known from many other Cretaceous–Paleogene boundary sites (• Figure 15.21a).

The significance of this **iridium anomaly** is that iridium is rare in crustal rocks but is found in much higher concentrations in some meteorites. Accordingly, some investigators propose a meteorite impact to explain the anomaly and further postulate that the meteorite, perhaps 10 km in diameter, set in motion a chain of events leading to extinctions. Some Cretaceous–Paleogene boundary sites also contain soot and shock-metamorphosed quartz grains, both of which are cited as additional evidence of an impact.

According to the impact hypothesis, about 60 times the mass of the meteorite was blasted from the crust high into the atmosphere, and the heat generated at impact started raging forest fires that added more particulate matter to the atmosphere. Sunlight was blocked for several months, temporarily halting photosynthesis; food chains collapsed; and extinctions followed. Furthermore, with sunlight greatly diminished, Earth's surface temperatures were drastically reduced, adding to the biologic stress. Another consequence of the impact was that vaporized rock and atmospheric gases produced sulfuric acid (H_2SO_4) and nitric acid (HNO_3). Both would have contributed to strongly acid rain that may have had devastating effects on vegetation and marine organisms.

Now some geologists point to a probable impact site centered on the town of Chicxulub on the Yucatán Peninsula of Mexico (Figure 15.21b). The 170 km diameter structure lies beneath layers of sedimentary rock and appears to be the right age. Evidence that supports the conclusion that the Chicxulub structure is an impact crater includes shocked quartz, the deposits of huge waves, and tektites—small pieces of rock melted during the impact and hurled into the atmosphere.

Even if a meteorite did hit Earth, did it lead to these extinctions? If so, both terrestrial and marine extinctions must have occurred at the same time. To date, strict time equivalence between terrestrial and marine extinctions has not been demonstrated. The selective nature of the extinctions is also a problem. In the terrestrial realm, large animals were the most affected, but not all dinosaurs were large, and crocodiles, close relatives of dinosaurs, survived, although some species died out. Some paleontologists think dinosaurs, some marine invertebrates, and many plants were already on the decline and headed for extinction before the end of the Cretaceous. A meteorite impact may have simply hastened the process.

• Figure 15.21 **End of Mesozoic Extinctions**

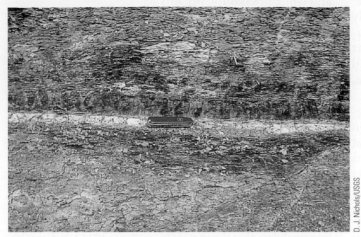

a Closeup view of the iridium-enriched, Cretaceous-Paleogene boundary clay (the white layer) in the Raton Basin, New Mexico.

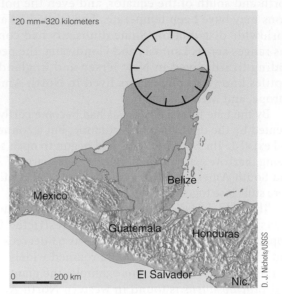

*20 mm=320 kilometers

b Proposed meteorite impact site centered on Chicxulub on the Yucatan Peninsula of Mexico.

In the final analysis, Mesozoic extinctions have not been explained to everyone's satisfaction. Most geologists now concede that a large meteorite impact occurred, but we also know that vast outpourings of lava were taking place in what is now India. Perhaps these brought about detrimental atmospheric changes. Furthermore, the vast, shallow seas that covered large parts of the continents had mostly withdrawn by the end of the Cretaceous, and the mild equable Mesozoic climates became harsher and more seasonal by the end of that era. Nevertheless, these extinctions were selective, and no single explanation accounts for all aspects of this crisis in life history.

SUMMARY

Table 15.3 summarizes the major Mesozoic evolutionary and climatic events.

- Invertebrate survivors of the Paleozoic extinctions diversified and gave rise to increasingly diverse marine communities.

- Some of the most abundant invertebrates were cephalopods, especially ammonoids, foraminifera, and the reef-building rudists.

- Land plant communities of the Triassic and Jurassic consisted of seedless vascular plants and gymnosperms. The angiosperms, or flowering plants, evolved during the Early Cretaceous, diversified rapidly, and were soon the most abundant land plants.

- Dinosaurs evolved from small, bipedal archosaurs during the Late Triassic, but they were most common during the Jurassic and Cretaceous periods.

- All dinosaurs evolved from a common ancestor but differ enough that two distinct orders are recognized: the Saurischia and the Ornithischia.

- Bone structure, predator–prey relationships, and other features have been cited as evidence of dinosaur endothermy. Although there is still no solid consensus, many paleontologists think some dinosaurs were indeed endotherms.

- That some theropods had feathers indicates they were warm-blooded and provides further evidence of their relationship to birds.

- Pterosaurs, the first flying vertebrates, varied from sparrow size to comparative giants. The larger pterosaurs probably depended on soaring to stay aloft, whereas smaller ones flapped their wings. At least one species had hair or hairlike feathers.

- The fish-eating, porpoise-like ichthyosaurs were thoroughly adapted to an aquatic environment, whereas the plesiosaurs with their paddle-like limbs could most likely come out of the water to lay their eggs. The marine reptiles known as mosasaurs were most closely related to lizards.

- During the Jurassic, crocodiles became the dominant freshwater predators. Turtles and lizards were present during most of the Mesozoic. By the Cretaceous, snakes had evolved from lizards.

- Jurassic-age *Archaeopteryx*, the oldest known bird, possesses so many theropod characteristics that it has convinced most paleontologists the two are closely related.

- Mammals evolved by the Late Triassic, but they differed little from their ancestors, the cynodonts. Minor differences in the lower jaw, teeth, and middle ear differentiate one group of fossils from the other.

- Several types of Mesozoic mammals existed, but most were small, and their diversity was low. Both marsupial and placental mammals evolved during the Cretaceous from a group known as eupantotheres.
- Because during much of the Mesozoic the continents were close together and climates were mild, plants and animals occupied much larger geographic ranges than they do now.
- Among the victims of the Mesozoic mass extinction were dinosaurs, flying reptiles, marine reptiles, and several groups of marine invertebrates. A huge meteorite impact may have caused these extinctions, but some paleontologists think other factors were important, too.

IMPORTANT TERMS

angiosperm, p. 308
Archaeopteryx, p. 317
archosaur, p. 310
bipedal, p. 310
cynodont, p. 319
dinosaur, p. 310
ectotherm, p. 314

endotherm, p. 314
ichthyosaur, p. 315
iridium anomaly, p. 323
marsupial mammal, p. 322
monotreme, p. 322
mosasaur, p. 316
Ornithischia, p. 310

placental mammal, p. 322
plesiosaur, p. 316
pterosaur, p. 315
quadrupedal, p. 310
Saurischia, p. 310
therapsid, p. 319

REVIEW QUESTIONS

1. The group of organisms known as angiosperms includes all
   a. _____ flowering plants; b. _____ ancestors of dinosaurs; c. _____ plankatonic bivalves; d. _____ mammal-like reptiles; e. _____ bipedal ectotherms.

2. All dinosaurs with a bird-like pelvis belong to the order
   a. _____ Therapsida; b. _____ Crossopterygii; c. _____ Pterosauria; d. _____ Ornithischia; e. _____ Ceratopsia.

3. The middle ear bones of mammals evolved from which of these bones in the mammal-like reptiles?
   a. _____ Palatine and vomer; b. _____ Articular and quadrate; c. _____ Prefrontal and parietal; d. _____ Dentary and incus; e. _____ Tibia and fibula.

4. Which one of the following is a Mesozoic marine reptile?
   a. _____ Teleost; b. _____ Plesiosaur; c. _____ Pterosaur; d. _____ Cynodont; e. _____ Monotreme.

5. Which one of the following statements is correct?
   a. _____ The first flying vertebrates were placoderms; b. _____ Ichthyosaurs look much like living opossums; c. _____ The first mammals evolved during the Cretaceous; d. _____ Rudists were important Mesozoic reef-building animals; e. _____ Most Triassic plants were angiosperms.

6. Modification of the hand yielding an elongated finger for wing support is found in
   a. _____ birds; b. _____ insects; c. _____ theropods; d. _____ bats; e. _____ pterosaurs.

7. Because of their rapid evolution and nektonic lifestyle, the _____ are good guide fossils.
   a. _____ amphibians; b. _____ cephalopods; c. _____

burrowing worms; d. _____ saurischians; e. _____ bivalves.

8. Which of the following were common during the Mesozoic, and are the major primary producers in the warm seas today?
   a. _____ Ammonoids; b. _____ Nautiloids; c. _____ Coleoids; d. _____ Dinoflagellates; e. _____ Rudists.

9. Theropods were
   a. _____ long-necked marine reptiles; b. _____ egg-laying mammals; c. _____ bipedal, carnivorous dinosaurs; d. _____ dome-headed, herbivorous mammals; e. _____ flying reptiles.

10. The eupantotheres were the
   a. _____ probable ancestor of placental and marsupial mammals; b. _____ most dolphin-like of all ichthyosaurs; c. _____ largest of all sauropods; d. _____ first vertebrate animals capable of flight; e. _____ most diverse bivalved gymnosperms.

11. During the transition from mammal-like reptiles to mammals what changes took place in the jaw and middle ear? Are there any other features that indicate these two groups of animals are closely related?

12. Why classify *Archaeopteryx* as a bird? After all, we now know that several dinosaurs had feathers.

13. Explain how plate position and climate influenced the geographic distribution of Mesozoic plants and animals.

14. Summarize the evidence that convinces many geologists that an asteroid impact took place at the end of the Mesozoic Era.

15. What are the three main groups of mammals and how do they differ from one another?

16. Describe the modifications for flight that occurred in pterosaurs.
17. What were the main communities of plants during the Triassic, Jurassic, and Cretaceous periods? Which one predominates now and why has it been so successful?
18. What are the two main groups of dinosaurs and how do they differ from one another?
19. Why do scientists think that at least some dinosaurs were warm-blooded?
20. Briefly summarize the overall trends among marine invertebrates during the Mesozoic Era.

APPLY YOUR KNOWLEDGE

1. In your high school science class a student notices that ichthyosaurs and porpoises are similar looking, and she speculates that the former is ancestral to the latter. After all, they have comparable shapes, both are marine predators, and ichthyosaurs lived before porpoises, so, she reasons, there must be a relationship between them. How would you explain that there is no evidence supporting her conclusion? (Hint: Remember the discussion in Chapter 7 and see the discussion on whales in Chapter 18.)
2. You observe limestone beds with fossil trilobites and brachiopods dipping at 50 degrees, but an overlying layer of volcanic ash followed upward by sandstone beds with dinosaur fossils dips at only 15 degrees. A basalt dike cuts though all of the strata. Explain the sequence of events that took place. What basic principles did you use to make your interpretation? Is it possible to determine the absolute ages of any of the events? If so, explain.
3. Construct a cladogram that shows the relationship among birds, saurischians, ornithischians, mammals, and cynodonts. (Hint: See Figures 15.8 and 15.21.)
4. Living crocodiles have a fairly well developed secondary palate and a 4-chambered heart like that in mammals. So why are crocodiles classified as reptiles rather than mammals?

Table 15.3
Major Evolutionary and Climatic Events of the Mesozoic Era

Age (millions of years)	Geologic Period	Invertebrates	Vertebrates	Plants	Climate
66 —	Cretaceous	Extinction of ammonites, rudists, and most planktonic foraminifera at end of Cretaceous. Continued diversification of ammonites and belemnoids. Rudists become major reef builders.	Extinctions of dinosaurs, flying reptiles, marine reptiles, and some marine invertebrates. Placental and marsupial mammals diverge.	Angiosperms evolve and diversify rapidly. Seedless plants and gymnosperms still common but less varied and abundant.	Climate becomes more seasonal and cooler at end of Cretaceous. North–south zonation of climates more marked but remains equable.
146 —	Jurassic	Ammonites and belemnoid cephalopods increase in diversity. Scleractinian coral reefs common. Appearance of rudist bivalves.	First birds (may have evolved in Late Triassic). Time of giant sauropod dinosaurs.	Seedless vascular plants and gymnosperms only.	Much like Triassic. Ferns with living relatives restricted to tropics live at high latitudes, indicating mild climates.
200 —	Triassic	The seas are repopulated by invertebrates that survived the Permian extinction event. Bivalves and echinoids expand into the infaunal niche.	Cynodonts become extinct. Mammals evolve from cynodonts. Ancestral archosaur gives rise to dinosaurs. Flying reptiles and marine reptiles evolve.	Land flora of seedless vascular plants and gymnosperms as in Late Paleozoic.	Warm temperate to tropical. Mild temperatures extend to high latitudes; polar regions may have been temperate. Local areas of aridity.
251 —					

Greatest Diversity of Dinosaurs (spanning Cretaceous through Triassic in Vertebrates column)

Sue Monroe

CHAPTER **16**

CENOZOIC EARTH HISTORY: THE PALEOGENE AND NEOGENE

The Oligocene Brule Formation of the White River Group in Badlands National Park, South Dakota, was deposited mostly in stream channels and on their floodplains. These rocks, and those of the underlying Chadron Formation, have one of the most complete successions of fossil mammals anywhere in the world. Notice the sharp angular slopes and ridges and numerous ravines that are typical of badlands topography.

[OUTLINE]

Introduction

Cenozoic Plate Tectonics—An Overview

Cenozoic Orogenic Belts
    Alpine–Himalayan Orogenic Belt
    Circum-Pacific Orogenic Belt

North American Cordillera
    Laramide Orogeny
    Cordilleran Igneous Activity
    Basin and Range Province
    Colorado Plateau
    Rio Grande Rift
    Pacific Coast

The Continental Interior

Cenozoic History of the Appalachian Mountains
    Perspective *The Great Plains*

North America's Southern and Eastern Continental Margins
    Gulf Coastal Plain
    Atlantic Continental Margin

Paleogene and Neogene Mineral Resources

Summary

At the end of this chapter, you will have learned that

- Geologists divide the 66-million-year-long Cenozoic Era into two periods, the Paleogene and the Neogene, each of which consists of several epochs.

- The breakup of Pangaea began during the Triassic Period and continues to the present, giving rise to the present distribution of land and sea.

- Cenozoic orogenies were concentrated in two belts, one that nearly encircles the Pacific Ocean basin, and another that trends east-west through the Mediterranean basin and on into southeast Asia.

- The Late Cretaceous to Eocene Laramide orogeny resulted in deformation of a large area in the west, called the North American Cordillera, which extends from Alaska to Mexico.

- Following the Laramide orogeny, the North American Cordillera continued to evolve as it experienced volcanism, uplift of broad plateaus, large-scale block faulting, and deep erosion.

- The Great Plains consist of huge quantities of sediments that were eroded from the Rocky Mountains and transported eastward.

- A subduction zone was present along the western margin of the North American plate until the plate collided with a spreading ridge, producing the San Andreas and Queen Charlotte transform faults.

- An epeiric sea briefly occupied North America's continental interior during the Paleogene.

- Thick deposits of sediment accumulated along the Gulf and Atlantic Coastal plains.

- Renewed uplift and erosion account for the present-day Appalachian Mountains.

- Paleogene and Neogene rocks contain mineral resources such as oil, oil shale, coal, phosphorus, and gold.

Introduction

At 66 million years long, the Cenozoic Era is only 1.4% of all geologic time, or just 20 minutes on our hypothetical 24-hour clock for geologic time (see Figure 8.1). So the Cenozoic Era was comparatively brief when considered in the context of geologic time, and yet it was extremely long by any other measure. It was certainly long enough for significant changes to occur as plates changed position, mountains and landscapes continued to develop, an ice age took place, and the biota evolved. In short, many events that began during the Cenozoic continue to the present, including the ongoing erosion of the Grand Canyon, continued uplift and erosion of the Himalayas in Asia and the Andes in South America, the origin and evolution of the San Andreas Fault, and the origin of the volcanoes that make up the Cascade Range.

Geologists divide the Cenozoic Era into two periods of unequal duration. The *Paleogene Period* (66 to 23 million years ago) includes the Paleocene, Eocene, and Oligocene epochs, and the *Neogene Period* (23 million years ago to the present) includes the Miocene, Pliocene, Pleistocene, and Holocene or Recent epochs. Although the terms *Tertiary Period* (66 to 1.8 million years ago) and *Quaternary Period* (for the last 1.8 million years) are still used by some geologists, they are no longer recommended as subdivisions of the Cenozoic Era (• Figure 16.1).

Geologists know more about Cenozoic Earth and life history than for any other interval of geologic time

• **Figure 16.1 Geologic Time Scale for the Cenozoic Era** In the past, Tertiary and Quaternary periods have been used, but in 2004 the International Commission on Stratigraphy recommended using Paleogene and Neogene, which we follow in this book. The status of the Quaternary is unresolved (see Chapter 18).

Holocene/Recent
.01

Era	Period			Epoch	Duration, millions of years (approx.)	Millions of years ago (approx.)
Cenozoic	Quaternary		Quat.	Pleistocene	1.79	
						1.8
	Tertiary	Neogene		Pliocene	3.2	
						5.0
				Miocene	18.0	
						23
		Paleogene		Oligocene	11	
						34
				Eocene	22	
						56
				Paleocene	10	
						66

Archean Eon	Proterozoic Eon	Phanerozoic Eon							
Precambrian		Paleozoic Era							
		Cambrian	Ordovician	Silurian	Devonian	Mississippian	Pennsylvanian	Permian	
						Carboniferous			

2500 MYA   *542 MYA*   *251 MYA*

because Cenozoic rocks, being the youngest, are the most accessible at or near the surface. Vast exposures of Cenozoic sedimentary and igneous rocks in western North America record the presence of a shallow sea in the continental interior, terrestrial depositional environments, lava flows, and volcanism on a huge scale in the Pacific Northwest (• Figure 16.2). Exposures of Cenozoic rocks in eastern North America are limited, except for Ice Age deposits, but notable exceptions are Florida, where fossil-bearing rocks of Middle to Late Cenozoic age are present, and Maryland.

One reason to study Cenozoic Earth history is that the present distribution of land and sea, climatic and oceanic circulation patterns, and Earth's present-day distinctive topography resulted from systems interactions during this time. In this chapter, our concern is Earth history of the Paleogene and the Neogene periods (except for the Pleistocene and Holocene epochs). The latter part of the Neogene was unusual because it was one of the few times in Earth history when widespread glaciers were present, so we consider the Pleistocene and Holocene epochs in Chapter 17.

Cenozoic Plate Tectonics—An Overview

The ongoing fragmentation of Pangaea, the supercontinent that existed at the end of the Paleozoic (see Figure 14.1), accounts for the present distribution of Earth's landmasses. Moving plates also directly affect the biosphere because the geographic locations of continents profoundly influence the atmosphere and hydrosphere. As we examine Cenozoic life history, you will see that some important biological events are related to isolation and/or connections between landmasses (see Chapter 18).

Notice from • Figure 16.3 that as the Americas separated from Europe and Africa, the Atlantic Ocean basin opened, first in the south and later in the north. Spreading ridges such as the Mid-Atlantic Ridge and East Pacific Rise were established, along which new oceanic crust formed and continues to form. However, the age of the oceanic crust in the Pacific is very asymmetric, because much of the crust in the eastern Pacific Ocean basin has been

• Figure 16.2 **Cenozoic Sedimentary and Volcanic Rocks in the Western United States**

James S. Monroe

a The Paleocene Cannonball Formation in Montana.

Sue Monroe

b The Eocene Ione Formation (light colored) and the overlying Miocene-age Lovejoy Basalt near Cherokee, California.

Phanerozoic Eon										
Mesozoic Era			Cenozoic Era							
Triassic	Jurassic	Cretaceous	Paleogene			Neogene			Quaternary	
			Paleocene	Eocene	Oligocene	Miocene	Pliocene		Pleistocene	Holocene

251 MYA

66 MYA

subducted beneath the westerly moving North and South American plates (see Figure 3.12).

Another important plate tectonic event was the northward movement of the Indian plate and its eventual collision with Asia (Figure 16.3b). Simultaneous northward movement of the African plate caused the closure of the Tethys Sea and initiated the tectonic activity that currently takes place throughout an east–west zone from the Mediterranean through northern India. Erupting volcanoes in Italy and Greece as well as seismic activity in Italy, Turkey, Greece, and Pakistan remind us of the continuing plate interactions in this part of the world.

Neogene rifting began in East Africa, the Red Sea, and the Gulf of Aden (see Figure 3.16a). Rifting in East Africa is in its early stages, because the continental crust has not yet stretched and thinned enough for new oceanic crust to form from below. Nevertheless, this area is

seismically active and has many active volcanoes. In the Red Sea, rifting and the Late Pliocene origin of oceanic crust followed vast eruptions of basalt, and in the Gulf of Aden, Earth's crust had stretched and thinned enough by Late Miocene time for upwelling basaltic magma to form new oceanic crust. Notice in Figure 3.16a that the Arabian plate is moving north, so it too causes some of the deformation taking place from the Mediterranean through India.

In the meantime, the North and South American plates continued their westerly movement as the Atlantic Ocean basin widened. Subduction zones bounded both continents on their western margins, but the situation changed in North America as it moved over the northerly extension of the East Pacific Rise and it now has a transform plate boundary, a topic we discuss more fully in a later section, although subduction continues in the Pacific Northwest.

• Figure 16.3 **Paleogeography of the World During the Cenozoic**

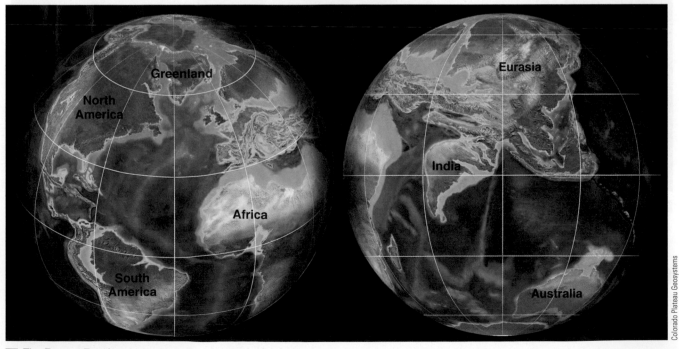

Colorado Plateau Geosystems

a The Eocene Epoch.

• Figure 16.3 (cont.)

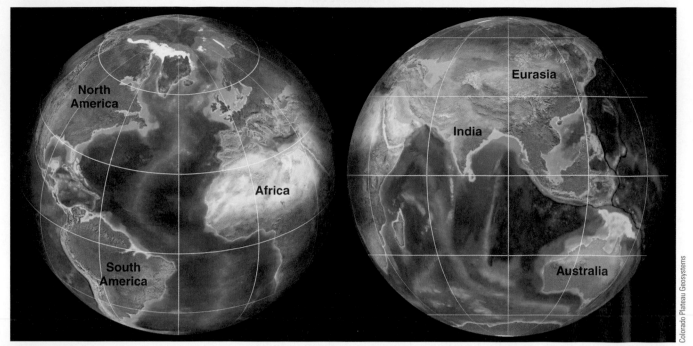

b The Miocene Epoch.

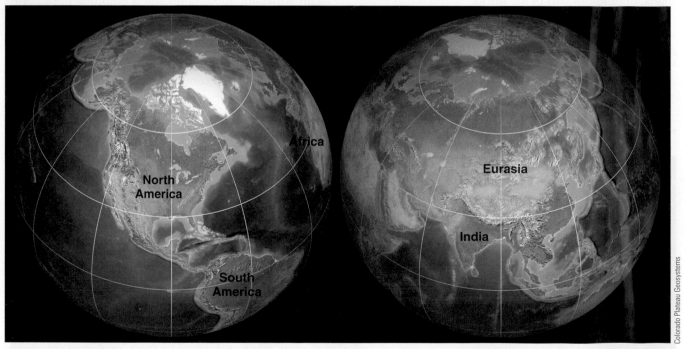

c The World Today.

Cenozoic Orogenic Belts

Remember that an *orogeny* is an episode of mountain building, during which deformation takes place over an elongate area. In addition, most orogenies involve volcanism, the emplacement of plutons, and regional metamorphism as Earth's crust is locally thickened and stands higher than adjacent areas. Cenozoic orogenic activity took place largely in two major zones or belts: the *Alpine–Himalayan orogenic belt* and the *circum-Pacific orogenic belt* (• Figure 16.4). Both belts are made up of smaller segments known as **orogens,** each of which shows the characteristics of an orogeny.

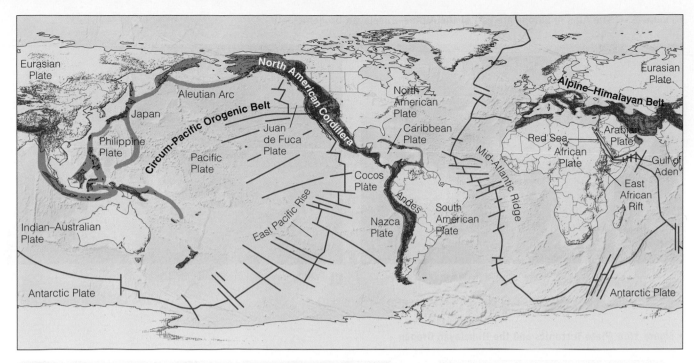

• **Figure 16.4 Earth's Present-Day Orogenic Belts** Most of Earth's geologically recent and present-day orogenic activity takes place in the circum-Pacific orogenic belt and the Alpine–Himalayan orogenic belt. Each belt is made up of smaller units known as orogens.

Alpine–Himalayan Orogenic Belt The **Alpine–Himalayan orogenic belt** extends eastward from Spain through the Mediterranean region as well as the Middle East and India and on into Southeast Asia (Figure 16.4). Remember that during Mesozoic time the Tethys Sea separated much of Gondwana from Eurasia. Closure of this sea took place during the Cenozoic as the African and Indian plates collided with the huge landmass to the north (Figure 16.3). Volcanism, seismicity, and deformation remind us that the Alpine–Himalayan orogenic belt remains active.

The Alps During the **Alpine orogeny,** deformation occurred in a linear zone in southern Europe extending from Spain eastward through Greece and Turkey. Concurrent deformation also occurred along Africa's northwest margin (Figure 16.4). Many details of this long, complex event are poorly understood, but the overall picture is now becoming clear.

Events leading to Alpine deformation began during the Mesozoic, yet Eocene to Late Miocene deformation was also important. Northward movements of the African and Arabian plates against Eurasia caused compression and deformation, but the overall picture is complicated by the collision of several smaller plates with Europe. These small plates were also deformed and are now found in the mountains in the Alpine orogen.

Mountain building produced the Pyrenees between Spain and France, the Apennines of Italy, as well as the Alps of mainland Europe (Figure 16.4). Indeed, the compressional forces generated by colliding plates resulted in complex thrust faults and huge overturned folds known as

nappes (• Figure 16.5). As a result, the geology of the Alps in France, Switzerland, and Austria is extremely complex. Plate convergence also produced an almost totally isolated sea in the Mediterranean basin, which had previously been part of the Tethys Sea. Late Miocene deposition in this sea, which was then in an arid environment, accounts for evaporite deposits up to 2 km thick (see Chapter 6 Perspective).

The collision of the African plate with Eurasia also accounts for the Atlas Mountains of northwest Africa, and further to the east, in the Mediterranean basin, Africa continues to force oceanic lithosphere northward beneath Greece and Turkey. Active volcanoes in Italy and Greece as well as seismic activity throughout this region indicate that southern Europe and the Middle East remain geologically active. In 2005, for instance, an earthquake with a magnitude of 7.6 on the Richter scale killed more than 86,000 people in Pakistan, and Mount Vesuvius in Italy has erupted 80 times since it destroyed Pompeii in A.D. 79; its most recent eruption was in 1944.

The Himalayas—Roof of the World During the Early Cretaceous, India broke away from Gondwana and began moving north, and oceanic lithosphere was consumed at a subduction zone along the southern margin of Asia (• Figure 16.6a). The descending plate partially melted, forming magma that rose to form a volcanic chain and large granitic plutons in what is now Tibet. The Indian plate eventually approached these volcanoes and destroyed them as it collided with Asia. As a result, two continental plates were sutured along a zone now recognized as the **Himalayan orogeny** (Figure 16.6b).

a View of the Alps near Interlaken, Switzerland.

b Folded rocks at Lütschental, Switzerland.

• Figure 16.6 **Plate Tectonics and the Himalayan Orogen**

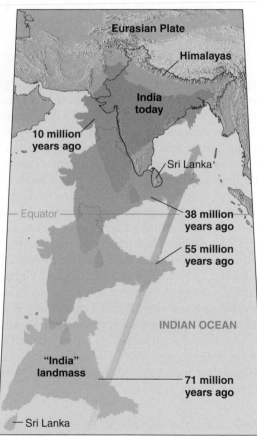

a The Indian plate moved northward for millions of years until it collided with Eurasia, causing crustal thickening and uplift of the Himalayas.

b The Karakoram Range seen here from Karimabad, Pakistan, is within the Himalayan origin. The range lies on the border of Pakistan, China, and India.

Sometime between 40 and 50 million years ago, India's drift rate decreased abruptly from 15 to 20 cm/year to about 5 cm/year. Because continental lithosphere is not dense enough to be subducted, this decrease most likely marks the time of collision and India's resistance to subduction. As a result of India's low density and resistance to subduction, it was underthrust about 2000 km beneath Asia, causing crustal thickening and uplift, a process that continues at about 5 cm/year. Furthermore, sedimentary rocks formed in the sea south of Asia were thrust northward into Tibet, and two huge thrust faults carried Paleozoic and Mesozoic rocks of Asian origin onto the Indian plate.

In the Himalayan origin there is no volcanism because the Indian plate does not penetrate deeply enough to generate magma, but seismic activity continues. Indeed, the entire Himalayan region including the Tibetan plateau and well into China is seismically active. The May 12, 2008 Sichuan earthquake in China in which about 70,000 people perished was a result of this collision between India and Asia.

Circum-Pacific Orogenic Belt

The **circum-Pacific orogenic belt** consists of orogens along the western margins of South, Central, and North America as well as the eastern margin of Asia and the islands north of Australia and New Zealand (Figure 16.4). Subduction of oceanic lithosphere accompanied by deformation and igneous activity characterize the orogens in the western and northern Pacific. Japan, for instance, is bounded on the east by the Japan Trench, where the Pacific plate is subducted, and the Sea of Japan, a **back-arc marginal basin,** lies between Japan and mainland Asia. According to some geologists, Japan was once part of mainland Asia and was separated when back-arc spreading took place (• Figure 16.7). Separation began during the Cretaceous as Japan moved eastward over the Pacific plate and oceanic crust formed in the Sea of Japan. Japan's geology is complex, and much of its deformation predates the Cenozoic, but considerable deformation, metamorphism, and volcanism occurred during the Cenozoic and continues to the present.

In the northern part of the Pacific Ocean, basin subduction of the Pacific plate at the Aleutian trench accounts for the tectonic activity in that region. Of the 80 or so potentially active volcanoes in Alaska, at least half have erupted since 1760, and of course, seismic activity is ongoing.

In the eastern part of the Pacific, the Cocos and Nazca plates moved east from the East Pacific Rise only to be consumed at subduction zones along the west coasts of Central and South America (Figure 16.3). Volcanism and seismic activity indicate these orogens in both Central and South America are active. One manifestation of ongoing tectonic activity in South America is the Andes Mountains, with more than 49 peaks higher than 6000 m. The Andes formed, and continue to do so, as Mesozoic-Cenozoic plate convergence resulted in crustal thickening as sedimentary rocks were deformed, uplifted, and intruded by huge granitic plutons (• Figure 16.8).

North American Cordillera

The **North American Cordillera,** a complex mountainous region in western North America, is a large segment of the circum-Pacific orogenic belt extending from Alaska to central Mexico. In the United States it widens to 1200 km, stretching east–west from the eastern flank of the Rocky Mountains to the Pacific Ocean (• Figure 16.9).

The geologic evolution of the North American Cordillera began during the Neoproterozoic when huge quantities of sediment accumulated along a westward-facing continental margin (see Figure 9.7). Deposition continued into the Paleozoic, and during the Devonian part of the region was deformed at the time of the Antler orogeny (see Chapter 11). A protracted episode of deformation known as the *Cordilleran orogeny* began during the Late Jurassic as the Nevadan, Sevier, and Laramide

• Figure 16.7 **Back-Arc Spreading and the Sea of Japan**

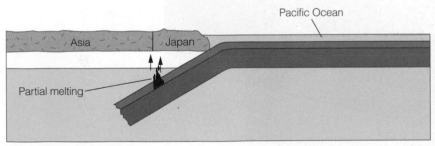

a Model showing the initial stage in the origin of the Sea of Japan.

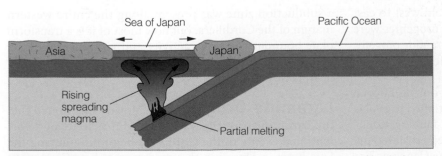

b A more advanced stage in which the Sea of Japan, a back-arc marginal basin, has opened.

• Figure 16.8 The Andes Mountains In South America

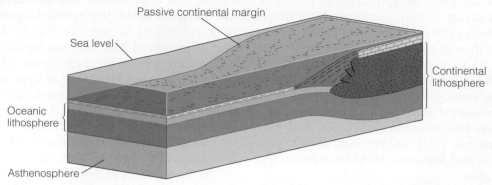

Passive continental margin

Sea level

Oceanic lithosphere

Continental lithosphere

Asthenosphere

a Prior to 200 million years ago, the western margin of South America was a passive continental margin.

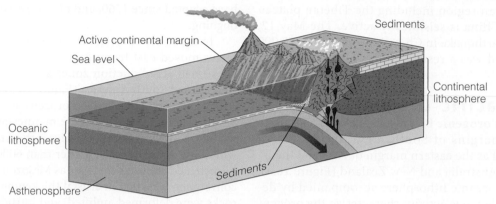

Active continental margin

Sea level

Sediments

Oceanic lithosphere

Continental lithosphere

Sediments

Asthenosphere

b Orogeny began when this area became an active continental margin as the South American plate moved to the west and collided with oceanic lithosphere.

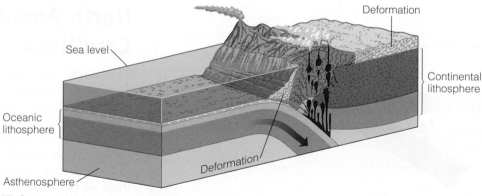

Deformation

Sea level

Oceanic lithosphere

Continental lithosphere

Deformation

Asthenosphere

c Continued deformation. plutonism, and volcanism.

orogenies progressively affected areas from west to east (see Figure 14.11). The first two of these orogenies were discussed in Chapter 14; we discuss the Laramide orogeny, a Late Cretaceous to Eocene episode of deformation in the following section.

After Laramide deformation ceased during Eocene time, the North American Cordillera continued to evolve as parts of it experienced large-scale block-faulting, extensive volcanism, and vertical uplift and deep erosion. Furthermore, during about the first half of the Cenozoic Era

a subduction zone was present along the entire western margin of the Cordillera, but now most of it is a transform plate boundary. Seismic activity and volcanism indicate plate interactions continue in the Cordillera, especially near its western margin.

Laramide Orogeny

We already mentioned that the **Laramide orogeny** was the third in a series of deformational events in the Cordillera beginning during the Late Jurassic. However, this orogeny was Late Cretaceous

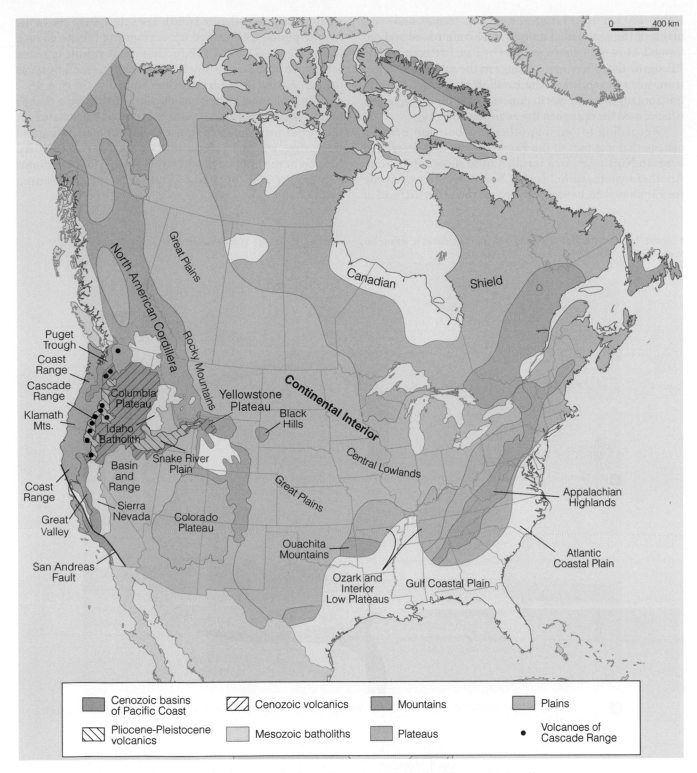

• Figure 16.9 **The North American Cordillera and the Major Provinces in the United States and Canada**

to Eocene and it differed from the previous orogenies in important ways. First, it occurred much further inland from a convergent plate boundary, and neither volcanism nor emplacement of plutons was very common. In addition, deformation mostly took the form of vertical, fault-bounded uplifts rather than the compression-induced folding and

thrust faulting typical of most orogenies. To account for these differences, geologists modified their model for orogenies at convergent plate boundaries. During the preceding Nevadan and Sevier orogenies, the Farallon plate was subducted at about a 50-degree angle along the western margin of North America. Volcanism and plutonism took place

150 to 200 km inland from the oceanic trench, and sediments of the continental margin were compressed and deformed. Most geologists agree that by Late Cretaceous Early Paleogene time there was a change in the subduction angle from steep to gentle and the Farallon plate moved nearly horizontally beneath North America, but they disagree on what caused the change in the angle of subduction.

According to one hypothesis, a buoyant oceanic plateau that was part of the Farallon plate that descended beneath North America resulted in shallow subduction. Another hypothesis holds that North America overrode the Farallon plate, beneath which was the deflected head of a mantle plume (• Figure 16.10). The lithosphere above the mantle plume was buoyed up, accounting for the change from steep to shallow subduction. As a result, igneous activity shifted farther inland and finally ceased because the descending plate no longer penetrated to the mantle.

This changing angle of subduction also caused a change in the type of deformation—the fold-thrust deformation of the Sevier orogeny gave way to large-scale buckling and fracturing, which yielded fault-bounded vertical uplifts. Erosion of the uplifted blocks yielded rugged mountainous topography and supplied sediments to the intervening basins.

• **Figure 16.10 Laramide Orogeny** The Late Cretaceous to Eocene Laramide orogeny took place as the Farallon plate was subducted beneath North America.

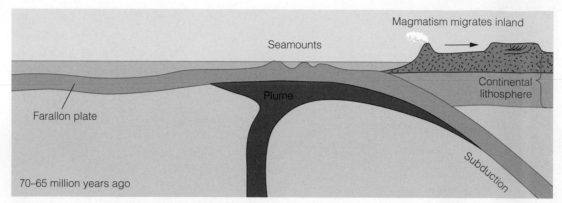

a As North America moved westward over the Farallon plate, beneath which was the deflected head of a mantle plume, the angle of subduction decreased and igneous activity shifted inland.

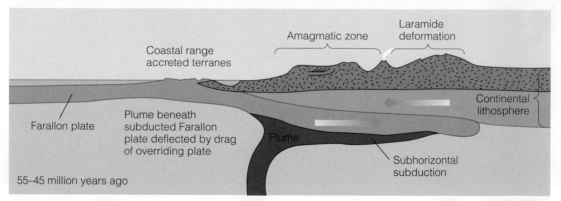

b With nearly horizontal subduction, igneous activity ceased and the continental crust was deformed mostly by vertical uplifts.

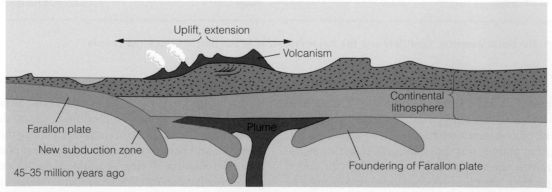

c Disruption of the oceanic plate by the mantle plume marked the onset of renewed igneous activity.

The Laramide orogen is centered in the middle and southern Rocky Mountains of Wyoming and Colorado, but deformation also took place far to the north and south. In the northern Rocky Mountains of Montana and Alberta, Canada, huge slabs of pre-Laramide strata moved eastward along overthrust faults.* On the Lewis overthrust in Montana, a slab of Precambrian rocks was displaced eastward about 75 km (• Figure 16.11), and similar deformation occurred in the Canadian Rocky Mountains. Far to the south of the main Laramide orogen, sedimentary rocks in the Sierra Madre Oriental of east-central Mexico are now part of a major fold-thrust belt.

By Middle Eocene time, Laramide deformation ceased and igneous activity resumed in the Cordillera when the mantle plume beneath the lithosphere disrupted the overlying oceanic plate (Figure 16.10c). The uplifted blocks of the Laramide orogen continued to erode, and by the Neogene, the rugged mountains had been nearly buried in their own debris, forming a vast plain across which

streams flowed. During a renewed cycle of erosion, these streams removed much of the basin fill sediments and incised their valleys into the uplifted blocks. Late Neogene uplift accounts for the present ranges, and uplift continues in some areas.

Cordilleran Igneous Activity
The enormous batholiths in Idaho, British Columbia, Canada, and the Sierra Nevada of California were emplaced during the Mesozoic (see Chapter 14), but intrusive activity continued into the Paleogene Period. Numerous small plutons formed, including copper- and molybdenum-bearing stocks in Utah, Nevada, Arizona, and New Mexico.

Volcanism was common in the Cordillera, but it varied in location, intensity, and eruptive style, and it ceased temporarily in the area of the Laramide orogen (• Figure 16.12a). In the Pacific Northwest, the Columbia Plateau (Figure 16.9) is underlain by 200,000 km³ of Miocene lava flows of the Columbia River basalts that have

• Figure 16.11 **Laramide Deformation in the Northern Rocky Mountains**

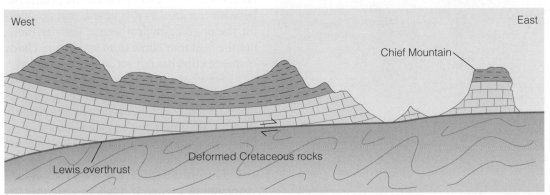

West   East

Chief Mountain

Deformed Cretaceous rocks

Lewis overthrust

a Cross section showing the Lewis overthrust fault in Glacier National Park, Montana. Meso- and Neoproterozoic rocks of the Belt Supergroup rest on deformed Cretaceous rocks.

Lewis overthrust

b The trace of the Lewis overthrust is visible in this image as a light-colored line on the mountain-side.

Chief Mountain

c Erosion has isolated Chief Mountain from the rest of the slab of overthrust rock.

*An overthrust fault is a large-scale, low-angle thrust fault with movement measured in kilometers.

• Figure 16.12 Cenozoic Volcanism

Predominantly andesitic ☐ Basalt and andesite ☐ Predominantly basaltic

a Distribution of Cenozoic volcanic rocks in the western United States.

an aggregate thickness of about 2500 m. These vast lava flows are now well exposed in the walls of the canyons eroded by the Columbia and Snake rivers and their tributaries (Figure 16.12b and c). The relationship of this huge outpouring of lava to plate tectonics remains unclear, but some geologists think it resulted from a mantle plume beneath western North America.

The Snake River Plain (Figure 16.9), which is mostly in Idaho, is actually a depression in the crust that was filled by Miocene and younger rhyolite, volcanic ash, and basalt (Figure 16.12c). These rocks are oldest in the southwest part of the area and become younger toward the northeast, leading some geologists to propose that North America has migrated over a mantle plume that now lies beneath Yellowstone National Park in Wyoming. Other geologists disagree, thinking that these volcanic rocks erupted along an intracontinental rift zone.

Bordering the Snake River Plain on the northeast is the Yellowstone Plateau (Figure 16.9), an area of Pliocene and Pleistocene volcanism. Perhaps a mantle plume lies beneath the area, as just noted, that accounts for the ongoing hydrothermal activity there, but the heat may come from an intruded body of magma that has not yet completely cooled.

Elsewhere in the Cordillera, andesite, volcanic breccia, and welded tuffs (ignimbrites), mostly of Oligocene age, cover more than 25,000 km² in the San Juan volcanic field in Colorado. Eruptions in the Coso volcanic field in California began during the Pliocene and continued until only a few thousands of years ago (• Figure 16.13a), in Arizona, the San Francisco volcanic field formed during

b The Columbia River basalts are exposed in the walls of this canyon eroded by the Columbia River in Oregon. Multnomah Falls plunge 189 m from a small tributary valley.

c Basalt lava flows of the Snake River Plain at Malad Gorge State Park, Idaho.

• Figure 16.13 **Cenozoic Volcanism in California and Oregon**

a Pliocene to Pleistocene volcanism took place in the Coso volcanic field in California. The cinder cone, called Red Hill, formed no more than a few tens of thousands of years ago.

b These rocks at Cape Foulweather in Oregon are outcrops of basalt. The rocks are the remnants of a Miocene volcano.

the Pliocene and Pleistocene, and volcanism took place along Oregon's coast (Figure 16.13b).

Some of the most majestic and highest mountains in the Cordillera are in the **Cascade Range** of northern California, Oregon, Washington, and southern British Columbia, Canada (• Figure 16.14). Thousands of volcanic vents are present, the most impressive of which are the dozen or so large composite volcanoes and Lassen Peak in California, the world's largest lava dome. Volcanism in this region is related to subduction of the Juan de Fuca plate beneath North America. Volcanism in the Cascade Range goes back at least to the Oligocene, but the most recent episode began during the Late Miocene or Early Pliocene. The eruption of Lassen Peak in California from 1914 to 1917 and the eruptions of Mount St. Helens in Washington

• **Figure 16.14 Cascade Range Volcanism** Volcanism in the Cascade Range dates back to at least the Oligocene, but the large volcanoes of the range formed more recently. This view shows Mount Hood in Oregon; its last large eruption was during the 1790s, but some minor explosive activity took place during the mid 1800s.

in 1980 and again in 2004 indicate that Cascade volcanoes remain active.

Basin and Range Province Earth's crust in the **Basin and Range Province** (Figure 16.9)—an area of nearly 780,000 km^2 centered on Nevada but extending into adjacent states and northern Mexico— has been stretched and thinned yielding north–south oriented mountain ranges with intervening valleys or basins (• Figure 16.15a). The 400 or so ranges are bounded on one or both sides by steeply dipping normal faults that probably curve and dip less steeply with depth. In any case, the faults outline blocks that show displacement and rotation.

Before faulting began, the region was deformed during the Nevadan, Sevier, and Laramide orogenies. Then, during the Paleogene, the entire area was highlands undergoing extensive erosion, but Early Miocene eruptions of rhyolitic lava flows and pyroclastic materials covered large areas. By the Late Miocene, large-scale faulting had begun, forming the basins and ranges. Sediment derived from the ranges was transported into the adjacent basins and accumulated as alluvial fan and playa lake deposits.

At its western margin, the Basin and Range Province is bounded by normal faults along the east flank of the Sierra Nevada (Figure 16.15b). Pliocene and Pleistocene uplift tilted the Sierra Nevada toward the west, and its crest now stands 3000 m above the basins to the east. Before this uplift took place, the Basin and Range had a subtropical climate, but the rising mountains created a rain shadow, and the climate became increasingly arid.

Geologists have proposed several models to account for basin-and-range structure but have not reached a consensus. Among these are back-arc spreading; spreading at the East Pacific Rise, the northern part of which is thought to now lie beneath this region; spreading above a mantle plume; and deformation related to movements along the San Andreas Fault.

• Figure 16.15 **The Basin and Range Province**

a The Basin and Range is mostly in Nevada, but it extends into adjacent states and northern Mexico. The ranges and the intervening basins trend north-south. Each range is bounded on one or both sides by normal faults.

b The Sierra Nevada at the western margin of the Basin and Range has risen along normal faults so that it is more than 3000 m above the valley to the east.

Colorado Plateau

The vast, elevated region in Colorado, Utah, Arizona, and New Mexico known as the **Colorado Plateau** (Figure 16.9) has volcanic mountains rising above it, brilliantly colored rocks, and deep canyons. In Chapters 11 and 14, we noted that during the Permian and Triassic the Colorado Plateau region was the site of extensive red bed deposition; many of these rocks are now exposed in the uplifts and canyons (• Figure 16.16).

Cretaceous-age marine sedimentary rocks indicate the Colorado Plateau was below sea level, but during the Paleogene Period Laramide deformation yielded broad anticlines and arches and basins, and a number of large normal faults. However, deformation was far less intense than elsewhere in the Cordillera. Neogene uplift elevated the region from near sea level to the 1200 to 1800 m elevations seen today, and as uplift proceeded, streams and rivers began eroding deep canyons.

Geologists disagree on the details of just how the deep canyons so typical of the region developed—such as the Grand Canyon. Some think the streams were *antecedent,* meaning they existed before the present topography developed, in which case they simply eroded downward as uplift proceeded. Others think the streams were *superposed,* implying that younger strata covered the area, on which streams were established. During uplift, the streams stripped away these younger rocks and eroded down into the underlying strata. In either case, the landscape continues to evolve as erosion of the canyons and their tributaries deepens and widens them.

Rio Grande Rift

The **Rio Grande rift** extends north to south about 1000 km from central Colorado through New Mexico and into northern Mexico. Recall our discussions of the Mesoproterozoic Midcontinent rift in the Great Lakes region (see Chapter 9) and the present-day rifting in the Gulf of Aden, the Red Sea, and East Africa (see Figure 3.16a). The Rio Grande rift is similar in that Earth's crust has been stretched and thinned, the rift is bounded on both sides by normal faults, seismic activity continues, and volcanoes and calderas are present (• Figure 16.17). Actually the Rio Grande rift consists of several basins through which the present-day Rio Grande flows, although the river simply exploited an easy route to the sea but was not responsible for the rift itself.

Rifting in this area began about 29 million years ago and persisted for 10 to 12 million years, from the Late Oligocene into the Early Miocene. A second period of rifting began during the Middle Miocene, about 17 million

• Figure 16.16 **Rocks of the Colorado Plateau**

a Agathla Peak is a volcanic neck that rises 457 m in Monument Valley Navajo Tribal Park in Arizona. It is composed of tuff breccia and formed in Late Oligocene time, about 25 million years ago.

b Mexican Hat in Utah is an erosional feature measuring about 18 m across. It is made up of Permian rocks, but its present form resulted from Cenozoic erosion.

• Figure 16.17 **The Rio Grande Rift**

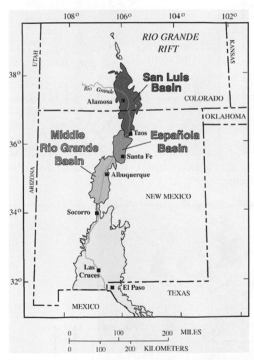

a Location of the basins making up the Rio Grande rift. A complex of normal faults is present on both sides of the rift.

b The Bandelier Tuff in Bandelier National Monument, New Mexico, erupted in the Jemez volcanic field 1.14 million years ago.

years ago, and it continues to the present. The displacement on some of the faults is as much as 8000 m, but concurrent with faulting, the basins within the rift filled with huge quantities of sediments and volcanic rocks. Some of the volcanic features such as Valles caldera, which measures 19 by 24 km, and the Bandelier Tuff are prominent features in New Mexico (Figure 16.17b). Rifting continues, but very slowly—only 2 mm or less per year. So even though ongoing rifting may eventually split the area so that it resembles the Red Sea, it will be in the far distant future.

Pacific Coast Before the Eocene, the entire Pacific Coast was a convergent plate boundary where the **Farallon plate** was consumed at a subduction zone that stretched from Mexico to Alaska. Now only two small remnants of the Farallon plate remain—the Juan de Fuca and Cocos plates (• Figure 16.18). Continuing subduction of these small plates accounts for the present seismic activity and volcanism in the Pacific Northwest and Central America, respectively.

Another consequence of these plate interactions was the westward movement of the North American plate and its collision with the **Pacific–Farallon ridge.** Because the Pacific–Farallon ridge was at an angle to the margin of North America, the continent–ridge collision took place first during the Eocene in northern Canada and only later during the Oligocene in southern California (Figure 16.18). In southern

California, two triple junctions formed, one at the intersection of the North American, Juan de Fuca, and Pacific plates, the other at the intersection of the North American, Cocos, and Pacific plates. Continued westward movement of the North American plate over the Pacific plate caused the triple junctions to migrate, one to the north and the other to the south, giving rise to the **San Andreas transform Fault** (Figure 16.18). A similar occurrence along Canada's west coast produced the *Queen Charlotte transform fault.*

Seismic activity on the San Andreas Fault results from continuing movements of the Pacific and North American plates along this complex zone of shattered rocks. Indeed, where the fault cuts through coastal California it is actually a zone as much as 2 km wide, and it has numerous branches. Movements on such complex fault systems subject blocks of rocks adjacent to and within the fault zone to

• Figure 16.18 **Origin of the San Andreas and Queen Chartotte Faults**

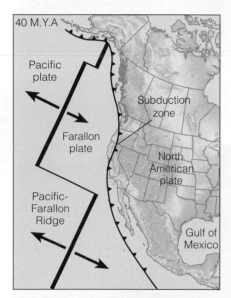
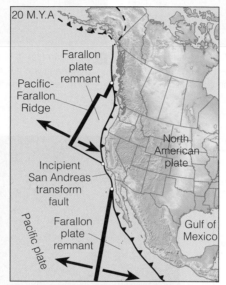

a Three stages in the westward movement of North America and its collision with the Pacific-Farallon ridge. As North America overrode the ridge, its margin became bounded by transform faults except in the Pacific Northwest.

b Aerial view of the San Andreas Fault today. On land, we call it a right-lateral strike-slip fault. The creek has been offset nearly 100 m.

extensional and compressive stresses forming basins and elevated areas, the latter supplying sediments to the former. Many of the fault-bounded basins in the southern California area have subsided below sea level and soon filled with turbidites and other deposits. A number of these basins are areas of prolific oil and gas production.

The Continental Interior

Notice in Figure 16.9 that much of central North America is a vast area called the **continental interior,** which in turn is made up of the *Great Plains* and the *Central Lowlands.* During the Cretaceous, the Great Plains were covered by the **Zuni epeiric sea,** but by Early Paleogene time, this sea had largely withdrawn except for a sizable remnant in North Dakota. Sediments eroded from the Laramide highlands were transported to this sea and deposited in transitional and marine environments. Following this brief episode of marine deposition, all other sedimentation in the Great Plains took place in terrestrial environments, especially fluvial systems. These formed eastward-thinning wedges of sediment that now underlie the entire region (• Figure 16.19).

The only local sediment source within the Great Plains was the Black Hills in South Dakota (see Perspective). The Great Plains have a history of marine deposition during the Cretaceous followed by the origin of terrestrial deposits derived from the Black Hills that are now well exposed in Badlands National Park, South Dakota (see the chapter opening photo). Judging from the sedimentary rocks and their numerous fossil mammals and other animals, the

area was initially covered by semitropical forest, but grasslands replaced the forests as the climate became more arid (see Chapter 18).

Igneous activity was not widespread in the continental interior but was significant in some parts of the Great Plains. For instance, igneous activity in northeastern New Mexico was responsible for volcanoes and numerous lava flows (• Figure 16.20a) and several small plutons were emplaced in Colorado, Wyoming, Montana, South Dakota, and New Mexico. Indeed, one of the most widely recognized igneous bodies in the entire continent, Devil's Tower in northeastern Wyoming, is probably an Eocene volcanic

• Figure 16.20 **Cenozoic Volcanism in the Great Plains**

a Sierra Grande is a shield volcano in northeastern New Mexico. It stands 600 m high and is 15 km in diameter. It was active from 2.6 to 4.0 million years ago.

• Figure 16.19 **Huge Amounts of Sediments Shed from the Laramide Highland Were Deposited on the Great Planes** Paleocene sedimentary rocks seen from Scoria Point in Theodore Roosevelt National Park in North Dakota. The scoria, the reddish rock, is not the volcanic rock, but rather scoria-like material that formed when an ancient coal bed burned and baked clay and silt in the surrounding beds.

b At 650 m high, Devil's Tower in Wyoming can be seen from 48 km away. It was emplaced as a small pluton during the Eocene, 45 to 50 million years ago.

neck, although some geologists think it is an eroded laccolith (Figure 16.20b).

Our discussion thus far has focused on the Great Plains, but what about the Central Lowlands to the east? Pleistocene glacial deposits are present in the northern part of this region, as well as in the northern Great Plains (see Chapter 17), but during most of the Cenozoic Era nearly all the Central Lowlands was an area of active erosion rather than deposition. Of course, the eroded materials had to be deposited somewhere, and that was on the Gulf Coastal Plain (Figure 16.9).

Cenozoic History of the Appalachian Mountains

Deformation and mountain building in the area of the present Appalachian Mountains began during the Neoproterozoic with the Grenville orogeny (see Figure 9.2c). The area was deformed again during the Taconic and Acadian orogenies, and during the Late Paleozoic closure of the Iapetus Ocean, which resulted in the Hercynian-Alleghenian orogeny (see Chapters 10 and 11). Then, during Late Triassic time, the entire region experienced block-faulting as Pangaea fragmented (see Figure 14.7). By the end of the Mesozoic, though, erosion had reduced the mountains to a plain across which streams flowed eastward to the ocean.

The present distinctive aspect of the Appalachian Mountains developed as a result of Cenozoic uplift and erosion (• Figure 16.21). As uplift proceeded, upturned resistant rocks formed northeast–southwest trending ridges with intervening valleys eroding into less resistant rocks. The preexisting streams eroded downward while uplift took place, were superposed on resistant rocks, and cut large canyons across the ridges, forming *water gaps* (• Figure 16.22), deep passes through which streams flow, and *wind gaps,* which are water gaps no longer containing streams.

Erosion surfaces at different elevations in the Appalachians are a source of continuing debate among geologists. Some are convinced these more or less planar surfaces show evidence of uplift followed by extensive erosion and then renewed uplift and another cycle of erosion. Others think that each surface represents differential response to weathering and erosion. According to this view, a low-elevation erosion surface developed on softer strata that eroded more or less uniformly, whereas higher surfaces represent weathering and erosion of more resistant rocks.

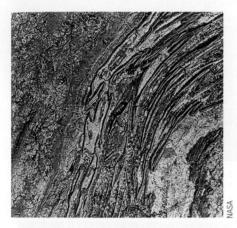

• **Figure 16.21 Landsat Image of the Appalachian Mountains** This view of the Appalachians shows the central part of Pennsylvania. Notice the long ridges with intervening valleys.

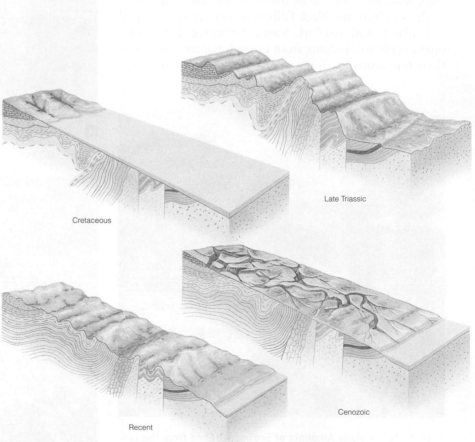

• **Figure 16.22 Evolution of the Present Topography of the Appalachian Mountains** Although these mountains have a long history, their present topographic expression resulted mainly from Cenozoic uplift and erosion.

The Great Plains

The Great Plains is a huge expanse measuring 3200 km north to south and 800 km west to east lying east of the Rocky Mountains covering parts of 10 states, 3 Canadian provinces, and a small part of northern Mexico (Figure 1). Many people think of the Great Plains as a rather monotonous landscape with little of interest, but such an assessment is incorrect. Indeed, there are many areas of geologic interest, scenic beauty, and historic importance including Dinosaur Provincial Park in Alberta, Canada; Agate Fossil Beds National Monument, Nebraska; Devil's Tower National Monument, Wyoming;

Capulin Volcano National Monument, New Mexico; and many others. Here, however, we will concentrate on only two areas—Badlands National Park in South Dakota, and Theodore Roosevelt National Park in North Dakota (Figure 16.19a).

Both national parks feature sedimentary rocks of Cenozoic age that originated as sediment shed from the Laramide highlands to the west, or, in the case of Badlands National Park, sediments derived from the Black Hills. Much of the sediment was transported to the east by meandering streams and deposited in their channels, and on their floodplains, and in

small lakes. Coal beds found in Theodore Roosevelt National Park formed when vegetation partially decayed in swamps (Figure 16.19). In addition, both areas were then populated by a variety of mammals, several of them now extinct. In fact, Badlands National Park has one of the most complete sequences of Late Eocene and Oligocene fossil mammals anywhere in the world. The Park Service has left a number of these fossils in place but protected for viewing. Not nearly as many fossils are found at Theodore Roosevelt National Park, but we can be sure that similar animals lived there.

Another feature of interest found in both parks is *badlands topography*. Badlands develop in dry areas with sparse vegetation yet easily eroded clay-rich rocks. Infrequent but intense rainfall on such unprotected rocks rapidly runs off and intricately dissects the surface forming numerous closely spaced, small gullies and deep ravines, thus yielding steep slopes, angular ridges and divides between gullies, and steep pinnacles (see the chapter opening photo). Obviously Badlands National Park has well-developed badlands, hence its name, but similar topography is found in a discontinuous band from Alberta, Canada to Texas. And, of course, the same kind of topography is also found on the other continents where similar conditions exist.

The relief is not great in areas of badlands, but the surface is so complex that it is difficult to traverse. Indeed, there are stories of gold miners and others attempting to sneak through the badlands into the Black Hills of South Dakota in defiance of treaties with the Sioux. Some of these people became hopelessly lost and eventually died of thirst.

Many sedimentary rocks and some pyroclastic rocks contain *concretions*, which are hard irregular to spherical masses that formed by precipitation of minerals

Figure 1 There are many places of interest in the Great Plains, but here we discuss only two—Theodore Roosevelt National Park in North Dakota and Badlands National Park in South Dakota. The Black Hills of South Dakota are also shown because they were the source of sediments that now make the rocks in Badlands National Park.

Figure 2 These spherical masses of rock known as cannonball concretions are in Theodore Roosevelt National Park in North Dakota. These concretions, which measure about 0.6 m across, are hard, so when the host rock weathers they collect at the surface.

around some nucleus, perhaps a shell or bone. In any case, they are much harder than the host rock and stand out as the host rock weathers, or collect at the surface as the parent rock materials are eroded. The remarkable cannonball concretions, so named because of their shape, in Theodore Roosevelt National Park are mostly small objects measuring less than one meter in diameter but some are as much as 3 m across (Figure 2).

We have focused on two areas within the Great Plains and mentioned a few others in passing in our opening paragraph. However, we close this section by noting that there are many other areas of interest. For example, Shiprock in New Mexico is a huge volcanic neck that is scenic and considered sacred by the local Native Americans, and Chimney Rock in Nebraska is a famous landmark that helped guide pioneers on their way along the Oregon Trail.

North America's Southern and Eastern Continental Margins

In a previous section, we mentioned that much of the Central Lowlands eroded during the Cenozoic. Even in the Great Plains where vast deposits of Cenozoic rocks are present, sediment was carried across the region and into the drainage systems that emptied into the Gulf of Mexico. Likewise, sediment eroded from the western margin of the Appalachian Mountains ended up in the Gulf, but these mountains also shed huge quantities of sediment eastward that was deposited along the Atlantic Coastal Plain. Notice in Figure 16.9 that the **Atlantic Coastal Plain** and the **Gulf Coastal Plain** form a continuous belt extending from the northeastern United States to Texas. Both areas have horizontal or gently seaward-dipping strata deposited mostly by streams. Seaward of the coastal plains lie the continental shelf, slope, and rise, also areas of notable Mesozoic and Cenozoic deposition.

Gulf Coastal Plain
After the withdrawal of the Cretaceous to Early Paleogene Zuni Sea, the Cenozoic **Tejas epeiric sea** made a brief appearance on the continent. But even at its maximum extent it was largely restricted to the Atlantic and Gulf Coastal plains and parts of coastal California. It did, however, extend up the Mississippi River Valley, where it reached as far north as southern Illinois.

The overall Gulf Coast sedimentation pattern was established during the Jurassic and persisted throughout the Cenozoic. Sediments derived from the Cordillera, western Appalachians, and the Central Lowlands were transported toward the Gulf of Mexico, where they were deposited in terrestrial, transitional, and marine environments. In general, the sediments form seaward-thickening wedges grading from terrestrial facies in the north to marine facies in the south (• Figure 16.23).

Sedimentary facies development was controlled mostly by regression of the Tejas epeiric sea. After its maximum extent onto the continent during the Paleogene, this sea began its long withdrawal toward the Gulf of Mexico. Its regression, however, was periodically reversed by minor transgressions—eight transgressive–regressive episodes are recorded in Gulf Coastal Plain sedimentary rocks, accounting for the intertonguing among the various facies.

Many sedimentary rocks in the Gulf Coastal Plain are either source rocks or reservoirs for hydrocarbons, a topic we discuss more fully in the section on Paleogene and Neogene mineral resources.

Most of the Gulf Coastal Plain was dominated by detrital deposition, but in the Florida section of the region and the

• Figure 16.23 **Cenozoic Deposition on the Gulf Coastal Plain**

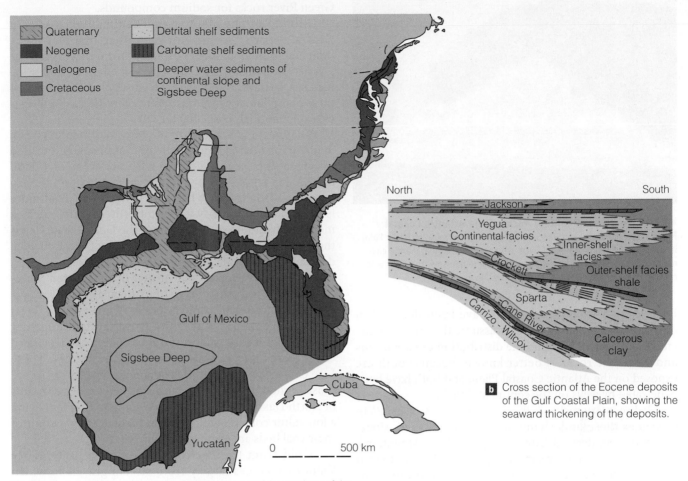

Quaternary
Neogene
Paleogene
Cretaceous
Detrital shelf sediments
Carbonate shelf sediments
Deeper water sediments of continental slope and Sigsbee Deep

a Areas of deposition in the Gulf of Mexico and surface geology of the Gulf and Atlantic Coastal plains.

b Cross section of the Eocene deposits of the Gulf Coastal Plain, showing the seaward thickening of the deposits.

Gulf Coast of Mexico significant carbonate deposition took place. Florida was a carbonate platform during the Cretaceous and continued as an area of carbonate deposition into the Early Paleogene; carbonate deposition continues even now in Florida Bay and the Florida Keys. Southeast of Florida, across the 85-km-wide Florida Strait, lies the Great Bahama Bank, an area of carbonate deposition from the Cretaceous to the present.

Atlantic Continental

Margin The east coast of North America includes the Atlantic Coastal Plain and extends seaward across the continental shelf, slope, and rise (• Figure 16.24). When Pangaea began fragmenting during the Triassic, continental crust rifted, and a new ocean basin began to form. Remem-

ber that the North American plate moved westerly, so its eastern margin was within the plate, where a passive continental margin developed.

The Atlantic continental margin has a number of Mesozoic and Cenozoic basins, formed as a result of rifting, in which sedimentation began by Jurassic time.

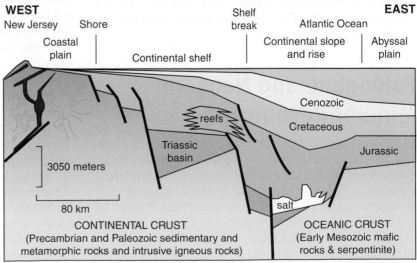

• Figure 16.24 **The Continental Margin in Eastern North America** The coastal plain and continental margin of New Jersey are covered mostly by Cenozoic sandstones and shales. Beneath these rocks lie Cretaceous and probably Jurassic sedimentary rocks.

• Figure 16.25 **Sedimentary Rocks of the Atlantic Coastal Plain** These Miocene- and Pliocene-age sedimentary rocks at Rocky Point in the Calvert Cliffs of Maryland, were deposited in marginal marine environments. In addition to fossils of marine microorganisms, the rocks also contain fossil invertebrates, sharks, and marine mammals.

Even though Jurassic-age rocks have been detected in only a few deep wells, geologists assume they underlie the entire continental margin. The distribution of Cretaceous and Cenozoic rocks is better known, because both are exposed on the Atlantic Coastal Plain, and both have been penetrated by wells on the continental shelf.

Sedimentary rocks on the broad Atlantic Coastal Plain as well as those underlying the continental shelf, slope, and rise, were derived from the Appalachian Mountains. Numerous rivers and streams transported sediments toward the east, where they were deposited in seaward thickening wedges (up to 14 km thick) that grade from terrestrial deposits on the west to marine deposits further east. For instance, the Calvert Cliffs in Maryland consist of rocks deposited in marginal marine environments (• Figure 16.25).

An interesting note regarding the geologic evolution of the Atlantic Coastal Plain is the evidence indicating that a 3- to 5-km-diameter comet or asteroid impact occurred in the present-day area of Chesapeake Bay. This event took place about 35 million years ago, during the Late Eocene, and left an impact crater measuring 85 km in diameter and 1.3 km deep. Now buried beneath 300 to 500 m of younger sedimentary rocks, it has been detected by drilling and geophysical surveys.

Paleogene and Neogene Mineral Resources

The Eocene Green River Formation of Wyoming, Utah, and Colorado, well known for fossils, also contains huge quantities of oil shale and evaporites of economic interest. Oil shale consists of clay particles, carbonate minerals, and an organic compound called *kerogen* from which liquid oil and combustible gases can be extracted. No oil is currently derived from these rocks, but according to one estimate 80 billion barrels of oil could be recovered with present technology. The evaporite mineral *trona* is mined from Green River rocks for sodium compounds.

Mining of phosphorus-rich sedimentary rocks in Central Florida accounts for more than half that state's mineral production. The phosphorus from these rocks has a variety of uses in metallurgy, preserved foods, ceramics, matches, fertilizers, and animal feed supplements. Some of these phosphate rocks also contain interesting assemblages of fossil mammals (see Chapter 18).

Diatomite is a soft, low-density sedimentary rock made up of microscopic shells of diatoms, single-celled marine and freshwater plants with skeletons of silicon dioxide (SiO_2) (see Figure 15.4b). In fact, diatomite is so porous and light that when dry it will float. Diatomite is used mostly to purify gas and to filter liquids such as molasses, fruit juices, and sewage. The United States leads the world in diatomite production, mostly from Cenozoic deposits in California, Oregon, and Washington.

Historically, most coal mined in the United States (Canada has very little coal) has been Pennsylvanian-age bituminous coal from mines in Pennsylvania, West Virginia, Kentucky, and Ohio. Now, though, huge deposits of lignite and subbituminous coal in the Northern Great Plains are becoming important resources. These Late Cretaceous to Early Paleogene-age coal deposits are most abundant in the Williston and Powder River basins of North Dakota, Montana, and Wyoming. Besides having a low sulfur content, which makes them desirable, some of these coal beds are more than 30 m thick!

Gold from the Pacific Coast states, particularly California, comes largely from stream gravels in which placer deposits are found. A placer is an accumulation resulting from the separation and concentration of minerals of greater density from those of lesser density in streams or on beaches. The gold in these placers was weathered and eroded from Mesozoic-age quartz veins in the Sierra Nevada batholith and adjacent rocks (see Chapter 14 Introduction).

Hydrocarbons are recovered from the Cenozoic fault-bounded basins in Southern California and from many rocks of the Gulf Coastal Plain. Many rocks in the latter region form reservoirs for petroleum and natural gas because of different physical properties of the strata, and are thus called *stratigraphic traps*. Hydrocarbons are also found in geologic structures, such as folds, particularly those adjacent to salt domes, and such reservoirs are accordingly called *structural traps*. Because rock salt is a low-density sedimentary rock, when deeply buried and under pressure it rises toward the surface, and in doing so it penetrates and deforms the overlying rocks.

Another potential resource is methane hydrate, which consists of single methane molecules bound up in networks formed by frozen water. Huge deposits of methane hydrate are present along the eastern continental margin of North America, but so far it is not known whether they can be effectively recovered and used as an energy source. According to one estimate, the amount of carbon in methane hydrates worldwide is double that in all coal, oil, and conventional natural gas reserves.

SUMMARY

- The Late Triassic rifting of Pangaea continued through the Cenozoic and accounts for the present distribution of continents and oceans.

- Cenozoic orogenic activity was concentrated in two major belts: the Alpine–Himalayan orogenic belt and the circum-Pacific orogenic belt. Each belt is composed of smaller units called orogens.

- The Alpine orogeny resulted from convergence of the African and Eurasian plates. Mountain building took place in southern Europe, the Middle East, and North Africa. Plate motions also caused the closure of the Mediterranean basin, which became a site of evaporite deposition.

- India separated from Gondwana, moved north, and eventually collided with Asia, causing deformation and uplift of the Himalayas.

- Orogens characterized by subduction of oceanic lithosphere and volcanism took place in the western and northern Pacific Ocean basin. Back-arc spreading produced back-arc marginal basins such as the Sea of Japan.

- Subduction of oceanic lithosphere occurred along the western margins of the Americas during much of the Cenozoic.

- Subduction continues beneath Central and South America, but the North American plate is now bounded mostly by transform faults, except in the Pacific Northwest.

- The North American Cordillera is a complex mountainous region extending from Alaska into Mexico. Its Cenozoic evolution included deformation during the Laramide orogeny, extensional tectonics that formed the Basin and Range structures, intrusive and extrusive igneous activity, and uplift and erosion.

- Shallow angle subduction of the Farallon plate beneath North America resulted in the vertical uplifts of the Laramide orogeny. The Laramide orogen is centered in the middle and southern Rockies, but deformation occurred from Alaska to Mexico.

- Cordilleran volcanism was more or less continuous in the Cordillera through the Cenozoic. The Columbia River basalts represent one of the world's greatest eruptive events. Volcanism continues in the Cascade Range of the Pacific Northwest.

- Crustal extension in the Basin and Range Province yielded north–south oriented, normal faults. Differential movement on these faults produced uplifted ranges separated by broad, sediment-filled basins.

- The Colorado Plateau was deformed less than other areas in the Cordillera. Late Neogene uplift and erosion were responsible for the present topography of the region.

- The westward drift of North America resulted in its collision with the Pacific–Farallon ridge. Subduction ceased, and the continental margin became bounded by major transform faults, except where the Juan de Fuca plate continues to collide with North America.

- The Rio Grande rift formed as north–south oriented rifting took place in an area extending from Colorado into Mexico. The basins within this rift filled with sediments and volcanic rocks.

- Sediments eroded from Laramide uplifts were deposited in intermontane basins, on the Great Plains, and in a remnant of the Cretaceous epeiric sea in North Dakota.

- Deposition on the Gulf Coastal Plain and Atlantic Coastal Plain took place throughout the Cenozoic, resulting in seaward-thickening wedges of rocks grading from terrestrial facies to marine facies.

- Cenozoic uplift and erosion were responsible for the present topography of the Appalachian Mountains. Much of the sediment eroded from the Appalachians was deposited on the Atlantic Coastal Plain.

- Paleogene and Neogene mineral resources include oil and natural gas, gold, and phosphorus-rich sedimentary rocks.

IMPORTANT TERMS

Alpine–Himalayan orogenic belt, p. 333
Alpine orogeny, p. 333
Atlantic Coastal Plain, p. 348
back-arc marginal basin, p. 335
Basin and Range Province, p. 341
Cascade Range, p. 341
circum-Pacific orogenic belt, p. 335

Colorado Plateau, p. 342
continental interior p. 345
Farallon plate, p. 344
Gulf Coastal Plain, p. 348
Himalayan orogeny p. 333
Laramide orogeny, p. 336
North American Cordillera, p. 335

orogen, p. 332
Pacific–Farallon ridge, p. 344
Rio Grande rift, p. 342
San Andreas transform Fault, p. 344
Tejas epeiric sea, p. 348
Zuni epeiric sea, p. 345

REVIEW QUESTIONS

1. A complex part of the circum-Pacific orogenic belt in the United States is the
   a. _____ Tejas sedimentary sequence; b. _____ North American Cordillera; c. _____ Rio Grande rift; d. _____ Atlantic coastal plain; e. _____ Pacific-Farallon ridge.

2. The Basin and Range Province in the United States is
   a. _____ a huge area of block faulting; b. _____ mainly in Kansas and Nebraska; c. _____ made up mostly of volcanic mountains; d. _____ characterized by compression and crustal thickening; e. _____ bordered on the east and west by the Appalachians and the Great Plains, respectively.

3. As North America moved westward, the _____ plate was largely consumed as it was subducted beneath the continent
   a. _____ Zuni; b. _____ Orogenic; c. _____ Cascade; d. _____ Alpine; e. _____ Farallon.

4. Geologic evidence indicates that the Larmide orogeny ceased during the
   a. _____ Miocene; b. _____ Quaternary; c. _____ Eocene; d. _____ Permian; e. _____ Mesozoic.

5. A vast area of overlapping lava flows mostly in Washington state is known as the
   a. _____ Coast Ranges; b. _____ San Juan volcanic field; c. _____ Columbia River basalts; d. _____ Gulf Coastal Plain; e. _____ Zuni epeiric sea.

6. The Himalayas formed when the _____ plate collided with the _____ plate.
   a. _____ Farallon/Pacific; b. _____ Nazca/Cacos; c. _____ African/European ;d. _____ Indian/Asian; e. _____ Australian/South American.

7. The Cenozoic Era consists of two periods: the _____ and _____ .
   a. _____ Paleogene and Neogene; b. _____ Permian and Cretaceous; c. _____ Proterozoic and Archean; d. _____ Mesozoic and Triassic; e. _____ Miocene and Eocene.

8. Most of the Cenozoic-age sediment on the Atlantic coastal plain was eroded from the:
   a. _____ Rocky Mountains; b. _____ Cascade Range; c. _____ Appalachian Mountains; d. _____ Ozark Plateau; e. _____ Farallon ridge.

9. Cenozoic deposition of limestone was common in the
   a. _____ Florida section of the Gulf Coastal Plain; b. _____ basins formed by faulting in southern California;
   c. _____ Cannonball Sea in North Dakota; d. _____ Pacific Northwest of the U.S. and Canada; e. _____ Basin and Range Province.

10. The San Andreas Fault formed when the North American plate overrode the
    a. _____ Ouachita subduction zone; b. _____ Zuni epeiric sea ; c. _____ Andes Mountains ; d. _____ Pacific-Farallon ridge; e. _____ Atlantic Coastal Plain.

11. How does the Cenozoic geologic history of the Colorado Plateau differ from the history of other parts of the North American Cordillera?

12. Where is the Cascade Range, what kinds of volcanoes are found there, and what accounts for ongoing volcanism in that area?

13. How did the Laramide orogeny differ from other orogenies at convergent plate boundaries?

14. What features are shared by the Gulf Coastal Plain and the Atlantic Coastal Plain? Are there any differences between these two areas?

15. When and what sequence of events led to the origin of the Himalayas in Asia?

16. Explain how plate interactions were responsible for the origin of the San Andreas Fault.

17. What and where is the Basin and Range Province? Any speculation on what caused the structures typical of this region?

18. What kinds of sedimentary rocks are found on the Great Plains, and what was the source of sediment?

19. How does a back-arc marginal basin form?

20. What happened as the African plate moved northward against Europe and Asia?

APPLY YOUR KNOWLEDGE

1. A curious high school student wants to know why western North America has volcanoes, earthquakes, many mountain ranges, and numerous small glaciers, whereas these features are absent or nearly so in the eastern part of the continent. How would you explain this disparity, and further, can you think of how the situation might be reversed? That is, what kinds of events would lead to these kinds of geologic events in the east?

2. According to one estimate, 80 billion barrels of oil could be produced from oil shale in the Eocene Green River Formation of Wyoming, Colorado, and Utah, yet none is currently being produced. Given that U.S. domestic oil production is 8.45 million barrels per day (March 2008) and imports total 12.36 million barrels per day, what contribution would the Green River Formation make assuming that its total potential could be realized?

Do you see any problems with your projections? Also, what kinds of economic, technologic, and environmental problems might be encountered?

3. The United States uses just over 1.0 billion metric tons of coal each year from a reserve of 250 billion metric tons. Assuming that all of this coal can be mined, how long will it last at the current rate of consumption? Is there any reason to think that the current rate of consumption will remain the same, and is it even likely that all of the coal reserve could actually be recovered?

4. Twenty or thirty years from now it is doubtful that you will remember many of the terms and names of events you learned in this course, unless you become a geologist or work in some related area. So, how would you explain the evolution of the North American cordillera in terms that a nongeologist could readily understand?

CHAPTER 17

CENOZOIC GEOLOGIC HISTORY: THE PLEISTOCENE AND HOLOCENE EPOCHS

James S. Monroe

View of the Jungfrau Firn in Switzerland which merges with two other valley glaciers to form the Aletsch Glacier, the largest glacier in the Alps. From this viewpoint the glacier's terminus is 23 km distant. The Aletsch Glacier covers 120 km², but here its tributary is only about 1.5 km wide. Valley glaciers similar to this one are present on all the continents except Australia, but during the Pleistocene Epoch (the Ice Age) they were more numerous and much larger than they are now.

[OUTLINE]

Introduction

Pleistocene and Holocene Tectonism and Volcanism

    Tectonism

    Volcanism

Pleistocene Stratigraphy

    Terrestrial Stratigraphy

    Deep-Sea Stratigraphy

Onset of the Ice Age

    Climates of the Pleistocene and Holocene

    Glaciers—How Do They Form?

Glaciation and Its Effects

    Glacial Landforms

    Changes in Sea Level

Perspective *Waterton Lakes National Park, Alberta, and Glacier National Park, Montana*

Glaciers and Isostasy

Pluvial and Proglacial Lakes

What Caused Pleistocene Glaciation?

    The Milankovitch Theory

    Short-Term Climatic Events

Glaciers Today

Pleistocene Mineral Resources

Summary

At the end of this chapter, you will have learned that

• The Pleistocene and the Holocene or Recent epochs encompass only the most recent 1.8 million years of Earth history.

• The Pleistocene Epoch, lasting from 1.8 million years to 10,000 years ago, is best known for widespread glaciers but was also a time of continuing orogeny and volcanism.

• Much of our information about Pleistocene climates comes from oxygen isotope ratios, pollen analyses, and the distribution and coiling directions of planktonic foraminifera.

• Pleistocene continental glaciers were present on the Northern Hemisphere continents as well as Antarctica, and thousands of small glaciers were in mountain valleys on all continents.

• Sea level fell and rose during the several Pleistocene advances and retreats of glaciers, depending on how much water from the ocean was frozen on land.

• The tremendous weight of continental glaciers caused Earth's crust to subside into the mantle and to rise again when the glaciers wasted away.

• Many now-arid regions far from glaciated areas supported large lakes as a result of greater precipitation and lower evaporation rates during the Pleistocene, and numerous other lakes formed along the margins of glaciers.

• A current widely accepted theory explaining the onset of ice ages relies on irregularities in Earth's rotation and orbit.

• Important Pleistocene mineral resources include sand and gravel, diatomite, peat, and placer deposits of gold.

Introduction

The *Pleistocene Epoch* and *Holocene* or *Recent Epoch* are the designations for the most recent 1.8 million years of geologic time (see Figure 16.1). Notice in Figure 16.1 that in past usage the Quaternary Period encompassed the Pleistocene and Holocene, but more recently the Quaternary is a subperiod within the Neogene. However, because the chronostratigraphic status of the Quaternary has not yet been resolved, we will dispense with any further reference to it. Accordingly, the Pleistocene and Holocene epochs are the last two epochs of the Neogene Period. Obviously the Pleistocene from 1.8 million years ago to 10,000 years ago constitutes far more time than the Holocene, so our discussion in this chapter focuses on that interval of geologic time.

Of course, *only* 1.8 million years is far longer than we can visualize, and yet it is brief when put in the perspective of geologic time. Recall our analogy of all geologic time represented by a 24-hour clock (see Figure 8.1). In this context, the Pleistocene is 38 seconds long, but these are certainly an important 38 seconds, at least from our perspective, because during this time our species (*Homo sapiens*) evolved, and it was one of the few times in Earth history when vast glaciers were present.

A **glacier** is a body of ice on land that moves as a result of *plastic flow* (internal deformation in response to pressure) and by *basal slip* (sliding over its underlying surface). **Continental glaciers** (also called ice sheets) are the most important for our consideration in this chapter. By definition, they cover at least 50,000 km² and they are unconfined by topography, meaning that they flow outward from a point or points of accumulation (• Figure 17.1a). An **ice cap** is similar to a continental glacier but covers less than 50,000 km² (Figure 17.1b). And lastly, **valley glaciers** are long, narrow tongues of ice confined to mountain valleys where they flow from higher to lower elevations (Figure 17.1c).

In hindsight, it is difficult to understand why many scientists of the 1830s refused to accept the evidence indicating that widespread glaciers were present on the Northern Hemisphere continents during the recent geologic past. Many of them invoked the biblical deluge to explain the large boulders throughout Europe far from their source, whereas others thought the boulders were rafted by ice to their present positions during vast floods. By 1837, the Swiss naturalist Louis Agassiz argued convincingly that these boulders, as well as polished and striated bedrock and U-shaped valleys in many areas, resulted from huge masses of ice moving over the land.

We now know that the Pleistocene Epoch, more popularly known as the Ice Age, was a time of several major episodes of glacial advances each separated by warmer interglacial intervals. In addition, during times of glacial expansion more precipitation fell in regions now arid, such as the Sahara Desert of North Africa and Death Valley in California, both of which supported streams, lakes, and lush vegetation. Indeed, cultures existed in what is now the Sahara Desert as recently as 4500 years ago. Although we now know much about the Ice Age, an unresolved question is whether the Ice Age is truly over or are we simply in an interglacial period that will be followed by renewed glaciation?

We focus on Pleistocene glaciers in this chapter because they had such a profound impact on the continents, but remember that even at their maximum extent glaciers covered only about 30% of Earth's land surface. Of course, the climatic conditions that led to glaciation had worldwide effects, but other processes were operating as usual in the nonglaciated areas. From the systems approach that we introduced in Chapter 1, glaciers are part of the hydrosphere, although some geologists prefer the term *cryosphere* for all of Earth's frozen water, which includes glaciers, sea ice, snow, and even permafrost (permanently frozen ground).

• Figure 17.1 Continental Glaciers, Ice Caps, and Valley Glaciers

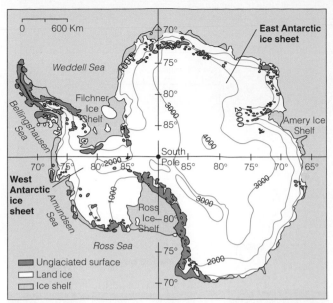

a The West and East Antarctic ice sheets merge to form a nearly continuous ice cover that averages 2160 m thick. The blue lines are lines of equal thickness.

b The Penny Ice Cap on Baffin Island, Canada, covers about 6000 km².

c A valley glacier such as this one in Alaska is a long, narrow tongue of moving ice confined to a mountain valley.

Pleistocene and Holocene Tectonism and Volcanism

The Pleistocene is best known for vast glaciers and their effects, but it was also a time of tectonism and volcanism, processes that continued through the Holocene to the present. Indeed, today plates diverge and converge, and, in places, slide past one another at transform plate boundaries. As a consequence, orogenic activity is ongoing as is seismic activity and volcanic eruptions.

Tectonism In Chapter 16, we discussed the continent-continent collision between India and Asia and the convergence of the Pacific plate with South America that formed the Andes. These areas of orogenic activity continue unabated, as do those in the Aleutian Islands, Japan, the Philippines, and elsewhere. Interactions between the North America and Pacific plates along the San Andreas transform plate boundary produced folding (• Figure 17.2a), faulting, and a number of basins and uplifted areas. Marine terraces covered with Pleistocene sediments attest to periodic uplift all along the Pacific Coast of the United States (Figure 17.2b).

Archean Eon	Proterozoic Eon	Phanerozoic Eon								
Precambrian		Paleozoic Era								
		Cambrian	Ordovician	Silurian	Devonian	Mississippian	Pennsylvanian	Permian
						Carboniferous		

2500 MYA 542 MYA 251 MYA

• Figure 17.2 **Pleistocene Uplift and Tectonism**

a These deformed sedimentary rocks are only a few hundred meters from the San Andreas Fault in southern California.

Marli Bryant Miller

Marine terraces

b Several marine terraces on San Clemente Island, California. Each terrace was at sea level when it formed. The highest is 400 m above sea level.

University of Washington Libraries Special Collections, Neg. no. KC8974

Volcanism Ongoing subduction of remnants of the Farallon plate beneath Central America and the Pacific Northwest accounts for volcanism in these areas. The Cascade Range of California, Oregon, Washington, and British Columbia has a history dating back to the Oligocene, but the large composite volcanoes and Lassen Peak, a large lava dome, formed mostly during the Pleitocene and Holocene (• Figure 17.3a). Indeed, Lassen Peak in California and Mount St. Helens in Washington State erupted during the 1900s, and in the case of the latter, it showed renewed activity in 2004.

Volcanism also took place in several other areas in the western United States, including Arizona, Idaho, and California (Figure 17.3b and c), and many of the volcanoes in Alaska have erupted in the recent past. Following colossal eruptions, huge calderas formed in the area of Yellowstone National Park, Wyoming. Vast eruptions took place 2.0 and 1.3 million years ago and again 600,000 years ago that left a composite caldera measuring 76 by 45 km. Since its huge eruption, part of the area has risen, presumably from magma below the surface, forming a *resurgent dome*. And finally, between 150,000 and 75,000 years ago, the Yellowstone Tuff was erupted and partially filled the caldera.

Elsewhere in the world, volcanoes erupted in South America, the Philippines, Japan, the east Indies, as well as Iceland, Spitzbergen, and the Azores. So, even though the amount of heat generated within Earth has decreased through time (see Chapter 8), volcanism and the other processes driven by internal heat remain significant.

Pleistocene Stratigraphy

Although geologists continue to debate which rocks should serve as the Pleistocene stratotype,* they agree the Pleistocene Epoch began 1.8 million years ago. The Pleistocene–Holocene boundary at 10,000 years ago is based on climatic change from cold to warmer conditions concurrent with the melting of the most recent ice sheets. Changes in vegetation, as well as oxygen isotope ratios determined from shells of marine organisms, provide ample evidence for this climatic change.

---

* Recall from Chapter 5 that the stratotype is a section of rocks where a named stratigraphic unit such as a system or series was defined—for example, the stratotype for the Cambrian System is in Wales.

Phanerozoic Eon											
Mesozoic Era			Cenozoic Era								
Triassic	Jurassic	Cretaceous	Paleogene			Neogene		Quaternary			
			Paleocene	Eocene	Oligocene	Miocene	Pliocene	Pleistocene	Holocene		

251 MYA

66 MYA

Terrestrial Stratigraphy

Soon after Louis Agassiz proposed his theory for glaciation, research focused on deciphering the history of the Ice Age. This work involved recognizing and mapping terrestrial glacial features and placing them in a stratigraphic sequence. From glacial features such as moraines, erratic boulders, and glacial striations, geologists have determined that Pleistocene glaciers at their greatest extent were up to 3 km thick and covered about three times as much of Earth's surface as they do now, or about 45,000,000 km^2 (• Figure 17.4). Furthermore, detailed mapping of glacial features reveals that several glacial advances and retreats occurred.

Geologists have mapped the distribution of glacial deposits, and determined that North America had at least four major episodes of Pleistocene glaciation. Each glacial advance was followed by a glacial retreat and warmer

• Figure 17.3 **Pleistocene and Recent Volcanism**

Marli Bryant Miller

a View of several volcanoes in the Cascade Range in central Oregon. Mount Bachelor at only 11,000 to 15,000 years old is the youngest volcano in the range. Volcanism in the range goes back at least to the Oligocene, but the large volcanoes are mostly Pleistocene to Recent in age.

• Figure 17.3 (Cont.)

b This 115-m-high cinder cone known as High Hole Crater in northern California lies on the flank of a huge shield volcano. The aa lava flow in the foreground was erupted about 1100 years ago.

c The Lower Falls of the McCloud River in California plunges 3.5 m over a precipice in a Pleistocene lava flow.

• Figure 17.4 **Centers of Ice Accumulation and Maximum Extent of Pleistocene Glaciers**

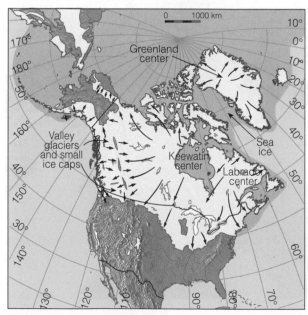

a North America.

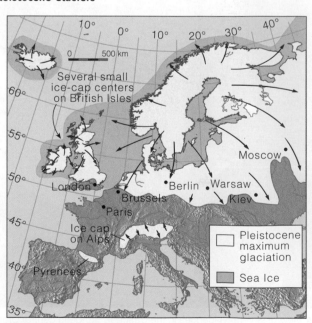

b Europe and part of Asia.

• Figure 17.5 **Pleistocene Glaciers in North America**

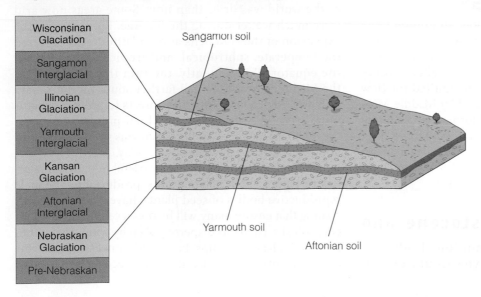

Wisconsinan Glaciation

Sangamon Interglacial

Illinoian Glaciation

Yarmouth Interglacial

Kansan Glaciation

Aftonian Interglacial

Nebraskan Glaciation

Pre-Nebraskan

Sangamon soil

Yarmouth soil

Aftonian soil

a Traditional terminology for Pleistocene glacial and interglacial stages in North America.

b Idealized succession of deposits and soils developed during the glacial and interglacial stages.

climates. The four **glacial stages,** the *Wisconsinan, Illinoian, Kansan,* and *Nebraskan,* are named for the states where the southernmost glacial deposits are well exposed. The three **interglacial stages,** the *Sangamon, Yarmouth,* and *Aftonian,* are named for localities of well-exposed interglacial soil and other deposits (• Figure 17.5). Recent detailed studies of glacial deposits indicate, however, that there were an as yet undetermined number of pre-Illinoian glacial events and that the history of glacial advances and retreats in North America is more complex than previously thought.

Six or seven major glacial advances and retreats are recognized in Europe, and at least 20 major warm–cold cycles have been detected in deep-sea cores. Why isn't there better correlation among the different areas if glaciation was so widespread? Part of the problem is that glacial deposits are typically chaotic mixtures of coarse materials that are difficult to correlate. Furthermore, glacial advances and retreats usually obscure or destroy the sediment left by the previous advances. Even within a single major glacial advance, several minor advances and retreats may have occurred. For example, careful study of deposits from the Wisconsinan glacial stage reveals at least four distinct fluctuations of the ice margin during the last 70,000 years in Wisconsin and Illinois.

Deep-Sea Stratigraphy
Until the 1960s, the traditional view of Pleistocene chronology was based on sequences of glacial sediments on land. However, new evidence from ocean sediment samples indicate numerous climatic fluctuations during the Pleistocene.

Evidence for these climatic fluctuations comes from changes in surface ocean temperature recorded in the shells of planktonic foraminifera, which after they die sink to the seafloor and accumulate as sediment.

One way to determine past changes in ocean surface temperatures is to resolve whether planktonic foraminifera were warm- or cold-water species. Many are sensitive to variations in temperature and migrate to different latitudes when the surface water temperature changes. For example, the tropical species *Globorotalia menardii* during periods of cooler climate, is found only near the equator, whereas during times of warming its range extends into the higher latitudes.

Some planktonic foraminifera species change the direction they coil during growth in response to temperature fluctuations. The Pleistocene species *Globorotalia truncatulinoides* coils predominantly to the right in water temperatures above 10°C but coils mostly to the left in water below 8°–10°C. On the basis of changing coiling ratios, geologists have constructed detailed climatic curves for the Pleistocene and earlier epochs.

Changes in the O^{18}-to-O^{16} ratio in the shells of planktonic foraminifera also provide data about climate. The abundance of these two oxygen isotopes in the calcareous ($CaCO_3$) shells of foraminifera is a function of the oxygen isotope ratio in water molecules and water temperature when the shell forms. The ratio of these isotopes reflects the amount of ocean water stored in glacial ice. Seawater has a higher O^{18}-to-O^{16} ratio than glacial ice, because water containing the lighter O^{16} isotope is more easily evaporated than water containing the O^{18} isotope. Therefore, Pleistocene glacial ice was enriched in O^{16} relative to O^{18}, whereas the heavier O^{18} isotope is concentrated in seawater. The declining percentage of O^{16} and consequent rise of O^{18} in seawater during times of glaciation is preserved in the shells of planktonic foraminifera. Consequently, oxygen isotope fluctuations indicate surface water temperature changes and thus climatic changes.

Unfortunately, geologists have not yet been able to correlate these detailed climatic changes with corresponding changes recorded in the sedimentary record on land. The time lag between the onset of cooling and any resulting glacial advance produces discrepancies between the marine and terrestrial records. Thus, it is unlikely that all the minor climatic fluctuations recorded in deep-sea sediments will ever be correlated with continental deposits.

Onset of the Ice Age

Glacial conditions actually set in about 40 million years ago when surface ocean waters at high southern latitudes rapidly cooled, and the water in the deep ocean became much colder than it was previously. The gradual closure of the Tethys Sea during the Oligocene limited the flow of warm water to higher latitudes, and by Middle Miocene time an Antarctic ice sheet had formed, accelerating the formation of very cold oceanic waters. After a brief Pliocene warming trend, continental glaciers began forming in the Northern Hemisphere about 1.8 million years ago—the Pleistocene Ice Age was underway.

Climates of the Pleistocene and Holocene

The climatic conditions leading to Pleistocene glaciation were, as you would expect, worldwide. Contrary to popular belief and depictions in cartoons and movies, Earth was not as cold as commonly portrayed. In fact, evidence of various kinds indicates that the world's climate cooled gradually from Eocene through Pleistocene time (• Figure 17.6). Oxygen isotope ratios (O^{18} to O^{16}) from deep-sea cores reveal that during the last 2 million years Earth has had 20 major warm–cold cycles during which the temperature fluctuated by as much as 10°C (see the section on deep-sea stratigraphy). And studies of glacial deposits attest to at least four major episodes of glaciation in North America and six or seven similar events in Europe.

During glacial growth, those areas covered by or near glaciers experienced short, cool summers and long, wet winters. Areas distant from glaciers had varied climates. When glaciers grew and advanced, lower ocean temperatures reduced evaporation rates, so most of the world was drier than now. Some areas now arid were much wetter during the Ice Age. For instance, the expansion of the cold belts at high latitudes compressed the temperate, subtropical, and tropical zones toward the equator. Consequently, the rain that now falls on the Mediterranean then fell farther south on the Sahara of North Africa, enabling lush forests to grow in what is now desert. In North America, a high-pressure zone over the northern ice sheets deflected storms south, so the arid Southwest was much wetter than today.

Pollen analysis is particularly useful in paleoclimatology (• Figure 17.7). Pollen grains, produced by the male reproductive bodies of seed plants, have a resistant waxy coating that ensure many will be preserved in the fossil record. Most seed plants disperse pollen by wind, so it settles in streams, lakes, swamps, bogs, and in nearshore marine environments. Once paleontologists recover pollen from

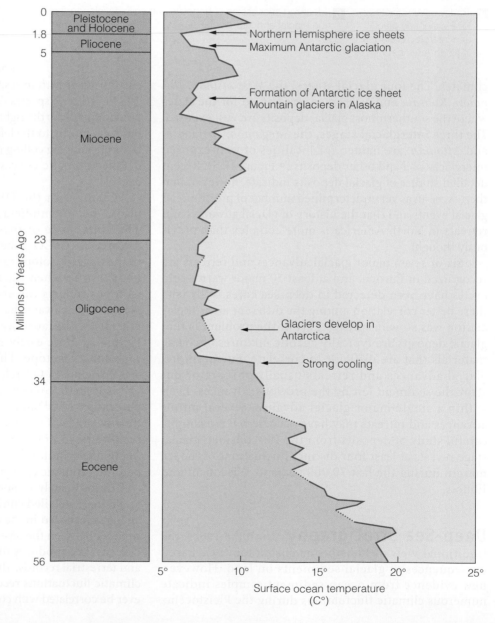

• Figure 17.6 **Climatic Changes During the Cenozoic** Oxygen isotope ratios from a sediment core in the western Pacific Ocean indicate that ocean surface temperatures changed during the last 56 million years. A change from warm surface waters to colder conditions took place about 32 million years ago.

• Figure 17.7 **Pollen Analyses and Climate**

Chatsworth Bog, Illinois

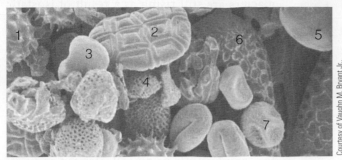

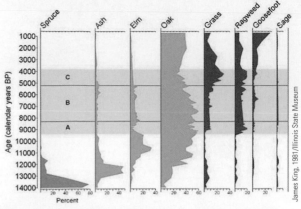

Courtesy of Vaughn M. Bryant Jr., Texas A and M University

a Scanning electron microscope view of present-day pollen: (1) sunflower, (2) acacia, (3) oak, (4) white mustard, (5) little walnut, (6) agave, and (7) juniper.

James King, 1981/Illinois State Museum

b Pollen diagram for the last 4000 years for Chatsworth Bog in Illinois. Spruce forest was replaced by ash and elm in a wetter climate. Oak, grasses, and ragweed then increased indicating prairie development.

sediments, they identify the type of plant it came from, determine the floral composition of the area, and make climatic inferences (Figure 17.7).

Pollen diagrams (Figure 17.7b), tree-ring analysis (see Chapter 4), and studies of the advances and retreats of valley glaciers have yielded a wealth of information about the Northern Hemisphere climate for the last 10,000 years—that is, since the time the last major continental glaciers retreated and disappeared. Data from pollen analysis indicate a continuous trend toward a warmer climate until about 6000 years ago. In fact, between 8000 to 6000 years ago temperatures were very warm. Then the climate became cooler and moister, favoring the growth of valley glaciers on the Northern Hemisphere continents. Three episodes of glacial expansion took place during this **neoglaciation,** as it is called. The most recent one, the **Little Ice Age** occurred between 1500 and the mid- to late 1800s. During the Little Ice Age, glaciers in mountain valleys expanded, an ice cap formed in Iceland, sea ice persisted much longer into the spring and summer at high northern latitudes, and rivers and canals in Europe regularly froze over. However, the greatest effect on humans came from the cooler, wetter summers and shorter growing seasons that resulted in famines as well as migrations of many Europeans to the New World. In England, the growing season was five weeks shorter from 1680 to 1730. In Europe and Iceland, glaciers reached their greatest historic extent by the early 1800s, and glaciers in the western United States, Alaska, and Canada also expanded.

Glaciers—How Do They Form?

We defined the terms *continental glacier, ice cap,* and *valley glacier* as moving bodies of ice on land (see the Introduction) (Figure 17.1). During the Pleistocene, all types of glaciers were much more widespread than now. For example, the only continental glaciers today are the ones in Antarctica and Greenland (• Figure 17.8), but during the Pleistocene they covered about 30% of Earth's land surface, especially on the Northern Hemisphere

continents. These continental glaciers formed, advanced, and then retreated several times, forming much of the present topography of the glaciated regions and nearby areas. The Pleistocene was also a time when small valley glaciers were more common in mountain ranges. Indeed, much of the spectacular scenery in such areas as Grand Teton National Park, Wyoming and Glacier National Park in Montana resulted from erosion by valley glaciers.

The question "How do glaciers form?" is rather easily answered, unlike "What causes the onset of an ice age?" Any area that receives more snow in the cold season than melts in the warm season has a net accumulation over the years. As accumulation takes place, the snow at depth is converted to glacial ice and when a critical thickness of about 40 m is reached, flow in response to pressure begins. Once a glacier forms, it moves from a zone of accumulation, where additions exceed losses, toward its zone of wastage, where losses exceed additions. As long as a balance exists between the two, the glacier has a *balanced budget,* but the budget may be *negative* or *positive,* depending on any imbalances that exist in these two zones. Consequently, a glacier's terminus may advance, retreat, or remain stationary, depending on its budget.

Glaciation and Its Effects

Huge glaciers moving over Earth's surface reshaped the previously existing topography and yielded many distinctive glacial landforms. As glaciers formed and wasted away, sea level fell and rose, depending on how much water was frozen on land, and the continental margins were alternately exposed and water-covered. In addition, the climatic changes that initiated glacial growth had effects far beyond the glaciers themselves. Another legacy of the

James S. Monroe

• **Figure 17.8 The Greenland Ice Sheet** Greenland is mostly covered by a continental glacier that is more than 3000 m thick. Notice that only a few high mountains are not ice covered. Continental glaciers were much more widespread during the Pleistocene, but today only Greenland and Antarctica have continental glaciers.

Pleistocene is that areas once covered by thick glaciers are still rising as a result of isostatic rebound.

Glacial Landforms
Remember that glaciers are moving masses of ice on land, and as such continental and valley glaciers yield a number of easily recognized erosional and depositional landforms. A large part of Canada and parts of some northern states have subdued topography, little or no soil, striated and polished bedrock exposures, and poor surface drainage, characteristics of an **ice-scoured plain** (• Figure 17.9a). Pleistocene valley glaciers also yielded several distinctive landforms such as bowl-shaped depressions on mountainsides known as **cirques** and broad valleys called **U-shaped glacial troughs** (Figure 17.9b, and see Perspective).

The deposits of continental and valley glaciers are **moraines,** which are chaotic mixtures of poorly sorted sediment deposited directly by glacial ice, and **outwash** consisting of stream-deposited sand and gravel (• Figure 17.10). Any *moraine* deposited at a glacier's terminus is an **end moraine,** but notice from Figure 17.10 that *terminal* and *recessional moraines* are types of end moraines. Terminal moraines and outwash in southern Ohio, Indiana, and Illinois mark the greatest southerly extent of Pleistocene continental glaciers in the midcontinent region. Recessional moraines indicate the positions where the ice front stabilized temporarily during a general retreat to the north (• Figure 17.11).

Glaciers are, of course, made up of frozen water and thus, constitute an important part of

• **Figure 17.9 Erosion by Continental and Valley Glaciers Yield Distinctive Landscapes**

University of Washington Libraries Special Collections

a A continental glacier eroded this ice-scoured plain, a subdued surface with extensive rock exposures, in the Northwest Territories of Canada.

USGS

b Valley glaciers erode mountains and leave sharp, angular peaks and ridges and broad, smooth valleys as seen here in the Chugach Mountains in Alaska.

• Figure 17.10 **Origin of End Moraines and Outwash**

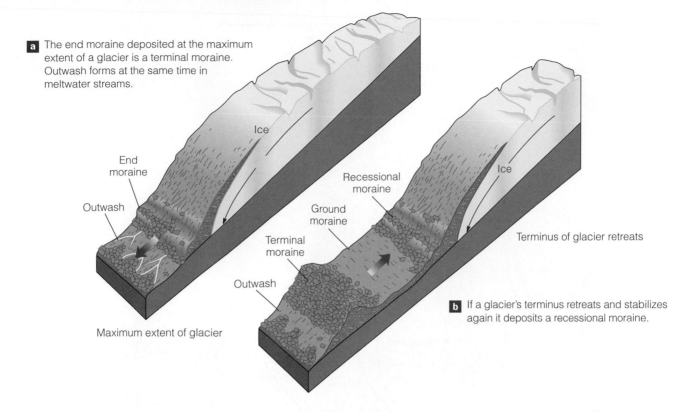

a The end moraine deposited at the maximum extent of a glacier is a terminal moraine. Outwash forms at the same time in meltwater streams.

End moraine

Outwash

Ice

Maximum extent of glacier

Recessional moraine

Ground moraine

Terminal moraine

Outwash

Ice

Terminus of glacier retreats

b If a glacier's terminus retreats and stabilizes again it deposits a recessional moraine.

c This terminal moraine in California is typical; it is unsorted and shows no stratification.

d Outwash deposited by streams that come from melting glaciers on Mount Rainier in Washington State.

the hydrosphere, one of Earth's major systems. In Chapter 1, we emphasized the systems approach to Earth history, and here we have an excellent opportunity to see interactions among systems at work. Cape Cod, Massachusetts, is a distinctive landform that resembles a human arm extending into the Atlantic Ocean. It and nearby Martha's Vineyard and Nantucket Island owe their existence to deposition by Pleistocene glaciers and modification of these deposits by wind-generated waves and nearshore currents (• Figure 17.12).

Changes in Sea Level Today, 28 to 35 million km³ of water is frozen in glaciers, all of which came from the oceans. During the maximum extent of Pleistocene glaciers, though, more than 70 million km³ of ice was present on the continents. These huge masses of ice themselves had a tremendous impact on the glaciated areas (see the next section), and they contained enough frozen water to lower sea level by 130 m. Accordingly, large areas of today's continental shelves were exposed and quickly blanketed by vegetation. In fact, the Bering

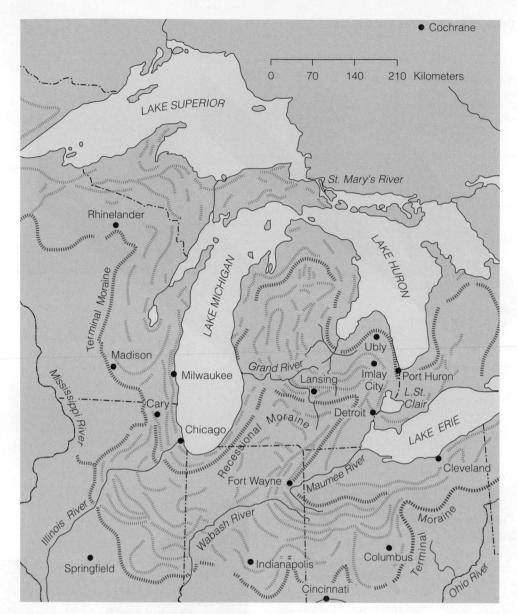

• **Figure 17.11 Terminal and Recessional Moraines in the Mid-Continent Region** These moraines were deposited during the latter part of the Wisconsinan glaciation. The oldest ones, those farthest south, are about 16,000 years old. Only one terminal moraine is present, marking the greatest extent of Wisconsinan glaciers; all the others on the map are recessional moraines.

Strait connected Alaska with Siberia via a broad land bridge across which Native Americans and various mammals such as the bison migrated (• Figure 17.13). The shallow floor of the North Sea was also above sea level so Great Britain and mainland Europe formed a single landmass. When the glaciers melted, these areas were flooded, drowning the plants and forcing the animals to migrate.

Lower sea level during the several Pleistocene glacial intervals also affected the *base level,* the lowest level to which running water can erode, of rivers and streams flowing into the oceans. As sea level dropped, rivers eroded deeper valleys and extended them across the emergent continental shelves. During times of lower sea level, rivers transported huge quantities of sediment across the exposed continental shelves and onto the continental slopes, where the sediment contributed to the growth of submarine fans. As the glaciers melted, however, sea level rose, and the lower ends of these river valleys along North America's East Coast were flooded, whereas those along the West Coast formed impressive submarine canyons.

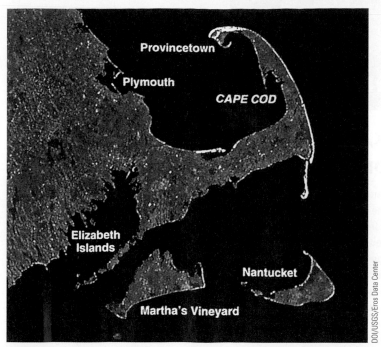

DOI/USGS/Eros Data Center

a Cape Cod and the nearby islands are made up of mostly end moraines. although the deposits have been modified by waves since they were deposited 23,000 to 14,000 years ago.

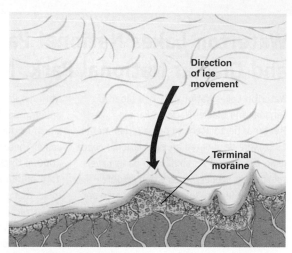

b Position of the glacier when it deposited a terminal moraine that would become Martha's Vineyard and Nantucket Island.

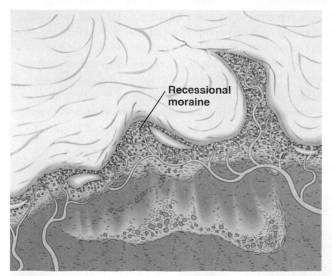

c Position of the glacier when it deposited a recessional moraine that now forms much of Cape Cod.

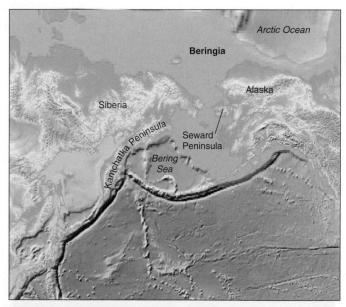

☐ Bering land bridge

• Figure 17.13 **The Bering Land Bridge** During the Pleistocene, sea level was as much as 130 m lower than it is now, and a broad area called the Bering land bridge (Beringia) connected Asia and North America. It was exposed above sea level during times of glacial advances and served as a corridor for the migration of people, animals, and plants.

Waterton Lakes National Park, Alberta, and Glacier National Park, Montana

Waterton Lakes National Park in Alberta, Canada, and Glacier National Park in Montana, lie adjacent to one another and in 1932 were designated an international peace park, the first of its kind. Both parks have spectacular scenery; interesting wildlife such as mountain goats, bighorn sheep, and grizzly bears; and an impressive geologic history. The present-day landscapes resulted from deformation and uplift from Cretaceous to Eocene times, followed by deep erosion by streams and glaciers. Park visitors can see the results of the phenomenal forces at work during an orogeny by visiting sites where a large fault is visible, and glacial landforms such as U-shaped glacial troughs, arêtes, cirques, and horns are some of the finest in North America (Figure 1).

Most of the rocks exposed in Glacier National Park belong to the Late Proterozoic-age Belt Supergroup* (see Figure 9.1b and 9.7a), whereas those in Waterton Lakes

National Park are assigned to the Purcell Supergroup. The names differ north and south of the border, but the rocks are the same. These Belt-Purcell rocks are nearly 4000 m thick and were deposited between 1.45 billion and 850 million years ago. The rocks themselves are interesting, and some are attractive, especially red and green rocks consisting mostly of mud and thick limestone formations. In addition, many of the rocks contain sedimentary structures such as mud cracks, ripple marks, and cross-bedding that help geologists interpret how they were deposited. Dark-colored mudstones and sandstones were deposited during the Cretaceous Period when a marine transgression took place that covered a large part of North America, including the area of the present-day parks.

The most impressive geologic structure in the parks is the Lewis overthrust,* a large fault along which Belt-Purcell rocks have

moved at least 75 km eastward so that they now rest on much younger Cretaceous age rocks (see Figure 16.11). If you take the trail from Marias Pass in Glacier National Park to get a closer look at the fault, you can see intense deformation of the rocks lying below the fault.

During the Pleistocene Epoch, glaciers formed and grew, overtopping the divides between valleys and thus forming an ice cap that nearly buried the entire area. In fact, several episodes of Pleistocene glaciation took place, but the evidence for the most recent one is most obvious. These glaciers flowed outward in all directions, and in the east they merged with the continental glacier covering most of Canada and the northern states. Much of the parks' landscapes developed during these glacial episodes as valleys were gouged deeper and widened, and cirques, arêtes, and horns developed (Figure 1).

Figure 1 These sharp angular peaks and broad rounded valleys resulted from erosion by valley glaciers

James S. Monroe

a St. Mary Lake in Glacier National Park, Montana.

© Ron Watts/CORBIS

b Wateron Lakes National Park in Alberta, Canada.

Supergroup is a geologic term for two or more groups that in turn are composed of two or more formations.

*An overthrust fault is simply a very low angle thrust fault along which movement is usually measured in kilometers.

Today, only about two dozen small glaciers remain active in the parks. But just like earlier glaciers, they continue to erode, transport, and deposit sediment, only at a considerably reduced rate. In fact, many of the 150 or so glaciers present in Glacier National Park in 1850 are now gone or remain simply as patches of stagnant ice. And even among the others it is difficult to determine exactly how many are active because they are so small and move so slowly, only a few meters per year. They did, however, expand markedly during the Little Ice Age, but have since retreated (Figure 2).

It seems that these small glaciers, which are very sensitive to climatic changes, are shrinking as a result of the 1°C increase in average summer temperatures in this region since 1900. According to one U.S. Geological Survey report, expected increased warming will eliminate the glaciers by 2030, and certainly by 2100, even if no additional warming takes place.

None of the active glaciers in either park can be reached by road, but several are visible from a distance. Nevertheless, Pleistocene glaciers and the remaining active ones were responsible for much of the striking scenery. Now weathering, mass wasting, and streams are modifying the glacial landscape.

Figure 2 Grinnell Glacier from Mount Gould from 1938 to 2006.

What would happen if the world's glaciers all melted? Obviously, the water stored in them would return to the oceans, and sea level would rise about 70 m. If this were to happen, many of the world's large population centers would be flooded.

Glaciers and Isostasy

In a manner of speaking, Earth's crust floats on the denser mantle below, a phenomenon geologists call **isostasy.** An analogy can help you understand this concept, which is certainly counterintuitive; after all, how can rock float in rock? Consider an iceberg. Ice is slightly less dense than water, so an iceberg sinks to its equilibrium position in water with only about 10% of its volume above the surface. Earth's crust is more complicated, but it sinks into the mantle, which behaves like a fluid, until it reaches its equilibrium position depending on its thickness and density. Remember, oceanic crust is thinner but denser than continental crust, which varies considerably in thickness.

If the crust has more mass added to it, as when thick layers of sediment accumulate or vast glaciers form, it sinks lower into the mantle until it once again achieves equilibrium. However, if erosion or melting ice reduces the load, the crust slowly rises by **isostatic rebound.** Think of the iceberg again. If some of it above sea level were to melt, it would rise in the water until it regained equilibrium.

No one doubts that Earth's crust subsided from the great weight of glaciers during the Pleistocene or that it has rebounded and continues to do so in some areas. Indeed, the surface in some places was depressed as much as 300 m below preglacial elevations. But as the glaciers melted and eventually wasted away, the downwarped areas gradually rebounded to their former positions. Evidence of isostatic rebound is seen in formerly glaciated areas such as Scandinavia and the North American Great Lakes region. Some coastal cities in Scandinavia have rebounded enough so that docks built only a few centuries ago are now far inland from the shore. And in Canada as much as 100 m of isostatic rebound has taken place during the last 6000 years (• Figure 17.14).

Pluvial and Proglacial Lakes

During the Wisconsinan glacial stage, many now arid parts of the western United States supported large lakes when glaciers were present far to the north. These **pluvial lakes,** as they are called, existed because of the greater precipitation and

• Figure 17.14 Glaciers and Isostasy

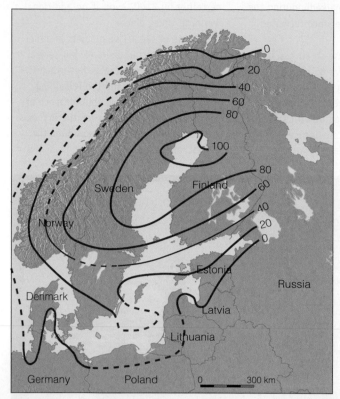

c Isostatic rebound in Scandinavia. The lines show rates of uplift in centimeters per century.

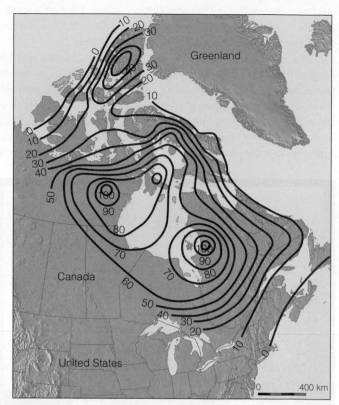

d Isostatic rebound in eastern Canada in meters during the last 6000 years.

overall cooler temperatures, especially during the summer, which lowered the evaporation rate. Wave-cut cliffs, beaches, deltas, and lake deposits along with fossils of freshwater organisms attest to the presence of these lakes (• Figure 17.15).

Death Valley on the California–Nevada border is the hottest, driest place in North America, yet during the Wisconsinan it supported Lake Manly, a large pluvial lake (Figure 17.15). It was 145 km long, nearly 180 m deep, and when it dried up dissolved salts precipitated on the valley floor. Borax, one of the minerals in these lake deposits, is mined for use in ceramics, fertilizers, glass, solder, and pharmaceuticals.

In contrast to pluvial lakes, which are far from areas of glaciation, **proglacial lakes** form where meltwater accumulates along a glacier's margin. Lake Agassiz, named in honor of the French naturalist Louis Agassiz, formed in this manner. It covered about 250,000 km² in North Dakota, Manitoba, Saskatchewan, and Ontario, and persisted until the ice along its northern margin melted, at which time it drained northward into Hudson Bay.

Deposits in lakes adjacent to or near glaciers vary considerably from gravel to mud, but of special interest are the finely laminated mud deposits consisting of alternating dark and light layers. Each dark-light couplet is a **varve** (see Figure 6.14d), representing an annual deposit. The light-colored layer of silt and clay formed during the spring and summer, and the dark layer is made up of smaller particles and organic matter formed during winter

when the lake froze over. These varved deposits may also contain gravel-sized particles known as *dropstones*, released from melting ice.

Glacial Lake Missoula In 1923, geologist J Harlan Bretz proposed that a Pleistocene lake in what is now western Montana periodically burst through its ice dam and flooded a large area in the Pacific Northwest. He further claimed that these huge floods were responsible for the giant ripple marks and other fluvial features seen in Montana and Idaho as well as the *scablands* of eastern Washington, an area in which the surface deposits were scoured, exposing underlying bedrock (• Figure 17.16a).

Bretz's hypothesis initially met with considerable opposition, but he marshaled his evidence and eventually convinced geologists these huge floods had taken place, the most recent one probably no more than 18,000 to 20,000 years ago. It now is well accepted that Lake Missoula, a large proglacial lake covering about 7800 km², was impounded by an ice dam in Idaho that periodically failed. In fact, the shorelines of this ancient lake are clearly visible on the mountainsides around Missoula, Montana. When the ice dam failed, the water rushed out at tremendous velocity, accounting for the various fluvial features seen in Montana and Idaho and the scablands in eastern Washington (Figure 17.15a).

Lake Bonneville Lake Bonneville, with a maximum size of about 50,000 km² and at least 335 m deep, was a large pluvial

• Figure 17.15 **Pleistocene Lakes in the Western United States**

• Figure 17.16 **Glacial Lake Missoula was a proglacial lake, whereas Lake Bonneville was a pluvial lake**

a This rumpled surface of gravel ridges, or so-called giant ripple marks, formed when Lake Missoula drained across this area near Camas Prairie, Montana.

b The Melon Gravel was deposited by the floodwaters from Lake Bonneville. This exposure is near Hagerman, Idaho.

a Pyramid Lake and Great Salt Lake are shrunken remnants of much larger ancient lakes. Of all the lakes shown, only Lake Columbia and Lake Missoula were proglacial lakes. When the 600-m-high ice dam impounding Lake Missoula failed, it drained westward and scoured out the scablands of eastern Washington.

b The area in this image east of Fallon, Nevada, was covered by Lake Lahontan. In fact, the image was taken from Grimes Point Archaeolgcal Site where Native Americans lived near the lakeshore.

lake mostly in what is now Utah, but parts of it extended into eastern Nevada and southern Idaho (Figure 17.15a). About 15,000 years ago, Lake Bonneville flooded catastrophically when it overflowed and rapidly eroded a natural dam at Red Rock Pass in Idaho. The flood waters followed the course of the Snake River, and, just as the Lake Missoula flood, it left abundant evidence of its passing. For example, the Melon Gravels in Idaho consist of rounded basalt boulders up to 3 m in diameter, and the gravel bars are as much as 90 m thick and 2.4 km long (Figure 17.16b). Although a catastrophic flood with an estimated discharge of 1.3 km³/hr, the Lake Missoula flood's discharge was about 30 times as great. The vast salt deposits of the Bonneville Salt Flats west of Salt Lake City, Utah, formed when parts of this ancient lake dried up, and the Great Salt Lake is a shrunken remnant of this once much larger lake.

A Brief History of the Great Lakes Before the Pleistocene, the Great Lakes region was a rather flat lowland with broad stream valleys. As the continental glaciers advanced southward from Canada, the entire area was ice-covered and deeply eroded. Indeed, four of the five Great Lakes basins were eroded below sea level; glacial erosion is not restricted by base level, as erosion by running water is. In any case, the glaciers advanced far to the south, but eventually began retreating north, depositing numerous recessional moraines as they did so (Figure 17.11).

By about 14,000 years ago, parts of the Lake Michigan and Lake Erie basins were ice free, and glacial meltwater began forming proglacial lakes (• Figure 17.17). As the ice

front resumed its retreat northward—although interrupted by minor readvances—the Great Lakes basins eventually became ice free, and the lakes expanded until they reached their present size and shape.

This brief history of the Great Lakes is generally correct, but oversimplified. The minor readvances of the ice front mentioned earlier caused the lakes to fluctuate widely, and as they filled, they overflowed their margins and partly drained. In addition, once the glaciers were gone, isostatic rebound took place, and this too has affected the Great Lakes.

What Caused Pleistocene Glaciation?

We know how glaciers move, erode, transport, and deposit sediment, and we even know the conditions necessary for them to originate—more winter snowfall than melts during the following warmer seasons. But this really does not address the broader question of what caused large-scale glaciation during the Ice Age, and why so few episodes of glaciation have occurred. Geologists, oceanographers, climatologists, and others have tried for more than a century to develop a comprehensive theory explaining all aspects of ice ages, but so far have not been completely successful. One reason for their lack of success is that the climatic changes responsible for glaciation, the cyclic occurrence of glacial–interglacial stages, and short-term events such as the Little Ice Age operate on vastly different time scales.

The few periods of glaciation recognized in the geologic record are separated by long intervals of mild climate. Slow geographic changes related to plate tectonic activity are probably responsible for such long-term climatic changes. Plate movements may carry continents into latitudes where glaciers are possible, provided they receive enough snowfall. Long-term climatic changes also take place as plates collide, causing uplift of vast areas far above sea level, and of course the distribution of land and sea has an important influence on oceanic and atmospheric circulation patterns.

One proposed mechanism for the onset of the cooling trend that began following the Mesozoic and culminated with Pleistocene glaciation is decreased levels of carbon dioxide (CO_2) in the atmosphere. Carbon dioxide is a greenhouse gas, so if less were present to trap sunlight, Earth's overall temperature would perhaps

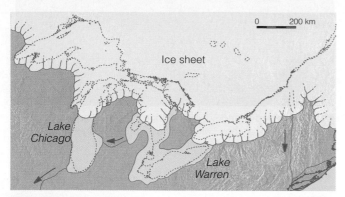

a 13,000 years ago.

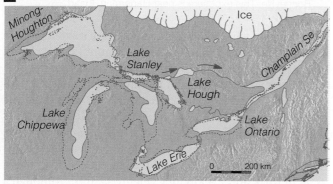

b 11,500 years ago.

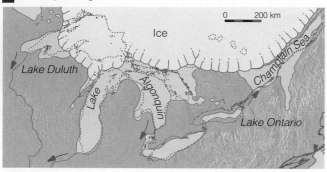

c 9,500 years ago.

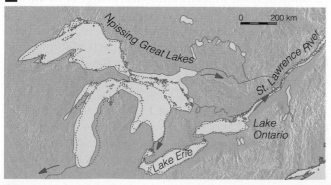

d 6,000 years ago.

• **Figure 17.17 Four Stages in the Evolution of the Great Lakes** As the glacial ice retreated northward, the lake basins began filling with meltwater. The dotted lines indicated the present-day shorelines of the lakes. Source: From V.K. Prest. *Economic Geology Report* 1, 5d, 1970, pp. 90–91 (fig. 7–6). Geological Survey of Canada, Department of Energy. Mines, and Resource, reproduced with permission of Minister of Supply and Services.

be low enough for glaciers to form. The problem is, no hard data exist to demonstrate that such a decrease in CO_2 levels actually occurred, nor do scientists agree on a mechanism to cause a decrease, although uplift of the Himalayas or other mountain ranges has been suggested.

Intermediate climatic changes lasting for a few thousand to a few hundred thousand years, such as the Pleistocene glacial–interglacial stages, have also proved difficult to explain, but the Milankovitch theory, proposed many years ago, is now widely accepted.

The Milankovitch Theory
Changes in Earth's orbit as a cause of intermediate-term climatic events was first proposed during the mid-1800s, but the idea was made popular during the 1920s by the Serbian astronomer Milutin Milankovitch. He proposed that minor irregularities in Earth's rotation and orbit are sufficient to alter the amount of solar radiation received at any given latitude and hence bring about climate changes. Now called the **Milankovitch theory,** it was initially ignored but has received renewed interest since the 1970s and is now widely accepted.

Milankovitch attributed the onset of the Pleistocene Ice Age to variations in three aspects of Earth's orbit. The first is *orbital eccentricity,* which is the degree to which Earth's orbit around the Sun changes over time (• Figure 17.18a). When the orbit is nearly circular, both the Northern and Southern Hemispheres have similar contrasts between the seasons. However, if the orbit is more elliptic, hot summers and cold winters will occur in one hemisphere, whereas warm summers and cool winters will take place in the other hemisphere. Calculations indicate a roughly 100,000-year cycle between times of maximum eccentricity, which corresponds closely to the 20 warm—cold climatic cycles that took place during the Pleistocene.

Milankovitch also pointed out that the angle between Earth's axis and a line perpendicular to the plane of Earth's orbit shifts about 1.5 degrees from its current value of 23.5 degrees during a 41,000-year cycle (Figure 17.18b). Although changes in *axial tilt* have little effect on equatorial latitudes, they strongly affect the amount of solar radiation received at high latitudes and the duration of the dark period at and near Earth's poles. Coupled with the third aspect of Earth's orbit, precession of the equinoxes, high latitudes might receive as much as 15% less solar radiation, certainly enough to affect glacial growth and melting.

Precession of the equinoxes, the last aspect of Earth's orbit that Milankovitch cited, refers to a change in the time of the equinoxes. At present, the equinoxes take place on about March 21 and September 21 when the Sun is directly over the equator. But as Earth rotates on its axis, it also wobbles as its axial tilt varies 1.5 degrees from its current value, thus changing the time of the equinoxes. Taken alone, the time of the equinoxes

• Figure 17.18 **According to the Milankovitch Theory, Minor Irregularities in Earth's Rotation and Orbit May Affect Climactic Changes**

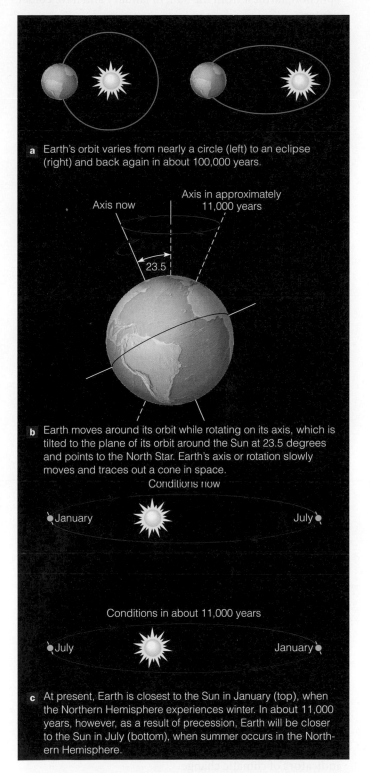

a Earth's orbit varies from nearly a circle (left) to an eclipse (right) and back again in about 100,000 years.

b Earth moves around its orbit while rotating on its axis, which is tilted to the plane of its orbit around the Sun at 23.5 degrees and points to the North Star. Earth's axis or rotation slowly moves and traces out a cone in space.

Conditions now

January   July

Conditions in about 11,000 years

July   January

c At present, Earth is closest to the Sun in January (top), when the Northern Hemisphere experiences winter. In about 11,000 years, however, as a result of precession, Earth will be closer to the Sun in July (bottom), when summer occurs in the Northern Hemisphere.

has little climatic effect, but changes in Earth's axial tilt also change the times of *aphelion* and *perihelion,* which are, respectively, when Earth is farthest from and closest to the Sun during its orbit (Figure 17.18c). Earth is now at perihelion, closest to the Sun, during

Northern Hemisphere winters, but in about 11,000 years perihelion will be in July. Accordingly, Earth will be at aphelion, farthest from the Sun, in January and have colder winters.

Continuous variations in Earth's orbit and axial tilt cause the amount of solar heat received at any latitude to vary slightly through time. The total heat received by the planet changes little, but according to Milankovitch, and now many scientists agree, these changes cause complex climatic variations and provided the triggering mechanism for the glacial—interglacial episodes of the Pleistocene.

Short-Term Climatic Events
Climatic events with durations of several centuries, such as the Little Ice Age, are too short to be accounted for by plate tectonics or Milankovitch cycles. Several hypotheses have been proposed, including variations in solar energy and volcanism.

Variations in solar energy could result from changes within the Sun itself or from anything that would reduce the amount of energy Earth receives from the Sun. The latter could result from the solar system passing through clouds of interstellar dust and gas or from substances in the atmosphere reflecting solar radiation back into space. Records kept over the past 90 years indicate that during this time the amount of solar radiation has varied only slightly. Although variations in solar energy may influence short-term climatic events, such a correlation has not been demonstrated.

During large volcanic eruptions, tremendous amounts of ash and gases are spewed into the atmosphere, where they reflect incoming solar radiation and thus reduce atmospheric temperatures. Small droplets of sulfur gases remain in the atmosphere for years and can have a significant effect on climate. Several large-scale volcanic events have occurred, such as the 1815 eruption of Tambora, and are known to have had climatic effects. However, no relationship between periods of volcanic activity and periods of glaciation has yet been established.

Glaciers Today

Glaciers today are much more restricted, but they nevertheless remain potent agents of erosion, sediment transport, and deposition. After all, even now they cover about 10% of Earth's land surface. Scientists monitor the behavior of glaciers to better understand the dynamics of moving bodies of ice, but they are also interested in glaciers as indicators of climatic change.

No doubt you have heard of *global warming*, a phenomenon of warming of Earth's atmosphere during the last 100 years or so. Many scientists are convinced that the cause of global warming is an increase of greenhouse gases, especially carbon dioxide, in the atmosphere as a result of burning fossil fuels. Others agree that surface temperatures have increased but attribute the increase to normal climatic variation. In either case, glaciers are good indicators of short-term climatic changes.

Any glacier's behavior depends on its budget—that is, its gains versus losses—which in turn is related to temperature and the amount of precipitation. According to one estimate there are about 160,000 valley glaciers and small ice caps outside Antarctica and Greenland, with Alaska alone having several tens of thousands. It is true that not many of these glaciers have been studied, but many of those that have show an alarming trend: They are retreating, ceased moving entirely, or have disappeared.

For example, in 1850 there were about 150 glaciers in Glacier National Park in Montana, but now only about two dozen remain, and nearly all the glaciers in the Cascade Range of the Pacific Northwest are retreating. Glacier Peak in Washington has more than a dozen glaciers, all of which are retreating, and Whitechuck Glacier will soon be inactive (• Figure 17.19a). When Mount St. Helens in Washington erupted in May 1980, all 12 of its glaciers were destroyed or considerably diminished. By 1982 the lava dome in the mountain's crater had cooled sufficiently for a new glacier to form; it is now about 190 m thick (Figure 17.19b). Remember, though, that Mount St. Helens already had the conditions for glaciers to exist, so the fact that one has become reestablished does not counter the evidence from virtually all other glaciers in the range.

There is one notable exception to shrinking glaciers in the Cascade Range. The seven glaciers on Mount Shasta in California are all growing, probably because of increased precipitation resulting from warming of the Pacific Ocean. Nevertheless, the glaciers are small and the trend is not likely to continue for long, because as warming continues it will soon overtake the increased snowfall.

The ice sheet in Greenland has lost about 162 km^3 of ice during each of the years from 2003 through 2005, and many of the glaciers that flow into the sea from the ice sheet have speeded up markedly. For instance, the Kangerdlugssuag Glacier is moving at about 14 km per year (38.4 m/day), and its terminus retreated 5 km in 2005 alone. The termini of many glaciers in Alaska are also retreating, particularly valley glaciers that flow into the sea (the so-called tidewater glaciers). Two factors account for these phenomena. First, the glaciers are moving faster because more meltwater is present that percolates downward and facilitates basal slip. The other factor is that warmer ocean temperatures melt the glaciers where they flow into the sea.

Most of Antarctica shows no signs of a decreasing volume of ice because the continent is at such high latitudes and so cold that little melting takes place. The greatest concern here is that some of the ice shelves—the parts of vast glaciers that flow into the sea—will collapse and allow the glaciers inland to flow more rapidly. In fact, huge sections of ice shelves have broken off in recent years, allowing land-based glaciers to surge into the ocean. The ice

• Figure 17.19 **Glaciers in the Cascade Range**

a Whitechuck Glacier on Glacier Peak in Washington State. The south branch (foreground) has a small accumulation area, but the north branch no longer has one.

b View of the lava dome (dark) and the newly formed Crater Glacier in the crater of Mount St. Helens on April 19, 20005. Notice the ash on the glacier's surface and the large crevasses.

shelves themselves are floating, so when they melt, that does not cause sea level to rise, but when the glacial ice on land flows into the ocean and melts, it does result in a rising sea level.

Pleistocene Mineral Resources

Many mineral deposits formed as a direct or indirect result of glacial activity during the Pleistocene and Holocene. We have already mentioned the vast salt deposits in Utah and the borax deposits in Death Valley, California, that originated when Pleistocene pluvial lakes evaporated. And some deposits of diatomite, rock composed of the shells of microscopic plants called *diatoms,* formed in the West Coast states during the Late Neogene.

In many U.S. states, as well as Canadian provinces, the most valuable mineral commodity is sand and gravel used in construction, much of which is recovered from glacial deposits, especially outwash. These same commodities are also recovered from deposits on the continental shelves and from stream deposits unrelated to glaciation. Silica sand is used in the manufacture of glass, and fine-grained glacial lake deposits are used to manufacture bricks and ceramics.

The California gold rush of the late 1840s and early 1850s was fueled by the discovery of Pleistocene and Holocene placer deposits of gold in the American River (see Chapter 14). Most of the $200 million in gold mined in California from 1848 to 1853 came from placer deposits. Discoveries of gold placer deposits in the Yukon Territory of Canada were primarily responsible for settlement of that area.

Peat consisting of semi-carbonized plant material in bogs and swamps is an important resource that has been exploited in Canada and Ireland. It is burned as a fuel in some areas but also finds other uses, as in gardening.

SUMMARY

- The most recent part of geologic time is the Pleistocene (1.8 million to 10,000 years ago) and the Holocene or Recent epochs (10,000 years ago to the present).
- Although the Pleistocene is best known for widespread glaciers, it was also a time of volcanism and tectonism.
- Pleistocene glaciers covered about 30% of the land surface, and were most widespread on the Northern Hemisphere continents.

- At least four intervals of extensive Pleistocene glaciation took place in North America, each separated by interglacial stages. Fossils and oxygen isotope data indicate about 20 warm–cold cycles occurred during the Pleistocene.
- Areas far beyond the ice were affected by Pleistocene glaciers: Climate belts were compressed toward the equator, large pluvial lakes existed in what are now arid regions, and when glaciers were present sea level was as much as 130 m lower than now.

- Moraines, striations, outwash, and various other glacial landforms are found throughout Canada, in the northern tier of states, and in many mountain ranges where valley glaciers were present.
- The tremendous weight of Pleistocene glaciers caused isostatic subsidence of Earth's crust. When the glaciers melted, isostatic rebound began and continues even now in some areas.
- Major glacial episodes separated by tens or hundreds of millions of years probably stem from changing positions of plates, which in turn profoundly affects oceanic and atmospheric circulation patterns.
- According to the Milankovitch theory, minor changes in Earth's rotation and orbit bring about climatic changes that produce glacial-interglacial intervals.
- The causes of short-term climatic changes such as occurred during the Little Ice Age are unknown; two proposed causes are variations in the amount of solar energy and volcanism.
- Pleistocene mineral resources include sand and gravel, placer deposits of gold, and some evaporite minerals such as borax.

IMPORTANT TERMS

cirque, p. 362
continental glacier, p. 354
end moraine, p. 362
glacial stage, p. 359
glacier, p. 354
ice cap, p. 354
ice-scoured plain, p. 362

interglacial stage, p. 359
isostasy, p. 367
isostatic rebound, p. 367
Little Ice Age, p. 361
Milankovitch theory, p. 371
moraine, p. 362
neoglaciation, p. 361

outwash, p. 362
pluvial lake, p. 367
pollen analysis, p. 360
proglacial lake, p. 368
U-shaped glacial trough, p. 362
valley glacier, p. 354
varve, p. 368

REVIEW QUESTIONS

1. A proposal to explain intermediate-term climatic fluctuations such as the Pleistocene glacial-interglacial stages is the
   a. _____ glacial moraine hypothesis; b. _____ Milankovitch theory; c. _____ Wegener's dictum; d. _____ principle of superposition; e. _____ Isostasy theory.
2. Lakes that existed during times of glaciation because of increased rainfall and decreased evaporation are called _____ lakes.
   a. _____ alluvial fan; b. _____ outwash; c. _____ varve; d. _____ pluvial; e. _____ cirque.
3. The time from 1500 to sometime during the 1800s when glaciers expanded markedly is referred to as the
   a. _____ Little Ice Age; b. _____ Medieval Cold Spell; c. _____ Wisconsinan; d. _____ Proglacial Lake Episode; e. _____ Milankovitch Interval.
4. Which one of the following statements is correct?
   a. _____ All of North America was ice covered during the Pleistocene; b. _____ Continental glaciers are long, narrow tongues of ice in mountain valleys; c. _____ The deposit that forms when a glacier is at its greatest extent is a recessional moraine; d. _____ Varves are gravel deposits that form during the catastrophic flooding of lakes; e. _____ The Pleistocene Epoch began 1.8 million years ago and ended 10,000 years ago.
5. An important area of Pleistocene and Holocene volcanism in North America is
   a. _____ the Cascade Range; b. _____ Cape Cod; c. _____ the Interior Lowlands; d. _____ Glacial Lake Missoula; e. _____ the Atlantic Coastal Plain.
6. The distinctive subdued topography resulting from erosion by continental glaciers with exposed, striated bedrock and little soil is a(n)
   a. _____ outwash plain; b. _____ U-shaped glacial trough; c. _____ ice-scoured plain; d. _____ ice cap; e. _____ pluvial landscape.
7. The phenomenon in which Earth's crust rises after unloading, as when a vast glacier melts, is called
   a. _____ precession; b. _____ orogenic deformation; c. _____ isostatic rebound; d. _____ neoglaciation; e. _____ postglacial adjustment.
8. The most recent episode of glaciation in North America is the
   a. _____ Montanan; b. _____ Wisconsinan; c. _____ Michiganian; d. _____ Oklahoman ; e. _____ New Jerseyan.
9. A large pluvial lake that was the forerunner of the Great Salt Lake was
   a. _____ Lake Ontario; b. _____ Lake Agassiz; c. _____ Lake Missoula; d. _____ Lake Bonneville; e. _____ Lake Champlain.
10. A ridge-like, unsorted, nonstratified sediment deposit at a glacier's terminus is a(n)
    a. _____ moraine; b. _____ varve; c. _____ delta; d. _____ cirque ; e. _____ point bar.
11. Explain how the Milankovitch theory accounts for the onset of glacial ages.
12. How does the landscape formed by erosion by continental glaciers differ from that eroded by valley glaciers?

13. What was the Little Ice Age, when did it take place, and what impact did it have on humans?
14. What are pluvial and proglacial lakes? Give an example of each.
15. How does an end moraine form? What criteria would you use to distinguish a terminal moraine from a recessional moraine?
16. What is isostatic rebound and what kinds of evidence indicates that it has occurred?
17. Give an account of the origin and subsequent history of glacial Lake Missoula.

18. How are oxygen isotope ratios and pollen analyses used to make inferences about ancient climates?
19. What accounts for the ongoing eruptions of volcanoes in the Cascade Range, and why are there no volcanoes along the rest of the U.S. Pacific coast other than Alaska?
20. Why is it so difficult to correlate the sequence of cold–warm intervals recorded in seafloor sediment with the glacial record on land?

APPLY YOUR KNOWLEDGE

1. After observing the same valley glacier for several years, you conclude (1) that the glacier's terminus has retreated 2 km, and yet (2) debris on the glacier's surface has clearly moved toward the terminus. How can you explain these observations?
2. Suppose that you live in western Nevada, and during one of your weekend excursions you notice a valley with a very flat floor. Where eroded, you see that the valley floor deposits are mostly laminated mud with sparse sand. Furthermore, you see several bench-like surfaces eroded into the hillsides around the valley. Your geologic map shows that the deposits are Pleistocene age, but it does not indicate how the deposits formed or what the bench-like surfaces are. How would you interpret the geology of this area?

3. When you give a talk on local geology in Plymouth, Massachusetts, how would you summarize how Cape Cod and nearby islands evolved to their present form?
4. You notice the following in a gorge. In the lower part of the gorge you see a bedrock surface with linear scratches and polish that is overlain by a 12-m-thick nonstratified mixture of mud, sand, and gravel; some of the gravel is up to 2 m in diameter. Next upward are moderately well-sorted layers of cross-bedded sand and horizontal layers of conglomerate. The uppermost part of the rock sequence is made up of thin (2–4 mm thick), alternating layers of light- and dark-colored clay, with a few boulders 10–15 cm across. Decipher the geologic history revealed by these rocks.

Chase Studio/Photo Researchers, Inc.

CHAPTER 18

LIFE OF THE CENOZOIC ERA

This scene depicts what life was like in Nebraska during the Miocene Epoch, about 21 million years ago. Some of the animals shown in this restoration are relatives of today's elephants, a chalicothere (right center), pig-like oreodonts (in the water), small camels (left front) being chased by bear dogs. A saber-tooted cat in the tree feasts on its prey and cranes circle overhead.

[OUTLINE]

Introduction

Marine Invertebrates and Phytoplankton
  Perspective *The Messel Pit Fossil Site in Germany*

Cenozoic Vegetation and Climate

Cenozoic Birds

The Age of Mammals Begins
  Monotremes and Marsupial Mammals
  Diversification of Placental Mammals

Paleogene and Neogene Mammals
  Small Mammals—Insectivores, Rodents, Rabbits, and Bats

A Brief History of the Primates

The Meat Eaters—Carnivorous Mammals

The Ungulates or Hoofed Mammals

Giant Land-Dwelling Mammals—Elephants

Giant Aquatic Mammals—Whales

Pleistocene Faunas
  Ice Age Mammals
  Pleistocene Extinctions

Intercontinental Migrations

Summary

At the end of this chapter, you will have learned that

- Survivors of the Mesozoic extinctions evolved and gave rise to the present-day invertebrate marine fauna.

- Angiosperms continued to diversify and to dominate land plant communities, but seedless vascular plants and gymnosperms are still common.

- Many of today's families and genera of birds evolved, and large flightless birds were important Early Cenozoic predators.

- If we could visit the Paleocene, we would not recognize many of the mammals, but more familiar ones evolved during the following epochs.

- Small mammals such as rodents, rabbits, insectivores, and bats adapted to the microhabitats unavailable to larger mammals.

- Carnivorous mammal fossils are not as common as those of herbivores, but there are enough to show their evolutionary trends and relationships to one another.

- The evolutionary histories of odd-toed and even-toed hoofed mammals are well documented by fossils.

- Today's giant land mammals (elephants) and giant marine mammals (whales) evolved from small Early Cenozoic ancestors.

- Extinctions at the end of the Pleistocene Epoch were most severe in the Americas and Australia, and the animals most affected were large land-dwelling mammals.

- As Pangaea continued to fragment during the Cenozoic, inter-continental migrations became increasingly difficult.

- A Late Cenozoic land connection formed between North and South America, resulting in migrations in both directions.

Introduction

We noted in Chapter 8 that when Earth formed, it was hot, barren, and waterless, the atmosphere was noxious; it was bombarded by meteorites and comets; and no organisms existed. During the Precambrian and following Paleozoic and Mesozoic eras, however, the planet and its biota evolved, and during the Cenozoic Era, Earth and its organisms took on their present-day appearance. So even though the Cenozoic makes up only 1.4% of all geologic time (see Figure 8.1), this comparatively brief period of *only* 66 million years was certainly long enough for considerable change to take place. In Chapter 16 we emphasized that Earth evolution continues unabated as plates move, volcanoes erupt, landscapes change, and mountains evolve. Here, our emphasis is on Earth's biota and it too continues to change, although most of the changes are minor from our perspective, but nevertheless important.

Remember from Chapter 15 that mammals evolved during the Late Triassic, so they were contemporaries with dinosaurs. Furthermore, some of the earliest mammals differed little from their ancestors, the cynodonts, but as they evolved they became increasingly familiar. During the Cenozoic, they diversified into numerous types that adapted to nearly all terrestrial habitats as well as aquatic environments and one group, the bats, became fliers. We emphasize the evolution of mammals in this chapter but there were other equally important biologic events taking place.

Angiosperms, or flowering plants, evolved during the Cretaceous and soon became the dominant land plants; now they constitute more than 90% of all land plant species. However, their geographic distribution varied during the Cenozoic, depending on changing climates. The first birds evolved during the Jurassic Period but the families common now appeared during the Paleogene and Neogene periods, reached their greatest diversity during the Pleistocene Epoch, and have declined slightly since then. The marine invertebrates that survived the Mesozoic extinctions diversified and gave rise to the present-day marine fauna. Overall, we can think of the Cenozoic as the time during which the fauna and flora became more and more like it is today.

We mentioned in Chapter 16 that Cenozoic rocks are the most easily accessible at or near the surface, and as a result we know more about life history for this time than for any of the previous eras. Widespread exposures of Cenozoic rocks are present in western North America, many of which were deposited in continental environments, but rocks of this age are also found along the Gulf and Atlantic coasts, although many of these were deposited in transitional or marine environments. In any case, we have particularly good fossil records for many Cenozoic organisms. In fact, several of our national parks and monuments as well as state parks in the west feature displays of fossil mammals, including Agate Fossil Beds National Monument in Nebraska (see the chapter opening photo), Badlands National Park in South Dakota, and John Day Fossil Beds National Monument in Oregon (• Figure 18.1).

Continental deposits with land-dwelling mammals are not nearly as common in the east, but Florida is a notable exception. Nevertheless, some eastern and southern states such as Maryland, South Carolina, and Alabama have deposits with fossils of marine mammals as well as fossil invertebrates and sharks. Indeed, the Alabama state fossil is *Basilosaurus cetoides,* a fossil whale that lived during the Eocene Epoch. Of course mammal fossil are found on the other continents, too, but certainly one of the most remarkable fossil sites anywhere in the world is the Messel fossil beds in Germany (see Perspective).

Archean Eon	Proterozoic Eon	Phanerozoic Eon							
					Paleozoic Era				
Precambrian		Cambrian	Ordovician	Silurian	Devonian	Mississippian	Pennsylvanian	Permian	
							Carboniferous		

2500 MYA   542 MYA   251 MYA

• **Figure 18.1** **Restoration of Fossils from the Eocene-age Clarno Formation in John Day Fossil Beds National Monument, Oregon**
The climate at this time was subtropical, and the lush forests of the region were occupied by **1** titanotheres standing 2.5 m high at the shoulder, **2** carnivores, **3** ancient horses, **4** tapirs, and **5** early rhinoceroses.

Roy Anderson

Marine Invertebrates and Phytoplankton

The Cenozoic marine ecosystem was populated mostly by plants, animals, and single-celled organisms that survived the terminal Mesozoic extinction. Gone were the ammonites, rudists, and most of the planktonic foraminifera. Especially prolific Cenozoic invertebrate groups were the foraminifera, radiolarians, corals, bryozoans, mollusks, and echinoids. The marine invertebrate community in general became more provincial during the Cenozoic because of changing ocean currents and latitudinal temperature gradients. In addition, the Cenozoic marine invertebrate faunas became more familiar in appearance.

Entire families of phytoplankton became extinct at the end of the Cretaceous with only a few species in each major group surviving into the Paleogene. These species diversified and expanded during the Cenozoic, perhaps because of decreased competitive pressures. Coccolithophores, diatoms, and dinoflagellates all recovered from their Late Cretaceous reduction in numbers to flourish during the Cenozoic. Diatoms were particularly abundant during

the Miocene, probably because of increased volcanism during this time. Volcanic ash provided increased dissolved silica in seawater, which diatoms used to construct their skeletons. Massive Miocene diatomite rocks, made up of diatom shells, are present in several western states (• Figure 18.2).

The foraminifera were a major component of the Cenozoic marine invertebrate community. Although dominated by relatively small forms (• Figure 18.3), it included some exceptionally large forms that lived in the warm waters of the Cenozoic Tethys Sea. Shells of these larger forms accumulated to form thick limestones, some of which the ancient Egyptians used to construct the Sphinx and the Pyramids of Giza (Figure 18.3c).

Corals were perhaps the main beneficiary of the Mesozoic extinctions. Having relinquished their reef-building role to rudists, which are mollusks, during the mid-Cretaceous, corals again became the dominant reef builders. They formed extensive reefs in the warm waters of the Cenozoic oceans and were especially prolific in the Caribbean and Indo-Pacific regions.

Other suspension feeders such as bryozoans and crinoids were also abundant and successful during the Paleogene and Neogene. Bryozoans, in particular, were very

Phanerozoic Eon								
Mesozoic Era			Cenozoic Era					
Triassic	Jurassic	Cretaceous	Paleogene			Neogene		
								Quaternary
			Paleocene	Eocene	Oligocene	Miocene	Pliocene	Pleistocene	Holocene

251 MYA

66 MYA

• **Figure 18.2 Miocene Diatomite and Diatoms**

Reed Wicander

a Outcrop of diatomite in the Monterey Formation at Newport Lagoon, California. The diatoms in **b** and **c** are from this formation.

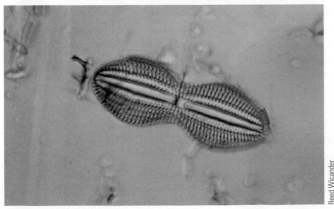

Reed Wicander

b A pinnate diatom.

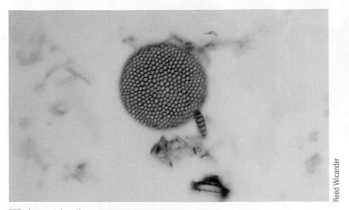

Reed Wicander

c A centric diatom.

• **Figure 18.3 Cenozoic Foraminifera**

Courtesy of B. A. masters

a *Cibicideds americanus* from the Early Miocene of California is a benthonic foraminifera.

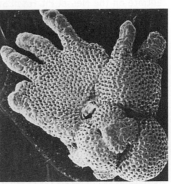

Courtesy of B. A. masters

b *Globigerinoides fistulosus* is a Pleistocene planktonic foraminifera from the south Pacific Ocean.

James S. Monroe

c The numerous disc-shaped objects in this image are specimens of Eocene-age *Nummulites,* a benthonic foraminifera in the limestone used to construct the Pyramids on the Giza Plateau in Egypt.

abundant. Perhaps the least important of the Cenozoic marine invertebrates were brachiopods, with fewer than 60 genera surviving today. Brachiopods never recovered from their reduction in diversity at the end of the Paleozoic (see Chapter 12).

Just as during the Mesozoic, bivalves and gastropods were two of the major groups of marine invertebrates

The Messel Pit Fossil Site In Germany

We have mentioned several outstanding fossil localities in this chapter such as John Day Fossil Beds National Monument in Oregon (Figure 18.1), but here we concentrate on the Messel Pit Fossil Site, near Frankfurt, Germany, which is one of the most remarkable fossil localities anywhere in the world. Indeed, it was designated a World Heritage Site, which is "an area or structure designated by UNESCO as being of global significance and conserved by a country that has signed a United Nations convention pledging its protection*." Of the 878 sites so designated, 679 are cultural (the Pyramids at Giza in Egypt), 174 are natural (the Messel Pit Fossil Site in Germany), and 25 are mixed (Mount Athos in Greece).

Actually, fossils have been known from the Messel Pit Fossil Site since 1875, but they have attracted the most attention only during the last few decades. The fossil site was originally a large pit excavated from 1886 to 1971 for oil shale that was used to produce crude oil. We noted in Chapter 16 that oil shale is a rock made up of mud and the organic compound kerogen from which liquid oil and combustible gases are extracted. These organic-rich mudrocks were deposited during the Middle Eocene, about 50 million years ago, in a small but deep lake in a fault valley. Fossils of palms, citrus fruits, laurels, beech, and other plants indicate that the climate was tropical or subtropical.

Streams carried mud and some sand into this small lake, which had oxygen-depleted deep waters where nothing but anaerobic bacteria could exist. As a consequence, the remains of organisms that sank into the deeper waters did not decompose completely and many have been preserved so well that even stomach contents and the outlines of internal organs are visible, as well as impressions of bird feathers, bat wings, and, most remarkably, the iridescent color of beetle wing covers. Although a wide variety of organisms existed in and around the lake (most of the vertebrate fossils are of fish), it is the mammals that are the most notable. For some of these animals every bone has been preserved, including all of the toe bones and every bony segment of their tails (Figure 1).

A complete inventory of all the animal fossils found at Messel is not practical, but we will mention that in addition to 10,000 or so fossil fish, thousands of insects have been found; as well as 12 genera of birds, including a "proto-ostrich" and a large flightless predatory bird similar to *Diatryma* (Figure 18.6); several types of amphibians and reptiles; and a wide assortment of mammals. Indeed, the mammals include anteaters, pangolins, marsupials, bats, carnivores, rodents, and an early relative of horses. The fossil in Figure 1 is known as *Messelobunodon*, which is a 78-cm-long early artiodactyl (even-toed hoofed mammal) that was probably close to the ancestry of pigs, cattle, and camels.

The fossils at Messel are abundant and of great scientific interest, but they present a particular problem for collecting and preservation. The host oil shale contains up to 40% water, so when the rock dries it crumbles and the fossils may be lost. Accordingly, scientists developed a technique of coating the rock surface with a synthetic resin in which the fossils are preserved.

Such a remarkable association of fossils gives us a unique glimpse of what the biota was like 50 million years ago in what is now Germany. Fortunately for us, the Messel fossil locality was designated a World Heritage Site, because when the oil shale excavation ceased, it was slated to become a sanitary landfill.

Figure 1 This fossil mammal known as *Messelobunodon* was recovered from one of the most remarkable fossil sites anywhere—the Eocene-age Messel Pit near Frankfurt, Germany. This fossil shows a small (78-cm-long) animal in a death pose with all of its bones preserved. *Messelobunodon* was one of the earliest known artiodactyls.

© Jonathan Blair/CORBIS

*http://encarta.msn.com/dictionary_1861714109/world_heritage_site.html

during the Cenozoic, and they had a markedly modern appearance (• Figures 18.4a, b).

After the extinction of ammonites and belemnites at the end of the Cretaceous, the Cenozoic cephalopod fauna consisted of nautiloids and shell-less cephalopods such as squids and octopuses.

Echinoids continued their expansion in the infaunal habitat and were very prolific during the Cenozoic. New forms such as sand dollars evolved during this time from biscuit-shaped ancestors (Figure 18.4c).

• Figure 18.4 **Cenozoic Fossil Invertebrates**

a The gastropod *Busycon contrariuim* from Pliocene rocks in Florida.

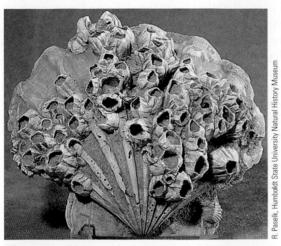

b The bivalve *Chlamya* sp. encrusted with barnacles from the Miocene of Virginia.

c A Pliocene echinoid (sand dollar) from Mexico.

Cenozoic Vegetation and Climate

Angiosperms continued to diversify during the Cenozoic, as more and more familiar varieties evolved, although seedless vascular plants and gymnosperms were also present in large numbers. In fact, many Paleogene plants would be familiar to us today, but their geographic distribution was not what it is now, because changing climatic conditions along with shifting plant distributions were occurring.

Some plants today are confined to the tropics, whereas others have adapted to drier conditions, and we have every reason to think that climate was a strong control on plant distribution during the past. Furthermore, leaves with entire or smooth margins, many with pointed drip-tips, are dominant in areas with abundant rainfall and high annual temperatures. Smaller leaves with incised margins are more typical of cooler, drier areas (• Figure 18.5a). Accordingly, fossil floras with mostly smooth-margined leaves with drip-tips indicate wet, warm conditions, whereas a cool, dry climate is indicated by a predominance of small leaves with incised margins and no drip-tips.

Paleocene rocks in the western interior of North America have fossil ferns and palms, both indicating a warm, subtropical climate. In a Paleocene flora in Colorado with about 100 species of trees, nearly 70 percent of the leaves are smooth margined, and many have drip-tips. The nature of these leaves coupled with the diversity of plants is much like that in today's rain forests. In fact, the Early Oligocene fossil plants at Florissant Fossil Beds National Monument in Colorado indicate that a warm, wet climate persisted then.

Seafloor sediments and geochemical evidence indicate that about 55 million years ago an abrupt warming trend took place. During this time, known as the **Paleocene-Eocene Thermal Maximum,** large-scale oceanic circulation was disrupted so that heat transfer from equatorial regions to the poles diminished or ceased. As a result, deep oceanic water became warmer, resulting in extinctions of many deep-water foraminifera. Some scientists think that this deep, warm oceanic water released methane from seafloor methane hydrates, contributing a greenhouse gas to the atmosphere and either causing or contributing to the temperature increase at this time.

Subtropical conditions persisted into the Eocene in North America, probably the warmest of all the Cenozoic epochs. Fossil plants in the Eocene John Day Beds in Oregon include ferns, figs, and laurels, all of which today live in much more humid regions, as in parts of Mexico and Central America. Yellowstone National Park in

• Figure 18.5 **Cenozoic Vegetation and Climate**

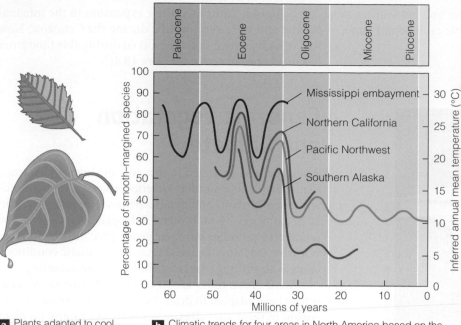

a Plants adapted to cool climates typically have small leaves with incised margins (top), but in humid, warm areas they have larger, smooth-margined leaves, many with drip-tips.

b Climatic trends for four areas in North America based on the percentages of plant species with smooth-margined leaves.

Wyoming has a temperate climate now, with warm, dry summers and cold, snowy winters, certainly not an area where you would expect avocado, magnolia, and laurel trees to grow. Yet their presence there during the Eocene indicates the area then had a considerably warmer climate than it does now.

A major climatic change took place at the end of the Eocene, when mean annual temperatures dropped as much as 7°C in about 3 million years (Figure 18.5b). Since the Oligocene, mean annual temperatures have varied somewhat worldwide, but overall they have not changed much in the middle latitudes except during the Pleistocene Epoch.

A general decrease in precipitation during the last 25 million years took place in the midcontinent region of North America. As the climate became drier, the vast forests of the Oligocene gave way first to *savannah* conditions (grasslands with scattered trees) and finally to *steppe* environments (short-grass prairie of the desert margin). Many herbivorous mammals quickly adapted to these new conditions by developing chewing teeth suitable for a diet of grass.

Cenozoic Birds

Birds today are diverse and numerous, making them the most easily observed vertebrates. The first members of many of the living orders, including owls, hawks, ducks, penguins, and vultures, evolved during the Paleogene. Beginning during the Miocene, a marked increase in the variety of songbirds took place, and by 5 to 10 million years ago, many of the existing genera of birds were present. Birds adapted to numerous habitats and continued to diversify into the Pleistocene, but since then their diversity has decreased slightly.

Today, birds vary considerably in diet, habitat, adaptations, and size. Nevertheless, their basic skeletal structure has remained remarkably constant throughout the Cenozoic. Given that birds evolved from a creature very much like *Archaeopteryx* (see Figure 15.16), this uniformity is not surprising, because adaptations for flying limit variations in structure. Penguins adapted to an aquatic environment, and in some large extinct and living flightless birds the skeleton became robust and the wings were reduced to vestiges.

Many authorities on prehistoric life are now convinced that birds are so closely related to dinosaurs that they refer to them as *avian dinosaurs* and all the others as *non-avian dinosaurs*. In fact, following the demise of the non-avian dinosaurs as the end of the Mesozoic Era, the dominant large, land-dwelling predators during the Paleocene and well into the Eocene were flightless birds, or the avian dinosaurs. Among these predators were giants such as *Diatryma,* which stood 2 m tall, had a huge head and beak, toes with large claws, and small, vestigial wings (• Figure 18.6). Its massive, short legs indicate that *Diatryma* was not very fast, but neither were the mammals it preyed upon. This extraordinary bird and related genera were widespread in North America and Europe, and in South America they were the dominant predators until replaced by carnivorous mammals during the Oligocene Epoch.

Two of the most notable large flightless birds were the now extinct moas of New Zealand and elephant birds of Madagascar. Moas were up to 3 m tall; elephant birds were shorter but more massive, weighing up to 500 kg. They are known only from Pleistocene-age deposits, and both went extinct shortly after humans occupied their respective areas.

Large flightless birds are truly remarkable creatures, but the real success among birds belongs to the fliers. Even though few skeletal modifications occurred during the Cenozoic, a bewildering array of adaptive types arose. If number of species and habitats occupied is any measure of success, birds have certainly been at least as successful as mammals.

• Figure 18.6
Restoration of
Diatryma Diatryma
was a flightless,
predatory bird that
stood more than
2 m tall. It lived
during the Paleo-
cene and Eocene in
North America and
Europe.

The Age of Mammals Begins

Mammals coexisted with dinosaurs for more than 140 million years, and yet, during this entire time they were not very diverse, and even the largest among them was only about 1 m long. Even at the end of the Cretaceous Period there were only a few families of mammals, a situation that was soon to change. With the demise of non-avian dinosaurs and their relatives, mammals quickly exploited the adaptive opportunities, beginning a diversification that continued throughout the Cenozoic Era. The Age of Mammals had begun.

We have already mentioned that Cenozoic deposits are easily accessible at or near the surface, and overall they show fewer changes resulting from metamorphism and deformation when compared with older rocks. In addition, because mammals have teeth fully differentiated into various types (see Figure 15.18), they are easier to identify and classify than members of the other classes of vertebrates. In fact, mammal teeth not only differ from front to back of the mouth, but they also differ among various mammalian orders and even among genera and species. This is especially true of chewing teeth, the **premolars** and **molars;** a single chewing tooth is commonly enough to identify the genus from which it came.

Monotremes and Marsupial Mammals

All warm-blooded vertebrates with hair and mammary glands belong to the class Mammalia, which includes the monotremes or egg-laying mammals—the platypus and spiny anteater of the Australian region; the marsupials commonly called the "pouched mammals"—kangaroos, opossums, wombats, and so on; and the placental mammals (about 18 orders). Female monotremes do secrete a milky substance that their young lick from the mother's hair, but both marsupials and placentals have true mammary glands and milk to nourish their young. Monotremes have the requisite features to be called mammals, but they appear to have had a completely separate evolutionary history from the marsupials and placentals (see Figure 15.19). Unfortunately, they have a very poor fossil record.

When the young of marsupial mammals are born, they are in an immature, almost embryonic state and then undergo further development in the mother's pouch. Marsupials probably migrated to Australia, the only area where they are common now, via Antarctica before Pangaea fragmented completely. They were also common in South America during much of the Cenozoic Era until only a few millions of years ago. However, when a land connection was finally established between the Americas, most of the South American marsupials died out as placental mammals from North America replaced them. Now the only marsupials outside of Australia and nearby islands are species of opossums.

Diversification of Placental Mammals

Like marsupials, placental mammals give birth to live young, but their reproductive method differs in important details. In placentals, the amnion of the amniote egg (see Figure 13.15) has fused with the walls of the uterus, forming a *placenta*. Nutrients and oxygen flow from mother to embryo through the placenta, permitting the young to develop much more fully before birth. Actually, marsupials also have a placenta, but it is less efficient, explaining why their newborn are not as fully developed.

A measure of the success of placental mammals is related in part to their method of reproduction—more than 90% of all mammals, fossil and extinct, are placentals.

In our following discussion of placental mammals, we emphasize the origin and evolution of several of the 18 or so living orders (• Figure 18.7). During the Paleocene Epoch, several orders of mammals were present, but some were simply holdovers from the Mesozoic or belonged to new but short-lived groups that have no living descendants. These so-called *archaic* mammals, including marsupials, insectivores, and the rodentlike multituburculates, occupied a world with several new mammalian orders (• Figure 18.8), such as the first rodents, rabbits, primates, and carnivores, and the ancestors of hoofed mammals. Most of these Paleocene mammals, even those belonging to orders that still exist, had not yet become clearly differentiated from their ancestors, and the differences between herbivores and carnivores were slight.

The Paleocene mammalian fauna was also made up mostly of small creatures. By Late Paleocene time, though, some rather large mammals were around, although giant terrestrial mammals did not appear until the Eocene. With the evolution of a now extinct order known as the Dinocerata, better known as uintatheres, and the strange creature known as *Arsinoitherium*, giant mammals of one kind or another have been present ever since (• Figure 18.9). In addition, there were also some large carnivores by Eocene time, such as *Andrewsarchus* that was probably a 3-m-long wolf-like predator.

Many mammalian orders that evolved during the Paleocene died out, but of the several that first appeared during the Eocene, only one has become extinct. Thus, by Eocene time many of the mammalian orders existing now were present, yet if we could go back for a visit we would not recognize most of these animals (Figure 18.8). Surely we would know they were mammals and some would be at least vaguely familiar, but the ancestors of horses, camels, elephants, whales, and rhinoceroses bore little resemblance to their living descendants.

Warm, humid climates persisted throughout the Paleocene and Eocene of North America, but by Oligocene time drier and cooler conditions prevailed. Most of the archaic Paleocene mammals, as well as several groups that originated during the Eocene, had died out by this time. The large, rhinoceros-like titanotheres died out, and the uintatheres just mentioned also went extinct. In addition, some other groups of mammals suffered extinctions, including several types of herbivores loosely united as condylarths, carnivorous mammals known as *creodonts*, most of the

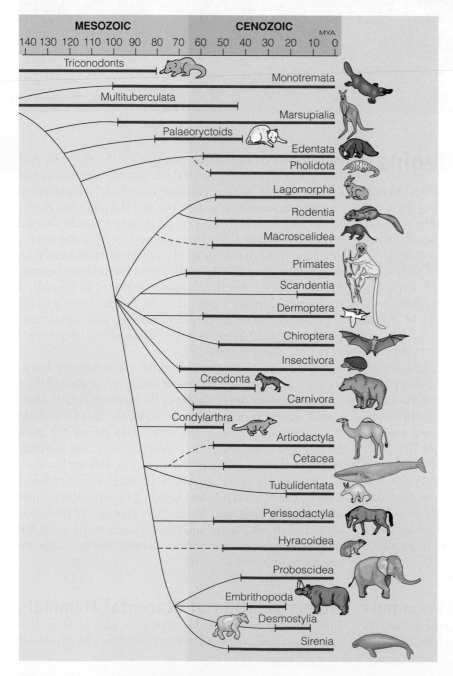

• **Figure 18.7 Mammals Existed During the Mesozoic, but Most Placental Mammals Diversified During the Paleocene and Eocene Epochs** Among the living orders of mammals, all are placentals except for the monotremes and marsupials. Several extinct orders are not shown. Bold lines indicate actual geologic ranges, whereas the thinner lines indicate the inferred branching of the groups.

• **Figure 18.8** **Archaic Mammals of the Paleocene Epoch** The mammals include **1** *Protictus,* an early carnivore, **2** insectivores, **3** the 19-cm-long, tree-dwelling, multituburculate *Ptilodus,* and **4** the pantodont known as *Pantolambda* that stood about 1 m high.

• Figure 18.9 **Some of the Earliest Large Mammals**

a Skull of *Arsinoitherium*, a rhinoceros-to elephant-sized Early Oligocene animal. Its paired, hollow horns were more than 0.5 m long.

Sue Monroe

b Scene from the Eocene showing the rhinoceros sized mammal known as *Uintatherium*. It had three pairs of bony protuberances on the skull and saberlike upper canine teeth.

remaining multituberculates, and some primates. All in all, this was a time of considerable biotic change.

By Oligocene time, most of the existing mammalian orders were present, but they continued to diversify as more and more familiar genera evolved. If we were to encounter some of these animals, we might think them a bit odd, but we would have little difficulty recognizing rhinoceroses (although some were hornless), elephants, horses, rodents, and many others. However, the large, horselike animals known as *chalicotheres*, with claws, and the large piglike entelodonts would be unfamiliar, and others would be found in areas where today we would not expect them—elephants in North America, for instance.

By Miocene and certainly Pliocene time, most mammals were quite similar to those existing now (• Figure 18.10). On close inspection, though, we would see horses with three toes, cats with huge canine teeth, deerlike animals with forked horns on their snouts, and very tall, slender camels.

And we would still see a few rather odd mammals, but overall the fauna would be quite familiar.

Paleogene and Neogene Mammals

We know mammals evolved from mammal-like reptiles called *cynodonts* during the Late Triassic, and diversified during the Cenozoic, eventually giving rise to the present-day mammalian fauna. Now more than 5000 species exist, ranging from tiny shrews to giants such as whales and elephants. Nevertheless, when one mentions the term *mammal*, what immediately comes to mind are horses, pigs, cattle, deer, dogs, cats, and so on, but most often we do not think much about small mammals, rodents, shrews, rabbits, and bats. Yet most species of mammals, probably 70%, are quite small, weighing less than 1 kg.

• **Figure 18.10 Pliocene Mammals of the Western North American Grasslands** The animals shown include **1** *Amebeledon*, a shovel-tusked mastodon, **2** *Teleoceras*, a short-legged rhinoceros, **3** *Cranioceras*, a horned, hoofed mammal, **4** a rodent, **5** a rabbit, **6** *Merycodus*, an extinct pronghorn, **7** *Synthetoceras*, a hoofed mammal with a horn on its snout, and **8** *Pliohippus*, a one-toed grazing horse.

Pliocene Mural by Jay H. Matternes © Copyright 1982

Small Mammals—Insectivores, Rodents, Rabbits, and Bats
Insectivores, rodents, rabbits, and bats share a common ancestor, but they have had separate evolutionary histories since they first evolved (Figure 18.7). With the exception of bats, the oldest of which is found in Eocene rocks, the others were present by the Late Mesozoic or Paleocene. The main reason we consider them together is that with few exceptions they are small and have adapted to the microhabitats unavailable to larger mammals. In addition to being small, bats are the only mammals capable of flight.

As you would expect from the name Insectivora, members of this group—today's shrews, moles, and hedgehogs—eat insects. Insectivores have probably not changed much since they appeared during the Late Cretaceous. In fact, an insectivore-like creature very likely lies at the base of the great diversification of placental mammals.

More than 40% of all living mammal species are members of the order Rodentia, most of which are very small animals. A few, though, including beavers and the capybara of South America, are sizable animals; the latter is more than 1 m long and weighs 45 kg. One Miocene beaver known as *Paleocastor* was not particularly large, but it constructed some remarkable burrows, and the Miocene rodent whimsically called "ratzilla" weighed an estimated 740 kg (• Figure 18.11). Rodents evolved during the Paleocene, diversified rapidly, and adapted to a wide range of habitats. One reason for their phenomenal success is that they can eat almost anything.

Rabbits (order Lagomorpha) superficially resemble rodents but differ from them in several anatomic details. Furthermore, since they arose from a common ancestor during the Paleocene, rabbits and rodents have evolved independently. Like rodents, rabbits are gnawing animals, although details of their gnawing teeth differ. The development of long, powerful hind limbs for speed is the most obvious evolutionary trend in this group.

The oldest fossil bat (order Chiroptera) comes from the Eocene-age Green River Formation of Wyoming, but well-preserved specimens are known from several other areas, too. Apart from having forelimbs modified into wings, bats differ little from their immediate ancestors among the insectivores. Indeed, with the exception of wings they closely resemble living shrews. Unlike pterosaurs and birds, bats use a modification of the hand in which four long fingers support the wings (see Figure 15.14c).

A Brief History of the Primates
The order **Primates** includes the "lower primates" (tarsiers, lemurs, and lorises), and the monkeys, apes, and humans, collectively referred to as "higher primates." Much of the primate story is more fully told in Chapter 19, where we consider human evolution, so in this chapter we will be brief. Primates may have evolved by Late Cretaceous time, but by the Paleocene they were undoubtedly present.

Small Paleocene primates closely resembled their contemporaries, the shrewlike insectivores. By the Eocene,

• Figure 18.11 **Fossil Rodents** Most species of rodents are small animals, about the size of mice and rats, but there were, and still are, some exceptions.

a *Paleocastor* was a 30-cm-long Miocene land-dwelling beaver that made these spiral burrows in Nebraska, which are locally called the Devil's Corkscrew.

b This restoration of *Phoberomys* from Miocene rocks in South America shows a huge rodent that was about 3 m long and weighed an estimated 740 kg.

though, larger primates had evolved, and lemurs and tarsiers that resemble their present descendants lived in Asia and North America. And by Oligocene time, primitive New World and Old World monkeys had developed in South America and Africa, respectively. The Hominoids, the group that includes apes and humans, evolved during the Miocene (see Chapter 19).

The Meat Eaters—Carnivorous Mammals

The order **Carnivora** is extremely varied, and includes bears, seals, weasels, skunks, dogs, and cats. All are predators and therefore meat eaters, but their diets vary considerably. For example, cats rarely eat anything but meat, whereas bears, raccoons, and skunks have a varied diet and are thus *omnivorous*. Most carnivores have well-developed, sharp, pointed canine teeth and specialized shearing teeth known as **carnassials** for slicing meat (• Figure 18.12). Some land-dwelling carnivores depend on speed, agility, and intelligence to chase down prey, but others employ different tactics. Badgers, for instance, are not very fast—they dig prey from burrows, and some small cats depend on stealth and pouncing to catch their meals.

Fossils of carnivorous mammals are not nearly as common as those of many other mammals—but why should this be so? First, in populations of warm-blooded (endothermic) animals, carnivores constitute no more than 5% of the total population, usually less. Second, many, but not all, carnivores are solitary animals, so the chance of large numbers of them being preserved together is remote. Nevertheless, fossil carnivores are common enough for us to piece together their overall evolutionary relationships with some confidence.

The order Carnivora began to diversify when two distinct lines evolved from *creodonts* and *miacids* during the Paleocene. Both had well-developed canines and carnassials, but they were rather short-limbed and flat-footed.

Certainly they were not very fast, but neither was their prey. The creodont branch became extinct by Miocene time, so need not concern us further, but the other branch, evolving from weasel-like miacids, led to all existing carnivorous mammals (• Figure 18.13).

© Jim Melli/San Diego Natural History Museum

• Figure 18.13 **Today's Carnivorous Mammals Evolved from a Primitive Group Known as Miacids** This miacid known as *Tapocyon* was a coyote-sized animal that lived during the Eocene. All weasels, otters, skunks, badgers, martins, wolverines, seals, sea lions, walruses, bears, raccoons, dogs, hyenas, mongooses, and cats evolved from an ancestor very much like this creature.

• Figure 18.12 **Teeth of Carnivorous Mammals**

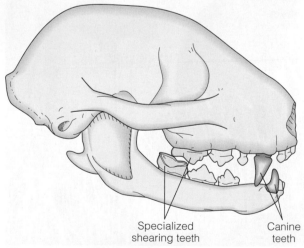

Specialized shearing teeth   Canine teeth

a This present-day skull and jaw of a large cat show the specialized sharp-crested shearing teeth, or carnassials, of carnivorous mammals.

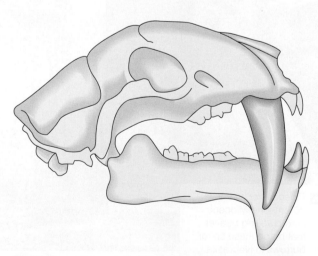

b Carnivorous mammals have sharp, pointed canine teeth, but in saber-toothed cats such as *Eusmilus* the upper canines were very large.

• Figure 18.14 **Characteristics and Evolutionary Trends in Hoofed Mammals**

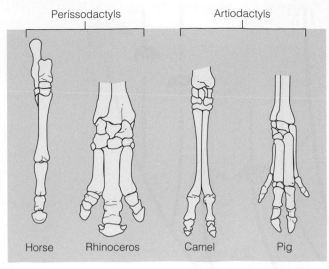

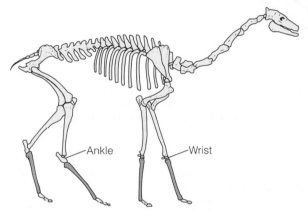

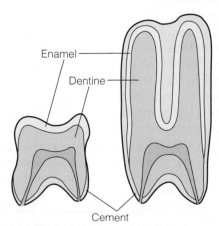

a Perissodactyls have one or three functional toes, whereas artiodactyls have two or four. In perissodactyls, the weight is borne on the third toe, but in artiodactyls, it is borne on toes three and four.

b In many hoofed mammals, long, slender limbs evolved as bones between the wrist and toes and the ankle and toes became longer.

c High-crowned, cement-covered chewing teeth (right) evolved in hoofed mammals that adapted to a diet of grass. Low-crowned chewing teeth are found in many other mammals, including primates and pigs, both of which have a varied diet.

An interesting note about carnivorous mammals is that cats, hyenas, and viverrids (civits and mongooses) share a common ancestor, but dogs are rather distantly related to the somewhat similar appearing hyenas. In fact, dogs (family Canidae) and hyenas (family Hyenadae) not only are similar in appearance but also, with few exceptions, are pack hunters. Nevertheless, the fossil record and studies of living animals clearly indicate hyenas are more closely related to cats and mongooses; their similarity to dogs is another example of convergent evolution.

One of the most remarkable developments in cats was the evolution of huge canines in the saber-tooth cats (Figure 18.12b). Saber-tooth cats existed throughout most of the Cenozoic Era and are particularly well known from Pleistocene-age deposits.

The aquatic carnivores, seals, sea lions, and walruses, are most closely related to bears, but unfortunately, their ancestry is less well known than for other families of carnivores. Aquatic adaptations include a somewhat streamlined body, a layer of blubber for insulation, and limbs modified into paddles. Most are fish eaters and have rather simple, single-cusped teeth, except walruses, which have flattened teeth for crushing shells.

The Ungulates or Hoofed Mammals

Ungulate is an informal term referring to several groups of living and extinct mammals, particularly the hoofed mammals of the orders **Artiodactyla** and **Perissodactyla.** The artiodactyls, or *even-toed hoofed mammals*, are the most diverse and numerous, with about 170 living species of cattle, goats, sheep, swine, antelope, deer, giraffes, hippopotamuses, camels, and several others. In marked contrast, the perissodactyls, or *odd-toed hoofed mammals*, have only 16 existing species of horses, rhinoceroses, and tapirs. During the Early Cenozoic, though, perissodactyls were more abundant than artiodactyls.

Some defining characteristics of these groups are the number of toes and how the animal's weight is borne on the toes. (Their teeth are also distinctive.) Artiodactyls have either two or four toes, and their weight is borne along an axis that passes between the third and fourth digits (• Figure 18.14a). For those artiodactyls with two toes, such as today's swine and deer, the first, second, and fifth digits have been lost or remain only as vestiges. Perissodactyls have one or three toes, although a few fossil species retained four toes on their forefeet. Nevertheless, their weight is borne on an axis passing through the third toe (• Figure 18.14a). Even today's horses have vestigial side toes, and rarely they are born with three toes.

Many hoofed mammals such as antelope and horses depend on speed to escape from predators in their open-grasslands habitat. As a result they have long, slender limbs, giving them a greater stride length. Notice from Figure 18.14b that the bones of the palm and sole have become very long. In addition, these speedy runners have fewer bones in their feet, mostly because they have fewer toes.

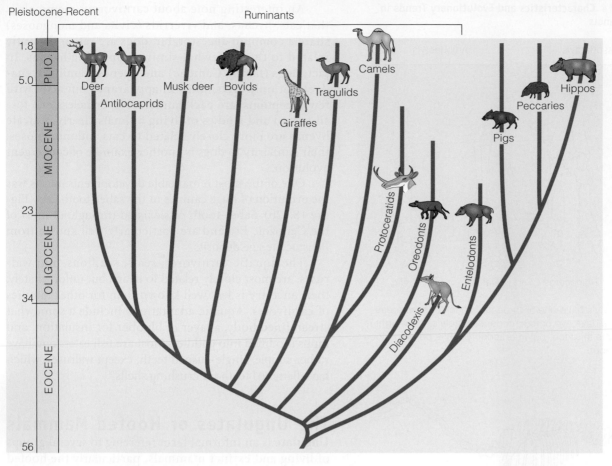

• **Figure 18.15 Relationships among Artiodactyls (Even-Toed Hoofed Mammals)** Most artiodactyls are ruminants—that is, cud-chewing animals with complex three or four-chambered stomachs. The bovids are the most diverse and numerous living artiodactyls, whereas some other groups proliferated during the past, the oreodonts for example. The oldest known artiodactyl, *Diacodexis*, was only about 50 cm long.

Not all hoofed mammals are long-limbed, speedy runners. Some are small and dart into heavy vegetation or a hole in the ground when threatened by predators. Size alone is adequate protection in some very large species such as rhinoceroses. In contrast to the long, slender limbs of horses, antelope, deer, and so on, rhinoceroses and hippopotamuses have developed massive, weight-supporting legs.

All artiodactyls and perissodactyls are herbivorous animals, with their chewing teeth—premolars and molars—modified for a diet of vegetation. One evolutionary trend in these animals was **molarization,** a change in the premolars so that they are more like molars, thus providing a continuous row of grinding teeth. Some ungulates—horses, for example—are characterized as **grazers** because they eat grass, as opposed to **browsers,** which eat the tender leaves, twigs, and shoots of trees and shrubs. Grasses are very abrasive, because as they grow through soil they pick up tiny particles of silt and sand that quickly wear teeth down. As a result, once grasses had evolved, many hoofed mammals became grazers and developed high-crowned, abrasion-resistant chewing teeth (Figure 18.14c).

Artiodactyls—Even-Toed Hoofed Mammals The oldest known artiodactyls were Early Eocene rabbit-sized

animals that differed little from their ancestors. Yet these small creatures were ancestral to the myriad living and several extinct families of even-toed hoofed mammals (Figure 18.15). Among the extinct families are the rather piglike oreodonts that were so common in North America until their extinction during the Pliocene and the peculiar genus *Synthetoceras* with forked horns on their snouts (chapter opener photo and Figure 18.10).

During much of the Cenozoic Era, especially in North America, camels of one kind or another were common. The earliest were small four-toed animals, but by Oligocene time all had two toes. Among the various types were very tall giraffe-like camels; slender, gazelle-like camels; and giants standing 3.5m high at the shoulder. Most camel evolution took place in North America, but during the Pliocene they migrated to Asia and South America, where the only living species exist now. North American camels went extinct near the end of the Pleistocene Epoch.

Among the artiodactyls, the family Bovidae is by far the most diverse, with dozens of species of cattle, bison, sheep, goats, and antelope. This family did not appear until the Miocene, but most of its diversification took place during Pliocene time on the northern continents. Bovids are

• Figure 18.16 **Evolution of Horses**

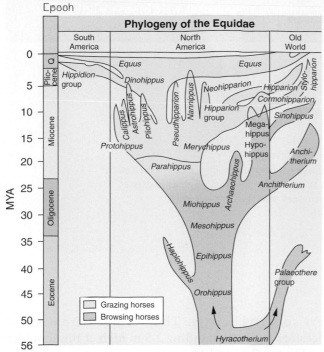

a Summary chart showing the relationships among the genera of horses. During the Oligocene two lines emerged, one leading to three-toed browsers and the other to one-toed grazers, including the present-day horse *Equus*.

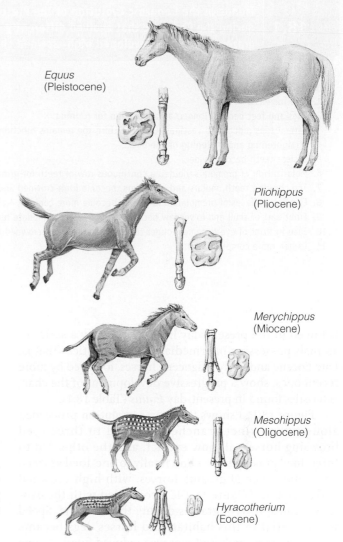

b Simplified diagram showing some trends in horse evolution. Trends include a size increase, a lengthening of the limbs and a reduction in the number of toes, and the development of high-crowned teeth with complex chewing surfaces.

now most numerous in Africa and southern Asia. North America still has its share of bovids such as bighorn sheep and mountain goats, but the most common ones during the Cenozoic were bison (which migrated from Asia), the pronghorn, and oreodonts, all of which roamed the western interior in vast herds.

Notice from Figure 18.15 that most living artiodactyls are **ruminants,** cud-chewing animals with complex three- or four-chambered stomachs in which food is processed to extract more nutrients. Perissodactyls lack such a complex digestive system. Perhaps the fact that artiodactyls use the same resources more effectively than do perissodactyls, explains why artiodactyls have flourished and mostly replaced perissodactyls in the hoofed mammal fauna.

Perissodactyls—Odd-Toed Hoofed Mammals In Table 7.1, we said, "If we examine the fossil record of related organisms such as horses and rhinoceroses, we should find that they were quite similar when they diverged from a common ancestor but became increasingly different as divergence continued." We also discussed how fossil records for horses, rhinoceroses, tapirs, and their extinct relatives, the chalicotheres and titanotheres, provide precisely this kind of evidence. In short, when these animals first appeared in the fossil record, they differed slightly in size and the structure of their teeth, but as they evolved, differences between them became more apparent. Perissodactyls evolved from a common ancestor during the Paleocene,

reached their greatest diversity during the Oligocene, and have declined markedly since then.

With the possible exception of camels, probably no group of mammals has a better fossil record than do horses. Indeed, horse fossils are so common, especially in North America where most of their evolution took place, that their overall history and evolutionary trends are well known. The earliest member of the horse family (family Equidae) is the fox-sized animal known as *Hyracotherium* (• Figure 18.16). This small, forest-dwelling animal had four-toed forefeet and three-toed hind feet, but each toe was covered by a small hoof. Otherwise it possessed few of the features of present-day horses. So how can we be sure it belongs to the family Equidae at all?

Horse evolution was a complex, branching affair, with numerous genera and species existing at various times during the Cenozoic (Figure 18.16a). Nevertheless, their exceptional fossil record clearly shows *Hyracotherium*

TABLE **18.1**

Trends in the Cenozoic Evolution of the Present-Day Horse _Equus_. A number of horse genera existed during the Cenozoic that evolved differently. For instance, some horses were browsers rather than grazers and never developed high-crowned chewing teeth, and retained three toes.

Trend

1. Size increase.
2. Legs and feet become longer, an adaptation for running.
3. Lateral toes reduced to vestiges. Only the third toe remains functional in _Equus_.
4. Straightening and stiffening of the back.
5. Incisor teeth become wider.
6. Molarization of premolars yielded a continuous row of teeth for grinding vegetation.
7. The chewing teeth, molars and premolars, become high-crowned and cement-covered for grinding abrasive grasses.
8. Chewing surfaces of premolars and molars become more complex—also an adaptation for grinding abrasive grasses.
9. Front part of skull and lower jaw become deeper to accommodate high-crowned premolars and molars.
10. Face in front of eye becomes longer to accommodate high-crowned teeth.
11. Larger, more complex brain.

is linked to the present-day horse, _Equus_, by a series of animals possessing intermediate characteristics. That is, Late Eocene and Early Oligocene horses, followed by more recent ones, show a progressive development of the characteristics found in present-day _Equus_ (Table 18.1).

Figure 18.16a shows that horse evolution proceeded along two distinct branches. One led to three-toed browsing horses, all now extinct, and the other led to three-toed grazing horses and finally to one-toed grazers. The appearance of grazing horses, with high-crowned chewing teeth (Figure 18.14c), coincided with the evolution and spread of grasses during the Miocene. Speed was essential in this habitat, and horses' legs became longer and the number of toes was reduced finally to one (Figure 18.16b). Pony-sized _Merychippus_ is a good example of the early grazing horses; it had three toes, but its teeth were high-crowned and covered by abrasion-resistant cement.

The other living perissodactyls, rhinoceroses and tapirs, increased in size from Early Cenozoic ancestors, and both became more diverse and widespread than they are now. Most rhinoceroses evolved in the Old World, but North American rhinoceroses were common until they became extinct at the end of the Pleistocene. At more than 5 m high at the shoulder and weighing perhaps 13 or 14 metric tons, a hornless Oligocene-Miocene rhinoceros in Asia was the largest land-dwelling mammal ever.

For the remaining perissodactyls, chalicotheres, and titanotheres, only the latter has a good fossil record. Chalicotheres, although never particularly abundant, are interesting, because the later members of this family, which were the size of large horses, had claws on their feet, rather than hooves. The prevailing opinion is that these claws were used to hook and pull down branches. Titanotheres existed only during the Eocene, giving them the distinction of being the shortest-lived

perissodactyl family. They evolved from small ancestors to giants standing 2.5 m high at the shoulder (see Figures 7.14 and 18.1).

Giant Land-Dwelling Mammals— Elephants

We just noted that the largest land-dwelling mammal ever was a hornless rhinoceros, but some of today's elephants, order **Proboscidea,** are also giants. The largest one on record weighed in at nearly 12 metric tons and stood 4.2 m at the shoulder, and some extinct mammoths were equally as large or perhaps slightly larger. In addition to their size, a distinctive feature of elephants is their long snout, or proboscis. During much of the Cenozoic Era, proboscideans of one kind or another (mastodons, mammoths, and today's elephants) were widespread on the northern continents, but now only three species exist, one in southeast Asia and two in Africa. The earliest member of the order was a 100- to 200-kg animal known as _Moeritherium_ from the Eocene that possessed few characteristics of elephants. It was probably aquatic.

By Oligocene time, elephants showed the trends toward large size and had developed a long proboscis and large tusks, which are enlarged incisors. Most elephants developed tusks in the upper jaw only, but a few had them in both jaws, and one, the deinotheres, had only lower tusks (• Figure 18.17).

The most familiar elephants, other than living ones, are the extinct _mastodons_ and _mammoths_. Mastodons evolved in Africa, but from Miocene to Pleistocene time they spread over the Northern Hemisphere continents and one genus even reached South America. These large browsing animals died out only a few thousands of years ago. During the Pliocene and Pleistocene, mammoths and living elephants diverged (Figure 18.17). Mammoths were

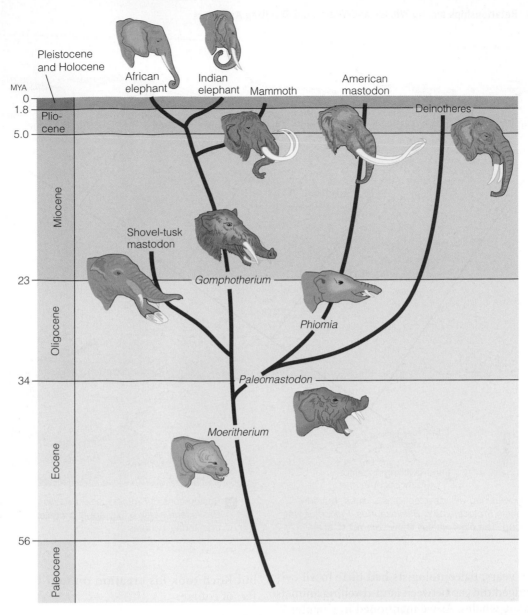

• **Figure 18.17 Phylogeny of Elephants and Some of Their Relatives** Large size, a long proboscis, and tusks were some of the evolutionary trends in proboscideans. Several fossil proboscideans are not shown here, so they were more diverse than indicated in this illustration.

about the size of elephants today, but they had the largest tusks of any elephant. In fact, mammoth tusks are common enough in Siberia that they have been and continue to be a source of ivory. Until their extinction near the end of the Pleistocene, mammoths lived on all Northern Hemisphere continents as well as in India and Africa.

Giant Aquatic Mammals—Whales Our
fascination with huge dinosaurs should not overshadow the fact that by far the largest animal ever is alive today. At more than 30 m long and weighing an estimated 130 metric tons, blue whales greatly exceed the size of any other living thing, except some plants such as redwood trees. But not all whales are large. Consider, for instance, dolphins and

porpoises—both are sizable but hardly giants. Nevertheless, an important trend in whale evolution has been increase in body size.

Several kinds of mammals are aquatic or semi-aquatic, but only sea cows (order Sirenia) and whales (order **Cetacea**), are so thoroughly aquatic that they cannot come out onto land. Fossils discovered in Eocene rocks in Southeast Asia indicate that the land-dwelling ancestors of whales were among the small dog-sized raoellids (• Figure 18.18). During the transition from land-dwelling animals to aquatic whales, the front limbs modified into paddlelike flippers; the rear limbs were lost; the nostrils migrated to the top of the head; and a large, horizontal tail fluke used for propulsion developed.

• Figure 18.18 **Relationships among Whales and Their Land-Dwelling Ancestors**

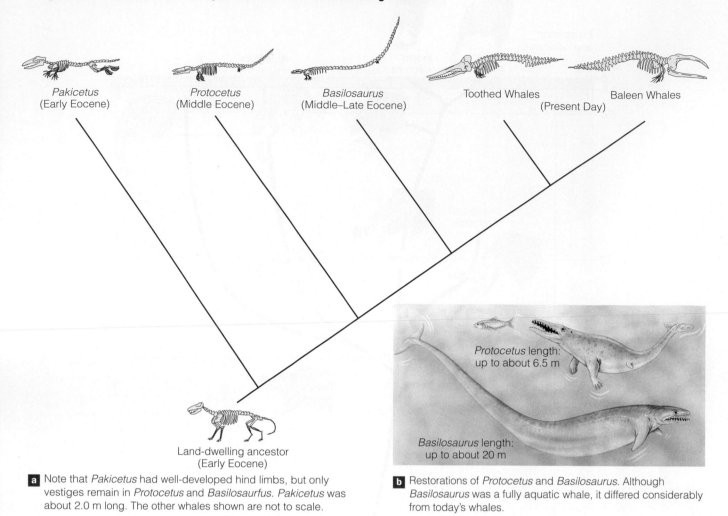

Pakicetus
(Early Eocene)

Protocetus
(Middle Eocene)

Basilosaurus
(Middle–Late Eocene)

Toothed Whales

Baleen Whales

(Present Day)

Land-dwelling ancestor
(Early Eocene)

Protocetus length:
up to about 6.5 m

Basilosaurus length:
up to about 20 m

a Note that *Pakicetus* had well-developed hind limbs, but only vestiges remain in *Protocetus* and *Basilosaurfus*. *Pakicetus* was about 2.0 m long. The other whales shown are not to scale.

b Restorations of *Protocetus* and *Basilosaurus*. Although *Basilosaurus* was a fully aquatic whale, it differed considerably from today's whales.

For many years, paleontologists had little fossil evidence that bridged the gap between land-dwelling animals and fully aquatic whales. As we mentioned in Chapter 7, though, this important transition took place in a part of the world where the fossil record was poorly known. Beginning about 20 years ago, paleontologists have made some remarkable finds that resolved this evolutionary enigma. For instance, the Early Eocene whale *Ambulocetus* still had limbs capable of support on land, whereas *Basilosaurus*, a 15-m-long Late Eocene whale, had only tiny, vestigial rear limbs (Figure 18.18). The latter had teeth similar to those of their ancestors, and its nostrils were on the snout, but it was truly a whale, although differently proportioned from those living now. By Oligocene time, both presently existing whale groups—baleen whales and toothed whales—had evolved.

An interesting note on fossil whales is that during the 1840s, Albert Koch found what he claimed was a fossil sea serpent in Eocene rocks in Alabama, which was actually a nearly complete skeleton of an extinct whale. In his restoration, Koch used the vertebrae of five different animals to render a "sea serpent" nearly 35 m long. No one in the scientific community was fooled,

but Koch took his creation on tour for viewing—for a fee, of course.

Pleistocene Faunas

Unlike the Paleocene fauna with its archaic mammals, unfamiliar ancestors of living mammalian orders, and large, predatory birds, the fauna of the Pleistocene consists mostly of familiar animals. Even so, their geographic distribution might surprise us, because rhinoceroses, elephants, and camels lived in North America, and a few unusual mammals, such as chalicotheres and the heavily armored glypotodonts, were present. In the avian fauna the giant moas and elephant birds were in New Zealand and in Madagascar, respectively.

Ice Age Mammals The most remarkable aspect of the Pleistocene mammalian fauna is that so many very large species existed. Mastodons, mammoths, giant bison, huge ground sloths, immense camels, and beavers 2 m tall at the shoulder were present in North America. South

• Figure 18.19 **The Irish Elk** Restoration of the giant deer *Megaloceros giganteus*, commonly called the Irish elk. It lived in Europe and Asia during the Pleistocene, and may have persisted in Eastern Europe until only a few thousands of years ago. Large males probably weighed 700 kg, about the same as a present-day moose, and had an antler spread of nearly 4.0 m.

America had its share of giants, too, especially sloths and glyptodonts. Elephants, cave bears, and giant deer known as Irish Elk lived in Europe and Asia (• Figure 18.19), and Australia had 3-m-tall kangaroos and wombats the size of rhinoceroses.

Of course, many smaller mammals also existed, but one obvious trend among Pleistocene mammals was large body size. Perhaps this was an adaptation to the cooler conditions that prevailed during that time. Large animals have less surface area compared to their volume and thus retain heat more effectively than do smaller animals.

Some of the world's best-known fossils come from Pleistocene deposits. You have probably heard of the frozen mammals found in Siberia and Alaska, such as mammoths, bison, and a few others. These extraordinary fossils, although very rare, provide much more information than most fossils do (see Figure 5.11b). Contrary to what you might hear in the popular press, all these frozen animals were partly decomposed, none were fresh enough to eat, and none were found in blocks of ice or icebergs. All were recovered from permanently frozen ground known as *permafrost*.

Paleontologists have recovered Pleistocene animals from many places in North America; two noteworthy areas are Florida and the La Brea tar pits at Rancho La Brea in southern California. In fact, Florida is one of the few places in the eastern United States where fossils of Cenozoic land-dwelling animals are common (• Figure 18.20a). At the La Brea tar pits, at least 230 kinds of vertebrate animals have been found trapped in the sticky residue where liquid petroleum seeped out at the surface and then evaporated.

The "tar" is really naturally formed asphalt, whereas tar is a product manufactured from peat or coal. Many of the fossils are carnivores, especially dire wolves, saber-tooth cats, and vultures, that gathered to dine on mammals that became mired in the tar (Figure 18.20b).

Pleistocene Extinctions

During the Pleistocene, the continental interior of North America was teeming with horses, rhinoceroses, camels, mammoths, mastodons, bison, giant ground sloths, glyptodonts, saber-toothed cats, dire wolves, rodents, and rabbits. Beginning about 14,000 years ago, however, many of these animals became extinct, especially the larger ones. These Pleistocene extinctions were modest compared to those at the end of the Paleozoic and Mesozoic eras, but they were unusual in that they had a profound impact on large, land-dwelling mammals (those weighing more than 44 kg). Particularly hard hit were the mammalian faunas of Australia and the Americas.

In Australia, 15 of the continent's 16 genera of large mammals died out. North America lost 33 of 45 genera of large mammals, and in South America 46 of 58 large mammal genera went extinct. In contrast, Europe had only 7 of 23 large genera die out, whereas Africa south of the Sahara lost only 2 of 44 genera. These data bring up three questions, none of which has been answered completely: (1) What caused Pleistocene extinctions? (2) Why did these extinctions eliminate mostly large mammals? (3) Why were extinctions most severe in Australia and the Americas? Scientists are currently debating two competing hypotheses for these extinctions. One, the *climatic change hypothesis*, holds that rapid changes in climate at the end of the Pleistocene caused extinctions, whereas the other, called *prehistoric overkill*, holds that human hunters were responsible.

Rapid changes in climate and vegetation did occur over much of Earth's surface during the Late Pleistocene, as glaciers began retreating. The North American and northern Eurasian open-steppe tundras were replaced by conifer and broadleaf forests as warmer and wetter conditions prevailed. The Arctic region flora changed from a productive herbaceous one that supported a variety of large mammals, to a relatively barren water-logged tundra that supported a much sparser fauna. The southwestern United States region also changed from a moist area with numerous lakes, where saber-tooth cats, giant ground sloths, and mammoths roamed, to a semiarid environment unable to support a diverse fauna of large mammals.

Rapid changes in climate and vegetation can certainly affect animal populations, but there are several problems with the climate change hypothesis. First, why didn't the large mammals migrate to more suitable habitats as the climate and vegetation changed? After all, many animal species did. For example, reindeer and the Arctic fox lived in southern France during the last glaciation and migrated to the Arctic when the climate became warmer. The second argument against the climate hypothesis is the apparent lack of correlation between extinctions and the earlier

• Figure 18.20 **Pleistocene Fossils from Florida and California**

Erika Simons/Florida Museum of Natural History

a Among the diverse Pliocene and Pleistocene mammals of Florida were 6-m-long ground sloths and armored glyptodonts that weighed more than 2 metric tons.

Sue Monroe

b Restoration of a mammoth trapped in the sticky tar (asphalt) at a present-day oil seep at the La Brea Tar Pits in Los Angeles, California.

extinctions occurred in Africa and most of Europe, because animals in those regions had long been familiar with humans.

One problem with the prehistoric overkill hypothesis is that archaeological evidence indicates the early human inhabitants of North and South America, as well as Australia, probably lived in small, scattered communities, gathering food and hunting. How could a few hunters decimate so many species of large mammals? However, it is true that humans have caused extinctions on oceanic islands. For example, in a period of about 600 years after arriving in New Zealand, humans exterminated several species of the large, flightless birds called moas.

A second problem is that present-day hunters concentrate on smaller, abundant, and less dangerous animals. The remains of horses, reindeer, and other smaller animals are found in many prehistoric sites in Europe, whereas mammoth and woolly rhinoceros remains are scarce.

Finally, few human artifacts are found among the remains of extinct animals in North and South America, and there is usually little evidence that the animals were hunted. Countering this argument is the assertion that the impact on the previously unhunted fauna was so swift as to leave little evidence.

The reason for the extinctions of large Pleistocene mammals is unresolved and probably will be for some time. It may turn out that the extinctions resulted from a combination of different circumstances. Populations that were already under stress from climate changes were perhaps more vulnerable to hunting, especially if small females and young animals were the preferred targets.

glacial advances and retreats throughout the Pleistocene Epoch. Previous changes in climate were not marked by episodes of mass extinctions.

Proponents of the prehistoric overkill hypothesis argue that the mass extinctions in North and South America and Australia coincided closely with the arrival of humans. Perhaps hunters had a tremendous impact on the faunas of North and South America about 11,000 years ago because the animals had no previous experience with humans. The same thing happened much earlier in Australia soon after people arrived about 40,000 years ago. No large-scale

Intercontinental Migrations

The mammalian faunas of North America, Europe, and northern Asia exhibited many similarities throughout the Cenozoic. Even today, Asia and North America are only narrowly separated at the Bering Strait, which at several

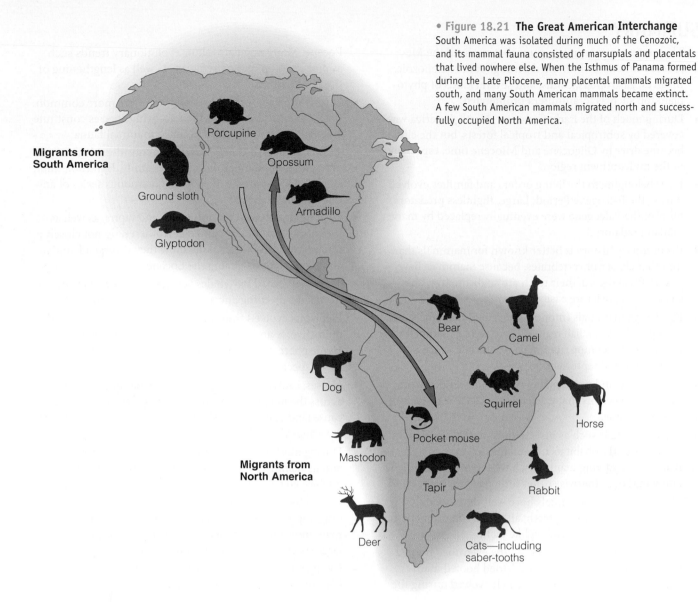

• Figure 18.21 **The Great American Interchange**
South America was isolated during much of the Cenozoic, and its mammal fauna consisted of marsupials and placentals that lived nowhere else. When the Isthmus of Panama formed during the Late Pliocene, many placental mammals migrated south, and many South American mammals became extinct. A few South American mammals migrated north and successfully occupied North America.

Migrants from South America

Porcupine

Opossum

Ground sloth

Armadillo

Glyptodon

Bear

Camel

Dog

Squirrel

Horse

Pocket mouse

Migrants from North America

Mastodon

Tapir

Rabbit

Deer

Cats—including saber-tooths

times during the Cenozoic formed a land corridor across which mammals migrated (see Figure 17.13). During the Early Cenozoic, a land connection between Europe and North America allowed mammals to roam across all the northern continents. Many did; camels and horses are only two examples.

However, the southern continents were largely separate island continents during much of the Cenozoic. Africa remained fairly close to Eurasia, and at times faunal interchange between those two continents was possible. For example, elephants first evolved in Africa, but they migrated to all the northern continents.

South America was isolated from all other landmasses from Late Cretaceous until a land connection with North America formed about 5 million years ago. Before the connection was established, the South American fauna was made up of marsupials and several orders of placental mammals that lived nowhere else. These animals thrived in isolation and showed remark-

able convergence with North American placental mammals (see Figure 7.11a). When the Isthmus of Panama formed, migrants from North America soon replaced many of the indigenous South American mammals, whereas fewer migrants from the south were successful in North America (• Figure 18.21). As a result of this *great American interchange*, today about 50% of South America's mammalian fauna came from the north, but in North America only 20% of its mammals came from the south. Even today, the coyote (*Canis latrans*) is extending its range from the north through Central America.

Most living species of marsupials are restricted to the Australian region. Recall from Chapter 15 that marsupials occupied Australia before its separation from Gondwana, but apparently placentals, other than bats and a few rodents, never got there until they were introduced by humans. So, unlike South America, which now has a connection with another continent, Australia has remained isolated, and its fauna is unique.

SUMMARY

- The marine invertebrate groups that survived the Mesozoic extinctions diversified throughout the Cenozoic. Bivalves, gastropods, corals, and several kinds of phytoplankton such as foraminifera proliferated.

- During much of the Early Cenozoic, North America was covered by subtropical and tropical forests, but the climate became drier by Oligocene and Miocene time, especially in the midcontinent region.

- Birds belonging to the living orders and families evolved during the Paleogene Period. Large, flightless predatory birds of the Paleogene were eventually replaced by mammalian predators.

- Evolutionary history is better known for mammals than for other classes of vertebrates, because mammals have a good fossil record, their teeth are so distinctive, and Cenozoic deposits are easily accessible.

- Egg-laying mammals (monotremes) and marsupials exist mostly in the Australian region. The placental mammals—by far the most common mammals—owe their success to their method of reproduction.

- All placental and marsupial mammals descended from shrewlike ancestors that existed from Late Cretaceous to Paleogene time.

- Small mammals such as insectivores, rodents, and rabbits occupy the microhabitats unavailable to larger mammals. Bats, the only flying mammals, have forelimbs modified into wings but otherwise differ little from their ancestors.

- Most carnivorous mammals have well-developed canines and specialized shearing teeth, although some aquatic carnivores such as seals have peglike teeth.

- The most common ungulates are the even-toed hoofed mammals (artiodactyls) and odd-toed hoofed mammals (perissodactyls), both of which evolved during the Eocene. Many ungulates show evolutionary trends such as molarization of the premolars as well as lengthening of the legs for speed.

- During the Paleogene, perissodactyls were more common than artiodactyls but now their 16 living species constitute less than 10% of the world's hoofed mammal fauna.

- Although present-day *Equus* differs considerably from the oldest known member of the horse family, *Hyracotherium*, an excellent fossil record shows a continuous series of animals linking the two.

- Even though horses, rhinoceroses, and tapirs, as well as the extinct titanotheres and chalicotheres, do not closely resemble one another, fossils show they diverged from a common ancestor during the Eocene.

- The fossil record for whales verifies that they evolved from land-dwelling ancestors during the Eocene.

- Elephants evolved from rather small ancestors, became quite diverse and abundant, especially on the Northern Hemisphere continents, and then dwindled to only three living species.

- Horses, camels, elephants, and other mammals spread across the northern continents during the Cenozoic because land connections existed between those landmasses at various times.

- During most of the Cenozoic, South America was isolated, and its mammal fauna was unique. A land connection was established between the Americas during the Late Cenozoic, and migrations in both directions took place.

- One important evolutionary trend in Pleistocene mammals and some birds was toward giantism. Many of these large species died out, beginning about 40,000 years ago.

- Changes in habitat and prehistoric overkill are the two hypotheses explaining Pleistocene extinctions.

IMPORTANT TERMS

Artiodactyla, p. 389
browser, p. 390
carnassials, p. 388
Carnivora, p. 388
Cetacea, p. 393
grazer, p. 390

Hyracotherium, p. 391
molar, p. 383
molarization, p. 390
Paleocene-Eocene Thermal
   Maximum, p. 381
Perissodactyla, p. 389

premolar, p. 383
Primates, p. 387
Proboscidea, p. 392
ruminant, p. 391
ungulate, p. 389

REVIEW QUESTIONS

1. Horses, rhinoceroses, and tapirs are all members of the mammal order Perissodactyla which is also known as the _____ mammals.
   a. _____ carnivorous/omnivorous; b. _____ odd-toed hoofed; c. _____ ruminant; d. _____ flightless predatory; e. _____ proboscidian.

2. One of the defining characteristics of the carnivorous mammals is
   a. _____ carnassial teeth; b. _____ molarization of the premolars; c. _____ hooves; d. _____ teeth adapted to grazing; e. _____ short, massive limbs.

3. During the Cenozoic Era, Earth's temperature was highest during the
   a. _____ Pleistocene; b. _____ Pliocene; c. _____ Eocene; d. _____ Cretaceous; e. _____ Neogene.
4. Cenozoic bryozoans were particularly abundant and successful
   a. _____ predatory birds; b. _____ marine reptiles; c. _____ carnivorous mammals; d. _____ suspension feeders; e. _____ artiodacyls.
5. Which one of the following statements is correct?
   a. _____ Africa had most extinctions at the end of the Pleistocene; b. _____ The oldest member of the horse family is *Hyracotherium*; c. _____ Ungulates includes all rodents, rabbits, and bats; d. _____ Browsers have high-crowned chewing teeth; e. _____ One trend in horse evolution was an increase in the number of toes.
6. The only living egg-laying mammals are
   a. _____ multituberculates; b. _____ megadonts; c. _____ marsupials d. _____ moerotheres; e. _____ monotremes.
7. One indication of a cool climate is
   a. _____ mammals with nostrils on the top of the head; b. _____ small leaves with incised margins; c. _____ evolution of large flightless birds; d. _____ increase in diversity of hoofed mammals; e. _____ extinction of coccolithophores.
8. One feature of Eocene whales that indicates they had land-dwelling ancestors is
   a. _____ low-crowned chewing teeth; b. _____ teeth modified for a diet of seaweed; c. _____ vestigial rear limbs; d. _____ a long slender body; e. _____ enlarged eyes.
9. One of the major components of the Cenozoic marine invertebrate fauna was
   a. _____ foraminifera; b. _____ cetaceans; c. _____ proboscidians; d. _____ ungulates, e. _____ ammonites.
10. Which one of the following has been proposed to account for Pleistocene extinctions?
    a. _____ Meteorite impact; b. _____ Climate change; c. _____ Evolution of grasses; d. _____ Migration of mammals from Australia; e. _____ Increased energy from the Sun.
11. Explain how and why the South American mammal fauna of the Cenozoic differed from faunas elsewhere and how has it changed?
12. What are the major evolutionary trends in hoofed mammals that adapted to a grazing, open plains habitat?
13. Why are so many fossil mammals found in the western United States but so few in the east?
14. Explain how fossil leaves and the composition of paleofloras give clues about ancient climates.
15. What were the dominant land-dwelling predators during the Early Cenozoic and what happened to them?
16. Discuss three evolutionary trends seen in whales.
17. What do fossils reveal about the evolution of today's horses from their ancient ancestor *Hyracotherium*?
18. Give a summary of the climatic changes that took place during the Cenozoic Era.
19. Why do paleontologists consider the Paleocene fauna as archaic?
20. Dogs and hyenas appear similar and yet they are distantly related. Explain how this came to be so.

APPLY YOUR KNOWLEDGE

1. While working in an area with well-exposed Pliocene, organic-rich mudstones and claytones you make a remarkable discovery: several fossil horses, camels, and mastodons, and dozens of skeletons of saber-toothed cats, dogs, and vultures. What is anomalous about this association of fossils, and what features of the rocks might help you resolve this apparent dilemma?
2. You are a science teacher who receives numerous unlabeled mammal and plant fossils. You are not too concerned with identifying genera and species, but you do want to show your students mammal adaptation for diet and speed. What features of the skulls, teeth, and bones would allow you to infer which animals were herbivores (grazers versus browsers) and carnivores, and which ones were speedy runners? Also, could you use the plant fossils to make any inferences about ancient climates?
3. Now that we have covered the evolution of vertebrates from fish to mammals, can you summarize the major events that occurred without using technical language? In other words, how would you explain the overall evolution of vertebrates to an interested by uninformed audience?
4. Can you think of evidence that would confirm or falsify the hypotheses of human overkill and climatic change for Pleistocene extinctions?

CHAPTER 19

PRIMATE AND HUMAN EVOLUTION

Olduvai Gorge on the eastern Serengeti Plain, Northern Tanzania, is often referred to as "The Cradle of Mankind" because of the many important hominid discoveries made there. The gorge, part of the East African Rift Valley, is 48 km long and 90 m deep and formed as a result of the tectonic forces shaping East Africa.

[OUTLINE]

Introduction

What Are Primates?

Prosimians

Anthropoids

Hominids
  Australopithecines

Perspective *Footprints at Laetoli*

The Human Lineage

Neanderthals

Cro-Magnons

Summary

At the end of this chapter, you will have learned that

- Primates are difficult to characterize as an order because they lack strong specializations found in most other mammalian orders.

- Primates are divided into two suborders: the prosimians, which include lemurs, lorises, tarsiers, and tree shrews, and the anthropoids, which include monkeys, apes, and humans.

- The hominids include present-day humans and their extinct ancestors.

- Human evolution is very complex and in a constant state of flux owing to new fossil and scientific discoveries.

- The most famous of all fossil humans are the Neanderthals, which were succeeded by the Cro-Magnons, about 30,000 years ago.

Introduction

Who are we? Where did we come from? What is the human genealogy? These are basic questions that we probably have all asked ourselves at some time or another. Just as many people enjoy tracing their own family history as far back as they can, paleoanthropologists are discovering, based on recent fossil finds, that the human family tree goes back much farther than we thought. In fact, a skull found in the African nation of Chad in 2002 and named *Sahelanthropus tchadensis* (but nicknamed *Tourmaï*, which means "hope of life" in the local Goran language) has pushed back the origins of humans to nearly 7 million years ago. Another discovery reported in 2006 provides strong evidence for an ancestor–descendant relationship between two early hominid lines, one of which leads to our own human lineage.

So where does this leave us, evolutionarily speaking? It leaves us at a very exciting time, as we seek to unravel the history of our species. Our understanding of our genealogy is presently in flux, and each new fossil hominid find sheds more light on our ancestry. Although some may find it frustrating, human evolution is just like that of other groups in that we have followed an uncertain evolutionary path. As new species evolved, they filled ecological niches and either gave rise to descendants better adapted to the changing environment or they became extinct. So it should not surprise us that our own evolutionary history has many "dead-end" side branches.

In this chapter, we examine the various primate groups, in particular the origin and evolution of the hominids, the group that includes our ancestors. However, we must point out that new discoveries of fossil hominids, as well as new techniques for scientific analysis, are leading to new hypotheses about our ancestry. By the time you read this chapter, it is possible that new discoveries may change some of the conclusions stated here. Such is the nature of paleoanthropology—and one reason why the study of hominids is so exciting.

What Are Primates?

Primates are difficult to characterize as an order because they lack the strong specializations found in most other mammalian orders. We can, however, point to several trends in their evolution that help define primates and are related to their *arboreal*, or tree-dwelling, ancestry. These include changes in the skeleton and mode of locomotion; an increase in brain size; a shift toward smaller, fewer, and less specialized teeth; and the evolution of stereoscopic vision and a grasping hand with an opposable thumb. Not all of these trends took place in every primate group, nor did they evolve at the same rate in each group. In fact, some primates have retained certain primitive features, whereas others show all or most of these trends.

The primate order is divided into two suborders, the Prosimii and Anthropoidea (● Figure 19.1, Table 19.1). The *prosimians*, or lower primates, include the lemurs, lorises, tarsiers, and tree shrews, whereas the *anthropoids*, or higher primates, include monkeys, apes, and humans.

Archean Eon	Proterozoic Eon	Phanerozoic Eon							
			Paleozoic Era						
Precambrian		Cambrian	Ordovician	Silurian	Devonian	Mississippian	Pennsylvanian	Permian	
						Carboniferous			

2500 MYA   542 MYA   251 MYA

a Tarsier

c Spider monkey

© Ron Austing/ Photo Researchers, Inc.

© Renee Lynn/ Photo Researchers, Inc.

b Baboon

© Tony Camacho / Photo Researchers, Inc.

d Chimpanzee

© Tim Davis/ Photo Researchers, Inc.

• Figure 19.1 **Primates** Primates are divided into two suborders: the prosimians **a** and the anthropoids **b**–**d**, which are further subdivided into three superfamilies: Old World monkeys **b**, New World monkeys **c**, and hominoids, which include the apes **d**, and humans.

TABLE **19.1**	**Classification of the Primates**

Order Primates: Lemurs, lorises, tarsiers, tree shrews, monkeys, apes, humans

  Suborder Prosimii: Lemurs, lorises, tarsiers, tree shrews (lower primates)

  Suborder Anthropoidea: Monkeys, apes, humans (higher primates)

    Superfamily Cercopithecoidea: Macaque, baboon, proboscis monkey (Old World monkeys)

    Superfamily Ceboidea: Howler, spider, and squirrel monkeys (New World monkeys)

    Superfamily Hominoidea: Apes, humans

      Family Pongidae: Chimpanzees, orangutans, gorillas

      Family Hylobatidae: Gibbons, siamangs

      Family Hominidae: Humans

Prosimians

Prosimians are generally small, ranging from species the size of a mouse up to those as large as a house cat. They are arboreal, have five digits on each hand and foot with either claws or nails, and are typically omnivorous. They have large, forwardly directed eyes specialized for night vision—hence, most are nocturnal (Figure 19.1a). As their name implies (*pro* means "before," and *simian* means "ape"), they are the oldest primate lineage, with a fossil record extending back to the Paleocene.

During the Eocene, prosimians were abundant, diversified, and widespread in North America, Europe, and Asia (• Figure 19.2). As the continents moved northward during the Cenozoic and the climate changed from warm

Phanerozoic Eon										
Mesozoic Era			Cenozoic Era							
Triassic	Jurassic	Cretaceous	Paleogene			Neogene			Quaternary	
			Paleocene	Eocene	Oligocene	Miocene	Pliocene		Pleistocene	Holocene

'51 MYA

99 MYA

• Figure 19.2 **Eocene Prosimian** *Notharctus,* a primitive Eocene prosimian from North America.

David L. Brill

• Figure 19.3 *Aegyptopithecus zeuxis* Skull of *Aegyptopithecus zeuxis,* one of the earliest known anthropoids.

tropical to cooler midlatitude conditions, the prosimian population decreased in both abundance and diversity.

By the Oligocene, hardly any prosimians were left in the northern continents as the once widespread Eocene populations migrated south to the warmer latitudes of Africa, Asia, and Southeast Asia. Presently, prosimians are found only in the tropical regions of Asia, India, Africa, and Madagascar.

Anthropoids

Anthropoids evolved from a prosimian lineage sometime during the Late Eocene, and by the Oligocene, they were a well-established group. Much of our knowledge about the early evolutionary history of anthropoids comes from fossils found in the Fayum district, a small desert area southwest of Cairo, Egypt. During the Late Eocene and Oligocene, this region of Africa was a lush, tropical rain forest that supported a diverse and abundant fauna and flora. Within this forest lived many different arboreal anthropoids as well as various prosimians. In fact, several thousand fossil specimens representing more than 20 primate species have been recovered from rocks of this region. One of the earliest anthropoids was *Aegyptopithecus,* a small, Late Eocene, fruit-eating, arboreal primate that weighed about 5 kg (• Figure 19.3). *Aegyptopithecus* had not only monkey characteristics, but also features that were more like those of apes as well. As such, it is presently the closest link we have to the Old World primates.

Anthropoids are divided into three superfamilies: Cercopithecoidea (Old World monkeys), Ceboidea (New World monkeys), and Hominoidea (apes and humans) (Table 19.1).

Old World monkeys (superfamily Cercopithecoidea) include the macaque, baboon, and proboscis monkey, and are characterized by close-set, downward-directed nostrils (like those of apes and humans), grasping hands, and a nonprehensile tail (Figure 19.1b). Present-day Old World monkeys are distributed throughout the tropical regions of Africa and Asia and are thought to have evolved from a primitive anthropoid ancestor, like *Aegyptopithecus,* sometime during the Oligocene.

New World monkeys (superfamily Ceboidea) are found only in Central and South America. They are characterized by a prehensile tail, flattish face, and widely separated nostrils, and include the howler, spider, and squirrel monkeys (Figure 19.1c). New World monkeys probably evolved from African monkeys that migrated across the widening Atlantic sometime during the Early Oligocene, and they have continued evolving in isolation to this day. No evidence exists of any prosimian or other primitive primates in Central or South America, or of any contact with Old World monkeys after the initial immigration from Africa.

Hominoids (superfamily Hominoidea) consist of three families: the *great apes* (family Pongidae), which include chimpanzees, orangutans, and gorillas (Figure 19.1d); the *lesser apes* (family Hylobatidae), which are gibbons and siamangs; and the *hominids* (family Hominidae), which are humans and their extinct ancestors.

The hominoid lineage diverged from Old World monkeys sometime before the Miocene, but exactly when is still being debated. It is generally accepted, however, that hominoids evolved in Africa, probably from the ancestral group that included *Aegyptopithecus*.

Recall that beginning in the Late Eocene the northward movement of the continents resulted in pronounced climatic shifts. In Africa, Europe, Asia, and elsewhere, a major cooling trend began (see Figure 18.5b), and the tropical and subtropical rain forests slowly began to change to a variety of mixed forests separated by savannas and open grasslands as temperatures and rainfall decreased. As the climate changed, the primate populations also changed. Prosimians and monkeys became rare, whereas hominoids diversified in the newly forming environments and became abundant. Ape populations became reproductively isolated from each other within the various forests, leading to adaptive radiation and increased diversity among the hominoids. During the Miocene, Africa collided with Eurasia, producing additional changes in the climate, as well as providing opportunities for migration of animals between the two landmasses.

Two apelike groups evolved during the Miocene that ultimately gave rise to present-day hominoids. Although there is still not agreement on the early evolutionary relationships among the hominoids, fossil evidence and molecular DNA similarities between modern hominoid families is providing a clearer picture of the evolutionary pathways and relationships among the hominoids.

The first group, the *dryopithecines*, evolved in Africa during the Miocene, and subsequently spread to Eurasia following the collision between the two continents. The dryopithecines were a group of hominoids that varied in size, skeletal features, and lifestyle. The best known of all later hominoids is *Proconsul,* an apelike, fruit-eating animal that led a quadrupedal arboreal existence with limited activity on the ground (• Figure 19.4). The dryopithecines were very abundant and diverse during the Miocene and Pliocene, particularly in Africa.

The second group, the *sivapithecids*, evolved in Africa during the Miocene, and then spread throughout Eurasia. The fossil remains of sivapithecids are plentiful and consist mostly of skulls, jaws, and isolated teeth. Body or limb bones are rare, limiting our knowledge about what they looked like and how they moved around. We do know that sivapithecids had powerful jaws and thick-enameled teeth with flat chewing surfaces, suggesting a diet of harder and coarser foods, including nuts.

It is clear from the fossil evidence that sivapithecids were not involved in the evolutionary branch leading to

• Figure 19.4 *Proconsul* Probable appearance of *Proconsul,* a dryopithecine.

humans, but they were probably the ancestral stock from which present-day orangutans evolved. In fact, one genus, *Gigantopithecus,* was a contemporary of early *Homo* in Eastern Asia.

Although many pieces are still missing, particularly during critical intervals in the African hominoid fossil record, molecular DNA as well as fossil evidence indicates the dryopithecines, African apes, and hominids form a closely related lineage. The sivapithecids and orangutans, as just discussed, form a different lineage that did not lead to humans.

Hominids

The **hominids** (family Hominidae), the primate family that includes present-day humans and their extinct ancestors (Table 19.1), have a fossil record extending back almost 7 million years. Several features distinguish them from other hominoids. Hominids are bipedal; that is, they have an upright posture, which is indicated by several modifications in their skeleton (• Figure 19.5a). In addition, they show a trend toward a large and internally reorganized brain (Figure 19.5b). Other features include a reduced face and reduced canine teeth, omnivorous feeding, increased manual dexterity, and the use of sophisticated tools.

Many anthropologists think that these hominid features evolved in response to major climatic changes that began during the Miocene and continued into the Pliocene. During this time, vast savannas replaced the African tropical rain forests where the lower primates and Old World monkeys had been so abundant. As the savannas and grasslands continued to expand, the hominids made the transition from true forest dwelling to life in an environment of mixed forests and grasslands.

At present, there is no clear consensus on the evolutionary history of the hominid lineage. This is due,

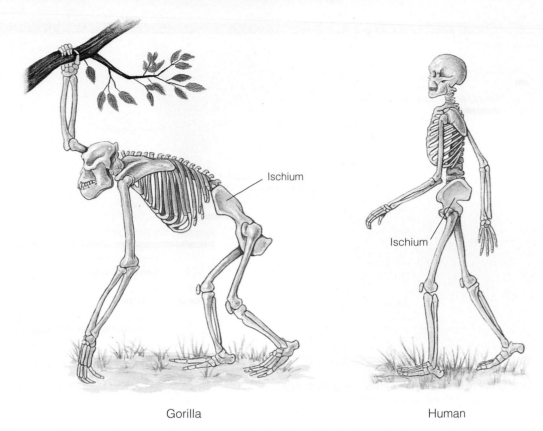

Gorilla   Human

a In gorillas, the ischium bone is long, and the entire pelvis is tilted toward the horizontal. In humans, the ischium bone is much shorter, and the pelvis is vertical.

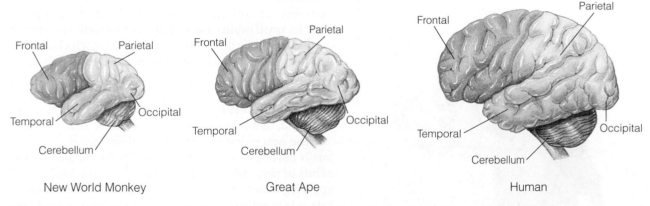

New World Monkey   Great Ape   Human

b An increase in brain size and organization is apparent in comparing the brains of a New World monkey (Superfamily Ceboidea), a great ape (Superfamily Hominoidea; Family Pongidae), and a present-day human (Superfamily Hominoidea; Family Hominidae).

in part, to the incomplete fossil record of hominids, as well as new discoveries, and also because some species are known only from partial specimens or fragments of bone.

Because of this, there is even disagreement on the total number of hominid species. A complete discussion of all the proposed hominid species and the various competing schemes of hominid evolution is beyond the scope of this chapter. However, we will briefly discuss the generally accepted taxa (• Figure 19.6) and present some of the current theories of hominid evolution.

Remember that although the fossil record of hominid evolution is not complete, what exists is well documented. Furthermore, the interpretation of that fossil record precipitates the often vigorous and sometimes acrimonious debates concerning our evolutionary history.

Discovered in northern Chad's Djurab Desert in July 2002, the nearly 7-million-year-old skull and dental remains of *Sahelanthropus tchadensis* (• Figure 19.7) make it the oldest known hominid yet unearthed and at or very near to the time when humans diverged from our closest-living relative, the chimpanzee. Currently, most paleoanthropologists

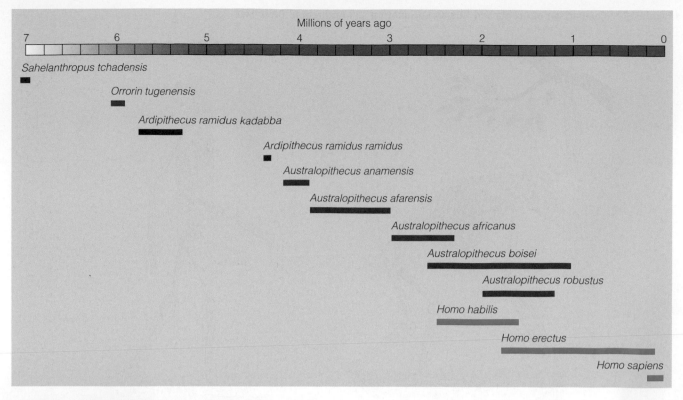

Millions of years ago

7   6   5   4   3   2   1   0

Sahelanthropus tchadensis

Orrorin tugenensis

Ardipithecus ramidus kadabba

Ardipithecus ramidus ramidus

Australopithecus anamensis

Australopithecus afarensis

Australopithecus africanus

Australopithecus boisei

Australopithecus robustus

Homo habilis

Homo erectus

Homo sapiens

• **Figure 19.6 The Stratigraphic Record of Hominids** The geologic ranges for the commonly accepted species of hominids (the branch of primates that includes present-day humans and their extinct ancestors).

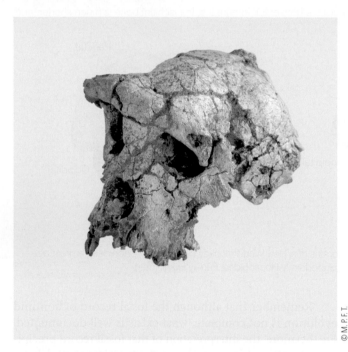

©M.P.F.T.

• **Figure 19.7 *Sahelanthropus tchadensis*** Discovered in Chad in 2002, and dated at nearly 7 million years, this skull of *Sahelanthropus tchadensis* is presently the oldest known hominid.

accept that the human–chimpanzee stock separated from gorillas approximately 8 million years ago and humans separated from chimpanzees about 5 million years ago.

Besides being the oldest hominid, *Sahelanthropus tchadensis* shows a mosaic of primitive and advanced features that has both excited and puzzled paleoanthropologists. Its small brain case and most of its teeth (except the canines) are chimplike. However, the nose, which is fairly flat, and the prominent brow ridges are features only seen, until now, in the human genus *Homo*. It is hypothesized that *Sahelanthropus tchadensis* was probably bipedal in its walking habits, but until bones from its legs and feet are found, that supposition remains conjecture.

The next oldest hominid is *Orrorin tugenensis*, whose fossils have been dated at 6 million years old and consist of bits of jaw, isolated teeth, and finger, arm, and partial upper leg bones (Figure 19.6). At this time, there is still debate as to exactly where *Orrorin tugenensis* fits in the hominid lineage.

Sometime between 5.8 and 5.2 million years ago, another hominid was present in eastern Africa. *Ardipithecus ramidus kadabba* is older than its 4.4-million-year-old relative *Ardipithecus ramidus ramidus* (Figure 19.6). *Ardipithecus ramidus kadabba* is very similar in most features to *Ardipithecus ramidus ramidus* but in specific features of its teeth, it is more apelike than its younger relative.

Although many paleoanthropologists think both *Orrorin tugenensis* and *Ardipithecus ramidus kadabba* were habitual bipedal walkers and thus on a direct evolutionary line to humans, others are not as impressed with the fossil evidence and are reserving judgment. Until more fossil evidence is found and analyzed, any single evolutionary scheme of hominid evolution presented here would be premature.

Australopithecines

Australopithecine is a collective term for all members of the genus *Australopithecus*. Currently, five species are recognized: *A. anamensis, A. afarensis, A. africanus, A. robustus,* and *A. boisei.* Many paleontologists accept the evolutionary scheme in which *A. anamensis,* the oldest known australopithecine, is ancestral to *A. afarensis,* who in turn is ancestral to *A. africanus* and the genus *Homo,* as well as the side branch of australopithecines represented by *A. robustus* and *A. boisei.*

The oldest known australopithecine is *Australopithecus anamensis.* Discovered at Kanapoi, a site near Lake Turkana, Kenya, by Meave Leakey of the National Museums of Kenya and her colleagues, this 4.2-million-year-old bipedal species has many features in common with its younger relative, *A. afarensis,* yet is more primitive in other characteristics, such as its teeth and skull. *A. anamensis* is estimated to have been between 1.3 and 1.5 m tall and weighed 33 to 50 kg.

A discovery, reported in 2006, of fossils of *Australopithecus anamensis* from the Middle Awash area in northeastern Ethiopia has shed new light on the transition between *Ardipithecus* and *Australopithecus.* Prior to this discovery, the origin of *Australopithecus* has been hampered by a sparse fossil record. The discovery of *Ardipithecus* in the same region of Africa and at the same time as the earliest *Australopithecus* provides strong evidence that *Ardipithecus* evolved into *Australopithecus* and links these two genera in the evolutionary lineage leading to humans.

Australopithecus afarensis (• Figure 19.8), who lived 3.9–3.0 million years ago, was fully bipedal (see Perspective) and exhibited great variability in size and weight. Members of this species ranged from just over 1 m to about 1.5 m tall and weighed between 29 and 45 kg. They had a brain size of 380–450 cubic centimeters (cc), larger than the 300–400 cc of a chimpanzee, but much smaller than that of present-day humans (1350 cc average).

The skull of *A. afarensis* retained many apelike features, including massive brow ridges and a forward-jutting jaw, but its teeth were intermediate between those of apes and humans. The heavily enameled molars were probably an adaptation to chewing fruits, seeds, and roots (• Figure 19.9).

A. afarensis was stratigraphically succeeded by *Australopithecus africanus,* who lived 3.0–2.3 million years ago (• Figure 19.10). The differences between the two species are relatively minor. They were both about the same size and weight, but *A. africanus* had a flatter face and somewhat larger brain. Furthermore, it appears the limbs

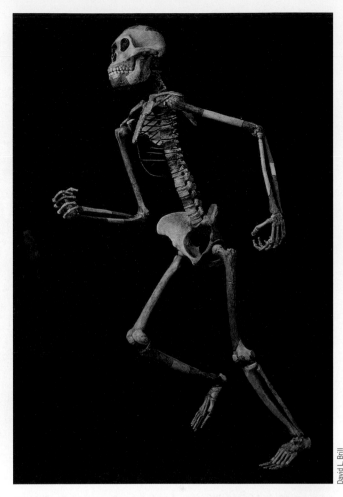

David L. Brill

• **Figure 19.8** **Skeleton of Lucy *(Australopithecus afarensis)*** A reconstruction of Lucy's skeleton by Owen Lovejoy and his students at Kent State University, Ohio. Lucy, whose fossil remains were discovered by Donald Johanson, is an approximately 3.5-million-year-old *Australopithecus afarensis* individual. This reconstruction illustrates how adaptations in Lucy's hip, leg, and foot allowed a fully bipedal means of locomotion.

of *A. africanus* may not have been as well adapted for bipedalism as those of *A. afarensis.*

Both *A. afarensis* and *A. africanus* differ markedly from the so-called robust species *A. boisei* (2.6–1.0 million years ago) and *A. robustus* (2.0–1.2 million years ago).

A. boisei was between 1.2 and 1.4 m tall and weighed between 34 and 49 kg. It had a powerful upper body, a distinctive bony crest on the top of its skull, a flat face, and the largest molars of any hominid. *A. robustus,* in contrast, was somewhat smaller (between 1.1 and 1.3 m tall) and lighter (32 to 40 kg). It had a flat face, and the crown of its skull had an elevated bony crest that provided additional area for the attachment of strong jaw muscles (• Figure 19.11). Its broad, flat molars indicated *A. robustus* was a vegetarian.

Most scientists accept the idea that the robust australopithecines form a separate lineage from the other australopithecines that went extinct 1 million years ago.

Darwen and Vally Hennings

• **Figure 19.9** **African Pliocene Landscape** Recreation of a Pliocene landscape showing members of *Australopithecus afarensis* gathering and eating various fruits and seeds.

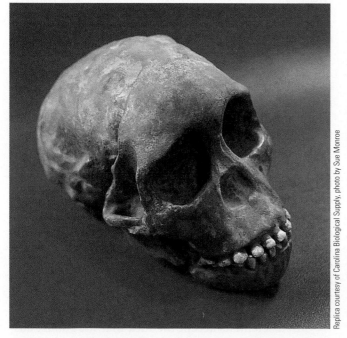

Replica courtesy of Carolina Biological Supply, photo by Sue Monroe

• **Figure 19.10** *Australopithecus africanus* A reconstruction of the skull of *Australopithecus africanus*. This skull, known as that of the Taung child, was discovered by Raymond Dart in South Africa in 1924 and marks the beginning of modern paleoanthropology.

Replica courtesy of Carolina Biological Supply, photo by Sue Monroe

• **Figure 19.11** *Australopithecus robustus* The skull of *Australopithecus robustus* had a massive jaw, powerful chewing muscles, and large, broad, flat chewing teeth apparently used for grinding up coarse plant food.

Footprints at Laetoli

During the summer of 1976, fossil footprints of such animals as giraffes, elephants, rhinoceroses, and several extinct mammals were found preserved in volcanic ash at Laetoli in northern Tanzania. Two years later, a member of Mary Leakey's archaeological team, which was searching for early hominid remains, found what appeared to be a human footprint in the same volcanic ash layer.

Dubbed the Footprint Tuff, a portion of this volcanic ash layer was excavated during the summers of 1978 and 1979, revealing two parallel trails of hominid footprints. This trackway stretched for 27 meters and consisted of 54 individual footprints (Figure 1). Radiometric dating of the ash indicates it was deposited between 3.8 and 3.4 million years ago

Figure 1 Hominid footprints preserved in volcanic ash at the Laetoli site, Tanzania. Discovered in 1978 by Mary Leakey, these footprints proved hominids were bipedal walkers at least 3.5 million years ago. The footprints of two adults and possibly those of a child are clearly visible in this photograph.

(Pliocene Epoch) during one of several eruptions of ash from the Sadiman volcano, located approximately 20 km east of Laetoli, therefore making it the oldest known hominid track-way.

In addition to the hominid footprints, there are approximately 18,000 other footprints, representing 17 families of mammals, found in the Laetoli area. Laetoli is part of the eastern branch of the Great Rift Valley of East Africa, a tectonically active area, which is separating from the rest of Africa along a divergent boundary (see Figure 3.16). During the Pliocene Epoch, Sadiman volcano erupted several times, spewing out tremendous quantities of ash that settled over the surrounding savannah. When light rains in the area moistened the ash, any animals walking over it left their footprints. As the ash dried, it hardened like cement, preserving whatever footprints had been made. Subsequent eruptions buried the footprint-bearing ash layer, thus further preserving the footprints.

What makes this find so exciting and scientifically valuable is that the footprints prove early hominids were fully bipedal and had an erect posture long before the advent of stone toolmaking or an increase in the size of the brain. Furthermore, the footprints showed that early hominids walked like modern humans by placing the full weight of the body on the ball of the heel. By examining how deep the impression in the ash is for various parts of the footprint, researchers can infer information about the soft tissue of the feet, something that can't be determined from fossil bones alone.

The question of who made the footprints and how many individuals were walking at the time the footprints were made has been debated since they were initially discovered. Most scientists think that the footprints were made by *Australopithecus afarensis,* one of the earlier known

hominids, whose fossil bones and teeth are found at Laetoli (Figure 19.9). *A. afarensis* lived from about 3.9 to 3.0 million years ago and exhibited great variability in size and weight. It is estimated that the largest of the hominids making the footprints, a male, was approximately 1.5 m tall, and the smallest hominid, either a female or a child, was approximately 1.2 m tall.

How many people made these parallel trails? There seems to be no argument that the people making these footprints were walking together, with one walking slightly behind the other. It was originally thought the footprints represented a male and female. However, closer examination of the footprints indicates there were probably three people. The larger footprints, which were probably made by a male, have features suggesting they are double prints. In this scenario, a second individual (possibly another male) followed the first one by deliberately stepping in its tracks, thus producing a double print. The smaller footprints are well defined and were probably made by a female or possibly a child. That three individuals rather than two made the trackway is now widely accepted.

To ensure that these footprints are not destroyed and will be available for future generations to study, the trackway has been reburied. The Antiquities Department of the Tanzanian government, in cooperation with the Getty Conservation Institute, completely reburied the site under five layers of sand, soil, and erosion-control matting. Some of the layers were treated with root inhibitors to prevent roots from destroying the footprints. The site is currently topped with a bed of lava boulders to provide additional protection against erosion and to mark its location. A sacred ceremony was held in 1996 in which the site was included as a place revered by the Masai people.

The Human Lineage

Homo habilis The earliest member of our own genus, ***Homo***, is *Homo habilis*, who lived 2.5 to 1.6 million years ago. Its remains were first found at Olduvai Gorge in Tanzania (see chapter opening photo) by Mary and Louis Leakey, but it is also known from Kenya, Ethiopia, and South Africa. *H. habilis* evolved from the *A. afarensis* and *A. africanus* lineage and coexisted with *A. africanus* for approximately 200,000 years (Figure 19.6). *H. habilis* had a larger brain (700 cc average) than its australopithecine ancestors, but smaller teeth (• Figure 19.12). It was between 1.2 and 1.3 m tall, had disproportionately long arms compared to modern humans, and only weighed 32 to 37 kg.

The evolutionary transition from *H. habilis* to *Homo erectus* appears to have occurred in a short period of time, between 1.8 and 1.6 million years ago. However, based on new findings published in 2007, which suggest that *H. habilis* and *H. erectus* apparently coexisted for approximately 500,000 years, some scientists think that *H. habilis* and *H. erectus* may have evolved from a common ancestor and represent separate lineages of *Homo*, rather than the traditional linear view of *H. erectus* evolving from *H. habilis*.

Homo erectus In contrast to the australopithecines and *H. habilis*, which are unknown outside Africa, *Homo erectus* was a widely distributed species, having migrated from Africa during the Pleistocene. Specimens have been found not only in Africa but also in Europe, India, China ("Peking Man"), and Indonesia ("Java Man"). *H. erectus* evolved in Africa 1.8 million years ago and by 1 million years ago was present in southeastern and eastern Asia, where it survived until about 100,000 years ago.

Although *H. erectus* developed regional variations in form, the species differed from modern humans in several ways. Its brain size of 800–1300 cc, although much larger than that of *H. habilis*, was still less than the average for *Homo sapiens* (1350 cc). The skull of *H. erectus* was thick-walled, its face was massive, it had prominent brow ridges, and its teeth were slightly larger than those of present-day humans (• Figure 19.13). *H. erectus* was comparable in size to modern humans, standing between 1.6 and 1.8 m tall and weighing between 53 and 63 kg.

The archaeological record indicates that *H. erectus* was a toolmaker. Furthermore, some sites show evidence that its members used fire and lived in caves, an advantage for those living in more northerly climates (• Figure 19.14).

Debate still surrounds the transition from *H. erectus* to our own species, *Homo sapiens*. Paleoanthropologists are split into two camps. On the one side are those who support the "out of Africa" view. According to this view, early modern humans evolved from a single woman in Africa, whose offspring then migrated from Africa, perhaps as recently as 100,000 years ago, and populated Europe and Asia, driving the earlier hominid populations to extinction.

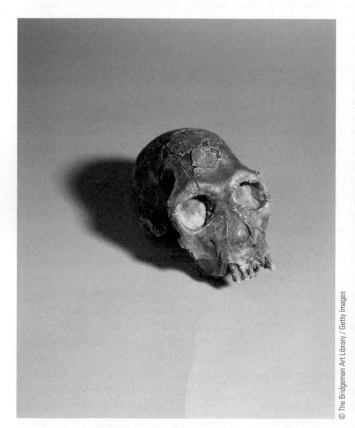

• **Figure 19.12** ***Homo habilis*** *Homo habilis* is the earliest species of the *Homo* lineage. Shown is an approximately 1.9-million-year-old skull from Kenya.

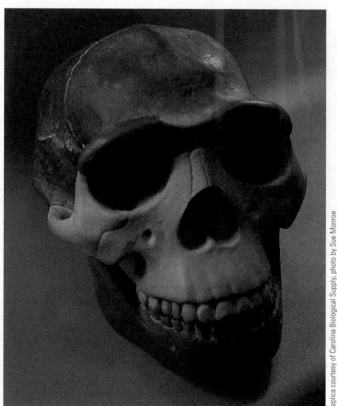

• **Figure 19.13** ***Homo erectus*** A reconstruction of the skull of *Homo erectus*, a widely distributed species whose remains have been found in Africa, Europe, India, China, and Indonesia.

• Figure 19.14 **European Pleistocene Landscape with *Homo erectus*** Recreation of a Pleistocene setting in Europe in which members of *Homo erectus* are using fire and stone tools.

The alternative explanation, the "multiregional" view, maintains that early modern humans did not have an isolated origin in Africa, but rather that they established separate populations throughout Eurasia. Occasional contact and interbreeding between these populations enabled our species to maintain its overall cohesiveness while still preserving the regional differences in people we see today. Regardless of which theory turns out to be correct, our species, *H. sapiens,* most certainly evolved from *H. erectus.*

Neanderthals Perhaps the most famous of all fossil humans are the **Neanderthals,** who inhabited Europe and the Near East from about 200,000 to 30,000 years ago, and according to the best estimates, never exceeded 15,000 individuals in western Europe. Some paleoanthropologists regard the Neanderthals as a variety or subspecies *(Homo sapiens neanderthalensis),* whereas others consider them as a separate species *(Homo neanderthalensis).* In any case, their name comes from the first specimens found in 1856 in the Neander Valley near Düsseldorf, Germany.

The most notable difference between Neanderthals and present-day humans is in the skull. Neanderthal skulls were long and low with heavy brow ridges, a projecting mouth, and a weak, receding chin (• Figure 19.15). Their brain was slightly larger on average than our own and somewhat differently shaped. The Neanderthal body was more massive and heavily muscled than ours, with a flaring rib cage, and rather short lower limbs, much like those of other cold-adapted people of today. Neanderthal males averaged between 1.6 and 1.7 m in height and about 83 kg in weight.

In 2007, scientists announced that they had isolated a pigmentation gene from a segment of Neanderthal DNA that indicated that at least some Neanderthals had red hair

• Figure 19.15 **Neanderthal Skull** The Neanderthals were characterized by prominent heavy brow ridges, a projecting mouth, and a weak chin as seen in this Neanderthal skull from Wadl Amud, Israel. In addition, the Neanderthal brain was slightly larger on average than that of modern humans.

• Figure 19.16 **Pleistocene Cave Setting with Neanderthals** Archaeological evidence indicates that Neanderthals lived in caves and participated in ritual burials, as depicted in this painting of a burial ceremony such as occurred approximately 60,000 years ago at Shanidar Cave, Iraq.

and light skin. Furthermore, the gene is different from that of modern red-haired people, suggesting that perhaps Neanderthals and present-day humans developed the trait independently, in response to similar higher northern latitude environmental pressures.

Based on specimens from more than 100 sites, we now know that Neanderthals were not much different from us, only more robust. Europe's Neanderthals were the first humans to move into truly cold climates, enduring miserably long winters and short summers as they pushed north into tundra country (• Figure 19.16).

The remains of Neanderthals are found chiefly in caves and hut-like rock shelters, which also contain a variety of specialized stone tools and weapons. Furthermore, archaeological evidence indicates that Neanderthals commonly took care of their injured and buried their dead, frequently with such grave items as tools, food, and perhaps even flowers.

As more fossil discoveries are made, and increasingly sophisticated techniques of DNA extraction and analysis are carried out, our view of Neanderthals and their society, as well as their place in human evolution is constantly changing.

Cro-Magnons
About 30,000 years ago, humans closely resembling modern Europeans moved into the region inhabited by the Neanderthals and completely replaced them. **Cro-Magnons,** the name given to the successors of the Neanderthals in France, lived from about 35,000 to 10,000 years ago. During this period the development of art and technology far exceeded anything the world had seen before.

Cro-Magnons were skilled nomadic hunters, following the herds in their seasonal migrations. They used a variety of specialized tools in their hunts, including perhaps the bow and arrow (• Figure 19.17). They sought refuge in caves and rock shelters and formed living groups of various sizes.

Cro-Magnons were also cave painters. Using paints made from manganese and iron oxides, Cro-Magnon people

• Figure 19.17 **Pleistocene Cro-Magnon Camp in Europe** Cro-Magnons were highly skilled hunters who formed living groups of various sizes.

Painting by Ronald Brown; photo courtesy of Robert Harding Picture Library

• Figure 19.18 **Cro-Magnon Cave Painting** Cro-Magnons were very skilled cave painters. Shown is a painting of a horse from the cave of Niaux, France.

painted hundreds of scenes on the ceilings and walls of caves in France and Spain, where many of them are still preserved today (• Figure 19.18).

With the appearance of Cro-Magnons, human evolution has become almost entirely cultural rather than biological. Humans have spread throughout the world by devising means to deal with a broad range of environmental conditions.

Since the evolution of the Neanderthals approximately 200,000 years ago, humans have gone from a stone culture to a technology that has allowed us to visit other planets with space probes and land astronauts on the Moon. It remains to be seen how we will use this technology in the future and whether we will continue as a species, evolve into another species, or become extinct as many groups have before us.

SUMMARY

- The primates evolved during the Paleocene. Several trends help characterize primates and differentiate them from other mammalian orders, including a change in overall skeletal structure and mode of locomotion, an increase in brain size, stereoscopic vision, and evolution of a grasping hand with opposable thumb.

- The primates are divided into two suborders: prosimians and anthropoids. The prosimians are the oldest primate lineage and include lemurs, lorises, tarsiers, and tree shrews. The anthropoids include the New and Old World monkeys, apes, and hominids, which are humans and their extinct ancestors.

- The oldest known hominid is *Sahelanthropus tchadensis*, dated at nearly 7 million years. It was followed by *Orrorin tugenensis* at 6 million years, then two subspecies of *Ardipithecus* at 5.8 and 4.4 million years, respectively. These early hominids were succeeded by the australopithecines, a fully bipedal group that evolved in Africa 4.2 million years ago. Recent discoveries indicate *Ardipithecus* evolved into *Australopithecus*. Currently, five australopithecine species are known: *Australopithecus anamensis*, *A. afarensis*, *A. africanus*, *A. robustus*, and *A. boisei*.

- The human lineage began approximately 2.5 million years ago in Africa with the evolution of *Homo habilis*,

which survived as a species until about 1.6 million years ago.

- *Homo erectus* evolved from *H. habilis* approximately 1.8 million years ago and was the first hominid to migrate out of Africa, spreading to Europe, India, China, and Indonesia, between 1.8 and 1 million years ago. The transition from *H. erectus* to *H. sapiens* is still unresolved because there is presently insufficient evidence to determine which hypothesis—the "out of Africa" or the "multiregional" hypothesis—is correct. Nonetheless, *H. erectus* used fire, made tools, and lived in caves.

- Neanderthals inhabited Europe and the Near East between 200,000 and 30,000 years ago and were not much different from present-day humans. They were, however, more robust and had differently shaped skulls. In addition, they made specialized tools and weapons, apparently took care of their injured, and buried their dead.

- The Cro-Magnons were the successors of the Neanderthals and lived from about 35,000 to 10,000 years ago. They were highly skilled nomadic hunters, formed living groups of various sizes, and were also skilled cave painters.

- Modern humans succeeded the Cro-Magnons about 10,000 years ago and have spread throughout the world, as well as having set foot on the Moon.

IMPORTANT TERMS

anthropoid, p. 403
australopithecine, p. 407
Cro-Magnon, p. 412

hominid, p. 404
hominoid, p. 404
Homo, p. 410

Neanderthal, p. 411
primate, p. 401
prosimian, p. 402

REVIEW QUESTIONS

1. The oldest currently known hominid is
   a. _____ *Sahelanthropus tchadensis*; b. _____ *Orrorin tugenensis*; c. _____ *Ardipithecus ramidus ramidus*; d. _____ *Australopithecus anamensis*; e. _____ *Homo erectus.*

2. Which extinct lineage of humans were skilled hunters and cave painters?
   a. _____ Acheuleans; b. _____ Archaic; c. _____ Neanderthals; d. _____ Cro-Magnons; e. _____ None of the previous answers.

3. Which of the following features distinguish hominids from other hominoids?
   a. _____ A large and internally reorganized brain; b. _____ A reduced face and reduced canine teeth; c. _____ Bipedalism; d. _____ Use of sophisticated tools; e. _____ All of the previous answers.

4. The human lineage began with the evolution of which species?
   a. _____ *Orrorin tugenensis*; b. _____ *Ardipithecus ramidus*; c. _____ *Sahelanthropus tchadensis*; d. _____ *Homo habilis*; e. _____ *Australopithecus boisei.*

5. The first hominids to migrate out of Africa and from which we evolved were
   a. _____ *Australopithecus robustus*; b. _____ *Homo erectus*; c. _____ *Homo sapiens*; d. _____ *Ardipithecus ramidus ramidus*; e. _____ *Homo habilis.*

6. To which of the following species do Java Man and Peking Man belong?
   a. _____ *Homo sapiens*; b. _____ *Australopithecus robustus*; c. _____ *Homo erectus*; d. _____ *Australopithecus boisei*; e. _____ *Homo habilis.*

7. Which is the oldest primate lineage?
   a. _____ Anthropoids; b. _____ Prosimians; c. _____ Insectivores; d. _____ Omnivores; e. _____ Hominids.

8. Which of the following evolutionary trends characterize primates?
   a. _____ Change in overall skeletal structure; b. _____ Grasping hand with opposable thumb; c. _____ Increase in brain size; d. _____ Stereoscopic vision; e. _____ All of the previous answers.

9. The oldest known australopithecine is *Australopithecus* _____.
   a. _____ *robustus*; b. _____ *afarensis*; c. _____ *anamensis*; d. _____ *boisei*; e. _____ *africanus.*

10. Which of the following is a hominid?
    a. _____ Chimpanzee; b. _____ Gibbon; c. _____ Prosimian; d. _____ Australopithecine; e. _____ Gorilla.

11. When did primates evolve?
    a. _____ Paleocene; b. _____ Eocene; c. _____ Oligocene; d. _____ Miocene; e. _____ Pliocene.

12. According to archaeological evidence, which were the first hominids to use fire?
    a. _____ *Australopithecus robustus*; b. _____ *Homo sapiens*; c. _____ *Homo erectus*; d. _____ *Homo habilis*; e. _____ *Australopithecus boisei.*

13. Discuss the importance of the discovery that *Ardipithecus* evolved into *Australopithecus* in terms of hominid evolution.

14. What major evolutionary trends characterize the primates and set them apart from the other orders of mammals?

15. What are the main differences between the Neanderthals and Cro-Magnons?

16. Discuss the evolutionary history of the genus *Homo.*

17. Discuss the evolutionary history of the anthropoids.

18. Discuss the evolutionary history of hominids.

19. Discuss the differences between the prosimians and anthropoids.

20. Discuss the merits of the "out of Africa" and "multiregional" views concerning the transition between *Homo erectus* and *Homo sapiens.*

APPLY YOUR KNOWLEDGE

1. Based on what you now know about human evolution, as well as what you've read and witnessed about the rapid technological advances in science and their impact on society and the environment, what factors do you think will influence the future course of human evolution? Do you think it is possible that we can control the direction that evolution takes?

2. Because of the recent controversy concerning the teaching of evolution in the public schools, your local school board has asked you to make a 30-minute presentation on the evolutionary history of humans and how the fossil record of humans and their ancestors is evidence that evolution is a valid scientific theory. With only 30 minutes to make your case, what evidence in the fossil record would you emphasize, and how would you go about convincing the school board that humans have indeed evolved from earlier hominids?

Epilogue

Introduction

Throughout this book, we have emphasized that Earth is a complex, dynamic planet that has changed continuously since its origin some 4.6 billion years ago. These changes, and the present-day features we observe, are the result of interactions between the various interrelated internal and external Earth systems, subsystems, and cycles. Furthermore, these interactions have also influenced the evolution of the biosphere.

The rock cycle (see Figure 2.7), with its recycling of Earth materials to form the three major rock groups, illustrates the interrelationships between Earth's internal and external processes. The hydrologic cycle explains the continuous recycling of water from the oceans, to the atmosphere, to the land, and eventually back to the oceans again. Changes within this cycle can have profound effects on Earth's topography as well as its biota. For example, a rise in global temperature will cause the ice caps to melt, contributing to rising sea level, which will greatly affect coastal areas where many of the world's large population centers are presently located (see Chapter 17). We have seen the effect of changing sea level on continents in the past, which resulted in large-scale transgressions and regressions. Some of these were caused by growing and shrinking continental ice caps when landmasses moved over the South Pole as a result of plate movements (see Chapter 11).

On a larger scale, the movement of plates has had a profound effect on the formation of landscapes, the distribution of mineral resources, and atmospheric and oceanic circulation patterns, as well as the evolution and diversification of life.

The launching in 1957 of *Sputnik 1*, the world's first artificial satellite, ushered in a new global consciousness in terms of how we view Earth and our place in the global ecosystem. Satellites have provided us with the ability to view not only the beauty of our planet, but also the fragility of Earth's biosphere and the role humans play in shaping and modifying the environment. The pollution of the atmosphere, oceans, and many of our lakes and streams; the denudation of huge areas of tropical forests; the scars from strip mining; the depletion of the ozone layer—all are visible in the satellite images beamed back from space and attest to the impact humans have had on the ecosystem.

Accordingly, we must understand that changes we make in the global ecosystem can have wide-ranging effects of which we may not even be aware. For this reason, an understanding of geology, and science in general, is of paramount importance so that disruption to the ecosystem is minimal. On the other hand, we must also remember that humans are part of the global ecosystem, and, like all other life-forms, our presence alone affects the ecosystem. We must therefore act in a responsible manner, based on sound scientific knowledge, so future generations will inherit a habitable environment.

One objective of this book, and much of your secondary education, is to develop your skills as a critical thinker. As opposed to simple disagreement, critical thinking involves evaluating the supporting evidence for a particular point of view. Although your exposure to geology is probably limited, you do have the fundamental knowledge needed to appraise why geologists accept plate tectonic theory, why they think that Earth is 4.6 billion years old, and why scientists are convinced that the theory of evolution is well supported by evidence. In addition, your abilities as a critical thinker will probably help you more effectively evaluate the arguments about global warming, ozone depletion, groundwater contamination, and many other environmental issues.

When such environmental issues as acid rain, the depletion of the ozone layer, and the greenhouse effect and global warming are discussed and debated, it is important to remember that they are not isolated topics but are part of a larger system that involves the entire Earth. Furthermore, it is important to remember that Earth goes through cycles of much longer duration than the human perspective of time.

Acid Rain

One result of industrialization is atmospheric pollution, which causes smog, possible disruption of the ozone layer, global warming, and acid rain. Acidity, a measure of hydrogen ion concentration, is measured on the pH scale (• Figure E.1a). A pH value of 7 is neutral, whereas acidic conditions correspond to values less than 7, and values greater than 7 denote alkaline, or basic, conditions. Normal rain has a pH value of slightly less than 6.0. Some areas experience acid snow and even acid fog with a pH as low as 1.7.

Several natural processes, including soil bacteria metabolism and volcanism, release gases into the atmosphere that contribute to acid rain. Human activities also produce added atmospheric stress, especially burning fossil fuels that release carbon dioxide and nitrogen oxide from internal combustion engines. Both of these gases add to acid rain, but the greatest culprit is sulfur dioxide released mostly by burning coal that contains sulfur that oxidizes to form sulfur dioxide (SO_2). As sulfur dioxide rises into the atmosphere, it reacts with oxygen and water droplets to form sulfuric acid (H_2SO_4), the main component of acid rain.

Robert Angus Smith first recognized acid rain in England in 1872, but not until 1961 did it become an environmental concern when scientists realized that acid rain is corrosive and irritating, kills vegetation, and has a detrimental effect on surface waters. Since then, the effects of acid rain are apparent in Europe (especially in Eastern Europe) and the eastern part of North America, where the problem has been getting worse for the last three decades (Figure E.1b).

The areas affected by acid rain invariably lie downwind from plants that emit sulfur gases, but the effects of acid rain in these areas may be modified by local conditions. For instance, if the area is underlain by limestone or alkaline soils, acid rain tends to be neutralized; however, granite has little or no modifying effect. Small lakes lose their ability to neutralize acid rain and become more and more acidic until some types of organisms disappear, and in some cases, all life-forms eventually die. However, just as discussed in Chapter 7, some organisms are able to adapt to new, and seemingly, hostile conditions, such as certain plants that have adapted to contaminated soils caused by coal mining operations.

Acid rain also causes increased chemical weathering of limestone and marble and, to a lesser degree, sandstone. The effects are especially evident on buildings, monuments, and tombstones, as in Gettysburg National Military Park in Pennsylvania.

The devastation caused by sulfur gases on vegetation near coal-burning power plants is apparent, and many forests in the eastern United States show signs of stress that cannot be attributed to other causes.

• Figure E.1 **Acid Rain**

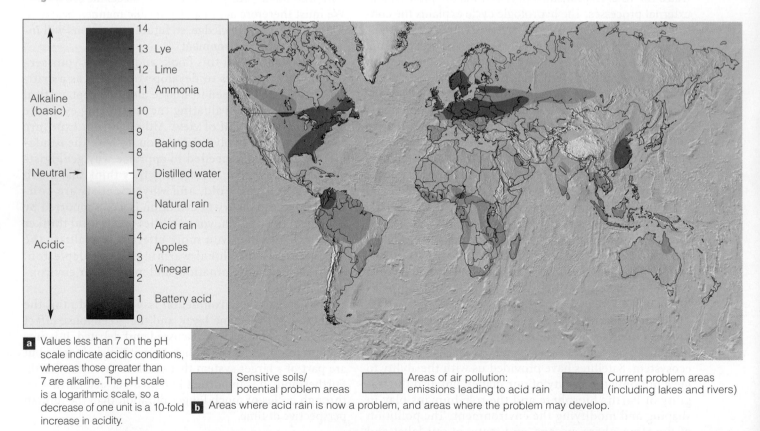

a Values less than 7 on the pH scale indicate acidic conditions, whereas those greater than 7 are alkaline. The pH scale is a logarithmic scale, so a decrease of one unit is a 10-fold increase in acidity.

Sensitive soils/ potential problem areas

Areas of air pollution: emissions leading to acid rain

Current problem areas (including lakes and rivers)

b Areas where acid rain is now a problem, and areas where the problem may develop.

Millions of tons of sulfur dioxide are released yearly into the atmosphere in the United States. Power plants built before 1975 have no emission controls, but the problems they pose must be addressed if emissions are to be reduced to an acceptable level. There are various methods that can be implemented to significantly reduce sulfur dioxide emissions by both older power plants, and more recently constructed ones. However, these methods are costly, and in some cases, it is simply too expensive to upgrade older power plants to reduce their emissions.

Acid rain, like global warming, is a worldwide problem that knows no national boundaries. Wind may blow pollutants from the source in one country to another where the effects are felt. For instance, much of the acid rain in eastern Canada actually comes from sources in the United States.

Ozone Depletion

Earth supports life because of its distance from the Sun, and the fact that it has abundant liquid water and an oxygen-rich atmosphere. An ozone layer (O_3) in the stratosphere (10 to 48 km above the surface) protects Earth because it blocks out most of the harmful ultraviolet radiation that bombards our planet (see Chapter 8). During the early 1980s, scientists discovered an ozone hole over Antarctica that has continued to grow. In fact,

depletion of the ozone layer is now also recognized over the Arctic region and elsewhere. Any ozone depletion is viewed with alarm because it allows more dangerous radiation to reach the surface, increasing the risk of skin cancer, among other effects (see Chapter 13 Perspective).

This discovery unleashed a public debate about the primary cause of ozone depletion, and how best to combat the problem. Scientists proposed that one cause of ozone depletion is chlorofluorocarbons (CFCs), which are used in several consumer products; for instance, in aerosol cans. According to this theory, CFCs rise into the upper atmosphere where reactions with ultraviolet radiation liberate chlorine, which in turn reacts with and depletes ozone (• Figure E.2). As a result of this view, an international agreement called the Montreal Protocol was reached in 1983, limiting the production of CFCs, along with other ozone-depleting substances.

However, during the 1990s, this view was challenged. Those opposed to the idea that CFCs were the cause of ozone depletion proposed an alternative idea that natural causes rather than commercial products (CFCs) were the main culprits. They pointed out that volcanoes release copious quantities of hydrogen chloride (HCl) gas that rise into the stratosphere and which could be responsible for ozone depletion. Furthermore, they claimed that because CFCs are heavier than air, they would not rise into the stratosphere.

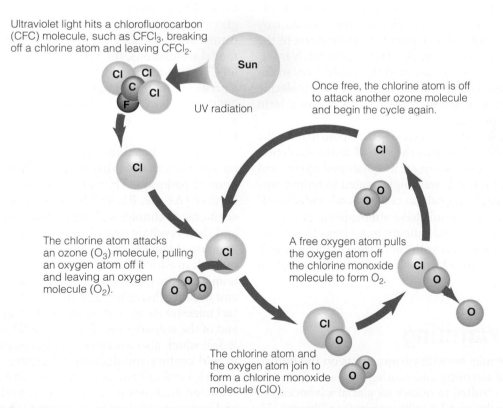

Ultraviolet light hits a chlorofluorocarbon (CFC) molecule, such as $CFCl_3$, breaking off a chlorine atom and leaving $CFCl_2$.

Sun

UV radiation

Once free, the chlorine atom is off to attack another ozone molecule and begin the cycle again.

The chlorine atom attacks an ozone (O_3) molecule, pulling an oxygen atom off it and leaving an oxygen molecule (O_2).

A free oxygen atom pulls the oxygen atom off the chlorine monoxide molecule to form O_2.

The chlorine atom and the oxygen atom join to form a chlorine monoxide molecule (ClO).

• **Figure E.2 Ozone Depletion** Ozone is destroyed by chlorofluorocarbons (CFCs). Chlorine atoms are continually regenerated, so one chlorine atom can destroy many ozone molecules.

It is true that volcanoes release HCl gas, as well as several other gases, some of which are quite dangerous. Nevertheless, most eruptions are too weak to inject gases of any kind high into the stratosphere. Even when it is released, HCl gas from volcanoes is very soluble and quickly removed from the atmosphere by rain and even by steam (water vapor) from the same eruption that released the HCl gas in the first place. Measurements of chlorine concentrations in the stratosphere show that only temporary increases occur following huge eruptions. For example, the largest volcanic outburst since 1912, the eruption of Mount Pinatubo in 1991, caused little increase in upper atmospheric chlorine. The impact of volcanic eruptions is certainly not enough to cause the average rate of ozone depletion taking place each year.

Conversely, as mentioned in the Chapter 13 Perspective, a 2007 paper, using a two-dimensional, atmospheric chemistry-transport model, discusses the possible role that the Siberian Traps flood basalt eruptions at the end of the Permian might have played in atmospheric ozone depletion. Such an eruption was far greater than that of Mount Pinatubo.

Although it is true that CFCs are heavier than air, this does not mean that they cannot rise into the stratosphere. Earth's surface heats differentially, meaning that more heat may be absorbed in one area than in an adjacent one. The heated air above a warmer area becomes less dense, rises by convection, and carries with it CFCs and other substances that are actually denser than air. Once in the stratosphere, ultraviolet radiation, which is usually absorbed by ozone, breaks up CFC molecules and releases chlorine that reacts with ozone. Indeed, a single chlorine atom can destroy 100,000 ozone molecules (Figure E.2). In contrast to the HCl gas produced by volcanoes, CFCs are absolutely insoluble: it is the fact that they are inert that made them so desirable for commercial products. Because a CFC molecule can last for decades, any increase in CFCs is a long-term threat to the ozone layer.

Another indication that CFCs are responsible for the Antarctic ozone hole is that the rate of ozone depletion decreased since 1989 when an international agreement (the Montreal Protocol) was implemented to reduce suspected ozone-depleting substances. A sound understanding of the science behind these atmospheric processes helped world leaders act quickly to address this issue. Now, scientists hope that with continued compliance with the Protocol, the ozone layer will recover by the middle of this century.

Global Warming

Although they may have disastrous effects on the human species, global warming and cooling are part of a larger cycle that has resulted in numerous glacial advances and retreats during the past 1.8 million years (see Chapter 17). In fact, geologists can make important contributions to the debate on global warming because of their geologic perspective (see Chapter 4 Perspective). Long-term trends can be studied by analyzing deep-sea sediments, ice cores, changes in sea level during the geologic past, and the distribution of plants and animals through time (see Chapter 4). As we have seen throughout this book, such studies have been done, and the results and synthesis of that information can be used to make intelligent decisions about how humans can better manage the environment and the effect we are having in altering the environment.

A good example of environmental imbalance is global warming caused by the greenhouse effect. Carbon dioxide is produced as a by-product of respiration and the burning of organic material. As such, it is a component of the global ecosystem and is constantly being recycled as part of the carbon cycle. The concern in recent years over the increase in atmospheric carbon dioxide has to do with its role in the greenhouse effect. The recycling of carbon dioxide between the crust and atmosphere is an important climatic regulator because carbon dioxide, as well as other gases such as methane, nitrous oxide, chlorofluorocarbons, and water vapor, allow sunlight to pass through them but trap the heat reflected back from Earth's surface. Heat is thus retained, causing the temperature of Earth's surface and, more importantly, the atmosphere, to increase, producing the greenhouse effect.

Because of the increase in human-produced greenhouse gases during the last 200 years, many scientists are concerned that a global warming trend has already begun and will result in severe global climatic shifts. Presently, most climate researchers use a range of scenarios for greenhouse gas emissions when making predictions for future warming rates. Based on state-of-the-art climate model simulations, the fourth Intergovernmental Panel on Climate Change (IPCC) report issued in 2007 showed a predicted increase in global average temperature from 2000 to 2100 of 1 to 3°C under the best conditions, to a 2.5 to 6.5°C rise under "business-as-usual" conditions (• Figure E.3).

These predicted increases in temperatures are based on various scenarios that explore different global development pathways. They are grouped into four scenario families (A1, A2, B1, and B2) that cover a wide range of economic, technological, and demographic possibilities and their resultant greenhouse gas emissions. The A1FI scenario (Figure E.3a) is based on a "business-as-usual" outlook in which the world experiences very rapid economic growth, global population peaks in the mid-century and declines thereafter, and the world continues a fossil-fuel intensive energy consumption strategy. At the opposite end of the scenario spectrum is the B1 scenario (Figure E.3a), which also assumes a global population peaking in mid-century and declining thereafter, a shift towards a service and information economy, and an emphasis on reducing materials usage and the introduction of clean and resource-efficient technologies. The A1B combination (Figure E.3a) is the "middle-of-the-road" scenario in which global population trends follow that for the A1FI

Estimated Temperature Rises From 2000 to 2100 and Projected Surface Temperature Changes for Early and Late 21st Century Source: Reprinted from Figure 3.2 in *Climate Change 2007: Synthesis Report*. Also known as the IPCC Fourth Assessment Report (AR4). Published by the Intergovernmental Panel on Climate Change. Adopted at IPCC Plenary XXVII, Valencia, Spain, November 12–17, 2007.

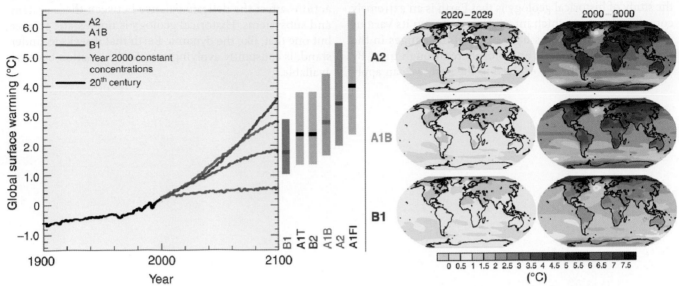

a Solid lines are multi-model global averages of surface warming (relative to 1980–1999) for the 2000 IPCC Special Report on Emissions Scenarios (SRES) A2, A1B and B1, shown as continuations of the 20th century simulations. The orange line is for the experiment where concentrations were held constant at year 2000 values. The bars in the middle of the figure indicate the best estimate (solid line within each bar) and the likely range assessed for the six SRES marker scenarios at 2090–2099 relative to 1980–1999. The assessment of the best estimate and likely ranges in the bars includes the Atmosphere-Ocean General Circulation Models (AOGCMs) in the left part of the figure, as well as results from a hierarchy of independent models and observational constraints.

b Projected surface temperature changes for the early and late 21st century relative to the period 1980–1999. The panels show the multi-AOGCM average projections for the A2 (top), A1B (middle) and B1 (bottom) SRES scenarios averaged over decades 2020–2029 (left) and 2090–2099 (right).

and B1 scenarios, but there is a balance across all energy sources and technologies used.

Regardless of which scenario is followed, the global temperature change will be uneven, with the greatest warming occurring in the higher latitudes of the northern hemisphere. Furthermore, there will be greater warming for the continents than for the oceanic regions because land areas tend to heat and cool more rapidly than oceans. Thus, there will be greater warming in the more land-dominated northern hemisphere than the ocean-dominated southern hemisphere (Figure E.3b). As a consequence of this warming, rainfall patterns will shift dramatically. This will have a major effect on the largest grain-producing areas of the world, such as the American Midwest. Drier and hotter conditions will intensify the severity and frequency of droughts, leading to more crop failures and higher food prices. With such shifts in climate, Earth may experience an increase in the expansion of deserts, which will remove valuable crop and grazing lands.

We cannot leave the subject of global warming without pointing out that some scientists are still not convinced that the global warming trend is the direct result of increased human activity related to industrialization. They point out that whereas there has been an increase in

greenhouse gases, there is still uncertainty about their rate of generation and rate of removal and about whether the 0.5°C rise in global temperature during the past century is the result of normal climatic variations through time or the result of human activity. Furthermore, they point out that even if there is a general global warming during the next 100 years, it is not certain that the dire predictions made by proponents of global warming will come true. Earth, as we know, is a remarkably complex system, with many feedback mechanisms and interconnections throughout its various subsystems. It is very difficult to predict all of the consequences that global warming would have for atmospheric and oceanic circulation patterns.

Since writing the above paragraph for the fifth edition of this book, it is interesting to note that many of the legitimate concerns raised about the role of human activity in causing global warming have largely been addressed, and the arguments countered. In fact, the evidence now strongly points to humans as the major driving force in global warming. We do not have the time to list all of the arguments and counterarguments here, and we refer the reader to the Fourth Assessment Report of the Intergovernmental Panel on Climate Change issued in 2007 for the full Synthesis Report.

A Final Word

In conclusion, the most important lesson to be learned from the study of historical geology is that Earth is an extremely complex planet in which interactions between its various systems and subsystems have resulted in changes in the atmosphere, lithosphere, and biosphere through time. By studying how Earth has evolved in the past, we can apply the lessons learned from this study to better understanding how the different Earth systems and subsystems work and interact with each other, and more importantly, how our actions affect the delicate balance between these systems and subsystems. Historical geology is not a static science, but one that, like the dynamic Earth that it seeks to understand, is constantly evolving as new information becomes available.

Appendix A
English-Metric Conversion Chart

English Unit	Conversion Factor	Metric Unit	Conversion Factor	English Unit
Length				
Inches (in)	2.54	Centimeters (cm)	0.39	Inches (in)
Feet (ft)	0.305	Meters (m)	3.28	Feet (ft)
Miles (mi)	1.61	Kilometers (km)	0.62	Miles (mi)
Area				
Square inches (in^2)	6.45	Square centimeters (cm^2)	0.16	Square inches (in^2)
Square feet (ft^2)	0.093	Square meters (m^2)	10.8	Square feet (ft^2)
Square miles (mi^2)	2.59	Square kilometers (km^2)	0.39	Square miles (mi^2)
Volume				
Cubic inches (in^3)	16.4	Cubic centimeters (cm^3)	0.061	Cubic inches (in^3)
Cubic feet (ft^3)	0.028	Cubic meters (m^3)	35.3	Cubic feet (ft^3)
Cubic miles (mi^3)	4.17	Cubic kilometers (km^3)	0.24	Cubic miles (mi^3)
Weight				
Ounces (oz)	28.3	Grams (g)	0.035	Ounces (oz)
Pounds (lb)	0.45	Kilograms (kg)	2.20	Pounds (lb)
Short tons (st)	0.91	Metric tons (t)	1.10	Short tons (st)
Temperature				
Degrees Fahrenheit (°F)	$-32° \times 0.56$	Degrees Celsius (Celsius) (°C)	$\times 1.80 + 32°$	Degrees Fahrenheit (°F)

Examples: 10 inches = 25.4 centimeters; 10 centimeters = 3.9 inches
100 square feet = 9.3 square meters; 100 square meters = 1080 square feet
50°F = 10.08°C; 50°C = 122°F

Appendix B
Classification of Organisms

Any classification is an attempt to make order out of disorder and to group similar items into the same categories. All classifications are schemes that attempt to relate items to each other based on current knowledge and therefore are progress reports on the current state of knowledge for the items classified. Because classifications are to some extent subjective, classification of organisms may vary among different texts.

The classification that follows is based on the five-kingdom system of classification of Margulis and Schwartz.* We have not attempted to include all known life forms, but rather major categories of both living and fossil groups.

Kingdom Monera

Prokaryotes
> Phylum Archaebacteria—(Archean–Recent)
> Phylum Cyanobacteria—Blue-green algae or blue-green bacteria (Archean–Recent)

Kingdom Protoctista

Solitary or colonial unicellular eukaryotes
> Phylum Acritarcha—Organic-walled unicellular algae of unknown affinity (Proterozoic–Recent)
> Phylum Bacillariophyta—Diatoms (Jurassic–Recent)
> Phylum Charophyta—Stoneworts (Silurian–Recent)
> Phylum Chlorophyta—Green algae (Proterozoic–Recent)
> Phylum Chrysophyta—Golden-brown algae, silico-flagellates and coccolithophorids (Jurassic–Recent)
> Phylum Euglenophyta—Euglenids (Cretaceous–Recent)
> Phylum Myxomycophyta—Slime molds (Proterozoic–Recent)
> Phylum Phaeophyta—Brown algae, multicellular, kelp, seaweed (Proterozoic–Recent)

> Phylum Protozoa—Unicellular heterotrophs (Cambrian–Recent)
>> Class Sarcodina—Forms with pseudopodia for locomotion (Cambrian–Recent)
>>> Order Foraminifera—Benthonic and planktonic sarcodinids most commonly with calcareous tests (Cambrian–Recent)
>>> Order Radiolaria—Planktonic sarcodinids with siliceous tests (Cambrian–Recent)
> Phylum Pyrrophyta—Dinoflagellates (Silurian?, Permian–Recent)
> Phylum Rhodophyta—Red algae (Proterozoic–Recent)
> Phylum Xanthophyta—Yellow-green algae (Miocene–Recent)

Kingdom Fungi

> Phylum Zygomycota—Fungi that lack cross walls (Proterozoic–Recent)
> Phylum Basidiomycota—Mushrooms (Pennsylvanian–Recent)
> Phylum Ascomycota—Yeasts, bread molds, morels (Mississippian–Recent)

Kingdom Plantae

Photosynthetic eukaryotes
> Division* Bryophyta—Liverworts, mosses, hornworts (Devonian–Recent)
> Division Psilophyta—Small, primitive vascular plants with no true roots or leaves (Silurian–Recent)
> Division Lycopodophyta—Club mosses, simple vascular systems, true roots and small leaves, including scale trees of Paleozoic Era (lycopsids) (Devonian–Recent)
> Division Sphenophyta—Horsetails (scouring rushes), and sphenopsids such as the Carboniferous *Calamites* (Devonian–Recent)
> Division Pteridophyta—Ferns (Devonian–Recent)

---

*Margulis, L., and K.V.S. Schwartz, 1982. *Five Kingdoms*. New York: W.H. Freeman and Co.

*In botany, division is the equivalent to phylum.

Division Pteridospermophyta—Seed ferns (Devonian–Jurassic)

Division Coniferophyta—Conifers or cone-bearing gymnosperms (Carboniferous–Recent)

Division Cycadophyta—Cycads (Triassic–Recent)

Division Ginkgophyta—Maidenhair tree (Triassic–Recent)

Division Angiospermophyta—Flowering plants and trees (Cretaceous–Recent)

Kingdom Animalia

Nonphotosynthetic multicellular eukaryotes (Proterozoic–Recent)

Phylum Porifera—Sponges (Cambrian–Recent)

Order Stromatoporoida—Extinct group of reef-building organisms (Cambrian–Oligocene)

Phylum Archaeocyatha—Extinct spongelike organisms (Cambrian)

Phylum Cnidaria—Hydrozoans, jellyfish, sea anemones, corals (Cambrian–Recent)

Class Hydrozoa—Hydrozoans (Cambrian–Recent)

Class Scyphozoa—Jellyfish (Proterozoic–Recent)

Class Anthozoa—Sea anemones and corals (Cambrian–Recent)

Order Tabulata—Exclusively colonial corals with reduced to nonexistent septa (Ordovician–Permian)

Order Rugosa—Solitary and colonial corals with fourfold symmetry (Ordovician–Permian)

Order Scleractinia—Solitary and colonial corals with sixfold symmetry. Most colonial forms have symbiotic dinoflagellates in their tissues. Important reef builders today (Triassic–Recent)

Phylum Bryozoa—Exclusively colonial suspension feeding marine animals that are useful for correlation and ecological interpretations (Ordovician–Recent)

Phylum Brachiopoda—Marine suspension feeding animals with two unequal sized valves. Each valve is bilaterally symmetrical (Cambrian–Recent)

Class Inarticulata—Primitive chitinophosphatic or calcareous brachiopods that lack a hinging structure. They open and close their valves by means of complex muscles (Cambrian–Recent)

Class Articulata—Advanced brachiopods with calcareous valves that are hinged (Cambrian–Recent)

Phylum Mollusca—A highly diverse group of invertebrates (Cambrian–Recent)

Class Monoplacophora—Segmented, bilaterally symmetrical crawling animals with cap-shaped shells (Cambrian–Recent)

Class Amphineura—Chitons. Marine crawling forms, typically with 8 separate calcareous plates (Cambrian–Recent)

Class Scaphopoda—Curved, tusk-shaped shells that are open at both ends (Ordovician–Recent)

Class Gastropoda—Single shelled generally coiled crawling forms. Found in marine, brackish, and fresh water as well as terrestrial environments (Cambrian–Recent)

Class Bivalvia—Mollusks with two valves that are mirror images of each other. Typically known as clams and oysters (Cambrian–Recent)

Class Cephalopoda—Highly evolved swimming animals. Includes shelled sutured forms as well as non-shelled types such as octopus and squid (Cambrian–Recent)

Order Nautiloidea—Forms in which the chamber partitions are connected to the wall along simple, slightly curved lines (Cambrian–Recent)

Order Ammonoidea—Forms in which the chamber partitions are connected to the wall along wavy lines (Devonian–Cretaceous)

Order Coleoidea—Forms in which the shell is reduced or lacking. Includes octopus, squid, and the extinct belemnoids (Mississippian–Recent)

Phylum Annelida—Segmented worms. Responsible for many of the Phanerozoic burrows and trail trace fossils (Proterozoic–Recent)

Phylum Arthropoda—The largest invertebrate group comprising about 80% of all known animals. Characterized by a segmented body and jointed appendages (Cambrian–Recent)

Class Trilobita—Earliest appearing arthropod class. Trilobites had a head, body, and tail and were bilaterally symmetrical (Cambrian–Permian)

Class Crustacea—Diverse class characterized by a fused head and body and an abdomen. Included are barnacles, copepods, crabs, ostracodes, and shrimp (Cambrian–Recent)

Class Insecta—Most diverse and common of all living invertebrates, but rare as fossils (Silurian–Recent)

Class Merostomata—Characterized by four pairs of appendages and a more flexible exoskeleton than crustaceans. Includes the extinct eurypterids, horseshoe crabs, scorpions, and spiders (Cambrian–Recent)

Phylum Echinodermata—Exclusively marine animals with fivefold radial symmetry and a unique water vascular system (Cambrian–Recent)

Subphylum Crinozoa—Forms attached by a calcareous jointed stem (Cambrian–Recent)

Class Crinoidea—Most important class of Paleozoic echinoderms. Suspension feeding forms that are either free-living or attached to sea floor by a stem (Cambrian–Recent)

Class Blastoidea—Small class of Paleozoic suspension feeding sessile forms with short stems (Ordovician–Permian)

Class Cystoidea—Globular to pear-shaped suspension feeding benthonic sessile forms with very short stems (Ordovician–Devonian)

Subphylum Homalozoa—A small group with flattened, asymmetrical bodies with no stems. Also called carpoids (Cambrian–Devonian)

Subphylum Echinozoa—Globose, predominantly benthonic mobile echinoderms (Cambrian–Recent)

Class Helioplacophora—Benthonic, mobile forms, shaped like a top with plates arranged in a helical spiral (Early Cambrian)

Class Edrioasteroidea—Benthonic, sessile or mobile, discoidal, globular- or cylindrical-shaped forms with five straight or curved feeding areas shaped like a starfish (Cambrian–Pennsylvanian)

Class Holothuroidea—Sea cucumbers. Sediment feeders having calcareous spicules embedded in a tough skin (Ordovician–Recent)

Class Echinoidea—Largest group of echinoderms. Globe- or disk-shaped with movable spines. Predominantly grazers or sediment feeders. Epifaunal and infaunal (Ordovician–Recent)

Subphylum Asterozoa—Stemless, benthonic mobile forms (Ordovician–Recent)

Class Asteroidea—Starfish. Arms merge into body (Ordovician–Recent)

Class Ophiuroidea—Brittle star. Distinct central body (Ordovician–Recent)

Phylum Hemichordata—Characterized by a notochord sometime during their life history. Modern acorn worms and extinct graptolites (Cambrian–Recent)

Class Graptolithina—Colonial marine hemichordates having a chitinous exoskeleton. Predominantly planktonic (Cambrian–Mississippian)

Phylum Chordata—Animals with notochord, hollow dorsal nerve cord, and gill slits during at least part of their life cycle (Cambrian–Recent)

Subphylum Urochordata—Sea squirts, tunicates. Larval forms with notochord in tail region.

Subphylum Cephalochordata—Small marine animals with notochords and small fish-like bodies (Cambrian–Recent)

Subphylum Vertebrata—Animals with a backbone of vertebrae (Cambrian–Recent)

Class Agnatha—Jawless fish. Includes the living lampreys and hagfish as well as extinct armored ostracoderms (Cambrian–Recent)

Class Acanthodii—Primitive jawed fish with numerous spiny fins (Silurian–Permian)

Class Placodermii—Primitive armored jawed fish (Silurian–Permian)

Class Chondrichthyes—Cartilaginous fish such as sharks and rays (Devonian–Recent)

Class Osteichthyes—Bony fish (Devonian–Recent)

Subclass Actinopterygii—Ray-finned fish (Devonian–Recent)

Subclass Sarcopterygii—Lobe-finned, air-breathing fish (Devonian–Recent)

Order Coelacanthimorpha—Lobe finned fish *Latimeria* (Devonian-Recent)

Order Crossoptergii—Lobe-finned fish that were ancestral to amphibians (Devonian–Permian)

Order Dipnoi—Lungfish (Devonian–Recent)

Class Amphibia—Amphibians. The first terrestrial vertebrates (Devonian–Recent)

Subclass Labyrinthodontia—Earliest amphibians. Solid skulls and complex tooth pattern (Devonian–Triassic)

Subclass Salientia—Frogs, toads, and their relatives (Triassic–Recent)

Subclass Condata—Salamanders and their relatives (Triassic–Recent)

Class Reptilia—Reptiles. A large and varied vertebrate group characterized by having scales and laying an amniote egg (Mississippian–Recent)

Subclass Anapsida—Reptiles whose skull has a solid roof with no openings (Mississippian–Recent)

Order Cotylosauria—One of the earliest reptile groups (Pennsylvanian–Triassic)

Order Chelonia—Turtles (Triassic–Recent)

Subclass Euryapsida—Reptiles with one opening high on the side of the skull behind the eye. Mostly marine (Permian–Cretaceous)

Order Protorosauria—Land living ancestral euryapsids (Permian–Cretaceous)

Order Placodontia—Placodonts. Bulky, paddle-limbed marine reptiles with rounded teeth for crushing mollusks (Triassic)

Order Ichthyosauria—Ichthyosaurs. Dolphin-shaped swimming reptiles (Triassic–Cretaceous)

Subclass Diapsida—Most diverse reptile class. Characterized by two openings in the skull behind the eye. Includes lizards, snakes, crocodiles, thecodonts, dinosaurs, and pterosaurs (Permian–Recent)

Infraclass Lepidosauria—Primitive diapsids including snakes, lizards, and the mosasaurs, a large Cretaceous marine reptile group (Permian–Recent)

Order Mosasauria—Mosasaurs (Cretaceous)

Order Plesiosauria—Plesiosaurs (Triassic–Cretaceous)

Order Squamata—Lizards and snakes (Triassic–Recent)

Order Rhynchocephalia—The living tuatara *Sphenodon* and its extinct relatives (Jurassic–Recent)

Infraclass Archosauria—Advanced diapsids (Triassic–Recent)

Order Thecodontia—Thecodontians were a diverse group that was ancestral to the crocodilians, pterosaurs, and dinosaurs (Permian–Triassic)

Order Crocodilia—Crocodiles, alligators, and gavials (Triassic–Recent)

Order Pterosauria—Flying and gliding reptiles called pterosaurs (Triassic–Cretaceous)

Infraclass Dinosauria—Dinosaurs (Triassic–Cretaceous)

Order Saurischia—Lizard-hipped dinosaurs (Triassic–Cretaceous)

Suborder Theropoda—Bipedal carnivores (Triassic–Cretaceous)

Suborder Sauropoda—Quadrupedal herbivores, including the largest known land animals (Jurassic–Cretaceous)

Order Ornithischia—Bird-hipped dinosaurs (Triassic–Cretaceous)

Suborder Ornithopoda—Bipedal herbivores, including the duck-billed dinosaurs (Triassic–Cretaceous)

Suborder Stegosauria—Quadrupedal herbivores with bony spikes on their tails and bony plates on their backs (Jurassic–Cretaceous)

Suborder Pachycephalosauria—Bipedal herbivores with thickened bones of the skull roof (Cretaceous)

Suborder Ceratopsia—Quadrupedal herbivores typically with horns or a bony frill over the top of the neck (Cretaceous)

Suborder Ankylosauria—Heavily armored quadrupedal herbivores (Cretaceous)

Subclass Synapsida—Mammal-like reptiles with one opening low on the side of the skull behind the eye (Pennsylvanian–Triassic)

Order Pelycosauria—Early mammal-like reptiles including those forms in which the vertebral spines were extended to support a "sail" (Pennsylvanian–Permian)

Order Therapsida—Advanced mammal-like reptiles with legs positioned beneath the body and the lower jaw formed largely of a single bone. Many therapsids may have been endothermic (Permian–Triassic)

Class Aves—Birds. Endothermic and feathered (Jurassic–Recent)

Class Mammalia—Mammals. Endothermic animals with hair (Triassic–Recent)

Subclass Prototheria—Egg-laying mammals (Triassic–Recent)

Order Docodonta—Small, primitive mammals (Triassic)

Order Triconodonta—Small, primitive mammals with specialized teeth (Triassic–Cretaceous)

Order Monotremata—Duck-billed platypus, spiny anteater (Cretaceous–Recent)

Subclass Allotheria—Small, extinct early mammals with complex teeth (Jurassic–Eocene)

Order Multituberculata—The first mammalian herbivores and the most diverse of Mesozoic mammals (Jurassic–Eocene)

Subclass Theria—Mammals that give birth to live young (Jurassic–Recent)

Order Symmetrodonta—Small, primitive Mesozoic therian mammals (Jurassic–Cretaceous)

Order Upantotheria—Trituberculates (Jurassic–Cretaceous)

Order Creodonta—Extinct ancient carnivores (Cretaceous–Paleocene)

Order Condylartha—Extinct ancestral hoofed placentals (ungulates) (Cretaceous–Oligocene)

Order Marsupialia—Pouched mammals. Opossum, kangaroo, koala (Cretaceous–Recent)

Order Insectivora—Primitive insect-eating mammals. Shrew, mole, hedgehog (Cretaceous–Recent)

Order Xenungulata—Large South American mammals that broadly resemble pantodonts and uintatheres (Paleocene)

Order Taeniodonta—Includes some of the most highly specialized terrestrial placentals of the Late Paleocene and Early Eocene (Paleocene–Eocene)

Order Tillodontia—Large, massive placentals with clawed, five-toed feet (Paleocene–Eocene)

Order Dinocerata—Uintatheres. Large herbivores with bony protuberances on the skull and greatly elongated canine teeth (Paleocene–Eocene)

Order Pantodonta—North American forms are large sheep to rhinoceros-sized. Asian forms are as small as a rat (Paleocene–Eocene)

Order Astropotheria—Large placental mammals with slender rear legs, stout forelimbs, and elongate canine teeth (Paleocene–Miocene)

Order Notoungulata—Largest assemblage of South American ungulates with a wide range of body forms (Paleocene–Pleistocene)

Order Liptoterna—Extinct South American hoofed mammals (Paleocene–Pleistocene)

Order Rodentia—Squirrel, mouse, rat, beaver, porcupine, gopher (Paleocene–Recent)

Order Lagomorpha—Hare, rabbit, pika (Paleocene–Recent)

Order Primates—Lemur, tarsier, loris, monkey, human (Paleocene–Recent)

Order Edentata—Anteater, sloth, armadillo, glyptodont (Paleocene–Recent)

Order Carnivora—Modern carnivorous placentals. Dog, cat, bear, skunk, seal, weasel, hyena, raccoon, panda, sea lion, walrus (Paleocene–Recent)

Order Pyrotheria—Large mammals with long bodies and short columnar limbs (Eocene–Oligocene)

Order Chiroptera—Bats (Eocene–Recent)

Order Dermoptera—Flying lemur (Eocene–Recent)

Order Cetacea—Whale, dolphin, porpoise (Eocene–Recent)

Order Tubulidentata—Aardvark (Eocene–Recent)

Order Perissodactyla—Odd-toed ungulates (hoofed placentals). Horse, rhinoceros, tapir, titanothere, chalicothere (Eocene–Recent)

Order Artiodactyla—Even-toed ungulates. Pig, hippo, camel, deer, elk, bison, cattle, sheep, antelope, entelodont, oreodont (Eocene–Recent)

Order Proboscidea—Elephant, mammoth, mastodon (Eocene–Recent)

Order Sirenia—Sea cow, manatee, dugong (Eocene–Recent)

Order Embrithopodoa—Known primarily from a single locality in Egypt. Large mammals with two gigantic bony processes arising from the nose area (Oligocene)

Order Desmostyla—Amphibious or seal-like in habit. Front and hind limbs well developed, but hands and feet somewhat specialized as paddles (Oligocene–Miocene)

Order Hyracoidea—Hyrax (Oligocene–Recent)

Order Pholidota—Scaly anteater (Oligocene–Recent)

Appendix C
Mineral Identification

To identify most common minerals, geologists use physical properties such as color, luster, crystal form, hardness, cleavage, specific gravity, and several others (Tables C1 and C2). Notice that the Mineral Identification Table (C3) is arranged with minerals having a metallic luster grouped separately from those with a nonmetallic luster. After determining luster, ascertain hardness and note that each part of the table is arranged with minerals in order of increasing hardness. Thus, if you have a nonmetallic mineral with a hardness of 6, it must be augite, hornblende, plagioclase, or one of the two potassium feldspars (orthoclase or microcline). If this hypothetical mineral is dark green or black, it must be augite or hornblende. Use other properties to make a final determination.

TABLE **C1**	Physical Properties Used to Identify Minerals
Mineral Property	**Comment**
Luster	Appearance in reflected light; if has appearance of a metal luster is metallic; those with nonmetallic luster do not look like metals
Color	Rather constant in minerals with metallic luster; varies in minerals with nonmetallic luster
Streak	Powdered mineral on an unglazed porcelain plate (streak plate) is more typical of a mineral's true color
Crystal form	Useful if crystals visible (see Figure 2.5)
Cleavage	Minerals with cleavage tend to break along a smooth plane or planes of weakness
Hardness	A mineral's resistance to abrasion (see Table C2)
Specific gravity	Ratio of a mineral's weight to an equal volume of water
Reaction with HCl (hydrochloric acid)	Calcite reacts vigorously, but dolomite reacts only when powdered
Other properties	Talc has a soapy feel; graphite writes on paper; magnetite is magnetic; closely spaced, parallel lines visible on plagioclase; halite tastes salty

TABLE **C2**	Moh's Hardness Scale

Austrian geologist Frederich Mohs devised this relative hardness scale for ten minerals. He assigned a value of 10 to diamond, the hardest mineral known, and lesser values to the other minerals. You can determine relative hardness of minerals by scratching one mineral with another or by using objects of known hardness.

Hardness	Mineral	Hardness of Some Common Objects
10	Diamond	
9	Corundum	
8	Topaz	
7	Quartz	
		Steel file ($6^{1}/_2$)
6	Orthoclase	
		Glass ($5^{1}/_2$–6)
5	Apatite	
4	Fluorite	
3	Calcite	Copper penny (3) Fingernail ($2^{1}/_2$)
2	Gypsum	
1	Talc	

Metallic Luster

Mineral	Chemical Composition	Color	Hardness Specific Gravity	Other Features	Comments
Graphite	C	Black	1–2 2.09–2.33	Greasy feel; writes on paper; 1 direction of cleavage	Used for pencil "leads." Mostly in metamorphic rocks.
Galena	PbS	Lead gray	$2^1/2$ 7.6	Cubic crystals; 3 cleavages at right angles	The ore of lead. Mostly in hydrothermal rocks.
Chalcopyrite	$CuFeS_2$	Brassy yellow	$3^1/2$–4 4.1–4.3	Usually massive; greenish black streak; iridescent tarnish	The most common copper mineral. Mostly in hydrothermal rocks.
Magnetite	Fe_3O_4	Black	$5^1/2$–$6^1/2$ 5.2	Strong magnetism	An ore of iron. An accessory mineral in many rocks.
Hematite	Fe_2O_3	Red brown	6 4.8–5.3	Usually granular or massive; reddish brown streak	Important iron ore. An accessory mineral in many rocks.
Pyrite	FeS_2	Brassy yellow	$6^1/2$ 5.0	Cubic and octahedral crystals	Found in some igneous and hydrothermal rocks and in sedimentary rocks associated with coal.

Nonmetallic Luster

Mineral	Chemical Composition	Color	Hardness Specific Gravity	Other Features	Comments
Talc	$Mg_3Si_4O_{10}(OH)_2$	White, green	1 2.82	1 cleavage direction; usually in compact masses; soapy feel	Formed by the alteration of magnesium silicates. Mostly in metamorphic rocks.
Clay minerals	Varies	Gray, buff, white	1–2 2.5–2.9	Earthy masses; particles too small to observe properties	Found in soils, mudrocks, slate, phyllite.
Chlorite	$(Mg,Fe)_3(Si,Al)_4O_{10}$ $(Mg,Fe)_3(OH)_6$	Green	2 2.6–3.4	1 cleavage; occurs in scaly masses	Common in low-grade metamorphic rocks such as slate.
Gypsum	$CaSO_4 \cdot 2H_2O$	Colorless, white	2 2.32	Elongate crystals; fibrous and earthy masses	The most common sulfate mineral. Found mostly in evaporite deposits.
Muscovite (Mica)	$KAL_2Si_3O_{10}(OH)_2$	Colorless	2–$2^1/2$ 2.7–2.9	1 direction of cleavage; cleaves into thin sheets	Common in felsic igneous rocks, metamorphic rocks, and some sedimentary rocks.
Biotite (Mica)	$K(Mg,Fe)_3AlSi_3O_{10}$ $(OH)_2$	Black, brown	$2^1/2$ 2.9–3.4	1 cleavage direction; cleaves into thin sheets	Occurs in both felsic and mafic igneous rocks, in metamorphic rocks, and in some sedimentary rocks.
Calcite	$CaCO_3$	Colorless, white	3 2.71	3 cleavages at oblique angles; cleaves into rhombs; reacts with dilute HCl	The most common carbonate mineral. Main component of limestone and marble.
Anhydrite	$CaSO_4$	White, gray	$3^1/2$ 2.9–3.0	Crystals with 2 cleavages; usually in granular masses	Found in limestones, evaporite deposits, and the cap rock of salt domes.

Nonmetallic Luster

Mineral	Chemical Composition	Color	Hardness Specific Gravity	Other Features	Comments
Halite	$NaCl$	Colorless, white	3–4 2.2	3 cleavages at right angles; cleaves into cubes; cubic crystals; salty taste	Occurs in evaporite deposits.
Dolomite	$CaMg(CO_3)_2$	White, yellow, gray, pink	$3\frac{1}{2}$–4 2.85	Cleavage as in calcite; reacts with dilute hydrochloric acid when powdered	The main constituent of dolostone. Also found associated with calcite in some limestones and marble.
Fluorite	CaF_2	Colorless, purple, green, brown	4 3.18	4 cleavage directions; cubic and octahedral crystals	Occurs mostly in hydrothermal rocks and in some limestones and dolostones.
Augite	$Ca(Mg,Fe,Al)(Al,Si)_2O_6$	Black, dark green	6 3.25–3.55	Short 8-sided crystals; 2 cleavages; cleavages nearly at right angles	The most common pyroxene mineral. Found mostly in mafic igneous rocks.
Hornblende	$NaCa_2(Mg,Fe,Al)_5$ $(Si,Al)_8O_{22}(OH)_2$	Green, black	6 3.0–3.4	Elongate, 6-sided crystals; 2 cleavages intersecting at 56° and 124°	A common rock-forming amphibole mineral in igneous and metamorphic rocks.
Plagioclase feldspars	Varies from $CaAl_2Si_2O_8$ to $NaAlSi_3O_8$	White, gray, brown	6 2.56	2 cleavages at right angles	Common in igneous rocks and a variety of metamorphic rocks. Also in some arkoses.
Potassium Feldspars — Microcline	$KAlSi_3O_8$	White, pink, green	6 2.56	2 cleavages at right angles	Common in felsic igneous rocks, some metamorphic rocks, and arkoses.
Potassium Feldspars — Orthoclase	$KAlSi_3O_8$	White, pink	6 2.56	2 cleavages at right angles	
Olivine	$(Fe,Mg)_2SiO_4$	Olive green	$6\frac{1}{2}$ 3.3–3.6	Small mineral grains in granular masses; conchoidal fracture	Common in mafic igneous rocks.
Quartz	SiO_2	Colorless, white, gray, pink, green	7 2.67	6-sided crystals; no cleavage; conchoidal fracture	A common rock-forming mineral in all rock groups and hydrothermal rocks. Also occurs in varieties known as chert, flint, agate, and chalcedony.
Garnet	$Fe_3Al_2(SiO_4)_3$	Dark red, green	7–$7\frac{1}{2}$ 4.32	12-sided crystals common; uneven fracture	Found mostly in gneiss and schist.
Zircon	Zr_2SiO_4	Brown, gray	$7\frac{1}{2}$ 3.9–4.7	4-sided, elongate crystals	Most common as an accessory in granitic rocks.
Topaz	$Al_2SiO_4(OH,F)$	Colorless, white, yellow, blue	8 3.5–3.6	High specific gravity; 1 cleavage direction	Found in pegmatites, granites, and hydrothermal rocks.
Corundum	Al_2O_3	Gray, blue, pink, brown	9 4.0	6-sided crystals and great hardness are distinctive	An accessory mineral in some igneous and metamorphic rocks.

Glossary

abiogenesis The origin of life from non-living matter.

Absaroka Sequence A widespread succession of Pennsylvanian and Permian sedimentary rocks bounded above and below by unconformities; deposited during a transgressive–regressive cycle of the Absaroka Sea.

absolute dating Assigning an age in years before the present to geologic events; absolute dates are determined by radioactive decay dating techniques (See *relative dating.*).

Acadian orogeny A Devonian episode of mountain building in the northern Appalachian mobile belt resulting from a collision of Baltica with Laurentia.

Alleghenian orogeny Pennsylvanian to Permian mountain building in the Appalachian mobile belt from New York to Alabama.

allele A variant form of a single gene. (See *gene.*)

allopatric speciation Model for the origin of a new species from a small population that became isolated from its parent population.

alluvial fan A cone-shaped accumulation of mostly sand and gravel where a stream flows from a mountain valley onto an adjacent lowland.

Alpine–Himalayan orogenic belt A linear zone of deformation extending from the Atlantic eastward across southern Europe and North Africa, through the Middle East and into Southeast Asia. (See *circum-Pacific orogenic belt.*)

Alpine orogeny A Late Mesozoic–Early Cenozoic episode of mountain building affecting southern Europe and North Africa.

amniote egg An egg in which an embryo develops in a liquid-filled cavity (the amnion); and a waste sac is present as well as a yolk sac for nourishment.

anaerobic Refers to organisms that do not depend on oxygen for respiration.

analogous structure Body part, such as wings of insects and birds, that serves the same function but differs in structure and development. (See *homologous structure.*)

Ancestral Rockies Late Paleozoic uplift in the southwestern part of the North American craton.

angiosperm Vascular plants having flowers and seeds; the flowering plants.

angular unconformity An unconformity below which strata generally dip at a steeper angle than those above. (See *disconformity, nonconformity,* and *unconformity.*)

anthropoid Any member of the primate suborder Anthropoidea; includes New World and Old World monkeys, apes, and humans.

Antler orogeny A Late Devonian to Mississippian episode of mountain building that affected the Cordilleran mobile belt from Nevada to Alberta, Canada.

Appalachian mobile belt A long narrow region of tectonic activity along the eastern margin of the North American craton extending from Newfoundland to Georgia; probably continuous to the southwest with the Ouachita mobile belt.

Archaeopteryx The oldest positively identified fossil bird; it had feathers but retained many reptile characteristics; from Jurassic rocks in Germany.

archosaur A term referring to the ruling reptiles—dinosaurs, pterosaurs, crocodiles, and birds.

artificial selection The practice of selectively breeding plants and animals with desirable traits.

Artiodactyla The mammalian order whose members have two or four toes; the even-toed hoofed mammals such as deer, goats, sheep, antelope, bison, swine, and camels.

asthenosphere Part of the upper mantle over which the lithosphere moves; it behaves as a plastic and flows.

Atlantic Coastal Plain The broad, low relief area of eastern North America extending from the Appalachian Mountains to the Atlantic shoreline.

atom The smallest unit of matter that retains the characteristics of an element.

atomic mass number The total number of protons and neutrons in an atom's nucleus.

atomic number The number of protons in an atom's nucleus.

australopithecine A collective term for all species of the extinct genus *Australopithecus* that existed in South Africa during the Pliocene and Pleistocene.

autotrophic Describes organisms that synthesize their organic nutrients from inorganic raw materials; photosynthesizing bacteria and plants are autotrophs. (See *heterotrophic* and *primary producer.*)

back-arc marginal basin A marine basin, such as the Sea of Japan, between a volcanic island arc and a continent; probably forms by back-arc spreading.

Baltica One of six major Paleozoic continents; composed of Russia west of the Ural Mountains, Scandinavia, Poland, and northern Germany.

banded iron formation (BIF) Sedimentary rocks made up of alternating thin layers of chert and iron minerals, mostly the iron oxides hematite and magnetite.

barrier island A long sand body more or less parallel with a shoreline but separated from it by a lagoon.

Basin and Range Province An area of Cenozoic block-faulting centered on Nevada but extending into adjacent states and northern Mexico.

benthos All bottom-dwelling marine organisms that live on the seafloor or within seafloor sediments.

Big Bang A theory for the evolution of the universe from a dense, hot state followed by expansion, cooling, and a less dense state.

biogenic sedimentary structure Any feature such as tracks, trails, and burrows in sedimentary rocks produced by the activities of organisms. (See *trace fossil.*)

biostratigraphic unit A unit of sedimentary rock defined solely by its fossil content.

bioturbation The churning of sediment by organisms that burrow through it.

biozone A general term referring to all biostratigraphic units such as range zones and concurrent range zones.

bipedal Walking on two legs as a means of locomotion as in birds and humans. (See *quadrupedal.*)

black smoker A submarine hydrothermal vent that emits a plume of black water colored by dissolved minerals. (See *submarine hydrothermal vent.*)

body fossil The shells, teeth, bones, or (rarely) the soft parts of organisms preserved in the fossil record. (See *fossil* and *trace fossil.*)

bonding The processes whereby atoms join with other atoms.

bony fish Members of the class Osteichthyes that evolved during the Devonian; characterized by a bony internal skeleton; includes the ray-finned fishes and the lobe-finned fishes.

braided stream A stream with an intricate network of dividing and rejoining channels.

browser An animal that eats tender shoots, twigs, and leaves. (See *grazer.*)

Caledonian orogeny A Silurian–Devonian episode of mountain building that took place along the northwestern margin of Baltica, resulting from the collision of Baltica with Laurentia.

Canadian shield The Precambrian shield in North America; mostly in Canada but also exposed in Minnesota, Wisconsin, Michigan, and New York.

carbon 14 dating An absolute dating technique relying on the ratio of C14 to C12 in organic substances; useful back to about 70,000 years ago.

carbonate mineral Any mineral with the negatively charged carbonate ion $(CO_3)^{-2}$ (e.g., calcite [$CaCO_3$] and dolomite [$CaMg(CO_3)_2$]).

carbonate rock Any rock composed mostly of carbonate minerals (such as limestone and dolostone).

carnassials A pair of specialized shearing teeth in members of the mammal order Carnivora.

Carnivora An order of mammals consisting of meat eaters such as dogs, cats, bears, weasels, and seals.

carnivore-scavenger Any animal that eats other animals, living or dead, as a source of nutrients.

cartilaginous fish Fish such as living sharks and their living and extinct relatives that have an internal skeleton of cartilage.

Cascade Range A mountain range made up of volcanic rock stretching from northern California through Oregon and Washington and into British Columbia, Canada.

cast A replica of an object such as a shell or bone formed when a mold of that object is filled by sediment or minerals. (See *mold.*)

catastrophism A concept proposed by Baron Georges Cuvier explaining Earth's physical and biologic history by sudden, worldwide catastrophes; also holds that geologic processes acted with much greater intensity during the past.

Catskill Delta A Devonian clastic wedge deposited adjacent to the highlands that formed during the Acadian orogeny.

Cetacea The mammal order that includes whales, porpoises, and dolphins.

chemical sedimentary rock Rock formed of minerals derived from materials dissolved during weathering.

China One of six major Paleozoic continents; composed of all Southeast Asia, including China, Indochina, part of Thailand, and the Malay Peninsula.

chordate Any member of the phylum Chordata, all of which have a notochord, dorsal hollow nerve cord, and gill slits at some time during their life cycle.

chromosome Complex, double-stranded, helical molecule of deoxyribonucleic acid (DNA); specific segments of chromosome are genes.

circum-Pacific orogenic belt One of two major Mesozoic-Cenozoic areas of large-scale deformation and the origin of mountains; includes orogens in South and Central America, the North American Cordillera, and the Aleutian, Japan, and Philippine arcs. (See *Alpine-Himalayan orogenic belt.*)

cirque A steep-walled, bowl-shaped depression formed on a mountainside by glacial erosion.

cladistics A type of analysis of organisms in which they are grouped together on the basis of derived as opposed to primitive characteristics.

cladogram A diagram showing the relationships among members of a clade, including their most recent common ancestor.

clastic wedge An extensive accumulation of mostly detrital sedimentary rocks eroded from and deposited adjacent to an area of uplift, as in the Catskill Delta or Queenston Delta.

Colorado Plateau A vast upland area in Colorado, Utah, Arizona, and New Mexico with only slightly deformed Phanerozoic rocks, deep canyons, and volcanic mountains.

compound A substance made up of different atoms bonded together (such as water [H_2O] and quartz [SiO_2]).

concurrent range zone A biozone established by plotting the overlapping geologic ranges of fossils.

conformable Refers to a sequence of sedimentary rocks deposited one after the other with no or only minor discontinuities resulting from nondeposition or erosion.

contact metamorphism Metamorphism taking place adjacent to a body of magma (a pluton) or beneath a lava flow from heat and chemically active fluids.

continental accretion The process whereby continents grow by additions of Earth materials along their margins.

continental–continental plate boundary A convergent plate boundary along which two continental lithospheric plates collide, such as the collision of India with Asia. (See *convergent plate boundary, oceanic–continental plate boundary,* and *oceanic–oceanic plate boundary.*)

continental drift The theory proposed by Alfred Wegener that all continents were once joined into a single landmass that broke apart with the various fragments (continents) moving with respect to one another.

continental glacier A glacier covering at least 50,000 km^2 and unconfined by topography. Also called an *ice sheet.*

continental interior An area in North America made up of the Great Plains and the Central Lowlands, bounded by the Rocky Mountains, the Canadian shield, the Appalachian Mountains, and parts of the Gulf Coastal Plain.

continental red bed Red-colored rock, especially mudrock and sandstone, on the continents. Iron oxides account for their color.

continental rise The gently sloping part of the seafloor lying between the base of the continental slope and the deep seafloor.

continental shelf The area where the seafloor slopes gently seaward between a shoreline and the continental slope.

continental slope The relatively steep part of the seafloor between the continental shelf and continental rise or an oceanic trench.

convergent evolution The origin of similar features in distantly related organisms as they adapt in comparable ways, such as ichthyosaurs and porpoises. (See *parallel evolution.*)

convergent plate boundary The boundary between two plates that move toward one another. (See *continental–continental plate boundary, oceanic–continental plate boundary,* and *oceanic–oceanic plate boundary.*)

Cordilleran mobile belt An area of extensive deformation in western North America bounded by the Pacific Ocean and the Great Plains; it extends north–south from Alaska into central Mexico.

Cordilleran orogeny A period of deformation affecting the western part of North America from Jurassic to Early Cenozoic time; divided into three phases known as the Nevadan, Sevier, and Laramide orogenies.

core The inner part of Earth from a depth of about 2900 km consisting of a liquid outer part and a solid inner part; probably composed mostly of iron and nickel.

correlation Demonstration of the physical continuity of stratigraphic units over an area; also matching up time-equivalent events in different areas.

craton Name applied to a stable nucleus of a continent consisting of a Precambrian shield and a platform of buried ancient rocks.

cratonic sequence A widespread association of sedimentary rocks bounded above and below by unconformities that were deposited during a transgressive-regressive cycle of an epeiric sea, such as the Sauk Sequence.

Cretaceous Interior Seaway A Late Cretaceous arm of the sea that effectively divided North America into two large landmasses.

Cro-Magnon A race of *Homo sapiens* that lived mostly in Europe from 35,000 to 10,000 years ago.

cross-bedding A type of bedding in which individual layers are deposited at an angle to the surface on which they accumulate, as in sand dunes.

crossopterygian A specific type of lobe-finned fish that had lungs.

crust The upper part of Earth's lithosphere, which is separated from the mantle by the Moho; consists of continental crust with an overall granitic composition and thinner, denser oceanic crust made up of basalt and gabbro.

crystalline solid A solid with its atoms arranged in a regular three-dimensional framework.

Curie point The temperature at which iron-bearing minerals in a cooling magma attain their magnetism.

cyclothem A sequence of cyclically repeated sedimentary rocks resulting from alternating periods of marine and nonmarine deposition; commonly contain a coal bed.

cynodont A type of therapsid (advanced mammal-like reptile); ancestors of mammals are among the cynodonts.

delta A deposit of sediment where a stream or river enters a lake or the ocean.

deoxyribonucleic acid (DNA) The chemical substance of which chromosomes are composed.

depositional environment Any area where sediment is deposited; a depositional site where physical, chemical, and biological processes operate to yield a distinctive kind of deposit.

detrital sedimentary rock Rock made up of the solid particles derived from preexisting rocks as in sandstone.

dinosaur Any of the Mesozoic reptiles belonging to the orders Saurischia and Ornithischia.

disconformity A type of unconformity above and below which the strata are parallel. (See *angular unconformity, nonconformity,* and *unconformity.*)

divergent evolution The diversification of a species into two or more descendant species.

divergent plate boundary The boundary between two plates that move apart; characterized by seismicity, volcanism, and the origin of new oceanic lithosphere.

drift A collective term for all sediment deposited by glacial activity; includes till deposited directly by ice, and outwash deposited by streams discharging from glaciers. (See *outwash.*)

dynamic metamorphism Metamorphism in fault zones where rocks are subjected to high differential pressure.

ectotherm Any of the cold-blooded vertebrates such as amphibians and reptiles; animals that depend on external heat to regulate body temperature. (See *endotherm.*)

Ediacaran fauna Name for all Late Proterozoic faunas with animal fossils similar to those of the Ediacara fauna of Australia.

element A substance composed of only one kind of atom (such as calcium [Ca] or silicon [Si]).

end moraine A pile or ridge of rubble deposited at the terminus of a glacier.

endosymbiosis A type of mutually beneficial symbiosis in which one symbiont lives within the other.

endotherm Any of the warm-blooded vertebrates such as birds and mammals who maintain their body temperature within narrow limits by internal processes. (See *ectotherm.*)

epeiric sea A broad shallow sea that covers part of a continent; six epeiric seas were present in North America during the Phanerozoic Eon, such as the Sauk Sea.

eukaryotic cell A cell with an internal membrane-bounded nucleus containing chromosomes and other internal structures such as mitochondria that are not present in prokaryotic cells. (See *prokaryotic cell.*)

evaporite Sedimentary rock formed by inorganic chemical precipitation from evaporating water (for example, rock salt and rock gypsum).

extrusive igneous rock An igneous rock that forms as lava cools and crystallizes or when pyroclastic materials are consolidated. (See *volcanic rock.*)

Farallon plate A Late Mesozoic–Cenozoic oceanic plate that was largely subducted beneath North America; the Cocos and Juan de Fuca plates are remnants.

fission-track dating The dating process in which small linear tracks (fission tracks) resulting from alpha decay are counted in mineral crystals.

fluvial Relating to streams and rivers and their deposits.

formation The basic lithostratigraphic unit; a mappable unit of strata with distinctive upper and lower boundaries.

fossil Remains or traces of prehistoric organisms preserved in rocks. (See *body fossil* and *trace fossil.*)

gene A specific segment of a chromosome constituting the basic unit of heredity. (See *allele.*)

geologic column A diagram showing a composite column of rocks arranged with the oldest at the bottom followed upward by progressively younger rocks. (See *geologic time scale.*)

geologic record The record of prehistoric physical and biologic events preserved in rocks.

geologic time scale A chart arranged so that the designation for the earliest part of geologic time appears at the bottom followed upward by progressively younger time designations. (See *geologic column.*)

geology The science concerned with the study of Earth; includes studies of Earth materials, internal and surface processes, and Earth and life history.

glacial stage A time of extensive glaciation that occurred several times in North America during the Pleistocene.

glacier A mass of ice on land that moves by plastic flow and basal slip.

***Glossopteris* flora** A Late Paleozoic association of plants found only on the Southern Hemisphere continents and India; named after its best-known genus, *Glossopteris.*

Gondwana One of six major Paleozoic continents; composed of South America, Africa, Australia, India, and parts of Southern Europe, Arabia, and Florida.

graded bedding A sediment layer in which grain size decreases from the bottom up.

granite-gneiss complex One of the two main rock associations found in areas of Archean rocks.

grazer An animal that eats low-growing vegetation, especially grasses. (See *browser.*)

greenstone belt A linear or podlike association of rocks particularly common in Archaean terranes; typically synclinal and consists of lower and middle volcanic units and an upper sedimentary unit.

Grenville orogeny An episode of deformation that took place in the eastern United States and Canada during the Neoproterozoic.

guide fossil Any easily identified fossil with a wide geographic distribution and short geologic range; useful for determining relative ages of strata in different areas.

Gulf Coastal Plain The broad low-relief area along the Gulf Coast of the United States.

gymnosperm A flowerless, seed-bearing plant.

half-life The time necessary for one-half of the original number of radioactive atoms of an element to decay to a stable daughter product; for example, the half-life of potassium 40 is 1.3 billion years.

herbivore An animal dependent on vegetation as a source of nutrients.

Hercynian orogeny Pennsylvanian to Permian deformation in the Hercynian mobile belt of southern Europe.

heterotrophic Organism such as an animal that depends on preformed organic molecules from its environment for nutrients. (See *autotrophic* and *primary producer.*)

hominid Abbreviated term for Hominidae, the family that includes bipedal primates such as *Australopithecus* and *Homo.* (See *hominoid.*)

hominoid Abbreviated term for Hominoidea, the superfamily that includes apes and humans. (See *hominid*.)

Homo The genus of hominids consisting of *Homo sapiens* and their ancestors *Homo erectus* and *Homo habilis*.

homologous structure Body part in different organisms with a similar structure, similar relationships to other organs, and similar development but does not necessarily serve the same function; such as forelimbs in whales, bats, and dogs. (See *analogous structure*.)

hot spot Localized zone of melting below the lithosphere; detected by volcanism at the surface.

hypothesis A provisional explanation for observations that is subject to continual testing and modification if necessary. If well supported by evidence, hypotheses may become theories.

Hyracotherium A small Early Eocene mammal that was ancestral to today's horses.

Iapetus Ocean A Paleozoic ocean between North America and Europe; it eventually closed as North America and Europe moved toward one another and collided during the Late Paleozoic.

ice cap A dome-shaped mass of glacial ice covering less than 50,000 km^2.

ice-scoured plain An area eroded by glaciers resulting in low-relief, extensive bedrock exposures with glacial polish and striations, and little soil.

ichthyosaur Any of the porpoise-like, Mesozoic marine reptiles.

igneous rock Rock formed when magma or lava cools and crystallizes and when pyroclastic materials become consolidated.

inheritance of acquired characteristics Jean-Baptiste de Lamarck's mechanism for evolution; holds that characteristics acquired during an individual's lifetime can be inherited by descendants.

interglacial stage A time of warmer temperatures between episodes of widespread glaciation.

intrusive igneous rock Igneous rock that cools and crystallizes from magma intruded into or formed within the crust. (See *plutonic rock*.)

iridium anomaly The occurence of a higher-than-usual concentration of the element iridium at the Cretaceous-Palogene boundary.

isostasy The concept of Earth's crust "floating" on the more dense underlying mantle. As a result of isostasy, thicker, less dense continental crust stands higher than oceanic crust.

isostatic rebound The phenomenon in which unloading of the crust causes it to rise, as when extensive glaciers melt, until it attains equilibrium.

Jovian planet Any planet with a low mean density that resembles Jupiter (Jupiter, Saturn, Uranus, and Neptune); the Jovian planets, or gas giants, are composed largely of hydrogen, helium, and frozen compounds such as methane and ammonia. (See *terrestrial planet*.)

Kaskaskia Sequence A widespread association of Devonian and Mississippian sedimentary rocks bounded above and below by unconformities; deposited during a transgressive–regressive cycle of the Kaskaskia Sea.

Kazakhstania One of six major Paleozoic continents; a triangular-shaped continent centered on Kazakhstan.

labyrinthodont Any of the Devonian to Triassic amphibians characterized by complex folding in the enamel of their teeth.

Laramide orogeny Late Cretaceous to Early Paleogene phase of the Cordilleran orogeny; responsible for many of the structural features in the present-day Rocky Mountains.

Laurasia A Late Paleozoic, Northern Hemisphere continent made up of North America, Greenland, Europe, and Asia.

Laurentia A Proterozoic continent composed mostly of North America and Greenland, parts of Scotland, and perhaps parts of the Baltic shield of Scandinavia.

lava Magma that reaches the surface.

lithification The process of converting sediment into sedimentary rock.

lithosphere The outer, rigid part of Earth consisting of the upper mantle, oceanic crust, and continental crust; lies above the asthenosphere.

lithostratigraphic unit A body of sedimentary rock, such as a formation, defined solely by its physical attributes.

Little Ice Age An interval from about 1500 to the mid- to late-1800s during which glaciers expanded to their greatest historic extent.

living fossil An existing organism that has descended from ancient ancestors with little apparent change.

lobe-finned fish Fish with limbs containing a fleshy shaft and a series of articulating bones; one of the two main groups of bony fish.

macroevolution Evolutionary changes that account for the origin of new species, genera, orders, and so on. (See *microevolution*.)

magma Molten rock material below the surface.

magnetic anomoly Any change, such as the average strength, in Earth's magnetic field.

magnetic reversal The phenomenon involving the complete reversal of the north and south magnetic poles.

mantle The inner part of Earth surrounding the core, accounting for about 85% of the planet's volume; probably composed of peridotite.

marine regression Withdrawal of the sea from a continent or coastal area resulting from emergence of the land with a resulting seaward migration of the shoreline.

marine transgression Invasion of a coastal area or much of a continent by the sea as sea level rises resulting in a landward migration of the shoreline.

marsupial mammal The pouched mammals such as wombats and kangaroos that give birth to young in a very immature state.

mass extinction Greatly accelerated extinction rates resulting in marked decrease in biodiversity, such as the mass extinction at the end of the Cretaceous.

meandering stream A stream with a single, sinuous channel with broadly looping curves.

meiosis Cell division yielding sex cells, sperm and eggs in animals, and pollen and ovules in plants, in which the number of chromosomes is reduced by half. (See *mitosis*.)

metamorphic rock Any rock altered in the solid state from preexisting rocks by any combination of heat, pressure, and chemically active fluids.

microevolution Evolutionary changes within a species. (See *macroevolution*.)

Midcontinent rift A Mesoproterozoic intracontinental rift in Lauentia in which volcanic and sedimentary rocks accumulated.

Milankovitch theory A theory that explains cyclic variations in climate and the onset of glacial episodes triggered by irregularities in Earth's rotation and orbit.

mineral Naturally occurring, inorganic, crystalline solid, having characteristic physical properties and a narrowly defined chemical composition.

mitosis Call division resulting in two cells with the same number of chromosomes as the parent cell; takes place in all cells except sex cells. (See *meiosis*.)

mobile belt Elongated area of deformation generally at the margins of a craton, such as the Appalachian mobile belt.

modern synthesis A combination of ideas of various scientists yielding a view of evolution that includes the chromosome theory of inheritance, mutations as a source of variation, and gradualism. It also rejects inheritance of acquired characteristics.

molar Any of a mammal's teeth that are used for grinding and chewing.

molarization An evolutionary trend in hoofed mammals in which the premolars become more like molars, giving the animals a continuous series of grinding teeth.

mold A cavity or impression of some kind of organic remains such as a bone or shell in sediment or sedimentary rock. (See *cast*.)

monomer A comparatively simple organic molecule, such as an amino acid, that can link with other monomers to form more complex polymers such as proteins. (See *polymer*.)

monotreme The egg-laying mammals; includes only the platypus and spiny anteater of the Australian region.

moraine A ridge or mound of unsorted, unstratified debris deposited by a glacier.

mosaic evolution The concept holding that not all parts of an organsim evolve at the same rate, thus yielding organisms with features retained from the ancestral condition as well as more recently evolved features.

mosasaur A term referring to a group of Mesozoic marine lizards.

mud crack A crack in clay-rich sediment that forms in response to drying and shrinkage.

multicelled organism Organism made up of many cells as opposed to a single cell; possesses cells specialized to perform specific functions.

mutation Any change in the genes of organisms; yields some of the variation on which natural selection acts.

Natural selection A mechanism accounting for differential survival and reproduction among members of a species; the mechanism proposed by Charles Darwin and Alfred Wallace to account for evolution.

Neanderthal A type of human that inhabited the Near East and Europe from 200,000 to 30,000 years ago; may be a subspecies (*Homo sapiens neanderthalensis*) of *Homo* or a separate species (*Homo neanderthalensis*).

nekton Actively swimming organisms, such as fish, whales, and squid. (See *plankton*.)

neoglaciation An episode in Earth history from about 6000 years ago until the mid to late 1800s during which three periods of glacial expansion took place.

neptunism The discarded concept held by Abraham Gottlob Werner and others that all rocks formed in a specific order by precipitation from a worldwide ocean.

Nevadan orogeny Late Jurassic to Cretaceous phase of the Cordilleran orogeny; most strongly affected the western part of the Cordilleran mobile belt.

nonconformity An unconformity in which stratified sedimentary rocks overlie an erosion surface cut into igneous or metamorphic rocks. (See *angular unconformity, disconformity,* and *unconformity*.)

North American Cordillera A complex mountainous region in western North America extending from Alaska into central Mexico.

oceanic–continental plate boundary A convergent plate boundary along which oceanic and continental lithosphere collide; characterized by subduction of the oceanic plate, seismicity, and volcanism.

(See *continental–continental plate boundary, convergent plate boundary,* and *oceanic–oceanic plate boundary*.)

oceanic–oceanic plate boundary A convergent plate boundary along which oceanic lithosphere collides with oceanic lithosphere; charcaterized by subduction of one of the oceanic plates, seismicity, and volcanism. (See *continental–continental plate boundary, convergent plate boundary,* and *oceanic–continental plate boundary*.)

organic evolution See *theory of evolution*.

organic reef A wave-resistant limestone structure with a framework of animal skeletons, such as a coral reef or stromatoporoid reef.

Ornithischia One of the two orders of dinosaurs; characterized by a birdlike pelvis; includes ornithopods, stegosaurs, ankylosaurs, pachycephalosaurs, and ceratopsians. (See *Saurischia*.)

orogen A linear part of Earth's crust that was or is being deformed during an orogeny; part of an orogenic belt.

orogeny An episode of mountain building involving deformation, usually accompanied by igneous activity, metamorphism, and crustal thickening.

ostracoderm The "bony-skinned" fish characterized by bony armor but no jaws or teeth; appeared during the Late Cambrian, making them the oldest known vertebrates.

Ouachita mobile belt An area of deformation along the southern margin of the North American craton; probably continuous to the northeast with the Appalachian mobile belt.

Ouachita orogeny A period of mountain building that took place in the Ouachita mobile belt during the Pennsylvanian Period.

outgassing The process whereby gases released from Earth's interior by volcanism formed an atmosphere.

outwash All sediment deposited by streams that issue from glaciers. (See *drift*.)

Pacific–Farallon ridge A spreading ridge that was located off the coast of western North America during part of the Cenozoic Era.

Paleocene–Eocene Thermal Maximum A warming trend that began abruptly about 55 million years ago.

paleogeography The study of Earth's ancient geography on a regional as well as a local scale.

paleomagnetism The study of the direction and strength of Earth's past magnetic field from remanent magnetism in rocks.

paleontology The use of fossils to study life history and relationships among organisms.

Pangaea Alfred Wegener's name for a Late Paleozoic supercontinent made up of most of Earth's landmasses.

Pannotia A supercontinent that existed during the Neoproterozoic.

Panthalassa A Late Paleozoic ocean that surrounded Pangaea.

parallel evolution Evolution of similar features in two separate but closely related lines of descent as a result of comparable adaptations. (See *convergent evolution*.)

pelycosaur Pennsylvanian to Permian reptile, many species with large fins on the back, that possessed some mammal characteristics.

period The fundamental unit in the hierarchy of time units; part of geologic time during which the rocks of a system were deposited.

Perissodactyla The order of odd-toed hoofed mammals; consists of present-day horses, tapirs, and rhinoceroses.

photochemical dissociation A process whereby water molecules in the upper atmosphere are disrupted by ultraviolet radiation, yielding oxygen (O_2) and hydrogen (H).

photosynthesis The metobolic process in which organic molecules are synthesized from water and carbon dioxide (CO_2), using the radiant energy of the Sun captured by chlorophyll-containing cells.

phyletic gradualism The concept that a species evolves gradually and continuously as it gives rise to new species. (See *puncuated equilibrium*.)

placental mammal All mammals with a placenta to nourish the developing embroyo, as opposed to egg-laying mammals (monotremes) and pouched mammals (marsupials).

placoderm Late Silurian through Permian "plate-skinned" fish with jaws and bony armor, especially in the head-shoulder region.

plankton Animals and plants that float passively, such as phytoplankton and zooplankton. (See *nekton*.)

plate A segment of Earth's crust and upper mantle (lithosphere) varying from 50 to 250 km thick.

plate tectonic theory Theory holding that lithospheric plates move with respect to one another at divergent, convergent, and transform plate boundaries.

platform The buried extension of a Precambrian shield, which together with a shield makes up a craton.

playa lake A temporary lake in an arid region.

plesiosaur A type of Mesozoic marine reptile; short-necked and long-necked plesiosaurs existed.

plutonic rock Igneous rock that cools and crystyllizes from magma intruded into or formed within the crust. (See *igneous rock*.)

pluvial lake Any lake that formed in nonglaciated areas during the Pleistocene as a result of increased precipitation and reduced evaporation rates during that time.

pollen analysis Identification and statistical analysis of pollen from sedimentary rocks; provides information about ancient floras and climates.

polymer A comparatively complex organic molecule, such as nucleic acids and proteins, formed by monomers linking together. (See *monomer.*)

Precambrian shield An area in which a continent's ancient craton is exposed, as in the Canadian shield.

premolar Any of a mammal's teeth between the canines and the molars; premolars and molars together are a mammal's chewing teeth.

primary producer Organism in a food chain, such as bacteria and green plants, that manufacture their own organic molecules, and on which all other members of the food chain depend for sustenance. (See *autotrophic.*)

Primates The order of mammals that includes prosimians (lemurs and tarsiers), monkeys, apes, and humans.

principle of cross-cutting relationships A principle holding that an igneous intrusion or fault must be younger than the rocks it intrudes or cuts across.

principle of fossil succession A principle holding that fossils, especially groups or assemblages of fossils, succeed one another through time in a regular and determinable order.

principle of inclusions A principle holding that inclusions or fragments in a rock unit are older than the rock itself, such as granite inclusions in sandstone are older than the sandstone.

principle of lateral continuity A principle holding that rock layers extend outward in all directions until they terminate.

principle of original horizontality According to this principle, sediments are deposited in horizontal or nearly horizontal layers.

principle of superposition A principle holding that sedimentary rocks in a vertical sequence formed one on top of the other so that the oldest layer is at the bottom of the sequence whereas the youngest is at the top.

principle of uniformitarianism A principle holding that we can interpret past events by understanding present-day processes, based on the idea that natural processes have always operated as they do now.

Proboscidea The order of mammals that includes elephants and their extinct relatives.

proglacial lake A lake formed of meltwater accumulating along the margin of a glacier.

progradation The seaward (or lakeward) migration of a shoreline as a result of nearshore sedimentation.

prokaryotic cell A cell lacking a nucleus and organelles such as mitochondria and plastids; the cells of bacteria and archaea. (See *eukaryotic cell.*)

prosimian Any of the so-called lower primates, such as tree shrews, lemurs, lorises, and tarsiers.

protorothyrid A loosely grouped category of small, lizardlike reptiles.

pterosaur Any of the Mesozoic flying reptiles that had a long finger to support a wing.

punctuated equilibrium A concept holding that new species evolve rapidly, in perhaps a few thousands of years, then remains much the same during its several million years of existence. (See *phyletic gradualism.*)

pyroclastic materials Fragmental materials such as ash explosively erupted from volcanoes.

quadrupedal A term referring to locomotion on all four limbs as in dogs and horses. (See *bipedal.*)

Queenston Delta A clastic wedge resulting from deposition of sediment eroded from the highland formed during the Taconic orogeny.

radioactive decay The spontaneous change in an atom by emission of a particle from its nucleus (alpha and beta decay) or by electron capture, thus changing the atom to a different element.

range zone A biostratigraphic unit defined by the occurrence of a single type of organism such as a species or a genus.

regional metamorphism Metamorphism taking place over a large but usually elongate area resulting from heat, pressure, and chemically active fluids.

relative dating The process of placing geologic events in their proper chronological order with no regard to when the events took place in number of years ago. (See *absolute dating.*)

relative geological time scale When it was first established, the geologic time scale as deduced from the geologic column showed only relative time; that is, Silurian rocks are younger than those of the Ordovician but older than those designated Devonian.

Rio Grande rift A linear depression made up of several interconnected basins extending from Colorado into Mexico.

ripple mark Wavelike structure on a bedding plane, especially in sand, formed by unidirectional flow of air or water currents, or by oscillating currents as in waves.

rock An aggregate of one or more minerals as in granite (feldspars and quartz) and limestone (calcite), but also includes rocklike materials such as natural glass (obsidian) and altered organic material (coal).

rock cycle A sequence of processes through which Earth materials may pass as they are transformed from one rock type to another.

rock-forming mineral Any of about two dozen minerals common enough in rocks to be important for their identification and classification.

Rodinia The name of a Neoproterozoic supercontinent.

rounding The process involving abrasion of sedimentary particles during transport so that their sharp edges and corners are smoothed off.

ruminant Any cud-chewing placental mammal with a complex three- or four-chambered stomach, such as deer, cattle, antelope, and camels.

San Andreas transform fault A major transform fault extending from the Gulf of Mexico through part of California to its termination in the Pacific Ocean off the north coast of California. (See *transform fault.*)

sand dune A ridge or mound of wind-deposited sand.

sandstone-carbonate-shale assemblage An association of sedimentary rocks typically found on passive continental margins.

Sauk Sequence A widespread association of sedimentary rocks bounded above and below by unconformities that was deposited during a Neoproterozoic to Early Ordovician transgressive–regressive cycle of the Sauk Sea.

Saurischia An order of dinosaurs; characterized by a lizardlike pelvis; includes theropods, prosauropods, and sauropods. (See *Ornithischia.*)

scientific method A logical, orderly approach involving data gathering, formulating and testing hypotheses, and proposing theories.

seafloor spreading The phenomenon involving the origin of new oceanic crust at spreading ridges that then moves away from ridges and is eventually consumed at subduction zones.

sedimentary facies Any aspect of sediment or sedimentary rocks that make them recognizably different from adjacent rocks of about the same age, such as a sandstone facies.

sedimentary rock Any rock composed of (1) particles of preexisting rocks, (2) or made up of minerals derived from solution by inorganic chemical processes or by the activities of organisms, and (3) masses of altered organic matter as in coal.

sedimentary structure All features in sedimentary rocks such as ripple marks, cross-beds, and burrows that formed as a result of physical or biological processes that operated in a depositional environment.

sediment-deposit feeder Animal that ingests sediment and extracts nutrients from it.

seedless vascular plant Plant with specialized tissues for transporting fluids and nutrients and that reproduces by spores rather than seeds, such as ferns and horsetail rushes.

sequence stratigraphy The study of rock relationships within a time-stratigraphic framework of related facies bounded by widespread unconformities.

Sevier orogeny Cretaceous phase of the Cordilleran orogeny that affected the continental shelf and slope areas of the Cordilleran mobile belt.

Siberia One of six major Paleozoic continents; composed of Russia east of the Ural Mountains, and Asia north of Kazakhstan and south of Mongolia.

silicate A mineral containing silica, a combination of silicon and oxygen, and usually one or more other elements.

solar nebula theory An explanation for the origin and evolution of the solar system from a rotating cloud of gases.

Sonoma orogeny A Permian–Triassic orogeny caused by the collision of an island arc with the southwestern margin of North America.

sorting The process whereby sedimentary particles are selected by size during transport; deposits are poorly sorted to well sorted depending on the range of particle sizes present.

species A population of similar individuals that in nature can interbreed and produce fertile offspring.

stratigraphy The branch of geology concerned with the composition, origin, areal extent, and age relationships of stratified (layered) rocks; concerned with all rock types but especially sediments and sedimentary rocks.

stratification (bedding) The layering in sedimentary rocks; layers less than 1 cm thick are laminations, whereas beds are thicker.

stromatolite A biogenic sedimentary structure, especially in limestone, produced by entrapment of sediment grains on sticky mats of photosynthesizing bacteria.

submarine hydrothermal vent A crack or fissure in the seafloor through which superheated water issues. (See *black smoker.*)

Sundance Sea A wide seaway that existed in western North America during the Middle Jurassic Period.

supercontinent A landmass consisting of most of Earth's continents (such as Pangaea).

suspension feeder Animal that consumes microscopic plants and animals or dissolved nutrients from water.

system The fundamental unit in the hierarchy of time-stratigraphic units, such as the Devonian System. A system is also a combination of related parts that interact in an organized manner. Earth's systems include the atmosphere, hydrosphere, biosphere, as well as Earth's lithosphere, mantle, and core.

Taconic orogeny An Ordovician episode of mountain building resulting in deformation of the Appalachian mobile belt.

Tejas epeiric sea A Cenozoic sea largely restricted to the Gulf and Atlantic Coastal Plains, coastal California, and the Mississippi Valley.

terrane A small lithospheric block with characteristics quite different from those of surrounding rocks. Terranes probably consist of seamounts, oceanic rises, and other seafloor features accreted to continents during orogenies.

terrestrial planet Any of the four, small inner planets (Mercury, Venus, Earth, and Mars) similar to Earth (*Terra*); all have high mean densities, indicating they are composed of rock. (See *Jovian planet.*)

theory An explanation for some natural phenomenon with a large body of supporting evidence; theories must be testable by experiments and/or observations, such as plate tectonic theory.

theory of evolution The theory holding that all living things are related and that they descended with modification from organisms that lived during the past.

therapsid Permian to Triassic mammal-like reptiles; the ancestors of mammals are among one group of therapsids known as cynodonts.

thermal convection cell A type of circulation of material in the asthenosphere during which hot material rises, moves laterally, cools and sinks, and is reheated and continues the cycle.

tidal flat A broad, extensive area along a coastline that is alternately water-covered at high tide and exposed at low tide.

till Sediment deposited directly by glacial ice, as in an end moraine.

time-stratigraphic unit A body of strata that was deposited during a specifc interval of geological time; for example, the Devonian System, a time-stratigraphic unit, was deposited during that part of geological time designated the Devonian Period.

time unit Any of the units such as eon, era, period, epoch, and age referring to specific intervals of geologic time.

Tippecanoe Sequence A widespread body of sedimentary rocks bounded above and below by unconformities; deposited during an Ordovician to Early Devonian transgressive-regressive cycle of the Tippecanoe Sea.

trace fossil Any indication of prehistoric organic activity such as tracks, trails, burrows, and nests. (See *biogenic sedimentary structure, body fossil,* and *fossil.*)

Transcontinental Arch Area extending from Minnesota to New Mexico that stood above sea level as several large islands during the Cambrian transgression of the Sauk Sea.

transform fault A type of fault that changes one kind of motion between plates into another type of motion; recognized on land as a strike-slip fault. (See *San Andreas transform fault.*)

transform plate boundary Plate boundary along which adjacent plates slide past one another and crust is neither produced nor destroyed.

tree-ring dating The process of determining the age of a tree or wood in a structure by counting the number of annual growth rings.

unconformity A break or gap in the geologic record resulting from erosion or nondeposition or both. Also the surface separating younger from older rocks where a break in the geologic record is present. (See *angular unconformity, nonconformity,* and *disconformity.*)

ungulate An informal term referring to a variety of mammals but especially the hoofed mammals of the orders Artiodactyla and Perissodactyla.

U-shaped glacial trough A valley with steep or nearly vertical walls and a broad, concave, or rather flat floor; formed by movement of a glacier through a stream valley.

valley glacier A glacier confined to a mountain valley.

varve A dark-light couplet of sedimentary laminations representing an annual deposit in a glacial lake.

vascular plant A plant with specialized tissues for transporting fluids in land plants.

vertebrate Any animal possessing a segmented vertebral column as in fish, amphibians, reptiles, birds, and mammals; members of the subphylum Vertebrata.

vestigial structure In an organism, any structure that no longer serves any or only a limited function, or a different function, such as dewclaws in dogs, wisdom teeth in humans, and middle ear bones in mammals.

volcanic rock An igneous rock that forms as lava cools and crystallizes or when pyroclastic materials are consolidated. (See *extrusive igneous rock.*)

Walther's law A concept holding that the facies in a conformable vertical sequence will be found laterally to one another.

Wilson cycle The relationship between mountain building (orogeny) and the opening and closing of ocean basins.

Zuni epeiric sea A widespread sea present in North America mostly during the Cretaceous, but it persisted into the Paleogene.

Answers to Multiple Choice Review Questions

Chapter 1
1. c; 2. c; 3. b; 4. c; 5. a; 6. e; 7. e; 8. b; 9. d; 10. d.

Chapter 2
1. b; 2. c; 3. e; 4. a; 5. b; 6. c; 7. d; 8. a; 9. a; 10. d.

Chapter 3
1. d; 2. a; 3. b; 4. b; 5. a; 6. c; 7. c; 8. e; 9. c; 10. a.

Chapter 4
1. e; 2. b; 3. e; 4. a; 5. e; 6. d; 7. a; 8. a; 9. b; 10. b.

Chapter 5
1. a; 2. b; 3. e; 4. c; 5. d; 6. d; 7. c; 8. c; 9. e; 10. c.

Chapter 6
1. a; 2. c; 3. e; 4. b; 5. d; 6. d; 7. a; 8. c; 9. d; 10. d.

Chapter 7
1. b; 2. c; 3. e ; 4. c; 5. a; 6. d; 7. e; 8. a; 9. c; 10. b.

Chapter 8
1. b; 2. c; 3. e; 4. a; 5. a; 6. d; 7. c; 8. a; 9. d; 10. b.

Chapter 9
1. a; 2. c; 3. e; 4. b; 5. c; 6. c; 7. a; 8. d; 9. c; 10. a.

Chapter 10
1. b; 2. d; 3. b; 4. b; 5. b; 6. e; 7. a; 8. c; 9. d; 10. b; 11. c; 12. e; 13. d.

Chapter 11
1. b; 2. d; 3. e; 4. b; 5. d; 6. a; 7. c; 8. c; 9. a; 10. b; 11. c; 12. e.

Chapter 12
1. a; 2. a; 3. d; 4. e; 5. b; 6. c; 7. e; 8. e; 9. c; 10. c; 11. a; 12. d; 13. a; 14. c.

Chapter 13
1. c; 2. b; 3. c; 4. c; 5. a; 6. e; 7. b; 8. e; 9. c; 10. d; 11. c; 12. e; 13. b; 14. b.

Chapter 14
1. c; 2. a; 3. c; 4. a; 5. e; 6. b; 7. e; 8. c; 9. c; 10. e.

Chapter 15
1. a; 2. d; 3. b; 4. b; 5. d; 6. e; 7. b; 8. d; 9. c; 10. a.

Chapter 16
1. b; 2. a; 3. e; 4. c; 5. c; 6. d; 7. a; 8. c; 9. a; 10. d.

Chapter 17
1. b; 2. d; 3. a; 4. e; 5. a; 6. c; 7. c; 8. b; 9. d; 10. a.

Chapter 18
1. b; 2. a; 3. c; 4. d; 5. b; 6. e; 7. b; 8. c; 9. a; 10. b.

Chapter 19
1. a; 2. d; 3. e; 4. d; 5. b; 6. c; 7. b; 8. e; 9. c; 10. d; 11. a; 12. c.

Index

A

Abiogenesis, 166
Absaroka Sea, 229
Absaroka Sequence, 226–227
Absolute dates
    and the relative geologic time scale, 105–106
Absolute dating, 67, 72–73, 75, 77
Acadian orogeny, 219–220, 234
Acanthostega, 268
Acasta Gneiss, 155, 162
Accessory minerals, 24
Accretion, 175
Accretion of terranes, 298–299
Acid rain, 323
Acquired characteristics, inheritance of, 133, 136, 138
Acritarchs, 186, 195, 224, 251–252, 255, 271
Actualism, 71
Adenosine triphosphate (ATP), 168–169
Afonin, S.A., 274
Africa
    Archean rocks, 157
    and continental drift, 39–42
    and divergent boundaries, 50–53
    hominids, 404
    human origins in, 411
    and meteorites, 178
    Paleoproterozoic glaciers in, 183
    plate tectonics and, 330–333
    rift valley system, 12, 47
Aftonian interglacial stage, 359
Agassiz, Louis, 354, 368
Agathla Peak, 104, 105
Ages, 100
Agnatha, 262–264
Alaska, 162, 298, 335, 355, 360–364, 372, 395
Aletsch glacier, 353
Aleutian Islands, 52, 355
Alleghenian orogeny, 222, 235
Alleles, 137
Allopatric speciation, 140
Alluvial fan deposits, 117
Alluvial rocks, 70
Alpha decay, 72
Alpine-Himalayan orogenic belt, 333–335
Altered remains, 95–96
Aluminum, 21, 161, 213, 236
Amino acids, 139, 166–167
Ammonoids, 144, 253–254, 267, 316
Amnion, 269–270, 384
Amniote egg, 269–270
Amphibians, 267–269, 307
    Mesozoic, 294–295, 307–309
    missing link and, 150
    Paleozoic, 265–270
    Permian, 256
Anaerobic organisms, 168
Analogous structures, 147, 148
Ancestral archosaur, 309
Ancestral Rockies, 229, 230
Angiosperms, 308–309

Angular unconformity, 70, 89–90
Animal fossils, Proterozoic, 191–192
Animals
    and continental drift, 39, 42, 60
    emergence of, 267–268
    evolution of, 140–144
    fossils of, 116
    and natural selection, 135–136
    Neoproterozoic, 190–191
    origin of, 319
Ankylosaurs, 314
Anning, Mary, 318
Antarctica
    continental drift and, 40–41
    glaciers in, 183, 360–362, 372
Anthracite, 34, 235
Anthropoids, 403–404
Antler orogeny, 220, 232
Aphelion, 371–372
Appalachian mobile belt, 198–199, 212–213, 223, 233–234
Appalachian Mountains, 59, 179, 287–289, 346–350
Aquatic mammals, 393–394
Aquatic vertebrates, 307
Archaeocyathids, 247
Archaeopteryx, 317–319
Archaic mammals, 384–385
Archean Eon, 154
    atmosphere and hydrosphere, 163–165
    continental foundations, 156–162
    microfossils, 169
    mineral resources, 169–170
    origin of life, 165–169
    plate tectonics, 163
    rocks, 157–159
    sulfide deposits, 170
Arches National Park, 109
Archosaurs, 310
Argillite, 159
Arkose, 29
Articulate brachiopods, 252
Artificial selection, 134
Artiodactyls, 390–391
Aseismic ridges, 56
Ash falls, 101, 103
Asia
    Alpine-Himalayan orogenic belt, 333
    human lineage, 410
    mammals, 391–393
    plate tectonics, 46, 52–54, 355
Asthenosphere
Atlantic Coastal Plain, 348–350
Atlantic continental margin, 349–350
Atlantic Ocean, 39, 59
Atlantic Ocean basin, 50
Atmosphere, 2–4, 163–165, 184–186
Atomic mass number, 20, 72
Atoms, 19–22, 72–76
Atrypa, 101
Australia, 225, 335, 383, 396–397
Australopithecines, 407
Autotrophic organisms, 168

Autotrophs, 246
Avalonia, 199, 235
Avian dinosaurs, 382
Axial tilt, 371

B

Back-arc marginal basins, 160–161
Back-arc spreading, 341
Background extinction, 145
Background radiation, 5, 6
Bacteria, 166–168, 186–187, 246
Badlands National Park, 328, 347
Badlands topography, 347
Baltica, 199
Banded-iron formations (BIF), 170, 184–185
Barrier islands, 121
Barrier reefs, 207
Basal rocks, 222
Basal slip, 354, 372
Basin and Range Province, 341, 342
Bats, 387
Bearpaw Formation, 106
Beartooth Mountains, 153
Bedding, 112–113
Bedding planes, 87
Belemnoids, 305, 327
Belt Supergroup, 173
Benthos, 244
Bering land bridge, 365
Beta decay, 72
Big Bang, 5
Biochemical sedimentary rocks, 30
Biogenic sedimentary structures, 115
Biosphere, 2–4
Biostratigraphic correlation, 252
Biostratigraphic units, 98–100
Biotic provinces, 60
Bioturbation, 115
Biozone, 100, 101
Bipedal carnivores, 311–313
Birds
    Cenozoic, 382
    Darwin and, 134
    evolution of, 309–311, 314, 317
Bituminous coal, 235
Bivalves, 305–306, 352, 379
Black Hills, 192, 345, 347
Black shales, 223–224
Black smokers, 167
Blanket geometry, 116
Body fossils, 94–95
Bonding, 20–21
Bony fish, 265–266
Bovids, 390–391
Brachiopods, 246
Brachiopods, inarticulate, 246, 251
Braided streams, 117, 118
Bretz, J Harlan, 368
Brongniart, Alexander, 98

Browsers, 390
Brule Formation, 328
Bryce Canyon National Park, 68
Buffon, Georges Louis de, 67
Burgess Shale, 241
Burgess Shale biota, 250–251
Bushveld Complex, 170, 192
Butte, 104

C

Calderas, 179, 342, 343, 356
Caledonian orogeny, 213, 219, 222, 233
Cambrian explosion, 241–242
Cambrian Period
    marine communities, 246–247, 252–253
    Paleozoic arthropods, 248
    rocks, 98, 156, 190, 205–206
    shelly fauna, 242–243
Canada
    Cordilleran igneous activity, 339–340
    evaporite deposits in, 125
    evolution of continents, 175
    glacial landforms, 362–363
    glaciers, 367–370
    mineral resources, 235–237, 373
    oldest known rocks, 162
    reef development in, 222–223
Canadian shield, 157
Canning Basin, 225
Cannonball concretions, 348
Cape Cod, Massachusetts, 363, 365
Capital Reef National Park, 101
Carbon, isotopes of, 21
Carbon-14 dating, 77–78
Carbonate depositional environments, 123
Carbonate desposition, 224–225
Carbonate minerals, 29
Carbonate rocks, 29
Carbonates, 23
Carboniferous Period, 220–221
    marine communities, 254–255
Carboniferous System, 98
Carnassials, 388
Carnivore-scavengers, 244
Carnivorous mammals, 388
Carnotite, 299
Cartilaginous fish, 265
Cascade Range, 341, 356, 372–373
Casts, 95, 97
Catastrophism, 69, 70
Catskill Delta, 234
Cave paintings, 412–413
Cedar Mesa Sandstone, 104
Cell cleavage, 262
Cellulose, 272
Cementation, 28
Cenozoic Era, 154
    Appalachian Mountains, 346
    birds, 382
    continental margins, 348–350
    life, 376–400
    marine invertebrates, 378–381
    orogenic belts, 332–335
    paleogeography of the world, 330–331
    plate tectonics, 330–332
    time scale, 329
    vegetation and climate, 381–382
    volcanism, 340, 345
Cenozoic geologic history, 353
Central Lowlands, 345, 346, 348
Cephalopods, 224, 263, 305
Cetacea, 384
Chaleuria cirrosa, 275–276
Chemical bonding, 221
Chemical sedimentary rocks, 28, 30
Chert, 29, 31
China, 199
Chinle Formation, 294–295, 297–298

Chondrichthyes, 263–264
Chordate, 261–262
Christian theology, 67, 133
Chrome deposits, 170
Chromosomes, 138
Chronostratigraphic units, 100
Circum-Pacific orogenic belt, 335
Cirques, 362, 366
Clack, Jennifer, 268
Clade, 143
Cladistics, 142–143
Cladograms, 142–143
Classification, 146
Classification of organisms, 187–189
Clastic wedge, 212, 214
Claystone, 29
Climate change, 81, 371–372
    data, 359
    and geologic time, 78–79
    hypothesis, 395
Climates
    Mesozoic, 322–324
    Pleistocene and Holocene, 360–361
Climatic events, short-term, 372
Closed systems, dating, 74
Coal, 29, 31, 196, 197, 235, 236
Cobalt, 19
Coccolithophores, 306, 378
Coelacanths, 266
Coleoids, 305
Colorado Plateau, 342, 343
Colorado River, 66
Columnar sections, 93
Compaction, 28
Compounds, 20–21
Concurrent range zones, 101, 103
Conformable strata, 89
Conglomerate, 28
Conifers, 294, 297, 308
Conodonts, 252
Consumers, primary and secondary, 246
Contact metamorphism, 32
Continental accretion, 163
Continental architecture, 197–199
Continental-continental plate boundaries,
    50–51, 53–54
Continental drift, 39–42, 39–43
Continental environments, 116–117
Continental fit, 40
Continental foundations
    cratons, 156–158, 175–176
    platforms, 156, 198
    shields, 198
Continental glaciers, 354, 355
Continental interior of North America, 345–346
Continental margins, North American, 348–350
Continental nuclei, 156
Continental red beds, 185–186
Continental rise, 122
Continental shelf, 122
Continental slope, 122
Continents, 182
    Proterozoic, 175
Convergent evolution, 141–142
Convergent plate boundaries, 47, 50, 53–54
Cooksonia, 272, 275
Copper deposits, 299
Coprolite, 94–96
Coral reefs, 208, 305
Corals, 22, 95, 116, 191, 207, 225
Cordaites, 277
Cordilleran Antler orogeny, 220
Cordilleran batholiths, 292
Cordilleran igneous activity, 339–340
Cordilleran mobile belt, 198–199, 203, 208,
    231–232, 290–294
Cordilleran orogeny, 286, 291–292,
    291–293, 335
Core, 3–5, 11
Correlation, 100–101
Cosmology, 5

Covalent bonds, 21, 22
Cranioceras, 386
Cratonic sequences, 201, 202–203, 222,
    226–227, 287
Cratonic uplift, 229
Cratons, 163
    Archean, 175–176
    and mobile belts, 197–198
    origin of, 156–158, 166
Creodonts, 384, 388
Cretaceous Interior Seaway, 296
Cro-Magnons, 412–413
Crocodiles, 317
Cross-bedding, 113
Cross-cutting relationships, Principle of, 69
Cross-dating, 77
Crossopterygians, 266
Crust, 11
Crustal evolution, 163, 174
Cryosphere, 354
Crystalline solids, 21
Crystalline texture, 29
Curie, Pierre and Marie, 72
Curie point, 42
Current ripples, 114
Cutin, 272
Cuvier, Georges, 70, 133
Cyclotherms, 227–229
Cynodonts, 319–321
Cynognathus, 41

D

Darwin, Charles, 132–133, 134
    *On the Origin of Species by Means of Natural
    Selection*, 13, 242
Darwin-Wallace theory of natural selection, 134
Dating techniques, 62, 72–73, 76–78
Daughter element, 73
Death Valley, 29, 118, 180, 373
Deep-sea lava flows, 212
Deep-sea stratigraphy, 359–360
Deep-seafloor sediments, 50, 54, 123, 126
Deep time, 2
Deinonychus, 311–312
Delicate Arch, 109
Deltas, 119, 121
Denver, Colorado weather of, 81
Depositional environments, 110, 116–124, 119,
    126–127
Detrital deposition, 223
Detrital marine environments, 122
Detrital sedimentary rocks, 28
Devoinan Period, 218–219
    marine communities, 252–253
    paleogeography of North America, 223
    reef complex, 224
    systems, 98
Diamonds, 299
Diatoms, 373
Differential pressure, 32
Differentiated Earth, 10, 155, 164
Dimetrodon, 270
Dinosaur fossils, 295–296
Dinosaurs, 308–315
Diplodocus, 311
Disconformity, 89–90
Divergent evolution, 141–142
Divergent plate boundaries, 47, 50
Diversity of life, 251
DNA (deoxyribonucleic acid), 138
Dolomite, 23, 24, 29, 111
Dolostone, 29, 111
Doppler effect, 5
Drift, 117
Dropstones, 368
du Toit, Alexander, 42
Dynamic metamorphism, 32

E

Early Paleozoic, 199
  mineral resources, 213–214
Earth
  age of, 67–68
  differentiation into a layered planet, 10
  early Paleozoic, 196–216
  internal heat source, 71
  oldest rock, 162
  surface water, 165
Earth–moon tidal interaction, 156
Earthquakes, 11–12, 38, 50–56, 71, 333
East African Rift Valley, 50, 52, 400
Eastern Coastal Region, 287–288
Echinoids, 305
Ectotherms, 314
Edaphosaurus, 270
Ediacaran fauna, 190–191, 241
Einstein, Albert, 5
El Capitan, 104
Electromagnetic force, 6
Electron capture decay, 72
Elements, 19–21, 72
Elephants, 392–393
Ellesmere orogeny, 220
Elongate geometry, 116
Emperor Seamount–Hawaiian Island
  chain, 58
End moraines, 362, 363, 365
Endosymbiosis, 187–189
Endotherms, 314–315
Eoarchean Era, 155–156
Eon, 100
Eonothem, 100
Eperic seas, 198
Epifauna, 224, 305–306
Epiflora, 245
Epoch, 100
Era, 100
Erathem, 100
Erosion, glacial, 362
Eukaryotic cells, 168, 186–189
Eurypterids, 253
Eusthenopteron, 266
Evaporites, 29, 31, 111, 125–126, 206–210
  deposits, 229
  environments, 123
  formation, 285, 289
  sedimentation, 212
Evolution, 132–133
  biological evidence for, 146–147, 150–151
  modern view of, 138
  and religion, 98
Evolutionary novelties, 142
Evolutionary trends, 144
Extinctions, 144–145, 395–396
Extrusive igneous rock, 24

F

Facies, 91
Farallon plate, 344, 356
Fault-block basins, 50–53, 287–290
Faunal succession, principle of, 98
Faunas, Pleistocene, 394–395
Feeding systems, 244–246
Felsic magma, 26
Fermentation, 169
Ferromagnesian silicates, 23–26, 33
Fishes, 244, 262–267, 307
Fission-track dating, 76
Fixity of species, 132
Florida, 299, 330, 348–349, 395
Fluid activity, 32
Fluvial deposits, 117
Flying reptiles, 315–316
Foliated metamorphic rocks, 33, 34

Foliated texture, 32
Foraminifera, 378
Formations, 100
Fossil evidence, 41
Fossil record, 95, 241–242
Fossils
  dinosaurs, 295–296
  and evolution, 149–150
  and fossilization, 93–94
  hominid, 409
  Mazon Creek, 218
  Messel Pit site, 380
  petrified wood, 294, 297–298
  Proterozoic, 191–192
  and sedimentary rocks, 116
  telling time, 97
Foster, C.B., 274
Fragmental texture, 25
Franciscan Complex, 292–293
Franklin mobile belt, 198–199
Fungi, 140, 166, 186, 271
Fusulinids, 255, 280

G

Galápagos Islands, 132, 133, 138
Garden of the Gods, 229
Gastropods, 252, 305, 379
Gemstones, 20, 192
Genes, 134, 137
Genesis, book of, 133
Genetic drift, 139
Genetics, 136–138
Geographic poles, 42, 43
Geologic column, 98
Geologic maps, 197
Geologic record, 19, 86
Geologic systems, 197
Geologic time, 13–14, 66, 67, 98
  and climate change, 78–79
Geology
  defined, 4
Geology, crisis in, 71–72
Geometric radioactive decay curve, 73
Glacial striations, 41
Glaciation
  effects of, 361–365
  landforms, 362–363
  Pleistocene, 370–371
  stages, 359
Glacier National Park, 173, 366–367
Glaciers, 183–184, 353, 354
  deposits, 120
  formation of, 361
  and isostasy, 367–368
  modern-day, 372–373
Glassy texture, 25
Glauconite, 74, 105
Global climates and ocean circulation,
  285–286
Global cooling, 253–254
Global warming, 372–373
Globorotalia, 359
Glossopteris, 41, 277
Glossopteris flora, 38
Glyptodonts, 395–396
Gneiss, 33, 34
Gold, 21, 61, 169, 282, 347, 350, 373
Gondwana, 38, 199, 220–221
  breakup of, 284–285
  continents, 40, 41
Gosse, Philip Henry, 86–87, 145
Graded bedding, 113
Grain size, 111
Grand Canyon region, 65, 66, 91, 204–206
Granite-gneiss complexes, 157
Graptolites, 252
Gravel, 26–28, 110, 296, 362
Gravity, 6

Gravity-driven mechanisms, 59
Grazers, 390
Great Barrier Reef, 225–226
Great Lakes region, 369–370
Great Plains, 345, 347–348
Great Rift Valley of Africa, 409
Great Salt Lake, 369
Great Valley Group, 292
Green River Formation, 350, 387
Greenstone, 33
Greenstone belts, 158–161, 175
Grenville orogeny, 179
Gressly, Armanz, 91
Groundmass, 25
Guadeloupe Mountains, 229
Guide fossils, 101, 103
Gulf Coastal Plain, 348–349
Gulf Coastal Region, 288–289
Gulf of California, 50
Gulf of Mexico, 121, 284, 344
Gymnosperms, 275
Gypsum, 23, 111, 210, 236, 294

H

Half-life, radioactive decay and, 72, 73
Halgaito Shale, 104
Heat transfer system, 44
Helicoplacus, 247
Hemicyclaspis, 263–264
Herbivores, 244
Hercynian orogeny, 222
Hercynian–Alleghenian orogeny, 234–235
Hess, Harry, 44
Heterospory, 275
Heterotrophic organisms, 168
Hiatus, 89
Himalayan orogeny, 333
Himalayas, 333
Historical geology
  benefits of studying, 14–15
  defined, 4
  theory formulation and, 4–5
History of life, 13
Holocene climate, 360–361
Holocene Epoch, 353
Hominids, 403–407
Hominoids, 404
Homo erectus, 410–411
Homo habilis, 410
Homologous structures, 147, 148
Hoofed mammals, 389–390
Hornfels, 34
Horses, 392
Hot spots and mantle plumes, 55–56, 57
Hox genes, 242
Human lineage, 410–411
Hutton, James, 70
Hydrocarbons, 192, 235, 299, 348
Hydrosphere, 3–5, 165–166
Hylonomus, 270
Hypotheses, 4
Hyracotherium, 391

I

Iapetus Ocean, 212
Iberia–Armorica, 235
Ice, 19
Ice Age, 354, 360–361
Ice Age mammals, 394–395
Ice caps, 354, 355
Ice-scoured plains, 362
Ice sheets, 354
Ichthyosaurs, 315–317
Ichthyostega, 268–269
Igneous activity, Cordilleran, 339–340

Igneous rocks, 24–26, 27
Illinoian glaciation, 359
Illinois, 218
Inarticulate brachiopods, 246
India, 284, 324, 331, 355
Inheritance of acquired characteristics,
    theory of, 133–134, 136
Insectivores, 387
Intercontinental migrations, 396–397
Interglacial stages, 359, 371
Intertonguing, 91
Interval zones, 101
Invertebrate groups and stratigraphic ranges, 245
Ionic bonding, 22
Iridium anomaly, 323
Isostasy, glaciers and, 367–368
Isostatic rebound, 367
Isotopes, 72
Isthmus of Panama, 60, 61

J

Jaw, vertebrate, 265
Johnson, Kirk R., 81
Joly, John, 68
Jovian planets, 6, 8–10

K

Kansan glaciation, 359
Kaskaskia Sequence, 222
Kayenta Formation, 295
Kazakhstania, 199
Kelvin, Lord, 71
Kerogen, 350, 380
Kimberella, 191–192
Koch, Albert, 394
Komatiites, 158

L

La Brea tar pits, 95, 395
Labyrinthodonts, 269
Laetoli (Tanzania), 409
Lake Agassiz, 357
Lake Bonneville, 368–369
Lake Manley, 368–369
Lake Missoula, 368–369, 369
Lakes, pluvial and proglacial, 367–368
Lamarck, Jean-Baptiste Pierre Antoine de,
    131–134
Laminations, 112, 113, 117
Land-living organisms
    amphibians, 267–269
    mammals, 363–394
    reptiles, 269–270
Lapilli, 26
Lapworth, Charles, 98, 99
Laramide orogeny, 293, 336–339
Laurasia, 219
Late Carboniferous floras, 276–277
Late Kaskaskia, 224–225
Late Paleozoic Earth history, 217–240
    Carboniferous Period, 220–221
    Devonian Period, 218–219
    evolution of North America, 222–223
    mineral resources, 235–236
    mobile belts, 231–232
    Permian Period, 222
Lateral continuity, principle of, 68, 91
Lateral gradation, 91
Lateral relationships–facies, 91
Laurasia, 42, 234
Laurentia, 175–176, 199

Lava flows, 42, 212
Leaves, evolution of, 272
Lepidodendron, 276
Life
    distribution of, 60
    origin of, 165–168
Lignin, 272
Limestone, 29, 111
Lingula, 101
Linnaeus, Carolus, 146
Lithification, 28
Lithosphere, 3–5
Lithostatic pressure, 32
Lithostratigraphic correlation, 100–102
Lithostratigraphic units, 98–100
Little Ice Age, 361
Living fossils, 144, 145
Living–nonliving distinction, 166
Lizards, 317
Lobe-finned fish, 266
Louann Salt, 299
Lovejoy, Owen, 407
Lower Silurian strata, 126–127
Lungfish, 266
Lycopsids, 276
Lyell, Charles, 70
Lysenko, Trofim Denisovich, 136
Lysenko affair, 136
Lystrosaurus, 41

M

Macroevolution, 142
Madagascar, 199, 266, 323, 382, 403
Mafic magma, 26
Magma, crystalization of, 74
Magnetic anomalies, 45, 56, 57
Magnetic field, 43
Magnetic poles, 42
Magnetic reversals, 43–44
Magnetosphere, 164
Malthus, 134
Mammals
    aquatic, 393–394
    bats, 387
    carnivores, 388
    elephants, 392–393
    Ice Age, 394–395
    insectivores, 387
    marsupial, 383
    Mesozoic, 321–322
    monotremes, 383
    Neogene, 386–387
    origin and evolution of, 319–320
    Paleogene, 386–387
    placental, 383–384
    primates, 387–388
    rabbits, 387
    rodents, 387
    ungulates (hoofed), 389–390
Mammoths, 82, 392
Manatees, 150
Mantle, 3–5, 11, 76
Mantle plumes, 55–56
Marble, 33, 34
Marine deltas, 121
Marine ecosystems
    Cambrian, 246–247
    Carboniferous and Permian, 254–255
    Ordovician, 251–252
    present day, 244–245
    Silurian and Devonian, 252–253, 252–254
Marine environments, 122–124
Marine food web, 246
Marine invertebrates, 304–305, 378–381
Marine regression, 92, 93–94
Marine reptiles, 315–316
Marine transgression, 91, 92, 93–94, 116, 366
Mars, 49

Marshall, James, 282
Marsupial mammals, 322–323
Mass extinctions, 145, 253, 255–257, 323–324
Mass spectrometer, 73
Mastodons, 82, 392
Matter, composition of, 19–20
Mazon Creek fossils, 218
Meandering stream, 117, 118
Meiosis, 138
Mélange, 54
Mendel, Gregor, 137–138
Mercury, 48
Mesa, 104
Mesoproterozoic Era
    accretion and igneous activity, 179
    orogeny and rifting, 179
    sedimentation, 180
Mesosaurus, 41, 42
Mesozoic Era
    breakup of Pangaea, 282–286
    climates and paleobiogeography, 322–324
    history of North America, 286–293
    life, 303–327
    mammals, 321–322
    mass extinctions, 323–324
    mineral resources, 298–299
    sedimentation, 293–294
    tectonics, 290–291
    vegetation, 308
Mesozoic Era, 154
Messel Pit fossil site, 330
Messinian salinity crisis, 126
Metallic mineral resources, 61–62, 236
Metamorphic rock, 31–34, 74–75
Metamorphism
    causes of, 32
    and radiometric dating, 75
Meteorite hypothesis, 256
Meteorites, 162
Meteors, 178
Methane hydrate, 350, 381
Miacids, 388
Michigan Basin, 211
Micrite, 123
Microcontinents, 235
Microevolution, 142
Microfossils, 116, 186, 273
Microplates, 235
Microspheres, 166
Mid-Atlantic Ridge, 39, 44, 45
Mid continent rift, 179
Migmatites, 33, 34
Migrations, intercontinental, 396–397
Milankovitch Theory, 371–372
Miller, Stanley, 166
Mineral deposits, 61–62
Mineral groups, 23
Mineral resources
    Early Paleozoic age, 213–214
    Late Paleozoic, 235–236
    Mesozoic, 298–299
    Paleogene and Neogene, 350
    Pleistocene, 373
    Proterozoic, 192
Minerals, 21–24
Minette iron ores, 299
Missing links, 150
Mississippian paleogeography of North
    America, 226
Mitosis, 138
Moas, 382, 394, 396
Mobile belts, 197–199, 231–232
Modern synthesis, 138
Moenkopi Formation, 293–294
Molarization, 390
Molds, 95, 97
Mollusks, 305, 378
Monomers, 166
Monotremes, 322
Monument Valley Navajo Tribal Park,
    104–105

Moraines, 117, 362
Morrison Formation, 295
Mosaic evolution, 144
Mosasaurs, 316
Mountain building and plate tectonics, 59
Mud cracks, 115
Mudrocks, 29
Multicelled organisms, 189–190
Murchison, Roderick, 98, 99
Mutagens, 139
Mutations, 139

N

Natural gas, 111, 235, 282, 350
Natural resources, distribution of, 61
Natural resources and plate tectonics, 61–62
Natural selection, 134–136, 139, 242
Navajo Sandstone, 295
Nazca plate, 335
Neanderthals, 411–412
Nebraskan glaciation, 359
Negaunee Iron Formation, 186
Nekton, 244
Neogene Period, 354
    mammals, 386–387
    mineral resources, 350
    rifting, 331
Neoglaciation, 361
Neoproterozoic Period
    animals, 190–191
    glaciers, 183
    sedimentation, 180
Neptune, 10
Neptunism, 69
Neutrons, 72
Nevadan orogeny, 292
New Zealand, 335, 382
Newark Group, 287–288
Noble gases, 21
Non-avian dinosaurs, 382
Nonconformity, 89–90
Nonferromagnesian silicates, 23
Nonfoliated metamorphic rocks, 33–34, 34
Nonfoliated texture, 32
North America
    black shales, 223–224
    continental margins, 348–350
    Devonian paleogeography of, 223
    Early Paleozoic evolution of, 201–203
    Late Paleozoic evolution of, 222–223
    Mesozoic history of, 286–293
    Mississippian paleogeography of, 226
    Pennsylvanian paleogeography of, 227
    Permian paleogeography of, 231
    Silurian paleogeography of, 210
North American Cordillera, 335–336, 337
Nuclear force, 6

O

Obsidian, 26
Ocean basins, 44, 46
Ocean circulation, global climates and,
    285–286
Oceanic-continental convergent plate
    boundary, 54
Oceanic-continental plate boundary,
    50–51, 52
Oceanic-oceanic convergent plate
    boundary, 53
Oceans, 165
    circulation patterns, 286
    currents, 60
    salinity, 68, 123, 165, 207
    surface temperature, 359
Oil, 192, 213, 380

Oil seeps, 95, 96
Oil shale, 350, 380
Old Red Sandstone, 234
Oldest known organisms, 168
Olduvai Gorge, 400, 410
On the Origin of Species by Means of Natural
    Selection (Darwin), 134, 242
Ooids, 123
Oolitic limestone, 123
Ooze, 123
Ophiolites, 54, 55, 59, 163, 182
Orbital eccentricity, 371
Ordovician marine community, 251–252
Ordovician System, 98, 99
Organ Rock Shale, 104
Organic evolution, 13, 98
Organic reefs, 206–209
Organisms, classification of, 187–189
Organisms, multicelled, 189–190
Origin of Continents and Oceans
    (Wegener), 39, 42
Origin of life
    experimental evidence, 166–167
    submarine hydrothermal vents, 167
Original horizontality, principle of, 68
Oriskany Sandstone, 222
Ornithischian dinosaurs, 310, 312–313
Orogenic belts, Cenozoic, 332–335
Orogens, 175
Orogeny, 59
Ostracoderms, 263
Ouachita mobile belt, 198–199, 233
Ouachita Mountains, 221
Ouachita orogeny, 222
Our Wandering Continents (du Toit), 42
Outgassing, 164
Outwash, 117, 362, 363
Owen, Richard, 310
Ozone layer, 155, 186, 274

P

Pacific Coast, 344
Pacific Ocean, 284–285, 360
Pacific–Farallon ridge, 344
Painted Desert, 297
Paleocene-Eocene thermal maximum, 381
Paleocene Epoch
    mammals, 386–387
    mineral resources, 350
Paleogeography, 127–128
    Late Paleozoic, 218–219
    Paleozoic, 199–200
Paleomagnetism, 42–43
Paleontology, 132
Paleoproterozoic glaciers, 183
Paleozoic Era, 154
    evolutionary and geologic events, 259
    global history, 199–200
    invertebrate marine life, 250–251
    invertebrates, 240–259
    paleogeography, 199–200
    plants, 260
    vertebrates, 260
Paleozoic invertebrate marine life
    Cambrian marine community, 246–247
    Carboniferous and Permian marine
        communities, 254–255
    Ordovician marine community, 251–252
    present marine ecosystem, 244–246
    Silurian and Devonian marine
        communities, 252–254
Palynology, 273–274
Panderichthys, 268
Pangaea, 39, 221, 222, 235, 282–286, 322–323
Pannotia, 182
Panthalassa Ocean, 222
Paradoxides, 101
Parallel evolution, 141–142

Parent element, 73
Peat, 29, 285, 373, 395
Pegmatite, 170
Pelagic clay, 123
Pelagic organisms, 244
Pelycosaurs, 270
Pennsylvanian paleogeography of North
    American, 227
Pennsylvanian Period, 276–277
Perihelion, 371–372
Period, 100
Perissodactyls, 391
Permafrost, 354, 395
Permian Delaware Basin, 235
Permian Period, 222
    coal deposits, 235
    floras, 276–277
    lava flows, 42
    marine communities, 254–255
    mass extinction, 256–257
    paleogeography of North America, 231
Permian System, 98
Permian–Triassic boundary, 274
Persian Gulf region, 61, 123, 299
Peru–Chile Trench, 53
Perunica, 235
Petrified Forest National Park, 297–298
Petroleum, 61
Phaneritic texture, 25
Phanerozoic Eon, 154
Photochemical dissociation, 164
Photosynthesis, 164, 169
Phyletic gradualism, 140
Phyllite, 33
Phylogenetic trees, 143
Phylogeny, 144
Physical geology, 4
Phytoplankton, 244, 304–305, 378–381
Pillow lavas, 50
Pinching out, 91
Pinnacle reefs, 209
Placental mammals, 322–323, 383–384
Placoderms, 265
Plankton, 244
Plant evolution, 271–272
Plants, 308–309
Plastic flow of glaciers, 354
Plate boundaries, 46–47
Plate movement, 38–39, 56–57
Plate tectonic theory, 12–13
Plate tectonics
    and the distribution of life, 60
    and the distribution of organisms, 61
    and mountain building, 59
    and natural resources, 61–62
    and other planets, 47
    and plate boundaries, 46–47
    and the rock cycle, 34–35
Plates, seven major, 46
Platforms, 156, 198
Playa lake deposits, 117
Pleistocene Epoch
    climate, 360–361
    extinctions, 395–396
    faunas, 394–395
    glaciation, 360, 370–371
    mineral resources, 373
    stratigraphy, 356–359
Plesiosaurs, 316
Pluto, downgraded, 7
Plutonic rocks, 24, 89, 178
Plutonism, 70
Plutons, 52
Pluvial lakes, 367–368
Point bar deposits, 117
Polar wandering, 42–43, 43
Pollen, analysis of fossil, 273–274
Pollen analysis and climate, 360–361
Polymers, 166
Polyploidy, 141
Populations, isolation of, 185

Porphyritic texture, 25
Porphyry copper deposits, 299
Powell, John Wesley, 65, 66, 205
Pre-Nebraskan glaciation, 359
Precambrian Earth, 154
    geologic time scale, 155
    and life history, 173–195
Precambrian shield, 156
Precession of the equinoxes, 371
Precious metals, 170
Prediction, 145
Prehistoric overkill, 394–396
Primates, 141, 387–388, 401–402
Primitive rocks, 70
Principle of crosscutting relationships, 69
Principle of faunal succession, 98
Principle of fossil succession, 97, 98
Principle of inclusions, 87–88
Principle of lateral continuity, 68–69
Principle of original horizontality, 68–69
Principle of superposition, 68
Principle of uniformitarianism, 14–15, 69
Principles of Geology (Lyell), 71
Producers, primary and secondary, 246
Proglacial lakes, 367–368
Prokaryotic cells, 168, 187
Prosimians, 402–403
Proterozoic Eon, 154, 173–195
    animal fossils, 191–192
    continents, 175–181
    life, 186–187
    mineral resources, 192
    supercontinents, 182
Protobionts, 166
Protocontinents, 156
Protons, 72
Protorothyrids, 270
Province boundaries, 60
Pterosaurs, 304, 309, 310, 315–316
Pumice, 26
Punctuated equilibrium, 140
Pyroclastic material, 24–26, 158, 341

Q

Quadrupedal animals, 310–314
Quartz, 21, 111, 297, 350
Quartz sandstone, 29, 33, 222
Quartzite, 33, 34
Quaternary, 354
Queen Charlotte Transform fault, 344
Queenston Delta, 212, 214

R

Rabbits, 387
Radioactive decay, 73
Radioactive isotope pairs, 76
Radioactivity, 71–72
Radiocarbon dating, 77–78
Radiogenic heat production, 165
Radiometric dating, 67
Range zone, 101
Ray-finned fish, 265–266
Recent Epoch, 354
Recessional moraines, 362, 364, 365
Recrystallization, 32
Red Sea, 50, 52
Reef development in western
    Canada, 222–223
Regional metamorphism, 32
Regressions, marine, 91–93, 116, 197, 256,
    287, 299
Reichert, Karl, 320
Relative dating, 66–69
Relative geologic time scale, 66, 98
Relativity, theory of, 5

Reptiles, 269–271, 304, 309–317
Resurgent dome, 356
Rhipidistians, 266
Rhizomes, 274
Ridge–push plate movement, 59
Rifting, 50, 53
    and Mesoproterozoic orogeny, 179
Rio Grande rift, 342–343
Ripple marks, 113–114
Rock cycle, 24
Rock defined, 19
Rock-forming minerals, 23–24
Rock groups, interrelated, 24
Rock gypsum, 29, 111
Rock salt, 29, 81, 111, 213
Rock sequences, 40
Rocky Mountains, 81, 157, 180, 230, 286, 335
Rodents, 387
Rodinia, 182
Roots, evolution of, 272
Rounding, 111
Rudistoid reefs, 289–290
Rudists, 305
Ruminants, 391

S

Sahara Desert, 81, 199, 360, 395
Salinity, ocean, 68, 123, 165, 207
Salt deposits, 368
Salt domes, 289, 299, 350
San Andreas fault, 56
San Andreas transform fault, 344
Sand dune deposits, 117
Sandstone, 29
Sandstone-carbonate-shale assemblage, 175
Sangamon interglacial stage, 359
Sauk Sequence, 202–203
Saurischian dinosaurs, 310, 311–312
Sauropods, 311–312
Savannah conditions, 382
Scablands, 368–369
Schist, 34
Schistose foliation, 33
Schistosity, 33
Scientific method, 4
Scleractinians, 305
Sea level, glacial changes in, 363–364
Seafloor spreading, 43–44, 45
Secondary rocks, 70
Sedgwick, Adam, 98, 99
Sediment-deposit feeders, 244
Sediment transport, 28, 70, 123, 372
Sedimentary breccia, 28
Sedimentary facies, 91
Sedimentary rock, 26–31
    biologic content of, 116
    composition and texture, 111–112
    geometry of, 116
    radiometric dating, 74
Sedimentary structures, 112
Sedimentation, Mesozoic, 293–294
Sediments, deep sea floor, 124
Seedless vascular plants, 272
Self-exciting dynamo, 42
Semiaquatic vertebrates, 307
Sequence stratigraphy, 201–202
Sevier orogeny, 293
Sex cells, 138
Shale, 29
Sheet geometry, 116
Shelly fauna, 242–244
Shelly fossils, 243
Shermer, Michael
    Why Darwin Matters, 145
Shields, 198
Shinarump Conglomerate, 294, 295
Shiprock, New Mexico, 76, 348
Shoestring geometry, 116

Shoreline movement, 93
Siberia, 199, 221, 299, 364, 393
Sierra Madre Oriental, 339
Sierra Nevada batholith, 282, 292, 350
Sigillaria, 276
Silica, 23
Silicate minerals, 23
Siltstone, 29
Silurian lava flows, 42
Silurian marine communities, 252–253
Slab–pull plate movement, 59
Slate, 34
Smith, William, 97, 196, 197
Snakes, 317
Snowball Earth hypothesis, 183
Solar nebula theory, 6–7
Solar System, origin and evolution of, 6–7
Sonoma orogeny, 291
Sonomia, 291
Sorting, 112
Southern Superior craton, 163
Space-time continuum, 5
Speciation, 140–141
Sphenopsids, 276–277
Spores, analysis of fossil, 273–274
Spreading ridges, 50
Sprigg, R.C., 190
Stalagmites and climate change, 79–80
Stars, life cycle of, 6
Stem reptiles, 309
Steno, Nicolas, 68–69, 91
Steppe environments, 382
Stössel, Iwan, 261
Stratification, 112
Stratigraphic relationships, vertical, 87–91
Stratigraphic terminology, 98–100
Stratigraphy, 87–93
Stratotype, 100
Stromatolites, 168, 184
Strong nuclear force, 6
Subduction complex, 50–52
Submarine hydrothermal vents, 167–168
Subsystems, interaction among Earth's, 3–5
Sudbury meteorite, 178
Suess, Edward, 39
Sundance Sea, 295
Supercontinents, 182–183
Supercontinents, Proterozoic, 182
Superposition, 87–88
Superposition, principle of, 68
Surface water, 165–166
Survival of the fittest, 135, 136
Suspension feeders, 244
Systems, 100
    defined, 4
    Earth as, 4–5

T

Taconic orogeny, 212
Tapeats Sandstone, 205
Taylor, Frank, 39
Tectonism, Pleistocene and Holocene, 355
Tejas epeiric sea, 348
Temperature gradients, 286
Terminal moraines, 362, 363, 364, 365
Terrane tectonics, 59–60
Terranes, 59–60, 235
Terranes, accretion of, 298–299
Terrestrial planets, 6, 8–10
    Earth unique among, 86
    tectonics of, 48
Terrestrial stratigraphy, 357–359
Tethys Sea, 284
Tetrapod trackway, 260, 261
Texture, 111
Theodore Roosevelt National Park, 347
Theory of evolution, 132
Therapsids, 270–271, 319

Thermal convection cells, 44, 46, 58
Theropods, 311
Tidal flats, 122
Tiktaalik roseae, 268
Till, 117
Time, linearity of, 67
Time-stratigraphic correlation, 101
Time-stratigraphic units, 100
Time transgressive, 92
Time units, 100
Tippecanoe reefs and evaporites, 206–208
Tippecanoe Sequence, 206–212, 222
Titanotheres, 144
Trace fossils, 95, 96, 115
Transcontinental Arch, 202–203
Transform faults, 55
Transform plate boundaries, 47, 55
Transitional rocks, 70
Transitional environments, 119–120
Tree-ring analysis and climate, 361
Tree-ring dating, 77–78
Triassic fault-block basins, 50, 53
Trilobites, 246, 248–249
Trophic levels, 244–245
Tsunami, 38
Tullimonstrum gregarium, 217, 218
Tully monster, 217, 218
Turbidity currents, 113
Turtles, 317

U

U-shaped glacial troughs, 362
Ultramafic magma, 26
Unaltered remains, 95–96
Uncertainty, sources of, 73–74
Unconformities, 70, 89–90

Ungulates, 389–390
Uniformitarianism, 13–14, 69, 70, 71
Universe, origin of, 5–6
Uranium, 58, 73, 162, 224
Uranus, 10
Ussher, James, 67

V

Valentia Island, Ireland, 260, 261
Valley glaciers, 354, 355
Variation, 136–138
Varve, 117, 368
Vascular plants, 271–272
Venus, 48
Vertebrates, 261–262, 267–269
Vertical running, 136
Vesicular rocks, 25
Vestigial structures, 147–149
Volcanic glass, 26
Volcanic island arc, 52
Volcanic rocks, 24–27, 59, 70, 87
Volcanism, Pleistocene and Holocene, 356
Volcano, 38

W

Walcott, Charles, 168, 241
Wallace, Alfred, 134
Walther's law, 92–93
Waterton Lakes National Park,
    Alberta, 366–367
Wave-formed ripples, 114
Weak nuclear force, 6

Wegener, Alfred, 39
Welded tuff, 26
Well cuttings, 116
Werner, Abraham Gottlob, 69
Western Region, 290–291
Westlothiana, 270
Whewell, William, 70
Whitechuck Glacier, 372, 373
Why Darwin Matters (Shermer), 145
Wilson cycle, 59, 175, 177
Wind and ocean currents, 60
Windjana Gorge, 225
Wingate Sandstone, 295
Wisconsin glaciation, 359
Wopmay orogen, 175–177
Worms, 166, 190–191, 242, 273, 306

Y

Yarmouth interglacial stage, 359
Yellowstone National Park, 56, 118, 340, 356, 381
Yellowstone Plateau, 337, 340
Yellowstone Tuff, 356

Z

Zion National Park, 110, 204, 295
Zircon, 162
Zooplankton, 244
Zuni eperic sea, 345, 348
Zuni Sequence, 287